Microbial Genomic Islands in Adaptation and Pathogenicity

Indra Mani • Vijai Singh •
Khalid J. Alzahrani • Dinh-Toi Chu
Editors

Microbial Genomic Islands in Adaptation and Pathogenicity

Springer

Editors
Indra Mani
Department of Microbiology, Gargi College
University of Delhi
New Delhi, India

Vijai Singh
Department of Biosciences
Indrashil University
Mehsana, Gujarat, India

Khalid J. Alzahrani
Department of Clinical Laboratories Sciences, College of Applied Medical Sciences
Taif University
Taif, Saudi Arabia

Dinh-Toi Chu
Center for Biomedicine and Community Health
Vietnam National University
Hanoi, Vietnam

ISBN 978-981-19-9341-1 ISBN 978-981-19-9342-8 (eBook)
https://doi.org/10.1007/978-981-19-9342-8

This Springer imprint is published by the registered company Springer Nature Singapore Pte Ltd.
The registered company address is: 152 Beach Road, #21-01/04 Gateway East, Singapore 189721, Singapore

Foreword

I am delighted to write the foreword for the book on *Microbial Genomic Islands in Adaptation and Pathogenicity* a well-timed contribution on a rapidly flourishing area of science. To this end, a number of topics from general characteristics of genomic islands (GEIs) and their subtypes are included. This book covers the introduction, tools used to analyze GEIs, and the involvement of GEIs in different microorganisms, such as *Escherichia coli, Helicobacter, Staphylococcus, Pseudomonas, Klebsiella,* and *Vibrio* species. It also covers the importance of GEIs in bacterial speciation, resistome gene acquisition, nutritional fitness, adaptation, genome plasticity, and stability. This book provides a comprehensive source of molecular basis understanding of GEIs and their subtypes.

The book also encompasses fundamental to advanced aspects of GEIs with regard to computation tools, discovery, visualization, analysis, and use of artificial intelligence in microbial GEI. It also covers GEIs and their subtypes, such as pathogenicity, resistance, fitness, and metabolic or symbiosis islands in pathogenic, environmental, commensal, and symbiotic bacteria. Chapters are included in this volume that are written by eminent scientists who are extremely learned in their respective areas of research in microbial genomics.

It gives me immense joy to recognize the valuable efforts of Dr. Indra Mani, Dr. Vijai Singh, Dr. Khalid J. Alzahrani, and Dr. Dinh-Toi Chu, who have tirelessly worked towards conceiving an excellent volume together with Springer Nature.

I believe that this book would be a valuable addition to the repertoire of information not just for beginners, but also for students, researchers, scientists, clinicians, practitioners, policymakers, and stakeholders interested in accessing the potential of microbial GEIs right from basic microbiology to its applications.

School of Life Sciences, University of Warwick
Coventry, UK

Antonia P. Sagona

Preface

This book elucidates the genomic islands (GEIs), which acquire through horizontal gene transfer (HGT). HGT is achieved through various ways, such as conjugation, transformation, and transduction. GEIs play an important role in the rapid and intense adaptation of bacterial species phenotypes by carrying clusters of gene, which can assist a cell with novel and useful phenotypes. In addition, GEIs have a crucial role to play in evolution of microbial genome and adaptation of organisms in changing environmental conditions. GEIs are ubiquitous in both pathogenic and nonpathogenic bacteria and are thought to contribute to accessory functions amongst bacterial populations. A broad category of GEIs is often seen, such as integrative and conjugative elements (ICEs), conjugative and prophages, resistance islands (REIs), metabolic islands, xenobiotic-degradation islands, and symbiosis islands. Remarkably before, it was being considered as a paradox, but due to the availability of vast genome sequencing data of various microorganisms, now it is being considered as a paradigm. Therefore, based on the composition, characteristics, and functions of GEIs, it has been further divided into various islands such as pathogenicity, resistance, fitness, and metabolic or symbiosis islands in pathogenic, environmental, commensal, and symbiotic bacteria.

The structure of this book covers fundamental to applied aspects. Chapter 1 offers an introduction to microbial GEIs for evolutionary adaptation and pathogenicity. Chapters 2 and 3 discuss computation tools for prediction and analysis of GEIs along with *Corynebacterium* pathogenic species. Chapter 4 explores microbial GEI discovery, visualization, and analysis. Chapter 5 confers GEIs and bacterial speciation. Chapter 6 offers GEIs in the gut microbiome. Chapters 7 and 8 explore GEIs in nutritional fitness and adaptation and are involved in iron uptake. From Chaps. 9–14 cover various aspects of GEIs in a number of bacteria such as *Escherichia coli*, *Helicobacter* species, *Staphylococcus*, *Pseudomonas* species, *Klebsiella pneumonia*, and *Vibrio cholerae*. Chapter 15 discusses GEIs in marine bacteria and Chap. 16 describes the challenges in HGT. Chapter 17 explores artificial intelligence and machine learning for prediction and analysis of GEIs.

Therefore, this book provides a comprehensive source of molecular basis understanding of GEIs and their subtypes. This book is a valuable source not only for beginners, but also for students, researchers, clinicians, stakeholders, and policymakers interested in the potential of the molecular level understanding of GEIs at different levels in their fields.

New Delhi, India — Indra Mani
Mehsana, Gujarat, India — Vijai Singh
Taif, Saudi Arabia — Khalid J. Alzahrani
Hanoi, Vietnam — Dinh-Toi Chu

Acknowledgements

I would like to express my sincere gratitude and deep appreciation to the Principal, Gargi College, the University of Delhi, India, for extending his outstanding support and motivation to complete this book. I am thankful to my friend, colleague, and co-editor Dr. Vijai Singh, for his great source of inspiration and motivation to complete this project. Further, I would like to thank my co-editors Dr. Dinh-Toi Chu and Dr. Khalid J. Alzahrani, for their encouragement and support to complete this project. I am delighted to thank all the authors for their excellent contributions to this book.

I would like to thank editorial staff members Dr. Bhavik Sawhney and Ms. Nandhini Viswanathan from Springer for their excellent management of this project. I would like to thank Prof. Janardan Yadav, Dr. Satya Prakash, Dr. Anurag Tiwari, Dr. Jay Prakash Verma, Dr. Ajay Kumar Singh, Dr. Santosh Kumar Mishra, Dr. Ashok Saini, Dr. Vimal C. Pandey, and who have directly or indirectly helped in shaping this project.

I wish to express my heartfelt gratitude to my parents, and all members of my family who always filled my mind with affirmations and crowded out the negative thoughts. Most of all, I thank my wife Reena Gond and sons Ankur Kumar Gond and Ankit Kumar Gond, who have helped me during this project. I would like to warmly thank the Faculty and Staff of the Department of Microbiology, Gargi College, University of Delhi, for providing a great working environment.

And finally, I express my sincere gratitude to all my friends and well-wishers for providing me moral support during the process of writing this book. Last but not least, my sincere thanks to GOD for his supreme POWER for endowing me to live with joy and victory in the shape of this book.

Dr. Indra Mani

I would like to express my sincere gratitude and deep appreciation to Dr. J. S. Yadav, Director (Research), Indrashil University, India, for extending his outstanding support and motivation to complete this book. I would like to give many thanks to co-editors, Dr. Indra Mani, Dr. Khalid J. Khalid J. Alzahrani, and Dr. Dinh-Toi Chu, of this book who gave me outstanding personal and professional support as well as inspiration to finish this book.

I am delighted to thank all the authors for their excellent contributions to this book. I would like to thank Dr. Bhavik Sawhney (Associate Editor—Biomedicine) and Ms. Nandhini Viswanathan (Production Editor) from Springer for their excellent management of this project.

I would like to thank Prof. Chaitanya G. Joshi, Prof. Rakesh Rawal, Prof. Bharat Maitreya. Prof. Pawan K. Dhar, Dr. Poonam Bhargava, Dr. Madhvi Joshi, Dr. Bhabatosh Das, Dr. Pablo Carbonell, Dr. Rupesh Maurya, Dr. Satya Prakash, Dr. Vimal C. Pandey, Dr. Suresh Ramakrishna, and those whose names do not feature here but have directly or indirectly contributed to shaping this project.

I greatly appreciate the support of my students Mr. Nisarg Gohil. Mr. Khushal Khambhati, and Ms. Gargi Bhattacharjee, whose discussion and comments helped to shape this book.

I wish to express my gratitude to my beloved wife Pritee Singh for her endless support, patience, and inspiration. Lots of affection for my kids Aaradhya and Ayush who missed me during this project. I would like to warmly thank the faculty and staff of Indrashil University for providing a great working environment.

I am aware that even despite our best efforts, the first version always comes with some error that may have crept in the compilation. I would be delighted to receive feedback from readers to further improve the future book.

Dr. Vijai Singh

I express my sincere thanks to all the authors for their excellent contributions to this book. This book would not have been possible without the support of many people. I am grateful to my parents, wife, and children for their love and support and for believing in me. They have been extremely supportive to me throughout this entire process and have made countless sacrifices in helping me get to this point.

I would like to express my deep and sincere gratitude to my research team at Taif University, Saudi Arabia, who have made my professional life such a pleasure. My sincere thanks goes to Dr. Khalaf Alsharif, Dr. Hosam Alzahrani, Dr. Ashraf Albarakati, Dr. Ibrahim Halawani, and Dr. Fuad Alzahrani for their camaraderie and unwavering support. Their levels of patience, knowledge, and ingenuity is something I will always keep aspiring and motivated.

Dr. Khalid J. Alzahrani

This book comes from the idea given not only by me but also by all editors Dr. Indra Mani, Dr. Vijai Singh, and Dr. Khalid J. Alzahrani. Therefore, I would like to express my sincere gratitude and deep appreciation to co-editors, especially Dr. Vijai Singh for this meaningful and great work.

I am delighted to thank all other contributors such as authors and production editors for their works to this project, without their contributions we could not have developed and finished this book.

I would like to thank my colleagues in the Faculty of Applied Sciences, International School, Vietnam National University, Hanoi, Vietnam, whose knowledge and experience have eased and developed the confidence to take up this work and finally finish the note.

I would like to thank the members at the Center for Biomedicine and Community Health, International School, Vietnam National University, Hanoi, Vietnam, for their contributions as the authors and the connectors to other scientists in the world who have their inputs directly or indirectly in the book.

Dr. Dinh-Toi Chu

Contents

Editors and Contributors

About the Editors

Indra Mani is an Assistant Professor in the Department of Microbiology, Gargi College, University of Delhi (DU), New Delhi, India. He received M.Sc. in Microbiology from Bundelkhand University (BU), Jhansi, UP, India, and Ph.D. in Biochemistry (2010) from Banaras Hindu University (BHU), Varanasi, UP, India. He completed postdoctoral training in the Department of Physiology, School of Medicine, Tulane University, New Orleans, Louisiana, USA. He has worked as a Senior Research Officer (SRO) in the Department of Medicine, All India Institute of Medical Sciences (AIIMS), New Delhi, India. He has more than 10 years of research and teaching experience in Microbiology, Molecular Genetics, Cell Signaling, and Bioinformatics. He has published more than 50 articles in peer-reviewed journals, 35 book chapters, and 4 books. He has presented several papers in the different national and international conferences. He has extensive experience in microbial characterization, molecular genetics, cell culture, fluorescence/confocal microscopy, and in silico analysis. He serves as a reviewer of peer-reviewed journals and a member of national and international societies.

Vijai Singh is an Associate Professor and Dean (Research and Innovation) in the Department of Biosciences, School of Science at Indrashil University, Rajpur, Mehsana, Gujarat, India. He was an Assistant Professor in the Department of Biological Sciences and Biotechnology at the Institute of Advanced Research, Gandhinagar, India, and also an Assistant Professor in the Department of Biotechnology at the Invertis University, Bareilly, India. Prior to that, he was a Postdoctoral Fellow in the Synthetic Biology Group at the Institute of Systems and Synthetic Biology, Paris, France, and at the School of Energy and Chemical Engineering at the Ulsan National Institute of Science and Technology, Ulsan, South Korea. He received his Ph.D. in Biotechnology (2009) from the National Bureau of Fish Genetic Resources, Uttar Pradesh Technical University, Lucknow, India. During his Ph.D., he has cloned, expressed, and characterized hemolysin from *Aeromonas hydrophila* in *Escherichia coli*. He has extensive experience in MAGE, small regulatory RNAs, pathway designing, CRISPR-Cas systems, and microfluidics. His research interests are focused on design of novel biosynthetic pathways for production of medically and industrially important biomolecules. Additionally, his laboratory is developing on CRISPR-Cas9 tools for genome editing and regulation. He has more than 10 years of research and teaching experience in synthetic biology, metabolic engineering, microbiology, and industrial microbiology. He has published more than 80 research articles, 51 book chapters, 16 books, and 4 patents. He serves as an associate editor, editorial board member, and reviewer of a number of peer-reviewed journals. He is also a member of the Board of Study and Academic Council of Indrashil University and is the Member Secretary of Institutional Biosafety Committee (IBSC) in the same University.

Khalid J. Alzahrani is an Associate Professor in the Department of Clinical Laboratories Sciences at Taif University in Taif, Saudi Arabia. He earned a B.Sc. in Laboratory medicine from Umm Al-Qura University, Saudi Arabia (2009), an M.Sc. in Clinical Microbiology from Glasgow Caledonian University, UK (2013), and a Ph.D. in Infection and Immunity from the University of Glasgow, UK (2017). His main areas of interest in research are tropical diseases, drug resistance, vaccine design, proteomics, and bioinformatics, and he has published more than 70 research articles and 11 chapters.

Dinh-Toi Chu is the Dean of the Faculty of Applied Sciences, International School, Vietnam National University, Hanoi, Vietnam, and he currently also serves as the Director at the Center for Biomedicine and Community Health, International School, Vietnam National University, Hanoi, Vietnam. He was a visiting professor at the Department of Cell Biology and Anatomy, College of Medicine, National Cheng Kung University, Tainan, Taiwan, from June 2021 to December 2021. Dr. Chu was a former researcher and Marie Curie Fellow at the Center for Molecular Medicine Norway and Institute of Cancer Research, Oslo University Hospital, Norway. Dr. Chu has a strong background in biomedical science and public health, especially in lipid metabolism, cancer immunity, and community health, with the Doctor of Veterinary Medicine degree focusing on infection immunity (Vietnam, 2006), a Master of Biological Science focusing on cancer and graft-vs-host-disease immunity (South Korea, 2011), and a Ph.D. in Medical Biology (Poland, 2015). He has more than 140 publications in biomedicine and epidemiology in peer-reviewed journals. Since 2014, he has been a reviewer for about 40 several high-qualified scientific journals such as Nature Communications, the FASEB Journal, Frontiers in Pharmacology, Scientific Reports, and Biomedicine and Pharmacotherapy. He has been a Section Editor for the Journal of Advanced Biotechnology and Experimental Therapeutics. He is also acting as an Associate Editor of the Bioengineered and an Academic Editor of the PLOS ONE.

Contributors

Flávia Figueira Aburjaile Preventive Veterinary Medicine Department, Veterinary School, Federal University of Minas Gerais (UFMG), Belo Horizonte, Brazil

Samarpita Adhikary Microbiology and Computational Biology Laboratory, Department of Botany, Raiganj University, Raiganj, West Bengal, India

Oluwatosin Akinwotu Department of Microbiology and Biotechnology Centre, Faculty of Science, The Maharaja Sayajirao University of Baroda, Vadodara, India
Environmental and Biotechnology Unit, Department of Microbiology, University of Ibadan, Ibadan, Nigeria

Zaaima AL-Jabri Department of Microbiology & Immunology, College of Medicine and Health Sciences, Sultan Qaboos University, Muscat, Sultanate of Oman

Nada AL-Mebairik Department of Clinical Laboratory Sciences, College of Applied Medical Sciences, King Saud University, Riyadh, Kingdom of Saudi Arabia

Carlos Augusto Almeida Postgraduate Program in Bioinformatics, Department of Genetics, Ecology, and Evolution, Institute of Biological Sciences, Federal University of Minas Gerais (UFMG), Belo Horizonte, Brazil
Preventive Veterinary Medicine Department, Veterinary School, Federal University of Minas Gerais (UFMG), Belo Horizonte, Brazil

Vasco Ariston de Carvalho Azevedo Postgraduate Program in Bioinformatics, Department of Genetics, Ecology, and Evolution, Institute of Biological Sciences, Federal University of Minas Gerais (UFMG), Belo Horizonte, Brazil

Rajib Bandopadhyay Microbiology Section, UGC-Centre for Advanced Studies, Department of Botany, The University of Burdwan, Golapbag, Bardhaman, West Bengal, India

Debmalya Barh Postgraduate Program in Bioinformatics, Department of Genetics, Ecology, and Evolution, Institute of Biological Sciences, Federal University of Minas Gerais (UFMG), Belo Horizonte, Brazil

Harshita Bhargava Department of Computer Science, IIS University, Jaipur, Rajasthan, India

Mahak Bhushan Department of Biological Sciences ISER Kolkata, Kolkata, West Bengal, India

Raju Biswas Microbiology Section, UGC-Centre for Advanced Studies, Department of Botany, The University of Burdwan, Golapbag, Bardhaman, West Bengal, India

Victor Augusto Sallum Ceballos Department of Microbiology, Immunology, and Parasitology, Institute of Biological and Natural Sciences, Federal University of Triângulo Mineiro (UFTM), Uberaba, Brazil

Ranadhir Chakraborty OMICS Laboratory, Department of Biotechnology, University of North Bengal, Siliguri, India

Stanzin Choksket CSIR-Institute of Microbial Technology, Chandigarh, India

Ankur Choudhary Department of Civil Engineering, Chitkara School of Engineering & Technology, Chitkara University, Baddi, Himachal Pradesh, India

Megha Choudhary CSIR-Institute of Microbial Technology, Chandigarh, India
All India Institute of Medical Sciences (AIIMS), New Delhi, India

Gourav Choudhir Centre for Rural Development and Technology, Indian Institute of Technology Delhi, New Delhi, India

Dinh Toi Chu Center for Biomedicine and Community Health, International School, Vietnam National University, Hanoi, Vietnam
Faculty of Applied Sciences, International School, Vietnam National University, Hanoi, Vietnam

Anu Dalal Chemistry Department, Indian Institute of Technology Delhi, New Delhi, India

Bhabatosh Das Molecular Genetics Laboratory, Department of Infection and Immunology, Translational Health Science and Technology Institute, Faridabad, Haryana, India

Ashkan Dashtban University College, London, UK

Thiago de Jesus Sousa Postgraduate Program in Bioinformatics, Department of Genetics, Ecology, and Evolution, Institute of Biological Sciences, Federal University of Minas Gerais (UFMG), Belo Horizonte, Brazil

Janaíne Aparecida de Paula Department of Microbiology, Immunology, and Parasitology, Institute of Biological and Natural Sciences, Federal University of Triângulo Mineiro (UFTM), Uberaba, Brazil

Sushanta Deb All India Institute of Medical Sciences (AIIMS), New Delhi, India
Department of Molecular Biology & Bioinformatics, Tripura University, Suryamaninagar, Tripura, India

Goutam Kumar Dhandh Bioclues.org, Hyderabad, India

Rodrigo Dias de Oliveira Carvalho Postgraduate Program in Bioinformatics, Department of Genetics, Ecology, and Evolution, Institute of Biological Sciences, Federal University of Minas Gerais (UFMG), Belo Horizonte, Brazil
Department of Biochemistry and Biophysics, Federal University of Bahia, Salvador, Brazil

Fabiana Vieira Dominici Department of Microbiology, Immunology, and Parasitology, Institute of Biological and Natural Sciences, Federal University of Triângulo Mineiro (UFTM), Uberaba, Brazil

Bhramar Dutta Microbiology Section, UGC-Centre for Advanced studies, Department of Botany, The University of Burdwan, Golapbag, Bardhaman, West Bengal, India

Prajesh Dutta Microbiology and Computational Biology Laboratory, Department of Botany, Raiganj University, Raiganj, West Bengal, India

Andrei Giacchetto Felice Department of Microbiology, Immunology, and Parasitology, Institute of Biological and Natural Sciences, Federal University of Triângulo Mineiro (UFTM), Uberaba, Brazil

Devarshi Gajjar Department of Microbiology and Biotechnology Centre, Faculty of Science, The Maharaja Sayajirao University of Baroda, Vadodara, India

Vishakha Grover Dr. HS Judge Institute of Dental Sciences and Hospital, Panjab University, Chandigarh, India

Urmi Halder Microbiology Section, UGC-Centre for Advanced studies, Department of Botany, The University of Burdwan, Golapbag, Bardhaman, West Bengal, India

Sowmya Hari Department of Bio-Engineering, School of Engineering, Vels Institute of Science, Technology and Advanced Studies, Chennai, Tamilnadu, India

Harshvardhan CSIR-Institute of Microbial Technology, Chandigarh, India

Meera Indracanti Department of Life Sciences, School of Allied Health Sciences, Malla Reddy University, Hyderabad, Telangana, India

Ashish Jain Dental Institute, Rajendra Institute of Medical Sciences, Ranchi, Jharkhand, India

Arun Kumar Jaiswal Postgraduate Program in Bioinformatics, Department of Genetics, Ecology, and Evolution, Institute of Biological Sciences, Federal University of Minas Gerais (UFMG), Belo Horizonte, Brazil

Purvi Joshi Department of Microbiology and Biotechnology Centre, Faculty of Science, The Maharaja Sayajirao University of Baroda, Vadodara, India

Alexandre Claudino Junior Postgraduate Program in Bioinformatics, Department of Genetics, Ecology, and Evolution, Institute of Biological Sciences, Federal University of Minas Gerais (UFMG), Belo Horizonte, Brazil

Varsha K Department of Bio-Engineering, School of Engineering, Vels Institute of Science, Technology and Advanced Studies, Chennai, Tamilnadu, India

Ashutosh Kabiraj Microbiology Section, UGC-Centre for Advanced Studies, Department of Botany, The University of Burdwan, Golapbag, Bardhaman, West Bengal, India

Dibyendu Khan Microbiology Section, UGC-Centre for Advanced Studies, Department of Botany, The University of Burdwan, Golapbag, Bardhaman, West Bengal, India

Ashutosh Khaswal Department of Biotechnology, IMS Engineering College, Ghaziabad, U.P, India

Maytiya Konkit Division of Microbiology, Faculty of Science and Technology, Nakhon-Pathom Rajabhat University, Nakhon-Pathom, Thailand

Suresh Korpole CSIR-Institute of Microbial Technology, Chandigarh, India

Anuj Kumar ICAR-Indian Agricultural Statistics Research Institute (IASRI), New Delhi, India

Mohit Kumar Department of Biotechnology, IMS Engineering College, Ghaziabad, U.P, India

Sushil Kumar Department of Botany, Shaheed Mangal Pandey Govt Girls PG College, Meerut, India

Moitri Let Microbiology Section, UGC-Centre for Advanced Studies, Department of Botany, The University of Burdwan, Golapbag, Bardhaman, West Bengal, India

Krishnendu Majhi Microbiology Section, UGC-Centre for Advanced Studies, Department of Botany, The University of Burdwan, Golapbag, Bardhaman, West Bengal, India

Indra Mani Department of Microbiology, Gargi College, University of Delhi, New Delhi, India

Pedro Henrique Marques Department of Microbiology, Immunology, and Parasitology, Institute of Biological and Natural Sciences, Federal University of Triângulo Mineiro (UFTM), Uberaba, Brazil

Yngrid Victória Cassiano Mascarenhas Department of Microbiology, Immunology, and Parasitology, Institute of Biological and Natural Sciences, Federal University of Triângulo Mineiro (UFTM), Uberaba, Brazil

Navnit Kumar Mishra University Institute of Biotechnology (UIBT), Chandigarh University, Mohali, Punjab, India

Santosh Kumar Mishra Department of Biotechnology, IMS Engineering College, Ghaziabad, Uttar Pradesh, India

Shriparna Mukherjee Department of Botany, Prasannadeb Women's College, Jalpaiguri, India

Lekshmi Narendrakumar Molecular Genetics Laboratory, Department of Infection and Immunology, Translational Health Science and Technology Institute, Faridabad, Haryana, India

Duy-Ha Nguyen Department of Microbiology, Vietnam Military Medical University, Hanoi, Vietnam

M. Nidheesh Amrita School of Biotechnology, Amrita Vishwavidyapeetham, Clappana, Kerala, India

Ayon Pal Microbiology and Computational Biology Laboratory, Department of Botany, Raiganj University, Raiganj, West Bengal, India

Sakthiselvan Punniavan Department of Bio-Engineering, School of Engineering, Vels Institute of Science, Technology and Advanced Studies, Chennai, Tamilnadu, India

Manpreet Kaur Rawal Department of Microbiology, Gargi College, University of Delhi, New Delhi, India

Rajendra Kr Roy Microbiology Section, UGC-Centre for Advanced Studies, Department of Botany, The University of Burdwan, Golapbag, Bardhaman, West Bengal, India

Vivek Roy Microbiology and Computational Biology Laboratory, Department of Botany, Raiganj University, Raiganj, West Bengal, India

Barnan Kr Saha Microbiology and Computational Biology Laboratory, Department of Botany, Raiganj University, Raiganj, West Bengal, India

Jayanti Saha Microbiology and Computational Biology Laboratory, Department of Botany, Raiganj University, Raiganj, West Bengal, India
Department of Botany, Acharya Prafulla Chandra Roy Government College, Siliguri, West Bengal, India

Ashok Saini Department of Microbiology, Institute of Home Economics, University of Delhi, New Delhi, India

Rodrigo Profeta Silveira Santos Postgraduate Program in Bioinformatics, Department of Genetics, Ecology, and Evolution, Institute of Biological Sciences, Federal University of Minas Gerais (UFMG), Belo Horizonte, Brazil

Shrabana Sarkar Microbiology Section, UGC-Centre for Advanced Studies, Department of Botany, The University of Burdwan, Golapbag, Bardhaman, West Bengal, India
Centro de Investigación de Estudios Avanzados del Maule, Vicerrectoría de Investigación y Posgrado, Universidad Católica del Maule, Talca, Chile

Amita Sharma Department of Computer Science, IIS University, Jaipur, Rajasthan, India

Muskan Sharma Department of Bioscience and Biotechnology, Banasthali Vidyapith, Tonk, Rajasthan, India

Sankalp Sharma Bioclues.org, Hyderabad, India

Ankita Shukla University Institute of Biotechnology (UIBT), Chandigarh University, Mohali, Punjab, India

Suraj Shukla Department of Microbiology and Biotechnology Centre, Faculty of Science, The Maharaja Sayajirao University of Baroda, Vadodara, India

Anju Singh Department of Chemistry, Shaheed Mangal Pandey Govt Girls PG College, Meerut, India

Vijai Singh Department of Biosciences, School of Science, Indrashil University, Rajpur, Gujarat, India

Siomar de Castro Soares Department of Microbiology, Immunology, and Parasitology, Institute of Biological and Natural Sciences, Federal University of Triângulo Mineiro (UFTM), Uberaba, Brazil

Nguyen Thai Son Department of Microbiology, Vietnam Military Medical University, Hanoi, Vietnam

Eduarda Guimarães Sousa Department of Microbiology, Immunology, and Parasitology, Institute of Biological and Natural Sciences, Federal University of Triângulo Mineiro (UFTM), Uberaba, Brazil

Meenambiga Setti Sudharsan Department of Bio-Engineering, School of Engineering, Vels Institute of Science, Technology and Advanced Studies, Chennai, Tamilnadu, India

Vijayaraghava S. Sundararajan School of Computing and Information Systems, Singapore Management University, Bras Basah, Singapore

Prashanth Suravajhala Bioclues.org, Hyderabad, India
Amrita School of Biotechnology, Amrita Vishwavidyapeetham, Clappana, Kerala, India

Van Thai Than Faculty of Applied Sciences, International School, Vietnam National University, Hanoi, Viet Nam
Center for Biomedicine and Community Health, International School, Vietnam National University, Hanoi, Viet Nam

Praveenya Tirunagari Department of Biotechnology, Indian Institute of Technology, Kharagpur, India

Sandeep Tiwari Postgraduate Program in Microbiology, Institute of Biology, Federal University of Bahia, Salvador, Brazil
Postgraduate Program in Immunology, Institute of Health Sciences, Federal University of Bahia, Salvador, Brazil

Pinal Trivedi Department of Microbiology and Biotechnology Centre, Faculty of Science, The Maharaja Sayajirao University of Baroda, Vadodara, India

Jayaraman Valadi Department of Computer Science, FLAME University, Pune, MH, India

Atul Vashist Molecular Genetics Laboratory, Department of Infection and Immunology, Translational Health Science and Technology Institute, Faridabad, Haryana, India

Abha Vashistha Department of Biosciences, University institute of Biotechnology, Chandigarh University, Mohali, Punjab, India

Chaitenya Verma Department of Pathology, Ohio State University, Columbus, OH, USA

Jyoti Verma Molecular Genetics Laboratory, Department of Infection and Immunology, Translational Health Science and Technology Institute, Faridabad, Haryana, India

Thaís Cristina Vilela Rodrigues Postgraduate Program in Bioinformatics, Department of Genetics, Ecology, and Evolution, Institute of Biological Sciences, Federal University of Minas Gerais (UFMG), Belo Horizonte, Brazil

Michele Min San Wu Department of Microbiology, Immunology, and Parasitology, Institute of Biological and Natural Sciences, Federal University of Triângulo Mineiro (UFTM), Uberaba, Brazil

An Introduction to Microbial Genomic Islands for Evolutionary Adaptation and Pathogenicity

1

Ashok Saini, Indra Mani, Manpreet Kaur Rawal, Chaitenya Verma, Vijai Singh, and Santosh Kumar Mishra

Abstract

Genomic islands (GEIs) play a vital role in the bacteria's evolutionary adaptation and pathogenicity. GEIs are the unusual region located in the bacterial genome acquired through horizontal gene transfer (HGT). HGT is achieved through various ways, such as conjugation, transformation, and transduction. General features of GEIs are different G + C content as compared to other parts of the genome (core genome), size about 10–200 kb, association with tRNA-encoding genes, mostly flanked by repeat structures and may carry other accessory elements such as insertion sequence (IS) elements, plasmids, bacteriophages, and other mobile elements. GEIs are present in both pathogenic and non-pathogenic bacteria. Therefore, based on the composition, characteristics, and functions of GEIs, it has further divided into various islands such as pathogenicity, resistance, fitness and metabolic or symbiosis islands in pathogenic, environmental, commensal, and symbiotic bacteria. This chapter briefly

A. Saini
Department of Microbiology, Institute of Home Economics, University of Delhi, New Delhi, India

I. Mani (✉) · M. K. Rawal
Department of Microbiology, Gargi College, University of Delhi, New Delhi, India
e-mail: indra.mani@gargi.du.ac.in

C. Verma
Department of Pathology, Ohio State University, Columbus, OH, USA

V. Singh
Department of Biosciences, School of Science, Indrashil University, Rajpur, Mehsana, Gujarat, India

S. K. Mishra
Department of Biotechnology, IMS Engineering College, Ghaziabad, Uttar Pradesh, India

I. Mani et al. (eds.), *Microbial Genomic Islands in Adaptation and Pathogenicity*,
https://doi.org/10.1007/978-981-19-9342-8_1

highlights the GEIs and their role in the bacteria's evolutionary adaptation and pathogenicity. In addition, the role of GEIs in pangenome is also discussed.

Keywords

Genomic Islands (GEIs) · Pathogenicity islands (PAIs) · Pangenome · Horizontal gene transfer (HGT) · Adaptation · Evolution · Antibiotic resistance

1.1 Introduction

Microbial genomes have evolved as a result of slow accumulation of mutations, largely by taking over new DNA either via the entry of single genes or the addition of large gene clusters, known as Genomic Islands (GEIs), via horizontal gene transfer (HGT). HGT is also known as lateral gene transfer (LGT). Generally, large blocks of horizontally acquired DNA sequences encode virulent determinant in pathogenic bacteria and help in the evolution of the microbial genomes. When there are multiple GEIs within a single genome, up to 28% of similarity has been found among them, indicating recurring possession from the same donor to the same acceptor (Roos and van Passel 2011). Further study will help to a better understanding of suggested output.

The GEIs are usually inserted into tRNA genes, transfer-messenger RNA (tmRNA) genes, and some small RNA genes. First, GEI was detected in the GMP synthase gene (*guaA*). Later on, around 34 GEIs have been detected in *guaA* genes of 987 completely sequenced genomes of archaea and bacteria (Proteobacteria, Firmicutes, and Actinobacteria) (Song et al. 2012). The mobility of these islands was further determined and found that *guaA*-associated islands are composed of integrases from P4 and PhiLC3 phages. Additionally, island-encoding proteins such as AlpA and XRE act as positive and negative transcriptional regulators for P4 and PhiLC integrases, respectively (Song et al. 2012). More studies will help to understand the distribution of GEIs in other microorganisms.

GEIs play many roles in various metabolic activities, adaptation during evolution, pathogenesis, and multiple drug resistance. Rao et al. predicted and functionally characterized around 46 GEIs found in the genome of livestock-associated *Staphylococcus aureus*. These GEIs are found to be consisting of various genes of metabolic operons like *leuABCD* and *folPK* genes suggesting their role and importance in niche adaptation. Mycoplasma and Rickettsia were found as the key donors of these elements to *S. aureus*, suggesting their evolutionary relatedness (Rao et al. 2020). The transfer of genes or part of genomes between species is usually done by transduction, conjugation, or transformation. A graphic representation of the mode of the HGT is given in Fig. 1.1. The transferred DNA modifies the microbial genome abruptly and provides some new properties such as pathogenicity or virulence. For pathogenicity, many virulence factors such as toxins, iron uptake system, adhesins, protein secretory system, etc., are required (Gal-Mor and Finlay 2006). The genes of these virulence factors are usually present in a particular region of the genome

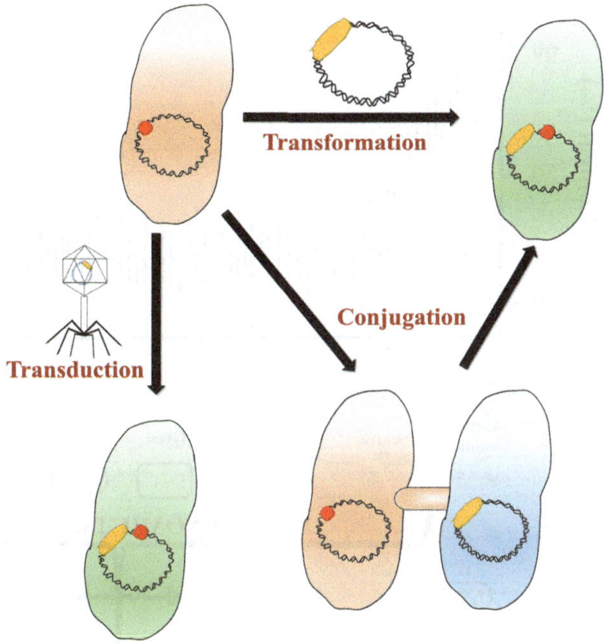

Fig. 1.1 Graphic representation of the mode of the horizontal gene transfer (HGT) via transformation, conjugation, and transduction

known as pathogenicity islands (PAIs) (Gal-Mor and Finlay 2006). Additional studies will help to recognize the different types of GEIs in numerous microorganisms.

The PAIs are only present in a wide range of pathogenic strains or species of both Gram-positive and Gram-negative bacteria which causes diseases in plants, animals, and humans but generally absent in non-pathogenic strains or species. The non-pathogenic strains sometimes acquired these pathogenicity islands via HGT mechanisms (Hacker and Kaper 2000). These PAIs were first identified in *Escherichia coli,* one of the most common human pathogenand later in the microbial genomes of other pathogenic bacteria causing diseases in plants, animals, and humans. Structurally, PAIs have many features such as different G + C content compared with other genomic regions, direct repeats at their ends, size of about 10–200 kb, associated with tRNA genes, genetic instability, presence of integrase enzyme and other mobile regions (Hacker and Kaper 2000). In addition, transferred DNA fragments usually contain clusters of genes associated with various functions such as microbial adaptation, pathogenicity, and multiple drug resistance known as GEIs. A schematic illustration of GEIs of bacteria is shown in Fig. 1.2. Thus, GEIs play an important role in microbial evolution, speciation, and disease outbreak.

A comprehensive analysis of genomic islands of 63 prokaryotic genomes was done by Hsiao et al. in 2005. They had observed that the novel genes are significantly present in the predicted genomic islands in comparison with the other genomic regions. Finding suggests that bacteria and archaea are adapted to a particular environmental condition, such as antibiotic resistance, metal resistance,

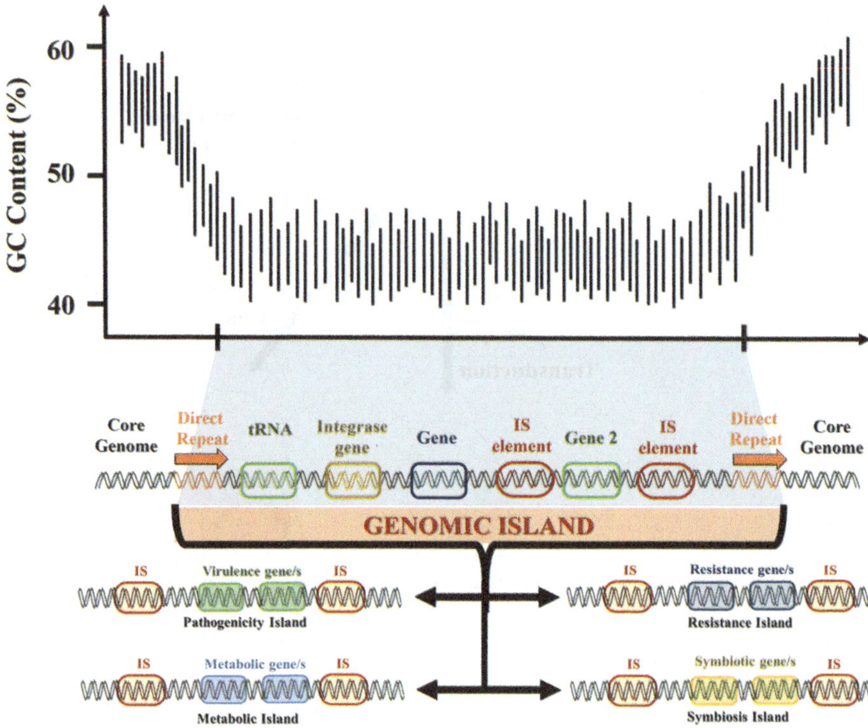

Fig. 1.2 Schematic illustration of genomic islands (GEIs) of bacteria. The GC content (%) of the GIEs is different compared to the core genome. Horizontally transferred DNA linked with tRNA gene and flanked by direct repeats. Functions encoded by acquired genes are shown in different islands such as pathogenicity, resistance, metabolic and symbiosis islands. IS, Insertion sequence element

pathogen virulence activity, etc., via acquiring these novel genes mainly found in genomic islands by the HGT mechanism (Hsiao et al. 2005). Therefore, the importance of PAIs and GEIs in the evolution of bacterial virulence, their identification and molecular characterization have become essential objectives. GC content distribution is a crucial characteristic of a genome. Traditionally, a window-based method is used for computing GC content distribution. However, it had a low sensitivity. In 2014, Zhang et al. developed another technique called the GC profile method, which was highly sensitive, gives a better resolution (Zhang et al. 2014) and therefore, is more favored when identifying horizontally transferred GEIs and other GC-content-related studies. The red recombination approach can be used to study the role of different loci. Cosmid libraries can be screened for clones harboring entire *Salmonella* pathogenic island (SPI) or GEI (Hensel 2007). Further, different bioinformatics approaches are used to identify putative SPI and GEI.

The functional diversity of closely related microorganisms can be studied using *Prochlorococcus* ecotypes as a system. It has been observed that 16S rRNA sequences of closely related strains of photoautotrophs and pathogenic *Prochlorococcus* cells living together were different from each other by ~1%, mainly in the region of GEIs present in them. The genes present in these GEIs are mainly acquired by phage-mediated HGT, which expresses differentially under nutrient and light stress and shares many features (Coleman et al. 2006). In a similar study, 21 GEIs had been identified in the *Burkholderia cenocepacia* J2315 strain using two independent prediction methods based on potential virulence factors. Out of 21 GEIs, 4 were predicted as PAIs, and the genes present in them are found to be responsible for bacterial invasion and replication (Guo et al. 2017).

1.2 Types of Genomic Islands

Based on genetic components and functions, various genomic islands are exhibited, such as pathogenicity, resistance, fitness and metabolic or symbiosis islands in the pathogenic, environmental, commensal, and symbiotic bacteria.

1.2.1 Pathogenicity Islands

Whole genome sequencing (WGS) data of various pathogenic microorganisms help to understand the mechanism of pathogenesis. For instance, the characterization of *S. aureus* genomic island *v*Saβ has led to the identification of the strain's virulence-associated genes (Kläui et al. 2019). WGS data of multi - and extensively drug-resistant (MDR and XDR) *Vibrio cholerae* revealed the presence of various mobile elements, as well as plasmids, play an important role in the fitness and adaptation (Verma et al. 2019). The presence of 4 GEIs in 7 tetracycline-resistant *Chlamydia suis* strains represents an important example of horizontally transferred new DNA into a natural isolate of *Chlamydiae* bacterium (Dugan et al. 2004). Furthermore, these islands have shown significant sequence similarity among several species.

Similarly, the presence of a novel variant of *Salmonella* genomic island 1 (SGI1) in avian pathogenic *E. coli* isolates presents another example of a HGT that enhances its capacity for antimicrobial resistance (Cummins et al. 2019). *Campylobacter* spp. like *Campylobacter coli* and *C. jejuni* cause acute intestinal diseases in humans. An ATP-binding cassette F (ABC-F) protein encoded by the *optrA* gene was found to confer resistance to oxazolidinones and phenicols in these pathogens. Moreover, sequence analysis of nine such *optrA*-positive strains of *Campylobacter* revealed the presence of the *optrA* gene on a chromosome-borne multidrug resistance genomic island (Tang et al. 2021). Similar is the case with another human pathogen, *Streptococcus pyogenes*. Virulence in this pathogen is closely linked to *S. pyogenes* phage-like chromosomal islands (SpyCI) (Nguyen and McShan 2014). These CIs are responsible for antibiotic resistance in *S. pyogenes* and other streptococcal species. Virulence genes present on GEIs and plasmids promoted bacterial pathogen emergence (Schoeniger et al. 2016). SPI in *Salmonella enterica* confers important virulence traits. Similarly, *Aeromonas salmonicida* genomic islands

(AsaGEIs) confer virulence in *Aeromonas salmonicida subsp. Salmonicida* (Vincent et al. 2021). Notably, a variation within the number and composition of GEIs was noticed, which was responsible for different antimicrobial patterns in many strains.

1.2.2 Resistance Islands

Genes responsible for multiple antibiotic resistance are usually found on plasmids. However, antibiotic-resistance genes are also located on mobile DNA elements such as transposons, integrative and conjugative elements (ICEs), GEIs, and prophages. The presence, geographical distribution, and genomic location of 7 mobile oxazolidinone resistance genes were studied recently in different Gram-positive and Gram-negative bacteria by Schwarz et al., in 2021. These genes usually code for 23S rRNA methylases and ABC-F proteins which confers resistance against antibiotics. These oxazolidinone resistance genes were located on transposons, prophages, ICEs, and GEIs, in addition to the plasmids. These mobile elements are mainly responsible for disseminating multiple drug resistance across various strains (Schwarz et al. 2021). Results suggest that different antibiotic-resistance genes may reside in several mobile elements, including plasmids.

A study has shown that the GEIs consist of antibiotic-resistance genes and thus play an important role in the transmission of multiple drug resistance between the species and strains. Mestrovic and Ljubin-Sternak (2018) studied the cause of generation of multiple drug resistance strains of *Chlamydia trachomatis* (*C. trachomatis*), a primary causative agent of sexually transmitted infections mainly in developed and undeveloped countries. They had found that *C. trachomatis* is resistant to multiple antibiotics such as azithromycin, tetracycline, fluoroquinolone, rifampicin, etc., and the development of resistance is linked to various mechanisms such as point mutation, nucleotide substitution, etc., and some of these genes are found in genomic islands present in *C. trachomatis* genome (Mestrovic and Ljubin-Sternak 2018). The transfer of multiple antibiotic-resistance genes from one microorganisms to other is a great concern to clinicians and researchers, and it is a major threat to society.

In another study, the factors responsible for antimicrobial resistance in *Trueperella pyogenes* (*T. pyogenes*) TP3 and TP4 isolates were studied, which causes a variety of suppurative infections. The comparative analysis showed that both isolates were resistant to multiple antibiotics such as erythromycin, azithromycin, tetracycline, amikacin, gentamicin and the genes responsible for the resistance were located on two different genomic islands. Many genes in these regions are homologous to each other and are mainly acquired via HGT (Dong et al. 2020). In a similar study, the role of HGT in *Riemerella anatipestifer,* a Gram-negative bacterium that causes contagious septicemia in birds, was studied to establish a relation of HGT with the acquirement of genetic diversity and antibiotic-resistance genes. An antibiotic-resistance gene cluster was identified in all studied genomes at the same loci, and these loci can acquire various virulence genes (Zhu et al. 2020). Such study needs to further expand in the plant, human, and animal diseases to understand the HGT.

Poultry-associated serovars of non-typhoidal *Salmonella enterica* (NTS) belonging to different epidemiological regions were characterized by the presence of resistome and multidrug resistance patterns. About 60% of all the isolates were found positive for multidrug resistance which is usually mediated by chromosomal single nuceotide polymorphisms (SNPs) and various mobile genetic elements (Cohen et al. 2020). Moreover, a novel genomic island (SGI1) having streptomycin-azithromycin resistance genes was identified in two isolates—Blockley and Kentucky which suggests that the autonomous genetic elements confer antibiotics and resistance to heavy metals and usually disseminate via the food chain to humans and poultry (Cohen et al. 2020). Methicillin-sensitive *Staphylococcus aureus* (MSSA) acquired methicillin-resistance gene and modified as Methicillin-resistance *Staphylococcus aureus* (MRSA) and it shows increased adaptability (Ito et al. 1999). The plasmid-encoded ciprofloxacin-modifying enzyme (CrpP) is responsible for increased resistance to fluoroquinolones in *Pseudomonas aeruginosa* isolates. Ortiz de la Rosa et al. 2020, observed similar crpP-like genes in chromosomes, a part of PAIs (Ortiz de la Rosa et al. 2020). These *crpP*-like genes confer variable levels of reduced susceptibility to fluoroquinolones in multidrug-resistant *P. aeruginosa* clinical isolates.

The structure and composition of the genomic island (AbaR-type genomic islands) responsible for the generation of antimicrobial resistance in *Acinetobacter baumannii* were studied. Conserved sequences have been identified at both ends of AbaRs, which were further used as signature sequences to identify AbaRs in all available genomes of *A. baumannii*. AbaRs were found in more than 2000 genomes and the insertion sites for these AbaRs had been mapped at various locations on the chromosomes, a few plasmids, prophages, transposons, and even other types of genomic islands. Moreover, around 1000 genes associated with antimicrobial resistance were located in AbaRs and some are unique to them, thus displaying a clonal-specific lineage pattern of distribution (Bi et al. 2019). The finding of this study can be helpful to design a primer and probe from the conserved region for rapid detection of microorganisms.

1.2.3 Fitness and Metabolic Islands

As HGT also contributes to the adaptation and diversification of bacteria mediated by GEIs. Thus, the evolutionary origins of *Haemophilus influenzae* antibiotic resistance island (ICEHin1056) were investigated. It has been found that these GEIs have core and accessory genes in almost equal part and the GC% of the core genes is also similar to the host bacteria (38–40%). But antibiotic-resistance genes change the GC % and number of variable sites indicating their recent acquirement (Juhas et al. 2007). New gene acquisition through HGT plays a significant role in the genomic variability of microbial species.

The effect of GEIs present in *Pseudomonas putida* KT2440 strains was analyzed which constitutes ~4.12% of the total genome size. The mutant strain (KTU-U13) was developed by deleting the genomic islands and further characterized. The mutant strain had shown many improved characteristics over original strains, such as higher plasmid transformation efficiency, high heterologous protein expression, effective utilization of various carbon sources, improved polyhydroxyalkanoate (PHA) content, high cell dry weight, and higher chromosomal integration efficiency of various biodegradation pathways. Thus, generating these kinds of mutants with better properties may be useful in various synthetic biology or biodegradation pathways (Liang et al. 2020). Finding suggests that a similar approach can be utilized in other microorganisms, and deleted GEIs mutants can explore for different purposes.

Food poisoning and infection are major threats to public health due to poor hygiene conditions. The major causative agent for food poisoning is *Salmonella enterica* QH (*S. enterica*). *S. enterica* QH strain displays different properties during pathogenesis and in vitro culture. In normal lab conditions *S. enterica* QH easily grows on Luria-Bertani (LB) medium and shows susceptibility toward various antibiotics whereas during infection it quickly adapts to the environmental/host niches and contributes to pathogenicity. Comparative genomic analysis of *S. enterica* QH with other *Salmonella* strains reveals that *S. enterica* QH strain has several large mobile sequences including GEIs which confer pathogenesis by modulating the host's immune system. These GEIs act as drivers for its adaptation in host niches requiring its colonization in intestine and further dissemination (Han et al. 2020). Comprehensive whole genome analyses are required for such niches for different pathogenic microorganisms.

The study has shown that *Pseudomonas putida* can degrade phenolic compounds and increase adaptability (Ravatn et al. 1998). Similarly, GEIs of fecal *E. coli*, *Klebsiella* spp., *Salmonella enterica* subgroups III and VI have been identified and their functions in iron uptake and increased adaptability (Bach et al. 2000; Oelschlaeger et al. 2003). In addition, the complete genome information of the *Wenzhouxiangella marina* supports multiple horizontal gene transfers with other marine bacteria living in the same habitat (Lee et al. 2015), thus expanding the repertoire of marine bacterial genomic diversity.

Fungi are considered model organisms for studying various processes such as adaptive divergence due to their properties like simple morphology, smaller genome size, contrasting and well-identified ecological niches, and shorter generation time, and most of them are responsive to various experimental methods. Fungi are adapted in a very diverse lifestyle ranging from saprotrophs to pathogens, thus playing very important roles such as saprophytes, mycorrhiza, lichens, pathogens, and various fermentation to produce drugs and other products, such as enzymes. Multiple mechanisms and patterns such as divergence source, variations in genomes and speciation are used by fungi to show adaptive divergence for various ecological niches. Gladieux et al. (2014) had reviewed the important processes such as genomic changes, amino acid substitution, gene duplication, gene loss, and changes in gene expression and genome architecture that have a role in adaptation to various ecological niches. They found that a variety of movable genomic segments such as transposable elements and genomic islands are responsible for the interspecific acquirement of genomic variations leading to the speciation of fungal pathogens (Gladieux et al. 2014). Further study will be helpful to understand the role of GEIs in fungal diversity.

1.2.4 Symbiotic Islands

Similar to genomic islands, symbiotic islands can be horizontally transferred. Recently, a putative symbiosis island facilitating symbioses between true bugs and several *Burkholderia* species was found (Stillson et al. 2022). First, symbiosis island was recognized as a transmittable 500-kb DNA element in *Mesorhizobium loti* ICMP3153. It was associated with a $tRNA^{Phe}$ gene (Sullivan and Ronson 1998). It carried the genes related to Nod factor synthesis, island transfer, and nitrogen fixation. In addition, IS elements and transposons are present in the symbiosis islands (Galibert et al. 2001; González et al. 2003). The study has characterized the functions of nitrogen fixation, and nodulation in Rhizobia that increased metabolic versatility and bacteria–host interactions (Sullivan et al. 2002). Similarly, the nitrogen fixation property has been characterized in *Wolinella succinogenes*, increasing the adaptation of microbes (Baar et al. 2003). Symbiotic islands are very helpful for the microorganisms to better survival in different environmental conditions. This area needs to be further explore.

1.3 In Silico Prediction Tools and Databases of Genomic Islands

Several tools have been developed for the identification of the different types of GEIs. Lee et al. developed a web server called—GI-POP (http://gipop.life.nthu.edu.tw) for predicting the genomic islands carrying genes for pathogenicity or antibiotic resistance while sequencing the genomes (Lee et al. 2013). Waack et al. in 2006 designed a very sensitive algorithm—SIGI-HMM to identify the genomic islands in the microbial genomes. This tool analyzes the microbial genomes based on codon usage of every gene of the genome under study in very detail in an interactive manner and finally generates a hypothesis regarding the origin of those islands and the genes present in them (Waack et al. 2006). In silico tools are very effective to predict the GEIs and their types based on different parameters.

There are various in silico tools available for the prediction of genomic islands from the microbial genome. Some of them are mentioned in ascending year of publication, such as PAI-IDA (Tu and Ding 2003), Islander (Mantri and Williams 2004), Wn-SVM (HGT) (Tsirigos and Rigoutsos 2005), SIGI-HMM (HGT) (Waack et al. 2006), MobilomeFINDER (Ou et al. 2007), Centroid (Aravamuthan and Mande 2007), PredictBias (Pundhir et al. 2008), MJSD (Arvey et al. 2009), INDeGenIUS (Shrivastava et al. 2010), EGID (Che et al. 2011), PIPS (Soares et al. 2012), GIST (Hasan et al. 2012), GI-POP (Lee et al. 2013), GC-profile (Zhang et al. 2014), Sighunt (Jaron et al. 2014), GIHunter (Han Wang et al. 2014), IslandViewer 3 (Dhillon et al. 2015), GI-SVM (Lu and Leong 2016), Zisland Explorer (Wei et al. 2017), IslandViewer (Bertelli et al. 2017), MTGIpick (Dai et al. 2018), IslandPath-DIMOB (Bertelli and Brinkman 2018), panRGP (Bazin et al. 2020), 2SigFinder (Kong et al. 2020), Shutter Island (Assaf et al. 2021), etc. In addition, there are various databases available for the genomic islands such as MOSAIC (Chiapello et al. 2005), PAIDB (Yoon et al. 2007), VFDB (Yang et al. 2008), Pre_GI (Pierneef et al. 2015), Islander (Hudson et al. 2015), etc. Databases provide a great source of biological data, which is very helpful at the molecular level of understanding.

1.4 Pangenome

Availability of high throughput sequencing technology (NGS), databases (NCBI, DDBJ, EMBL), and in silico tools helps to comprehensively analyze genomics and metagenomics data (Mani 2020; Gupta et al. 2021; Mani 2021; Gangotia et al. 2021). Comparative genomic analysis of bacteria has shown remarkable diversity in their genome. Further, it was hypothesized that due to the acquisition of new genes through HGT, the genome size of particular bacterial species has shown great diversity. The bacterial genome can acquire new genes anywhere, between 1.6–32% using HGT (Choi and Kim 2007). Hence, GEIs create an additional challenge to executing phylogenetic relationship among diverse bacterial species (Arndt et al. 2016). However, identification and characterization of different GEIs in microbial genomes certainly will help to develop diagnostic markers (molecular and immunological), antibiotics, new vaccines, or cancer therapies (Coates and Hu 2007; Bar et al. 2008). The study has shown the presence of virulence genes and pathogenicity genes in PAIs (Schmidt and Hensel 2004; Ho Sui et al. 2009). It has been identified that potential vaccine candidates are located within the PAIs (Moriel et al. 2010). Based on the rate of genomic variability of the microorganisms, the pangenome has been divided into open and closed pangenome.

1.4.1 Open and Closed Pangenome

Some species have unique sequences which may or may not share by other species. Such a region is considered an accessory/flexible genome. So, pangenome consists of a core genome and an accessory/flexible genome. The core genome will present in all the species, while the accessary/flexible genome is strain-specific. After analysis of strain-specific genome sequencing data, pangenome has been categorized as closed and open pangenome. Open pangenome occurred whereas most likely various bacterial species present and contain large numbers of bacteria such as rhizosphere, biofilms, the rumens, and guts of animals. However, microbes which show closed pangenome are not shown more genomic diversity. This is because such microbes reside in a very isolated and specific environment, such as seawater or the intracellular niches of obligate symbionts and parasites (Dobrindt et al. 2004). Due to this, there are fewer chances of the occurrence of HGT.

1.5 Concluding Remarks

The whole genome shotgun and metagenome sequencing data of various microbial isolates have shown great diversity in size and unusual features in the genome. Remarkably before, it was considered a paradox, but due to the availability of vast genome sequencing data of various microorganisms, now it is being considered a paradigm. To understand the genomic islands and their subtypes, multi-omics approaches such as metagenomics, metatranscriptomics, metaproteomics, and metabolomics can be helpful. To characterize the species/strain, a comprehensive analysis of genome sequencing data of various isolates is required. Due to emergence and re-emergence of pathogenic microorganisms in environments, there is a great opportunity to explore the culture-independent (Metagenomic) method to characterize and identify various genomic islands and their types. In addition, new gene acquisition can be used to develop molecular and immunological tools to diagnose pathogenic microbes and to develop a vaccine against emerging pathogens.

Conflict of Interest None.

References

Aravamuthan S, Mande SS (2007) Identification of compositionally distinct regions in genomes using the centroid method. Bioinformatics 23:2672–2677

Arndt D, Grant J, Marcu A, Sajed T, Pon A, Liang Y, Wishart D (2016) PHASTER: a better, faster version of the PHAST phage search tool. Nucleic Acids Res 44(W1):W16–W21

Arvey AJ, Azad RK, Raval A et al (2009) Detection of genomic islands via segmental genome heterogeneity. Nucleic Acids Res 37:5255–5266

Assaf R, Xia F, Stevens R (2021) Identifying genomic islands with deep neural networks. BMC Genomics 22(Suppl 3):281. https://doi.org/10.1186/s12864-021-07575-5

Baar C, Eppinger M, Raddatz G, Simon J, Lanz C et al (2003) Complete genome sequence and analysis of Wolinella succinogenes. Proc Natl Acad Sci U S A 100(20):11690–11695. https://doi.org/10.1073/pnas.1932838100

Bach S, de Almeida A, Carniel E (2000) The Yersinia high-pathogenicity island is present in different members of the family Enterobacteriaceae. FEMS Microbiol Lett 183(2):289–294. https://doi.org/10.1111/j.1574-6968.2000.tb08973.x

Bar H, Yacoby I, Benhar I (2008) Killing cancer cells by targeted drug-carrying phage nanomedicines. BMC Biotechnol 8:37

Bazin A, Gautreau G, Médigue C, Vallenet D, Calteau A (2020) panRGP: a pangenome-based method to predict genomic islands and explore their diversity. Bioinformatics 36(Suppl_2): i651–i658

Bertelli C, Brinkman FSL (2018) Improved genomic island predictions with IslandPath-DIMOB. Bioinformatics. https://doi.org/10.1093/bioinformatics/bty095

Bertelli C, Laird MR, Williams KP et al (2017) IslandViewer 4: expanded prediction of genomic islands for larger-scale datasets. Nucleic Acids Res 45(W1):W30–W35

Bi D, Xie R, Zheng J, Yang H, Zhu X, Ou HY, Wei Q (2019) Large-scale identification of AbaR-type genomic islands in acinetobacter baumannii reveals diverse insertion sites and clonal lineage-specific antimicrobial resistance gene profiles. Antimicrob Agents Chemother 63(4): e02526–e02518

Che D, Hasan MS, Wang H, Fazekas J, Huang J, Liu Q (2011) EGID: an ensemble algorithm for improved genomic island detection in genomic sequences. Bioinformation 7(6):311–314

Chiapello H, Bourgait I, Sourivong F, Heuclin G, Gendrault-Jacquemard A, Petit MA, El Karoui M (2005) Systematic determination of the mosaic structure of bacterial genomes: species backbone versus strain-specific loops. BMC Bioinf 12(6):171

Choi IG, Kim SH (2007) Global extent of horizontal gene transfer. Proc Natl Acad Sci U S A 104(11):4489–4494. https://doi.org/10.1073/pnas.0611557104

Coates AR, Hu Y (2007) Novel approaches to developing new antibiotics for bacterial infections. Br J Pharmacol 152:1147–1154

Cohen E, Davidovich M, Rokney A, Valinsky L, Rahav G, Gal-Mor O (2020) Emergence of new variants of antibiotic resistance genomic islands among multidrug-resistant Salmonella enterica in poultry. Environ Microbiol 22(1):413–432

Coleman ML, Sullivan MB, Martiny AC, Steglich C, Barry K, Delong EF, Chisholm SW (2006) Genomic islands and the ecology and evolution of Prochlorococcus. Science 311(5768): 1768–1770

Cummins ML, Roy Chowdhury P, Marenda MS, Browning GF, Djordjevic SP (2019) *Salmonella* genomic island 1B variant found in a sequence type 117 avian pathogenic Escherichia coli isolate. mSphere 4(3):e00169–e00119

Dai Q, Bao C, Hai Y et al (2018) MTGIpick allows robust identification of genomic islands from a single genome. Brief Bioinform 19:361–373

Dhillon BK, Laird MR, Shay JA, Winsor GL, Lo R, Nizam F, Pereira SK, Waglechner N, McArthur AG, Langille MG, Brinkman FS (2015) IslandViewer 3: more flexible, interactive genomic island discovery, visualization and analysis. Nucleic Acids Res 43(W1):W104–W108. https://doi.org/10.1093/nar/gkv401. Epub 2015 Apr 27

Dobrindt U, Hochhut B, Hentschel U, Hacker J (2004) Genomic islands in pathogenic and environmental microorganisms. Nat Rev Microbiol 2(5):414–424

Dong WL, Xu QJ, Atiah LA, Odah KA, Gao YH, Kong LC, Ma HX (2020) Genomic island type IV secretion system and transposons in genomic islands involved in antimicrobial resistance in Trueperella pyogenes. Vet Microbiol 242:108602

Dugan J, Rockey DD, Jones L, Andersen AA (2004) Tetracycline resistance in Chlamydia suis mediated by genomic islands inserted into the chlamydial inv-like gene. Antimicrob Agents Chemother 48(10):3989–3995

Galibert F, Finan TM, Long SR, Puhler A, Abola P, Ampe F et al (2001) The composite genome of the legume symbiont Sinorhizobium meliloti. Science 293(5530):668–672. https://doi.org/10.1126/science.1060966

Gal-Mor O, Finlay BB (2006) Pathogenicity islands: a molecular toolbox for bacterial virulence. Cell Microbiol 8(11):1707–1719

Gangotia D, Gupta A, Mani I (2021) Role of bioinformatics in biological sciences. In: Singh V, Kumar A (eds) Advances in bioinformatics. Springer Nature, Singapore, pp 37–57. ISBN: 978-981-336-191-1

Gladieux P, Ropars J, Badouin H, Branca A, Aguileta G, de Vienne DM, Rodríguez de la Vega RC, Branco S, Giraud T (2014) Fungal evolutionary genomics provides insight into the mechanisms of adaptive divergence in eukaryotes. Mol Ecol 23(4):753–773

González V, Bustos P, Ramírez-Romero MA, Medrano-Soto A, Salgado H, Hernandez-Gonzalez IL et al (2003) The mosaic structure of the symbiotic plasmid of Rhizobium etli CFN42 and its relation to other symbiotic genome compartments. Genome Biol 4(6):R36. https://doi.org/10.1186/gb-2003-4-6-r36

Guo FB, Xiong L, Zhang KY, Dong C, Zhang FZ, Woo PC (2017) Identification and analysis of genomic islands in Burkholderia cenocepacia AU 1054 with emphasis on pathogenicity islands. BMC Microbiol 17(1):73

Gupta A, Gangotia D and Mani I (2021) Bioinformatics tools and softwares Advances in bioinformatics. V. Singh, A Kumar (eds). Springer Nature, Singapore. 15–34. ISBN: 978-981-336-191-1

Hacker J, Kaper JB (2000) Pathogenicity islands and the evolution of microbes. Annu Rev Microbiol 54:641–679

Han S, Hu W, Kan W, Ge Z, Song X, Li L, Shang Y, Zeng Q, Zhou JH (2020) Analyses of genetics and pathogenesis of Salmonella enterica QH with narrow spectrum of antibiotic resistance isolated from yak. Infect Genet Evol 82:104293

Han Wang DC, Wang H, Fazekas J et al (2014) An accurate genomic island prediction method for sequenced bacterial and archaeal genomes. J Proteomics Bioinform 7:214–221

Hasan MS, Liu Q, Wang H, Fazekas J, Chen B, Che D (2012) GIST: Genomic Island suite of tools for predicting genomic islands in genomic sequences. Bioinformation 8(4):203–205

Hensel M (2007) Genome-based identification and molecular analyses of pathogenicity islands and genomic islands in Salmonella enterica. Methods Mol Biol 394:77–88

Ho Sui SJ, Fedynak A, Hsiao WWL, Langille MGI, Brinkman FSL (2009) The association of virulence factors with genomic islands. PLoS One 4:e8094

Hsiao WW, Ung K, Aeschliman D, Bryan J, Finlay BB, Brinkman FS (2005) Evidence of a large novel gene pool associated with prokaryotic genomic islands. PLoS Genet 1(5):e62

Hudson CM, Lau BY, Williams KP (2015) Islander: a database of precisely mapped genomic islands in tRNA and tmRNA genes. Nucleic Acids Res 43:D48–D53

Ito T, Katayama Y, Hiramatsu K (1999) Cloning and nucleotide sequence determination of the entire mec DNA of pre-methicillin-resistant Staphylococcus aureus N315. Antimicrob Agents Chemother 43(6):1449–1458. https://doi.org/10.1128/AAC.43.6.1449

Jaron KS, Moravec JC, Martı'nkova´ N (2014) SigHunt: horizontal gene transfer finder optimized for eukaryotic genomes. Bioinformatics 30(8):1081–1086

Juhas M, Power PM, Harding RM, Ferguson DJ, Dimopoulou ID, Elamin AR, Mohd-Zain Z, Hood DW, Adegbola R, Erwin A, Smith A, Munson RS, Harrison A, Mansfield L, Bentley S, Crook DW (2007) Sequence and functional analyses of Haemophilus spp. genomic islands. Genome Biol 8(11):R237

Kläui AJ, Boss R, Graber HU (2019) Characterization and comparative analysis of the Staphylococcus aureus Genomic Island *ν*Saβ: an *in silico* approach. J Bacteriol 201(22):e00777–e00718

Kong R, Xu X, Liu X, He P, Zhang MQ, Dai Q (2020) 2SigFinder: the combined use of small-scale and large-scale statistical testing for genomic island detection from a single genome. BMC Bioinform 21(1):159

Lee CC, Chen YP, Yao TJ, Ma CY, Lo WC, Lyu PC, Tang CY (2013) GI-POP: a combinational annotation and genomic island prediction pipeline for ongoing microbial genome projects. Gene 518(1):114–123

Lee SH, Choe H, Kim BK, Nasir A, Kim KM (2015) Complete genome of the marine bacterium Wenzhouxiangella marina KCTC 42284(T). Mar Genomics 24(Pt 3):277–280

Liang P, Zhang Y, Xu B, Zhao Y, Liu X, Gao W, Ma T, Yang C, Wang S, Liu R (2020) Deletion of genomic islands in the Pseudomonas putida KT2440 genome can create an optimal chassis for synthetic biology applications. Microb Cell Factories 19(1):70

Lu B, Leong HW (2016) GI-SVM: a sensitive method for predicting genomic islands based on unannotated sequence of a single genome. J Bioinforma Comput Biol 14(01):1640003

Mani I (2020) Current status and challenges of DNA sequencing. In: Singh V (ed) Advances in synthetic biology. Springer Nature, Singapore, pp 71–80. ISBN: 978-981-15-0081-7

Mani I (2021) Role of bioinformatics in microRNA analysis. In: Singh V, Kumar A (eds) Advances in bioinformatics. Springer Nature, Singapore, pp 365–373. ISBN: 978-981-336-191-1

Mantri Y, Williams KP (2004) Islander: a database of integrative islands in prokaryotic genomes, the associated integrases and their DNA site specificities. Nucleic Acids Res 32:D55–D58

Mestrovic T, Ljubin-Sternak S (2018) Molecular mechanisms of *Chlamydia trachomatis* resistance to antimicrobial drugs. Front Biosci 23(4):656–670

Moriel DG, Bertoldi I, Spagnuolo A, Marchi S, Rosini R et al (2010) Identification of protective and broadly conserved vaccine antigens from the genome of extraintestinal pathogenic Escherichia coli. Proc Natl Acad Sci U S A 107:9072–9077

Nguyen SV, McShan WM (2014) Chromosomal islands of Streptococcus pyogenes and related streptococci: molecular switches for survival and virulence. Front Cell Infect Microbiol 12(4): 109

Oelschlaeger TA, Zhang D, Schubert S, Carniel E, Rabsch W, Karch H, Hacker J (2003) The high-pathogenicity island is absent in human pathogens of Salmonella enterica subspecies I but present in isolates of subspecies III and VI. J Bacteriol 185(3):1107–1111. https://doi.org/10.1128/JB.185.3.1107-1111.2003

Ortiz de la Rosa JM, Nordmann P, Poirel L (2020) Pathogenicity genomic island-associated CrpP-like fluoroquinolone-modifying enzymes among Pseudomonas aeruginosa clinical isolates in Europe. Antimicrob Agents Chemother 64(7):e00489–e00420

Ou HY, He X, Harrison EM, Kulasekara BR, Thani AB, Kadioglu A, Lory S, Hinton JC, Barer MR, Deng Z, Rajakumar K (2007) MobilomeFINDER: web-based tools for in silico and experimental discovery of bacterial genomic islands. Nucleic Acids Res 35(Web Server issue):W97–W104

Pierneef R, Cronje L, Bezuidt O, Reva ON (2015) Pre_GI: a global map of ontological links between horizontally transferred genomic islands in bacterial and archaeal genomes. Database (Oxford) 17(2015):bav058. https://doi.org/10.1093/database/bav058

Pundhir S, Vijayvargiya H, Kumar A (2008) PredictBias: a server for the identification of genomic and pathogenicity islands in prokaryotes. In Silico Biol 8:223–234

Rao RT, Sharma S, Sivakumar N, Jayakumar K (2020) Genomic islands and the evolution of livestock-associated Staphylococcus aureus genomes. Biosci Rep 40(11):BSR20202287

Ravatn R, Studer S, Springael D, Zehnder AJ, van der Meer JR (1998 Sep) Chromosomal integration, tandem amplification, and deamplification in Pseudomonas putida F1 of a 105-kilobase genetic element containing the chlorocatechol degradative genes from Pseudomonas sp. Strain B13. J Bacteriol 180(17):4360–4369. https://doi.org/10.1128/JB.180.17.4360-4369.1998

Roos TE, van Passel MW (2011) A quantitative account of genomic island acquisitions in prokaryotes. BMC Genomics 24(12):427

Schmidt H, Hensel M (2004) Pathogenicity Islands in bacterial pathogenesis. Clin Mcrobiol Rev 17:14–56

Schoeniger JS, Hudson CM, Bent ZW, Sinha A, Williams KP (2016) Experimental single-strain mobilomics reveals events that shape pathogen emergence. Nucleic Acids Res 44(14): 6830–6839

Schwarz S, Zhang W, Du XD, Krüger H, Feßler AT, Ma S, Zhu Y, Wu C, Shen J, Wang Y (2021) Mobile oxazolidinone resistance genes in gram-positive and gram-negative bacteria. Clin Microbiol Rev 34(3):e0018820

Shrivastava S, Reddy CVSK, Mande SS (2010) INDeGenIUS, a new method for high-throughput identification of specialized functional islands in completely sequenced organisms. J Biosci 35: 351–364

Soares SC, Abreu VA, Ramos RT, Cerdeira L, Silva A, Baumbach J, Trost E, Tauch A, Hirata R Jr, Mattos-Guaraldi AL, Miyoshi A, Azevedo V (2012) PIPS: pathogenicity island prediction software. PLoS One 7(2):e30848

Song L, Pan Y, Chen S, Zhang X (2012) Structural characteristics of genomic islands associated with GMP synthases as integration hotspot among sequenced microbial genomes. Comput Biol Chem 36:62–70

Stillson PT, Baltrus DA, Ravenscraft A (2022) Prevalence of an insect-associated genomic region in environmentally acquired *Burkholderiaceae* symbionts. Appl Environ Microbiol 88(9): e0250221

Sullivan JT, Ronson CW (1998) Evolution of rhizobia by acquisition of a 500-kb symbiosis island that integrates into a phe-tRNA gene. Proc Natl Acad Sci U S A 95(9):5145–5149. https://doi.org/10.1073/pnas.95.9.5145. Erratum in: Proc Natl Acad Sci U S A 1998 Jul 21;95(15):9059

Sullivan JT, Trzebiatowski JR, Cruickshank RW, Gouzy J, Brown SD et al (2002 Jun) Comparative sequence analysis of the symbiosis island of Mesorhizobium loti strain R7A. J Bacteriol 184(11):3086–3095. https://doi.org/10.1128/JB.184.11.3086-3095.2002

Tang S, Yang R, Wu Q, Ding Y, Wang Z, Zhang J, Lei T, Wu S, Zhang F, Zhang W, Xue L, Zhang Y, Wei X, Pang R, Wang J (2021) First report of the optrA-carrying multidrug resistance genomic island in Campylobacter jejuni isolated from pigeon meat. Int J Food Microbiol 16(354):109320

Tsirigos A, Rigoutsos I (2005) A sensitive, support-vector-machine method for the detection of horizontal gene transfers in viral, archaeal and bacterial genomes. Nucleic Acids Res 33(12): 3699–3707

Tu Q, Ding D (2003) Detecting pathogenicity islands and anomalous gene clusters by iterative discriminant analysis. FEMS Microbiol Lett 221(2):269–275

Verma J, Bag S, Saha B, Kumar P, Ghosh TS, Dayal M, Senapati T, Mehra S, Dey P, Desigamani A, Kumar D, Rana P, Kumar B, Maiti TK, Sharma NC, Bhadra RK, Mutreja A, Nair GB, Ramamurthy T, Das B (2019) Genomic plasticity associated with antimicrobial resistance in *Vibrio cholerae*. Proc Natl Acad Sci U S A 116(13):6226–6231

Vincent AT, Intertaglia L, Loyer V, Paquet VE, Adouane É, Martin P, Bérard C, Lami R, Charette SJ (2021) AsaGEI2d: a new variant of a genomic island identified in a group of Aeromonas salmonicida subsp. salmonicida isolated from France, which bears the pAsa7 plasmid. FEMS Microbiol Lett 368(4):fnab021

Waack S, Keller O, Asper R, Brodag T, Damm C, Fricke WF, Surovcik K, Meinicke P, Merkl R (2006) Score-based prediction of genomic islands in prokaryotic genomes using hidden Markov models. BMC Bioinf 16(7):142

Wei W, Gao F, Du MZ et al (2017) Zisland Explorer: detect genomic islands by combining homogeneity and heterogeneity properties. Brief Bioinform 18:357–366

Yang J, Chen L, Sun L, Yu J, Jin Q (2008) VFDB 2008 release: an enhanced web-based resource for comparative pathogenomics. Nucleic Acids Res 36(Database issue):D539–D542. https://doi.org/10.1093/nar/gkm951

Yoon SH, Park YK, Lee S, Choi D, Oh TK, Hur CG, Kim JF (2007) Towards pathogenomics: a web-based resource for pathogenicity islands. Nucleic Acids Res 35(Database issue):D395–D400. https://doi.org/10.1093/nar/gkl790

Zhang R, Ou HY, Gao F, Luo H (2014) Identification of horizontally-transferred genomic islands and genome segmentation points by using the GC profile method. Curr Genomics 15(2): 113–121

Zhu D, Yang Z, Xu J, Wang M, Jia R, Chen S, Liu M, Zhao X, Yang Q, Wu Y, Zhang S, Liu Y, Zhang L, Yu Y, Chen X, Cheng A (2020) Pan-genome analysis of Riemerella anatipestifer reveals its genomic diversity and acquired antibiotic resistance associated with genomic islands. Funct Integr Genomics 20(3):307–320

Computation Tools for Prediction and Analysis of Genomic Islands

2

Muskan Sharma, Sushil Kumar, Gourav Choudhir, Anju Singh, Anu Dalal, and Anuj Kumar

Abstract

The genomic island (GI) is a discrete cluster of genes in prokaryotic genomes acquired through horizontal gene transfer (HGT). GI includes genes having significant roles in genome evolution and adaptation to the changed environment. Genomic islands may be pathogenicity islands (PAIs), metabolic islands (MIs), secretion islands (SIs), symbiotic islands (SymIs), etc., according to the traits they carry. Major characteristics of GIs exhibit are sporadic distribution, large size, sequence composition, inserted adjacent or near to tRNA genes, and mobility genes. Prediction of genomic islands can be performed via experimental and/or computational methods. Sequence composition and comparative genomics are the two major approaches to predict GIs at the level of single genome and related genomes, respectively. Several computational tools have been designed and developed by researchers to predict the presence, type, transmission, and origin of genomic islands. Representatives of window-based programs are AlienHunter,

M. Sharma
Department of Bioscience and Biotechnology, Banasthali Vidyapith, Tonk, Rajasthan, India

S. Kumar
Department of Botany, Shaheed Mangal Pandey Govt Girls PG College, Meerut, India

G. Choudhir
Centre for Rural Development and Technology, Indian Institute of Technology Delhi, New Delhi, India

A. Singh
Department of Chemistry, Shaheed Mangal Pandey Govt Girls PG College, Meerut, India

A. Dalal
Chemistry Department, Indian Institute of Technology Delhi, New Delhi, India

A. Kumar (✉)
ICAR-Indian Agricultural Statistics Research Institute (IASRI), New Delhi, India

I. Mani et al. (eds.), *Microbial Genomic Islands in Adaptation and Pathogenicity*, https://doi.org/10.1007/978-981-19-9342-8_2

Centroid, INDeGenIUS, Design-Island, and GI-SVM, whereas windowless methods include GC profile and MJSD. Among the available computational predictors GIPSy produced best results followed by Alien_Hunter, IslandViewer, predict Bias, GI Hunter, and Zisland Explorer. Till date except PAIs little information is available about other GIs like REIs, MIs, symbiotic islands, secretary islands, etc. PAIDB, IslandViewer, and InDeGenIUS software partially address these islands and need further exploration.

Keywords

Horizontal gene transfer · Pathogenicity islands · Mobile genetic elements · IslandViewer · AlienHunter · Zisland Explorer

2.1 Introduction

Horizontal gene transfer (HGT) or lateral gene transfer (LGT) is the acquisition and integration of foreign DNA in prokaryotes. Bacteria and Archaea multiply through binary fission and are facile to annex DNA from other organisms via HGT. HGT played a significant role in microbial adaptation, genome evolution, and diversification (Bertelli et al. 2019). In 1990, researchers discovered some virulence gene clusters in the genomes of some *E. Coli* strains that were absent in others (Hacker et al. 1990). Such alien gene clusters were initially termed pathogenicity islands (PAIs). Research revealed the presence of such gene clusters and referred them as secretion islands, antimicrobial resistance islands, and metabolic islands. These facilitate microbial evolution, survival, adaptation, and pathogenicity (Lu and Leong 2016a, 2016b). A more general term "genomic island" (GI) has been given to these horizontally acquired gene clusters. Hence, GIs are discrete segments of DNA established through HGT to the host genome (Jaron et al. 2014). Newly discovered mobile genetic elements (MGEs) are first classified as GIs until their possible mode of transfer, origin, and insertion site is established, then specific names are assigned. Many times the genes responsible for transmission and other functions undergo mutations resulting in the loss of gene functions. After integration in the genome genomic islands evolve through mutations, recombination, and loss or gain of MGEs. The Genomic islands broadly include prophages, integrons, conjugative transposons, and integrative conjugative elements (Hacker and Kaper 2000).

2.1.1 Major Features of GIs

Regardless of their kind putative horizontally transferred gene clusters or GIs share the following major features—

Sporadic Distribution GIs are found sporadically distributed in a few strains of a species and their phyletic patterns differ from the genome patterns. Researches indicate that GIs show instability and an ability to excise spontaneously even in the same strains (Middendorf et al. 2004). A basic local alignment search tool (BLAST) can be used to identify the group of genes that are present in some strains or species but not represented in other related species using sequence similarity (Altschul et al. 1997). *Mauve*, a whole genome sequence alignment tool can recognize conserved regions found around freshly inserted DNA and provide clues that the inserted region may be a GI (Darling et al. 2004).

Size This is challenging to identify horizontally transferred single genes with different phyletic patterns in a genome. But the nearby genes if showing abnormal phyletic patterns provide support to assume that the particular region may be a genomic island. Although, there is no clear-cut size limit to GIs many methods use a minimum size of 8 genes or 8 kb. The MGEs below this size are termed genomic islets.

Sequence Composition Genomic island predictors employing sequence composition greatly depend upon the fact that the sequence composition of genomes from different lineages is different. The sequence composition of GIs certainly differs from the host's genome. Percent GC content, oligonucleotide frequency, CU, and flanked with direct repeats (DRs) are the best signals indicating the presence of non-host gene contents (Juhas et al. 2009). These strategies can be employed for recently sequenced genomes waiting for annotation. The sequence composition bias may lead to false positive results under certain circumstances such as highly expressed ribosomal genes, GIs originating from a species having similar genome composition, and the mutational pressure. This indicates that the results should be confirmed with some other methods (Karlin 2001).

Inserted Adjacent or Near to tRNA Genes Several research reports indicated that the tRNA genes are preferred phage integration sites and direct repeats resulted due to phage insertion (Williams 2002). The BLAST and tRnAscan-SE analysis of genomes revealed that the GIs are generally found inserted in or adjacent to tRNA genes and flanked by direct repeats. Certain types of tRNA genes like tmRNA (transfer messenger RNA) and genes encoding tRNA for amino acids serine, arginine, leucine, threonine, etc., are preferentially used for insertion of GIs (Langille and Brinkman 2009). However, this shows only a small fraction of GIs because many more were found inserted at other locations in the genome.

Mobility Genes MGEs including GIs may be equipped with certain protein-coding genes such as integrases and transposases. The products of such genes provide autonomy (self-mobility) to these horizontally transferred GIs. A specific group of genes encoding surface antigens, host interaction proteins, and other phage-related proteins over-express in GIs (Vernikos and Parkhill 2008). This indicates that phages are probable ancestors of many GIs (Hsiao et al. 2005). All the mentioned features are not necessary to be present in a single genomic island but the presence of a few provides strong evidence in favor of HGT. Hence, we can say that GIs are a superfamily of mobile genetic elements with a core of variable characteristics (Vernikos and Parkhill 2008). These allowed the computational methods to be developed for the prediction of genomic islands in the genome of bacterial strains. Genomic islands can be classified based on their mobility as some are mobile and move horizontally to new locations, for example, integrative and conjugative elements, conjugative transposons, and prophages while others cannot.

2.1.2 Subclasses of GIs

Based on their functions GIs may be grouped as—

Pathogenicity Islands (PAIs) Horizontally transferred DNA fragments of 10–100 kb size encoding virulence factors and allied proteins are termed pathogenicity islands (PAIs). PAIs are associated with pathogenesis. These are generally found inserted in bacterial chromosomes and rarely from plasmids or phages. PAIs are flanked with DRs at both ends and show homology with tRNA genes which enable them to insert into tRNA genes through recombination. These are equipped with integrase, transposase along with virulence factors. The virulence genes include but are not limited to adhesins, secretion systems, toxins, invasions, modulins, effectors, superantigens, iron uptake systems, immunoglobulin A proteases, apoptosis, capsule synthesis, etc. PAIs are highly unstable due to their susceptibility to deletion or mobilization (Hacker and Kaper 2000).

Resistance Islands (REIs) The genomic islands that provide resistance against antibiotics are referred to as resistance islands (REIs). In *Staphylococcus aureus*, SCCmec islands provide methicillin resistance and function like the hotspot to integrate extra genetic elements. REIs size varies from 10 kb to more than 60 kb and incorporates more as well (Hiramatsu et al. 2002). These are generally found inserted in plasmid, integrons, super-integrons, and transposons mobile DNA, these are comparatively more mobile than those occur in bacterial chromosomes.

Metabolic Islands (MIs) The variability in genes related to primary metabolism leads to enhanced adaptability and competitiveness to the bacteria under changed niches. Enterobacteria, *Salmonella senftenberg* in general do not grow on media having only sucrose as a source of carbohydrate (sucrose negative) but a few isolates found sucrose positive. After analysis, this was observed that the responsible genes were located in a ~100 kb fragment inserted into *pheVtRNA* designated as CTn*scr94*. CTn*scr94* encodes an integrase gene and is flanked by 50 bp DRs. Possession of *scr* genes expands the metabolic versatility of the recipient (Dobrindt et al. 2004). Studies on *Thermotoga maritima*, a thermophile revealed that some strains show large DNA fragments transferred through HGT leading to increased genomic diversity. This enabled them to survive at higher temperatures (Nesbo and Doolittle

2003). Similarly, *Wolinella succinogenes* were reported to have horizontally transferred DNA inserted at a $tRNA^{Met}$ and codes for T4SS. The T4SS is found in *Helicobacter pylori* and *Campylobacter jejuni*, significant human pathogens (Baar et al. 2003).

Secretion Islands (SIs) Several microorganisms establish close associations like parasitism and symbiosis with their eukaryotic hosts. Specialized vectorial signal exchange from to host and vice versa through their plasma membranes is required. For example, some animal pathogens consist of PAIs encoding T3SSs and T4SSs responsible for the delivery of effectors that directly interfere with the functions of host cells. T4SSs also help in DNA exchange and led to adaptations to the changed niches. PAI based secretary systems modulate their interactions with a host by interfering with signal transduction pathways, boosting apoptosis, easing out intracellular growth, and initiating the synthesis of nutrients for pathogen multiplication (Büttner and Bonas 2002; Cornelis 2002).

Symbiosis Islands (SymIs) Symbiotic organisms like *Rhizobia* are soil inhabiting, they remain in contact with other soil microorganisms and frequently exchange genetic information through HGT. Shreds of evidence reflect that present diversity in rhizobial strains is mainly due to the acquisition of extra genetic information. This results in adaptation to soil life as well as high metabolic flexibility. This allows the evolution of symbiosis-related complex cellular programs. For instance, strain 1021 of *Sinorhizobium meliloti* the composite structure with three replicons had arisen by the acquisition of pSymA and pSymB megaplasmids and extended metabolic capacities and environmental adaptabilities. pSymA carries genes needed for nodule formation, multiplication at low oxygen concentrations, and nitrogen metabolism while pSymB enabled it to metabolize a broad range of polysaccharides (Galibert et al. 2001). USDA110 strain of *Bradyrhizobium japonicum* comprises the properties of both pSymA and pSymB in a single replicon located in the genome near $tRNA^{Val}$ (Kaneko et al. 2002). In *Mesorhizobium loti* strain ICMP3153 a symbiotic island of 500 kb associated with $tRNA^{Phe}$ has also been identified (Sullivan and Ronson 1998). The number of GIs has increased tremendously with the advancements in technology related to screening and comparing whole genomes.

2.1.3 Detection of Genomic Islands

GIs can be predicted via experimental or computational methods. The two approaches used for the detection of GIs are based on sequence composition and comparative genomics. The former is desirable and easy to apply because they require a single genome but even face the problem of false positive and false negative results while the latter method needs related genomes to be compared (Langille et al. 2010). There are many sequenced genomes without complete annotations and require GI prediction using DNA sequences alone. Based on genome segmentation approaches the sequence composition methods are divided into window-based and windowless methods (Lu and Leong 2016a, 2016b). Commonly used single threshold window-based methods using the single sliding window for segmentation of genome into small regions. Few representative programs under this approach are AlienHunter, Centroid, INDeGenIUS, Design-Island, and GI-SVM. Some windowless methods include GC profile and MJSD (Zhang et al. 2014). Both DNA sequence and gene sequence composition based methods have similar strengths and flaws.

2.2 Tools for Prediction and Analysis of Genomic Islands

An overview of computational tools with their functions is given in Table 2.1.

Virulence Factor Database (VFDB)

VFDB is a computational tool originally built to provide a better idea of the presence of virulence factors from bacterial pathogens. It is used to store entire data of bacterial pathogenesis on one platform. In further studies, comparative pathogenesis was introduced to this database to increase its efficiency. This allowed researchers to explore bacterial genome diversity and also the common virulence factors present in various bacterial species/strains. Manual analysis of results was considered a serious drawback to this database. To overcome this issue, a VFanalyzer was introduced into this database. This allowed automatic inspection of virulence factors present in complete bacterial genomes. The introduction of the VFanalyzer increased the efficiency of the original VFDB and provides divergent virulence factors, even with fewer sequence similarities. This eludes the presence of false positives in the virulence factors related to GI results (Liu et al. 2018).

Zisland Explorer

Keeping in mind the homogeneity and heterogeneity of genomic sequences, a novel tool was built that worked on segmental cumulative GC profiles. This tool is called Zisland Explorer. Zisland Explorer for the first time combines both homogeneity and heterogeneity of a sequence, which increases the dependence of GI prediction on the genomic sequence, making the process less time-consuming. Wei et al. (2016) predicted that Zisland Explorer showed more genuine GIs than many other widely used computational tools.

Pathogenic Island Database (PAIDB)

The PAIDB is the only database that is used to provide comprehensive information on predicted PAIs in prokaryotic genomes. It also automatically identifies homologous PAIs to previously known PAIs. The PAIDB released in the year 2007 had 112 types of PAIs and 889 GenBank accessions in around 497 pathogenic bacterial strains. In 2014, another version of PAIDB called PAIDB v2.0 was introduced,

Table 2.1 Overview of computational tools with their functions

S. N.	Computational tools	Functions
1	VFDB—virulence factor database	Identify PAIs on the basis of virulence, now available with an additional feature of VFanalyzer
2	Zisland Explorer	Works on the basis of heterogeneity and homogeneity of genomic sequence
3	PAIDB—Pathogenic Island Database	Provides comprehensive information on pathogenicity islands (PAIs), new version PAIDBv2.0 includes resistance islands (REIs)
4	GC profile	Works on identifying G-C content of the genome
5	GI-POP—Genomic Island prediction by genomic profile scanning	Use genomic profile scanning(GI-GPS) for the prediction of genomic islands
6	GIST—Genomic Island Suite of Tools	Combines 5 main genomic tools and the result is optimized using EGID (Ensemble Algorithm for Genomic Island Detection)
7	PIPS—Pathogenicity Island Prediction Software	Combines various pathogenicity island detecting methods at one platform
8	GI-SVM—Genomic Island-Support Vector Machine	Uses one-class SVM to detect GI of unannotated sequences.
9	LiSSI—Lifestyle Specific Islands	Prediction of GIs on the basis of bacterial lifestyles like oxygen consumption and pathogenicity.
10	GIPSy—Genomic Island Prediction Software	Modification of PIPS allows the identification of four types of genomic islands.
11	IslandViewer	Web interface that is a compilation of SIGI-HMM, IslandPath-DIMOB, and IslandPick
12	IslandPath-DIMOB	Uses dinucleotide bias including a mobility gene.
13	SIGI-HMM	Score-based prediction based on HMM.
14	IslandPick	Does a comparative study on genomic inputs
15	MobilomeFinder	Comparative study tool to identify tRNA associated genomic islands
16	Centroid	Uses k-mer frequency to identify genomic islands
17	Alien_Hunter	GI prediction using interpolated variable order motifs (IVOM)
18	INDeGenIUS	Identifies 6 different types of genomic islands, similar to the centroid method
19	PredictBias	Prediction by codon bias, percent GC, or dinucleotide bias, VFPB is used to identify PAIs
20	EuGI	Database to predict genomic islands of eukaryotes.
21	panRGP	Pangenome based prediction of region of genomic plasticity (RGPs)

which has 223 types of PAIs and 1331 GenBank accessions. Not only that, this new version also has over 88 types of REIs from 108 accessions. This version provides more accurate island detection than the previous one, and it has more diversity of analyzed genomes (Yoon et al. 2015).

GC Profile

To check the global availability of GC content on a genome, a GC profile is used. This is a high resolution method to predict GIs. The presence of GI is detected by a drop in the profile, the main reason for which is the decrease in the GC content. One of the major drawbacks of this method is its lack of discriminating power of GC content, therefore it is necessary to use other features with this method (Gao and Zhang 2006).

Genomic Island Prediction by Genomic Profile Scanning (GI-POP)

Prediction of GI of the whole microbial genome is a time-consuming process. Also, the presence of REIs an antibiotic gene on these GIs makes it very important to find a better GI predictor. Keeping this in mind, GI-POP was made. GI-POP is a web server that contains tools to assemble the genomic sequences and provides an annotation pipeline and GI module with high quality.

It uses SVM based method that is called genomic profile scanning (GI-GPS) and hence the name of the server GI-POP. When an ongoing genomic project is submitted to GI-POP, it provides both functional annotation and GI prediction. This feature allows researchers to find potential GIs which can further help in completing a genome sequence project in a time-efficient manner (Lee et al. 2013).

Pathogenicity Island Prediction Software (PIPS)

Different computational tools use different methods to predict PAIs on genomic sequences. Detection of PAIs is based on various methods like GC content deviation in CU, virulence factors, hypothetical protein, etc. To combine all these approaches on one platform to provide more accurate detection of PAIs, PIPS was introduced. PIPS is an easy to install (installation independent), more accessible tool which has a web-based interface and the potential to do fast analysis. Soares et al. (2012) used PIPS to study PAIs of *Corynebacterium pseudotuberculosis* and it was observed that PIPS provides more accurate GI prediction results as compared to other computational tools.

Genomic Island Suite of Tools (GIST)

GIST as the name suggest is a software that hosts five main genomic tools. These tools are: Columbo, SIGI-HMM, PAI-IDA, IslandPath, Alien_Hunter, and INDeGenIUS. Along with these tools, an optimization tool called EGID is also added. The main function of this tool is to overcome the lacking quality of results shown by the former five tools individually. EGID allows a better GI prediction by compiling the result of existing tools. Another advantage of GIST is its automatic genome download feature. This provides a platform to download the genomic files via the FTP server of the National Centre for Biotechnology Information (NCBI) (Hasan et al. 2012).

Genomic Island-Support Vector Machine (GI-SVM)

GI-SVM is a computational tool that allows the detection of GI of unannotated sequences to a single genome. The use of one-class SVM (Support Vector Machine) in GI-SVM allows detection of laterally transferred regions, also known as outliers. This method also uses a string kernel to describe the k-mer spectrum, which makes GI prediction more precise. Lu and Leong (2016a, 2016b) studied a comparative GI prediction was done between GI-SVM and other computational tools that use unannotated sequences (like Alien_Hunter and MJSD). The results of this study concluded that GI-SVM provides a better and improved result for GI prediction.

Lifestyle Specific Islands (LiSSI)
This bioinformatics tool helps in GI prediction based on different lifestyles of bacteria like oxygen consumption and pathogenicity. It works on a combination of evolutionary sequence analysis and a statistical approach. Unlike other tools, it works on lifestyle based comparative analysis between genomic islands of two sets of species. The major limitation of this tool is that the non-conserved genetic elements are usually difficult to detect with this approach (Barbosa et al. 2017).

Genomic Island Prediction Software (GIPSy)
GIPSy is a modification of PIPS. While PIPS was used to detect only one type of genomic island, i.e., PAIs, GIPSy helps in detecting all four types of GIs, which are MIs, SymIs, REIs, and PAIs. Therefore, it is not just a pathogenicity or virulence oriented tool. Developed in Java, GIPSy has no dependence on a platform for installation. But this tool does have its limitations. Due to the lack of pre-existing information on different islands other than PAIs, the validation of GIPSy is hindered. Still, the increased diversity of genome analysis using a comprehensive approach to target all types of genomic islands makes GIPSy one of a kind of tool that helps understand the bacterial genome in a better way (Soares et al. 2015).

IslandViewer
The unavailability of a web interface made researchers develop IslandViewer. Earlier the researchers used to download the program on their computer and then work on it. IslandViewer provides a web interface that compiles three tools for GI prediction (SIGI-HMM, IslandPath-DIMOB, and IslandPick). SIGI-HMM and IslandPath-DIMOB use a sequence composition based approach for genomic island prediction and IslandPick uses a comparative genome approach. Pre-computation of the GI dataset makes this web source easy to use for researchers. The validation of GI prediction is done manually. Prediction is available in various formats which can be uploaded to the genomic browser like Artemis; this provides multiple links to GI resources (Langille and Brinkman 2009).

IslandPath-DIMOB
One of the main GI prediction tools of the IslandViewer web server is IslandPath-DIMOB. It predicts GI and uses IslandPick which does comparative genomic analysis and builds a test set. The genomic regions are detected with biased dinucleotide composition. This composition encodes at least eight genes, one of which is a mobility gene. The presence of mobility gene along with dinucleotide bias increases the accuracy of IslandPath-DIMOB. Mobility gene identification is carried out by

using genome annotation for mobility gene research and an HMMer search is conducted for the predicted gene against PFAM mobility gene profiles. This tool has higher recall than SIGI-HMM (Langille et al. 2010). A new modified and more recall offering a version of IslandPath-DIMOB is developed called IslandPath-DIMOB v1.0.0 which provides more precise results for genomic island prediction (Bertelli et al. 2018).

SIGI-HMM

It is a score-based graphical tool under the IslandViewer web server that allows the identification of genomic islands using HMM (Hidden Markov Model). SIGI-HMM works on CU analysis of each alien gene present on the genome input. A comparison of this CU from each gene is done with selected CU tables that contain donor or highly expressed genes. This comparison allows researchers to predict the putative donor of a gene and thus helps detect the origin of a gene. Formulae are used to check if each gene resembles the donor or host. If it is more close to the host, then it is called a putative foreign gene. HMM is used to make putative GIs by combining non-contagious clusters of these putative foreign genes. This provides distinguished data between a normal and HGT associated CU derivation.

SIGI-HMM has higher accuracy than IslandPath-DIMOB. The value of less recall of this tool is compensated by the high recall of IslandPath-DIMOB as both of these tools are used in combination by IslandViewer. The higher accuracy of SIGI-HMM is because of its ability to remove ribosomal regions from the genome (Waack et al. 2006).

IslandPick

This tool is also a part of IslandViewer web sources like SIGI-HMM and IslandPath-DIMOB. It predicts overlapping GIs of SIGI-HMM and IslandPath-DIMOB. IslandPick uses a comparative approach to query genome input. The pre-computation of GI prediction here eliminates the presence of any manual selection-based bias. Alignment of the selected comparative genomes is done by Mauve; this allows refinement of the boundary of a predicted GI by the correction in flanking regions. BLAST is used to ensure that duplication is not present in the predicted region. Default is set to update available predicted data for a genome monthly. Researchers can also do a separate analysis of the unpublished genome as input (Langille et al. 2010).

MobilomeFinder

Like islandPick, this tool also works on a comparative approach. It predicts tRNA gene associated genomic island. A comparison among genomic inputs is made for similar tRNA genes and Mauve is used to align and predict GI in these tRNA regions. The presence of GI associated with the tRNA gene makes this a much more robust tool, but with a limitation that some GI can be missed because not all GIs are necessarily linked with tRNA (Langille et al. 2010).

Centroid
This tool uses the k-mer size or nonoverlapping window size of a genome. It breaks the genome into fragments of a certain size, using the centroid method which converts the k-mers into words and k-mer frequency is used for predicting Islands by detecting outliers. This identifies sequences embedded in the genome that have foreign genes. Working without annotations makes this method efficient. The centroid method for genome island prediction is considered better than the percent GC and Karlin method. Alien_Hunter and INDeGenIUS use the same approach of k-mer frequency for GI prediction (Rajan et al. 2007).

Alien_Hunter
It works on interpolated variable order motifs (IVOM) approach to predict GIs. Variable lengths of k-mers are used. Although the weight of longer k-mers is higher in the scoring system than the shorter k-mers, it is still preferred over shorter k-mers because it has more information and specificity. The predictions made are exposed to 2-state second-order HMM that helps set the boundaries of the predicted genomic island. GI prediction can be uploaded into Artemis genome viewer automatically in embl format (Da Silva Filho et al. 2018).

INDeGenIUS
The Improved N-mer Based Detection of GIs using Sequence-Clustering (INDeGenIUS) is similar to the centroid method, except that centroids are computed by hierarchical clustering. This method allows the study of diverse bacterial genomes by prediction of new GIs that were not mentioned by another prediction method. A study carried out by Shrivastava et al. (2010) mentioned that the application of INDeGenIUS on 400 sequenced species of proteobacteria lead to the identification of 6 different types of genomic islands (SymIs, PAIs, ReIs, MIs, Motility Islands, and Secretion Islands). INDeGenIUS algorithm allows the identification of GIs in a large number of genomes.

PredictBias
PredictBias is a web application used to predict GI along with PAIs and it also allows the study of various features of these islands. Genomic islands are predicted in regions that have a codon bias, percent GC, or dinucleotide bias. PredictBias helps to differentiate PAIs from GIs. If in case a sequence with more than three genes has VFPD hits but does not show composition bias, it is still considered a PAI. Similarly, if even a single virulence protein is encoded by a region, then it is categorized as PAIs and if no such protein is encoded then it is marked as GI. BLASTP provides a graphical view of protein BLAST hits for a GI and comparison can be done for a genome by using a comparison tool (Langille et al. 2010).

EuGI
Prediction of GI is mostly done for the prokaryotic genome. But the work on the eukaryotic genome is still in its juvenile phase. EuGI database is the first GI prediction database for the prediction and comparison of eukaryotes. This database

uses SWGIS 2.0 algorithm on 66 different eukaryotic species and is put together on the EuGI web resource. A total of 10,550 GIs were predicted using this tool, out of which 5299 GIs codes functional protein. SWGIS 2.0 is the first prediction tool for the eukaryotic genome. It is a modification of pre-existing SWGIS (SeqWord Gene Island Sniffer). EuGI provides a computational platform that allows researchers to freely access GI identification in an easy-to-use way with a visual interface as well (Clasen et al. 2018).

panRGP

Most recently developed computational tool that allows detection of Region of Genomic Plasticity (RGP) using pangenome. This tool is a part of PPanGGOLin software suite. panRGP uses pangenome graphs and makes a comparative genome approach for the prediction of Genomic Islands. It provides access to large-scale studies by identifying GI among thousands of genomes in a time-efficient way. panRGP algorithm is considered to be more reliable for detecting points of genome insertion and hence it gives more accurate results for genomic prediction. The large-scale study done by this tool allows researchers to explore a greater diversity of genome sequences (Bazin et al. 2020).

2.3 Conclusion and Future Perspectives

The evolution of genomes especially in bacteria occurs through various mechanisms like mutations, recombination, and HGT. HGT by comparison recently discovered a mechanism leading the fitness of bacterial strains in their niche. The major consequence of genomic islands is that they provide a large number of genes or full operons conferring new traits. The traits acquired in this way became the regular part of the genome by natural selection and provide adaptability to specific and constantly changing growth conditions, maintain genetic flexibility, and enhance competitive ability in their ecological niches. It is speculated that GIs originated from different mobile genetic elements such as integrative plasmids, phages, and insertion sequences and evolved through loss or gain processes. The loss of genetic information from a genome sometimes results in genetic diversification. This might help us in the assessment of risks related to antibiotic resistance in potential human pathogens. This will be interesting to know the mechanisms involved in the maintenance of the genome during the addition and deletion of genetic elements despite the disruption of operons.

The presence and mobility of GIs can be experimentally or computationally predicted. Computational tools are regarded as more convenient to use and manage the retrieved information. With the development in computer sciences and the accumulation of information regarding GIs, several tools have been made based on different characteristics of mobile genetic elements. Genomic comparison analysis and sequence composition analysis are two basic approaches for the development of GI prediction tools. The first strategy compares the related genomes for unique regions whereas the second compares composition of specific regions with other regions in the genome. The next-generation sequencing technologies and high throughput techniques have made large genome drafts that are waiting for annotation. Many software tools have been developed to tackle the problem of genomic island prediction but still, there is a need to develop new GI prediction tools in conjunction with artificial intelligence, machine learning, and pan genomic-based analysis. Software tools such as GI-SVM, IGIPT, PAI-IDA, and SIGI-HMM generally predict GIs acquired recently because these use the signature sequences only. EGID, Islander, and IslandPath help to identify the GIs of homogeneous genomic signatures by using flanking tRNAs, mobility genes, insertion sequences, etc. Based on the literature survey, it is concluded that among the available computational predictors GIPSy produced the best results followed by Alien_Hunter, IslandViewer, predict Bias, GI Hunter, and Zisland Explorer. To date except PAIs little information is available about other GIs like REIs, MIs, symbiotic islands, secretary islands, etc. PAIDB, IslandViewer, and InDeGenIUS software partially address these islands and need further exploration. Prediction of origin needs more attention as GIs adapt their signature with time and their origin prediction is not always possible by comparing the genomic signature of other organisms.

References

Altschul SF, Madden TL, Schaffer AA, Zhang J, Zhang Z, Miller W, Lipman DJ (1997) Gapped BLAST and PSI-BLAST: a new generation of protein database search programs. Nucleic Acids Res 25:3389–3402

Baar C, Eppinger M, Raddatz G, Simon J, Lanz C, Klimmek O, Nandakumar R, Gross R, Rosinus A, Keller H, Jagtap P, Linke B, Meyer F, Lederer H, Schuster SC (2003) Complete genome sequence and analysis of *Wolinella succinogenes*. Proc Natl Acad Sci U S A 100: 11690–11695. https://doi.org/10.1073/pnas.1932838100

Barbosa E, Rottger R, Hauschild AC, de Castro SS, Bocker S, Azevedo V, Baumbach J (2017) LifeStyle-specific-islands (LiSSI): integrated bioinformatics platform for genomic island analysis. J Integr Bioinform 14:20170010. https://doi.org/10.1515/jib-2017-0010

Bazin A, Gautreau G, Medigue C, Vallenet D, Calteau A (2020) panRGP: a pangenome-based method to predict genomic islands and explore their diversity. Bioinformatics 36:i651–i658. https://doi.org/10.1093/bioinformatics/btaa792

Bertelli C, Brinkman FSL, Valencia A (2018) Improved genomic island predictions with IslandPath-DIMOB. Bioinformatics 34:2161–2167. https://doi.org/10.1093/bioinforBioinformatic

Bertelli C, Tilley KE, Brinkman FSL (2019) Microbial genomic island discovery, visualization, and analysis. Brief Bioinform 20:1685–1698

Büttner D, Bonas U (2002) Getting across-bacterial type III effector proteins on their way to the plant cell. EMBO J 21:5313–5322

Clasen FJ, Pierneef RE, Slippers B, Reva O (2018) EuGI: a novel resource for studying genomic islands to facilitate horizontal gene transfer detection in eukaryotes. BMC Genom 19:323

Cornelis GR (2002) The *Yersinia* Ysc–Yop virulence apparatus. Int J Med Microbiol 291:455–462

Da Silva Filho AC, Raittz RT, Guizelini D, De Pierri CR, Augusto DW, dos Santos-Weiss ICR, Marchaukoski JN (2018) Comparative analysis of genomic island prediction tools. Front Genet 9:619. https://doi.org/10.3389/fgene.2018.00619

Darling ACE, Mau B, Blattner FR, Perna NT (2004) Mauve: multiple alignments of conserved genomic sequence with rearrangements. Genome Res 14:1394–1403

Dobrindt U, Hochhut B, Hentschel U, Hacker J (2004) Genomic islands in pathogenic and environmental microorganisms. Nat Rev Microbiol 2:414–425

Galibert F, Finan TM, Long SR, Puhler A, Abola P, Ampe F, Barloy-Hubler F, Barnett MJ, Becker A, Boistard P, Bothe G, Boutry M, Bowser L, Buhrmester J, Cadieu E, Capela D, Chain P, Cowie A, Davis RW, Dreano S, Federspiel NA, Fisher RF, Gloux S, Godrie T, Goffeau A, Golding B, Gouzy J, Gurjal M, Hernandez-Lucas I, Hong A, Huizar L, Hyman RW, Jones T, Kahn D, Kahn ML, Kalman S, Keating DH, Kiss E, Komp C, Lelaure V, Masuy D, Palm C, Peck MC, Pohl TM, Portetelle D, Purnelle B, Ramsperger U, Surzycki R, Thebault P, Vandenbol M, Vorholter FJ, Weidner S, Wells DH, Wong K, Yeh KC, Batut J (2001) The composite genome of the legume symbiont *Sinorhizobium meliloti*. Science 293(5530):668–672. https://doi.org/10.1126/science.1060966

Gao F, Zhang CT (2006) GC-Profile: a web-based tool for visualizing and analyzing the variation of GC content in genomic sequences. Nucleic Acids Res 34:W686–W691. https://doi.org/10.1093/nar/gkl040

Hacker J, Bender L, Ott M, Wingender J, Lund B, Marre R, Goebel W (1990) Deletions of chromosomal regions coding for fimbriae and hemolysins occur *in vitro* and *in vivo* in various extra intestinal *Escherichia coli* isolates. Microb Pathog 8(3):213–225

Hacker J, Kaper JB (2000) Pathogenicity islands and the evolution of microbes. Annu Rev Microbiol 54:641–679

Hasan MS, Liu Q, Wang H, Fazekas J, Chen B, Che D (2012) GIST: genomic island suite of tools for predicting genomic islands in genomic sequences. Bioinformation 8:203–205

Hiramatsu K, Katayama Y, Yuzawa H, Ito T (2002) Molecular genetics of methicillin-resistant *Staphylococcus aureus*. Int J Med Microbiol 292:67–74

Hsiao WW, Ung K, Aeschliman D, Bryan J, Finlay BB, Brinkman FS (2005) Evidence of a large novel gene pool associated with prokaryotic genomic islands. PLoS Genet 1:e62. https://doi.org/10.1371/journal.pgen.0010062

Jaron KS, Moravec JC, Martínková N (2014) SigHunt: horizontal gene transfer finder optimized for eukaryotic genomes. Bioinformatics 30:1081–1086

Juhas M, Van der Meer JR, Gaillard M, Harding RM, Hood D, Crook DW (2009) Genomic islands: tools of bacterial horizontal gene transfer and evolution. FEMS Microbiol Rev 33:376–393. https://doi.org/10.1111/j.1574-6976.2008.00136.x

Kaneko T, Nakamura Y, Sato S, Minamisawa K, Uchiumi T, Sasamoto S, Watanabe A, Idesawa K, Iriguchi M, Kawashima K, Kohara M, Matsumoto M, Shimpo S, Tsuruoka H, Wada T, Yamada M, Tabata S (2002) Complete genomic sequence of nitrogen-fixing symbiotic bacterium *Bradyrhizobium japonicum* USDA110. DNA Res 9:189–197. https://doi.org/10.1093/dnares/9.6.189

Karlin S (2001) Detecting anomalous gene clusters and pathogenicity islands in diverse bacterial genomes. Trends Microbiol 9:335–343

Langille MG, Brinkman FS (2009) IslandViewer: an integrated interface for computational identification and visualization of genomic islands. Bioinformatics 25:664–665

Langille MGI, Hsiao WWL, Brinkman FSL (2010) Detecting genomic islands using bioinformatics approaches. Nat Rev Microbiol 8:373–382

Lee CC, Chen YPP, Yao TJ, Ma CY, Lo WC, Lyu PC, Tang CY (2013) GI-POP: a combinational annotation and genomic island prediction pipeline for ongoing microbial genome projects. Gene 518:14–123. https://doi.org/10.1016/j.gene.2012.11.063

Liu B, Zheng D, Jin Q, Chen L, Yang J (2018) VFDB 2019: a comparative pathogenomic platform with an interactive web interface. Nucleic Acids Res 47:D687. https://doi.org/10.1093/nar/gky1080

Lu B, Leong HW (2016a) Computational methods for predicting genomic islands in microbial genomes. Comput Struct Biotechnol J 14:200–206

Lu B, Leong HW (2016b) GI-SVM: a sensitive method for predicting genomic islands based on unannotated sequence of a single genome. J Bioinforma Comput Biol 14:1640003. https://doi.org/10.1142/s0219720016400035

Middendorf B, Hochhut B, Leipold K, Dobrindt U, Blum-Oehler G, Hacker J (2004) Instability of pathogenicity islands in uropathogenic *Escherichia coli* 536. J Bacteriol 186(10):3086–3096
Nesbo CL, Doolittle WF (2003) Targeting clusters of transferred genes in *Thermotoga maritima*. Environ Microbiol 5:1144–1154
Rajan I, Aravamuthan S, Mande SS (2007) Identification of compositionally distinct regions in genomes using the centroid method. Bioinformatics 23(20):2672–2677. https://doi.org/10.1093/bioinformatics/btm405
Shrivastava S, Reddy Ch VSK, Mande SS (2010) INDeGenIUS, a new method for high-throughput identification of specialized functional islands in completely sequenced organisms. J Biosci 35: 351–364. https://doi.org/10.1007/s12038-010-0040-4
Soares SC, Geyik H, Ramos RTJ, de Sá PHCG, Barbosa EGV, Baumbach J, Figueiredo HCP, Miyoshi A, Tauch A, Silva A, Azevedo V (2015) GIPSy: genomic island prediction software. J Biotechnol 232:2. https://doi.org/10.1016/j.jbiotec.2015.09.008
Soares SC, Vinícius AC, Ramos RTJ, Cerdeira L, Silva A, Baumbach J, Trost E, Tauch A, Hirata R, Mattos-Guaraldi AL, Miyoshi A, Azevedo V, Mokrousov I (2012) PIPS: pathogenicity island prediction software. PLoS One 7:e30848. https://doi.org/10.1371/journal.pone.0030848
Sullivan JT, Ronson CW (1998) Evolution of rhizobia by acquisition of a 500-kb symbiosis island that integrates into a phe-tRNA gene. Proc Natl Acad Sci U S A 95:5145–5149
Vernikos GS, Parkhill J (2008) Resolving the structural features of genomic islands: a machine learning approach. Genome Res 18:331–342
Waack S, Keller O, Asper R, Brodag T, Damm C, Fricke WF, Surovcik K, Meinicke P, Merkl R (2006) Score-based prediction of genomic islands in prokaryotic genomes using hidden Markov models. BMC Bioinform 7:142. https://doi.org/10.1186/1471-2105-7-142
Wei W, Gao F, Du M-Z, Hua H-L, Wang J, Guo FB (2016) Zisland Explorer: detect genomic islands by combining homogeneity and heterogeneity properties. Brief Bioinform 18(3): 357–366. https://doi.org/10.1093/bib/bbw019
Williams KP (2002) Integration sites for genetic elements in prokaryotic tRNA and tmRNA genes: sublocation preference of integrase subfamilies. Nucleic Acids Res 30:866–875
Yoon SH, Park YK, Kim JF (2015) PAIDB v2.0: exploration and analysis of pathogenicity and resistance islands. Nucleic Acids Res 43(D1):D624–D630. https://doi.org/10.1093/nar/gku985
Zhang R, Ou HY, Gao F, Luo H (2014) Identification of horizontally-transferred genomic islands and genome segmentation points by using the GC profile method. Curr Genomics 15:113–121

An Overview of Genomic Islands' Main Features and Computational Prediction: The CMNR Group of Bacteria As a Case Study

3

Thaís Cristina Vilela Rodrigues, Arun Kumar Jaiswal, Flávia Figueira Aburjaile, Carlos Augusto Almeida, Rodrigo Dias de Oliveira Carvalho, Janaíne Aparecida de Paula, Rodrigo Profeta Silveira Santos, Fabiana Vieira Dominici, Andrei Giacchetto Felice, Pedro Henrique Marques, Michele Min San Wu, Yngrid Victória Cassiano Mascarenhas, Alexandre Claudino Junior, Thiago de Jesus Sousa, Victor Augusto Sallum Ceballos, Eduarda Guimarães Sousa, Debmalya Barh, Vasco Ariston de Carvalho Azevedo, Sandeep Tiwari, and Siomar de Castro Soares

Sandeep Tiwari and Siomar de Castro Soares contributed to this work equally and shared the position.

T. C. Vilela Rodrigues · A. K. Jaiswal · R. P. S. Santos · A. C. Junior · T. de Jesus Sousa · D. Barh · V. A. d. C. Azevedo (✉)
Postgraduate Program in Bioinformatics, Department of Genetics, Ecology, and Evolution, Institute of Biological Sciences, Federal University of Minas Gerais (UFMG), Belo Horizonte, Brazil

F. F. Aburjaile
Preventive Veterinary Medicine Department, Veterinary School, Federal University of Minas Gerais (UFMG), Belo Horizonte, Brazil

C. A. Almeida
Postgraduate Program in Bioinformatics, Department of Genetics, Ecology, and Evolution, Institute of Biological Sciences, Federal University of Minas Gerais (UFMG), Belo Horizonte, Brazil

Preventive Veterinary Medicine Department, Veterinary School, Federal University of Minas Gerais (UFMG), Belo Horizonte, Brazil

R. Dias de Oliveira Carvalho
Postgraduate Program in Bioinformatics, Department of Genetics, Ecology, and Evolution, Institute of Biological Sciences, Federal University of Minas Gerais (UFMG), Belo Horizonte, Brazil

Department of Biochemistry and Biophysics, Federal University of Bahia, Salvador, Brazil

I. Mani et al. (eds.), *Microbial Genomic Islands in Adaptation and Pathogenicity*, https://doi.org/10.1007/978-981-19-9342-8_3

Abstract

DNA sequencing, one of the biggest dilemmas of the age is solved with the next-generation sequencing technologies. The challenge now is to develop accurate tools to analyze the massive quantity of data that encompass invaluable information regarding all kinds of organisms. Genomic Islands (GEIs) contain genes and other elements that evidence the horizontal transfer process, having a huge impact on the evolution and adaptability of bacterial species. Several aspects are important to discriminate genomic regions that harbor GEIs from the regions inherited through vertical gene transfer. GC content, codon deviation, flanking tRNA, and transposase genes are some of the most important features. Besides that, distinct categories of GEIs are classified by the nature and function of the genes which it harbors. For bacterial pathogens, Pathogenicity and Resistance Islands are key features for the understanding of the virulence factors and the mechanism of the infection and disease development of a pathogen. Further, in the era of resistant microorganisms, knowing the behavior and the pattern of gene migration through different species and strains is of huge importance. In this book chapter, we described the leading points to predict GEIs, demonstrating the main available tools and some plasticity features regarding bacterial species from the CMNR group, which contains specific highly resistant species.

Keywords

Genomic Islands · GEIs · Comparative genomics · CMNR group bacteria

3.1 Introduction to the Genomic Age and Comparative Genomics of Microorganisms

The DNA structure was solved in 1953 by Watson and Crick when they also proposed that the molecule was a kind of a genetic repository discovering that revealing the sequence was crucial (Watson and Crick 1953). In the same year, the first sequence of a biological was achieved by Sanger through sequencing the insulin chains (Sanger and Thompson 1953). However, the first DNA molecule was only

J. A. de Paula · F. V. Dominici · A. G. Felice · P. H. Marques · M. M. S. Wu · Y. V. C. Mascarenhas · V. A. S. Ceballos · E. G. Sousa · S. d. C. Soares (✉)
Department of Microbiology, Immunology, and Parasitology, Institute of Biological and Natural Sciences, Federal University of Triângulo Mineiro (UFTM), Uberaba, Brazil

S. Tiwari
Postgraduate Program in Microbiology, Institute of Biology, Federal University of Bahia, Salvador, Brazil

Postgraduate Program in Immunology, Institute of Health Sciences, Federal University of Bahia, Salvador, Brazil sequenced in 1968 (Wu and Kaiser 1968). A few years later, Sanger and Coulson achieved the mark of sequencing the first genome, a phage of 5,368 bp, through the method they developed, named: plus, and minus (Sanger et al. 1977a). Finally, a similar procedure named after Sanger was proposed in 1977 (Sanger et al. 1977b).

Along with technological progress, larger genomes could be sequenced in a shorter time. Through Shotgun techniques, DNA fragments were cloned in bacterial vectors, and through bioinformatics, the overlapping of reads was firstly used to assemble the sequences (Staden 1979). The whole genome of the Epstein–Barr virus, with 172,282 bp, was one of the main accomplishments using the Sanger method at the time (Baer et al. 1984). Also, because bacterial and archaeal genomes are much smaller and less complex than eukaryotic genomes, a massive amount of data from these groups began to be generated (Setubal et al. 2018). The increase of data required the creation of specific repositories, so in 1982 the database GenBank from the US National Institute of Health (NIH) was created, achieving at the end of the 80s more than 40 million bases. Nowadays, it has more than 1 billion and 7 million sequences deposited (GenBank and WGS Statistics 2022). The last years had notable growth in genomic sequencing, with a decrease in the processing errors and an improvement in throughput, mediated mainly by the Human Genome Project, which aimed to obtain the entire genetic information of the complete human genome (Craig Venter et al. 1979).

The first bacterial genomes to be sequenced originated from two species with less than 2Mb genomes: *Haemophilus influenzae* and *Mycoplasma genitalium* in 1995. In the following years, the collection of genomes increased exponentially, and, simultaneously, the comparative analysis allowed us to understand phylogenetic relationships among them. This was the beginning of microbial genomics (Koonin et al. 2021). The number of computational methods developed to assess the microbial genomic data and the number of sequenced genomes increased in a direct proportion. Analysis regarding orthologous gene clusters supports evolutionary studies and functional gene annotation (Tatusov et al. 1997; Huerta-Cepas et al. 2017; Jensen et al. 2008). Besides, those approaches allow the comparison of multiple genomes yielding evolutionary insights beyond the development of models, which illustrates the rules that guide the prokaryote evolution (Koonin et al. 2021).

Comparative genomics allows the understanding of the genetic similarities and divergences among strains in species or several species in a genus. Thus, combined with phenotypic knowledge, that information plays a pivotal role in better comprehending the overall microbial species' behavior (Setubal et al. 2018). Orthology comparisons can be performed to verify the presence/absence of genes in genomes, assisting in further phylogenetic comprehension and genome annotation, which is highly important in the biology research field (Nichio et al. 2017). The pan-genome analysis is another section of this field defined by, briefly, the comparison of all genomes of an interesting set (species or genus, for instance), which is divided into the subsections: Core genome, including only the genes harbored by all genomes and, the accessory genome, which englobes gene families that are not present in all genomes (Tettelin et al. 2005). Those subsets can be used to discriminate essential genes between species- or strain-specific genes. Further, the phylogenetic analysis may be integrated into comparative genomics, in a phylogenomic analysis, to provide essential observations about the ancestrality, divergence, and taxonomy studies of microorganisms (Washburne et al. 2018). Those methodologies have brought new insights regarding the microbiologic area, mainly with the recent advances in metagenomics. One of the main challenges dealt with the microbiological research is the difficulty to obtain the DNA of some species once many of them are demand laborious growth techniques or are still unculturable. Nowadays, with the new advances in the area, the characterization of entire microbial populations and the identification of new species and strains are feasible (Segata 2018).

Hence, the combination of different approaches related to comparative microbiological data supports significant advances in interpreting the behavior of many microorganisms, providing highlights about the diversity of the species, genus, or strains and features related to its cellular mechanisms as cell pathogenic and metabolic characteristics.

3.2 Main Characteristics of Genomic Islands

3.2.1 Genomic Plasticity and Genomic Features of Horizontally Acquired Regions

Genome plasticity is one feature of the DNA that allows it to evolve dynamically. This alterable property results from several mechanisms, such as point mutations, rearrangements (inversion and translocation), gene conversion, deletions, and foreign DNA insertions (plasmids, bacteriophages, transposons, insertion elements, and GEIs). This genome plasticity mainly results in genomic changes where the evolutionary forces may act, taking the bacteria to evolve by leaps. Also, it is linked to bacterial diseases, in which highly resistant strains can share their genes, being one threat and demanding the attention of the scientific community (Anastasi et al. 2016).

In the first years of the studies of genomics of microorganisms, bacterial species were defined partly by the G+C (Guanine and Cytosine) content of their genomes. The overall G+C content reflects the base composition across the genome, where the bacterial genomes slowly change over time. Most genes in a given genome can evolve over a considerable period when subjected to selective pressures (Deng et al. 2002). Some genome sequence analyses have indicated that the nucleotide compositions from the genomes of individual replicates are not homogeneous, so a significant degree of gene flow involving the acquisition of DNA sequences from different sources at different times has been observed (Welch et al. 2002). Concluding, the bacterial genome is not static, presenting gene gain and loss over time, where those associated with selective advantage are preserved, and those not advantageous to the organism may be lost (Lawrence and Roth 2014).

Horizontal gene transfer (HGT) is the central process of gene sequence acquisition. HGT was described in 1928 by Frederick Griffith, which he demonstrated a bacterial transformation mechanism with external DNA (Griffith 1928). Mobile genetic elements (MGEs), such as plasmids, bacteriophages, and transposons, are frequent vectors of genes obtained through HGT. In the bacterial life cycle, HGT acts among the biological processes classified as Transformation, Conjugation, and Transduction. Transformation, the uptake of nucleic acids, is mediated by proteins natively present in the chromosome of bacteria. Conjugation, the plasmid-mediated gene transfer, requires independent replication of genetic components such as plasmids or transposons. Finally, transduction allows the displacement of genetic material mediated by bacterial viruses or bacteriophages (Frost et al. 2005). The gene regions acquired by HGT normally contain evidence of their functions, such as MGE transposases and site-specific recombinases that catalyze the intracellular movement of MGEs. In addition, host homologous recombination systems allow for chromosomal deletions and other rearrangements.

Several elements may be involved in HGT events, where they portray a decisive role in evolution by leaps in virtue of gene incorporations, such as plasmids, bacteriophages, transposons, insertion elements, and GEIs. As a result of source-specific characteristics and mechanisms used in incorporation, regions that have been acquired through HGT share some characteristics as deviation in genomic signature related to G+C content and codon usage, which reflect the genomic signature of the donor organism; the existence of insertion sequences (IS) and/or flanking tRNAs that may present a specific IS in their 3'-terminal regions; and harbor transposases (Soares et al. 2012).

Techniques to identify HGT have upgraded with time, revealing the notably extent and relevance of HGT to viral, prokaryotic, and eukaryotic genomes (Swithers et al. 2012; Treangen and Rocha 2011). For a transferred gene endure in the new organism for a prolonged time, it demands to contribute with a selective advantage for the receptor. However, many of the HGT genes identified through comparative genomics among near ancestors have indifferent or nearly neutral achievements for the receptor in both prokaryotic organisms (Gogarten and Townsend 2005). According to Zhou et al. (2022), the successfully integrated gene must not harm the receptor, be expressed at small rates, and encode an important function (Zhou et al. 2021). Phylogenetic distance, shared ecology, and economic restrictions are considered factors in HGT identification, although their relative contributions are not as clear (Williams et al. 2011).

On the other hand, vertical gene transfer is also described; however, it exerts less influence on genome evolution. It involves the transfer of DNA sequences from the parents to the progeny, like the asexual reproduction of bacteria. Although the horizontal transfer is expected to be succeed mainly between related than distant species, it also occurs with distant species as divergent as those found in the diverse domains of life (Sulaiman et al. 2018). In summary, the presence or absence of any gene between two phylogenetically related bacteria could be the result of a relevant HGT event. Studies in bacteria with open pan-genomes show higher variability in gene content compared to bacteria with closed pan-genomes and, thus, have great potential for discovering new genes. Therefore, comparative genomics is crucial to unravel the existence and absence of a genetic profile in certain strains. These procedures can also help to discriminate between phenotypic and genotypic patterns of pathogenic strains.

In comparative bacterial genomics, the characterization of the presence and absence of genes that are active in antibiotic resistance, pathogenesis, or related to metabolic ability is expected. This dynamic nature of the DNA is called genome plasticity, which is machinery adopted by the organism that causes it to genetically mold itself providing better adaptability to the environment. Finally, the presence–absence variation of genes (PAV) between genomes contributes fundamentally in the environmental- and host-adaptation, allowing the maintenance of the microorganism in new stress situations and the innovation and evolution of the bacterial genome (Wan et al. 2020; Sollitto et al. 2022). Generally, this variation of gene content within comparative genomics is related to accessory genes between species (Medini et al. 2005).

3.2.2 Codon Usage

61 combinations of the 4 nucleotide bases in groups of three result in coding codons, which are involved in amino acid recruitment, therefore encoding proteins and playing a vital role in biochemical information for the organism (Nirenberg 1963). However, the genetic code is degenerated, once the same amino acid may be coded by more than one codon (Gonzalez et al. 2019). The selection of a codon for one specific amino acid is not a random process, as some preferred codons influence the expression level, impacting the whole translational procedure (Zhoua et al. 2016).

The GEIs are gene clusters with compelling evidence that they were acquired by HGT and one such feature used in the identification and characterization of this process is the codon usage, as each species have its characteristic preferred codons (Azevedo et al. 2011) due to: tRNA availability, transcriptional fidelity, and efficiency; and selective and non-selective substitutional bias (Karlin et al. 1998). The level of expression of a gene acquired by horizontal transfer will depend on the compatibility of the receptor with the expression process. Due to codon preference, translation efficiency due to the unavailability of carrier RNA, for example, can negatively affect the production of gene products (Sharp and Matassi 1994; Callens et al. 2021).

Various theories about the use of these optimal codons have been created throughout history. Some bring that codon bias contributes for the equity of the natural selection, favoring the permanence of codons advantageous to the organism and genetic drift favoring the likelihood of fixing the less advantageous ones, but the mutation-selection-drift (MSD) theory is the most widely accepted (Bulmer 1988; Akashi 1994).

As a general rule for codon usage, the most efficient translations according to the strength of the relationship between codon and anticodon will be benefited, where an intermediate strength prevails over a strong or weak one and codons for medium and larger tRNA are preferred, with the smaller ones being avoided (Grosjean and Fiers 1982). Based on this, codon usage may be used as a feature of GEIs precisely because they carry a pattern/signature from the donor organism (Langille et al. 2008), which makes it easier to trace these regions.

3.2.3 GC Content

Chargaff's rule states that the total number of adenine (A) nucleotides equals the number of thymine nucleotides (T) due to double-stranded DNA base pairing. Similarly, the total number of guanine nucleotides (G) is equivalent to the number of cytosine nucleotides (C) (Kresge et al. 2005). For that reason, it is possible to refer to the genomic nucleotide composition as AT or GC content.

Because of the miscellaneous genome combination of distinct bacterial strains, GEIs will generally have a significantly different sequence composition from the new host genome (Langille et al. 2010). As shown for the codon usage, regarding the nucleotide composition, the GC content of a given island is also different from the overall GC of the genome, which is represented by the percentage of G and C it has, which is complementary to the fraction of A and T, as represented by the formula:

$$\%\mathrm{GC} = \left(\frac{G+C}{A+T+C+G}\right) \text{ x } 100$$

The GC content among species is highly variable, where it can range in prokaryotes from 13.5% in *Candidatus Zinderia insecticola* strain CARI, a betaproteobacterial symbiont (Langille et al. 2010), to 77.4% in *Actinomycetales bacterium* strain S29, a high *G* + *C* Gram-positive bacteria (Genome List 2022). In general, the GC content tends to increase as the genome size increases, with longer genomes having higher %GC (Almpanis et al. 2018). The reasons for this correlation are yet to be completely understood, although both environmental pressure and phylogenetic relationships seem to play a core influence (Reichenberger et al. 2015).

3.2.4 Transposases and Insertion Sequences

Insertion sequences (ISs) are little/single units of transposons, capable of moving through DNA and were widely distributed in bacterial genomes (Mahillon and Chandler 1998). ISs carry genes that encode enzymes as transposases, in which recognizes the terminal repeated and inverted sequences (RIs) that flank the transposon gene, enabling transposition to occur while transferring itself to the host's DNA, for instance (Cho et al. 2014; Siguier et al. 2014; Rice and Baker 2001).

Transposition largely contributes to the development of genomic diversity and plasticity as it may induce mutations, inversions, duplications, and deletions, with genomic rearrangements generating mosaic regions where the genes present in the inserted DNA may be differentially modulated (Preston et al. 2004; Lysnyansky et al. 2009). In this scenario, the evolution of the bacterial genome also stems from many results of transpositions, since bacteria can start to acquire ISs through the mechanisms of conjugation, transformation, and transduction. These are important factors that can confer evolutive advantages but also can generate instability of the bacterial genome, as deleterious mutations with dysfunctional phenotype can lead to drastic consequences for the organism (Williams 2016; Schlüter et al. 2007; Colonna Romano and Fanti 2022).

There are two transposition mechanisms: non-replicative and replicative. The non-replicative mechanism of transposition is carried out by composite transposons, those consisting of two insertion sequences that flank a gene, which is often antibiotic resistance genes. The integron-cassette system is one of the common genetic elements responsible for the spread of antibiotic resistance genes (Schlüter et al. 2007; Teuber et al. 1999). Replicative transposition is performed by non-composite transposons. In this mechanism, one copy of the sequence of the transposon is transferred to the target site and another to the original bacterial site (Hickman and Dyda 2015).

The progress of bioinformatics tools aimed at the identification of transposon elements (TEs) has been widely studied in recent years, although the choice of the best tools for specific cases is under discussion (Nelson et al. 2017; Hoen et al. 2015). There are two main approaches available for detecting these elements in whole genomes: mapping discordant reading pairs and "split" reads that share common alignment junctions (Vendrell-Mir et al. 2019). The combination of both approaches seems to be a good strategy, however, improvements in the accuracy of the predictions are still needed.

3.2.5 Specific Factors (Virulence, Resistance, Metabolic, and Symbiotic Factors)

Genomic Islands (GEIs) can be categorized differently according to their gene content: Symbiotic Islands, implicated in the association of bacteria to Leguminosae (Hadjilouka et al. 2018); Resistance Islands, harboring genes related to antibiotic resistance (Krizova and Nemec 2010); Metabolic Islands, composed by genes related to the biosynthesis of metabolites of the second class (Tumapa et al. 2008); and Pathogenicity Islands (PAIs), presenting a prominent concentration of virulence factors, which appear linked to pathogenic bacteria and are involved in the reemergence of several pathogens (Dobrindt et al. 2000; Soares et al. 2016). Among the specific factors, there are virulence factors, resistance genes, and metabolism- and symbiosis-associated genes, respectively, that are predominant in their respective islands.

Virulence factors (VFs) are gene products that can favor the pathogenicity of a microorganism by expanding its potential to cause disease and may be classified into at least 5 types of distinct mechanisms: motility, adhesion, invasion, immune system evasion, and toxin production (vanden Broeck et al. 2007; de Jong et al. 2019; Veerachamy et al. 2014; Chaban et al. 2015). An example of a classic pathogenicity island that contains essential VF's, is LIPI-1 from *Listeria monocytogenes*. It is composed of 6 genes (*prfA*, *plcA*, *hly*, *mpl*, *actA*, *plcB*) whose products are fundamental to the intracellular lifestyle of the bacterium, allowing the progression of infection (Hadjilouka et al. 2018).

Once the microorganism has already infected the host, resisting treatments is crucial for the life-maintenance of the pathogen. In this sense, antibiotic resistance genes (ARGs) encode proteins with functions that confer resistance to a specific drug. For instance, a new RI "AbGRI4" was recently identified in multidrug-resistant clinical isolates of *A. baumannii*, which contained the genes aacC1 and aadA1, conferring resistance to gentamicin and streptomycin, respectively (Chan et al. 2020). In this sense, selective pressure on ARGs and their spread via horizontal transfer is a serious public health problem, which makes the area of RIs studies impactful.

The bacterial metabolism can be a determining factor in whether or not it adapts to an environment (or in a host). MIs harbor genes associated with energy generation and conversion, metabolism, and transport of carbohydrates, amino acids, nucleotides, inorganic ions, lipids, coenzymes, and secondary metabolites. For example, in marine *Acinetobacter*, GEIs associated with secondary metabolites may provide a better functional adaptation of these microorganisms (Penn et al. 2009). In *Vibrio spp*., there are arguments that support the hypothesis that iron transport and acquisition systems may have been disseminated via horizontal gene transfer(Payne et al. 2016). Finally, genes associated with symbiosis are those that promote a symbiotic relationship. Therefore, the repertoire of a GEI that provides support for the symbiotic host–bacterium relationship to the point of sustaining it, classifies the island as a Symbiosis Island (SI). One gene with evidence of horizontal transmission is the *nifH* encoding nitrogenase present in species such as *Bradyrhizobium japonicum*, which fixes nitrogen, making it useful for plants-associated bacteria (Barcellos et al. 2007).

3.3 Software/Databases for Prediction and Visualization of Genomic Islands

GEIs affect the genome plasticity through their ability to transfer and incorporate a huge number of genes in the block, i.e. operons and groups of genes coding correlate functions. They may cause drastic modifications, taking the bacteria to evolve by leaps compared to the parent strain. GEIs are portions of DNA obtained from a different organism that sharing in common: the size ranging from 10 to 200 Kb; and, the presence of sequences derived from phage and/or plasmid, the existence of transfer genes or integrases and insertion sequences. Also, they are normally flanked by tRNA genes or insertion sequences that may be involved in their instability (Hacker and Carniel 2001), resulting in the deletion and transfer events, besides rendering the region mosaic (Letek et al. 2008). Tools for predicting GEIs have followed the advances in the next-generation sequencing technologies. Thus, comparative genomics and sequence composition are the main approaches for predicting GEIs, combined with the support of different databases (da Silva Filho et al. 2018).

Several analysis tools are currently available for these predictions (Table 3.1) and visualization of data.

GIPSy is a software program that predicts the 4 types of GEIs explained above. It has been developed in Java including different databases such as PFAM (protein families database), CARD (Comprehensive Antibiotic Resistance Database), and Mvirdb (microbial database of protein toxins, virulence factors, and antibiotic resistance genes), among others, to perform the predictions in an independent platform. The program requires the genomes of the target and a reference organism, where the data may be in EMBL (.embl) or Genbank (.genbank, .gb, .gbk) format. The prediction of GEIs will be based on: genomic signature deviation such as G + C content; presence/absence of transposase genes; virulence, metabolism, antibiotic resistance, and symbiosis factors; and, presence/absence of flanking tRNAs (Soares et al. 2016).

The Alien Hunter software created by researchers at the Sanger Institute in the UK predicts GEIs through Interpolated Variable Order Motifs (IVONs), detecting atypical regions of the genome such as G + C content, dinucleotides, and recurrence of the codons. The result is given through the IVON score, where the higher the score, the more accurate is the prediction, meaning the portion that corresponds to the island differs from the remaining genome. In addition, a threshold is also given based on the average of the complete genome according to the similarity. In this way, the atypical regions that are candidates for GIs are constructed (Vernikos and Parkhill 2006).

Another tool is GI Hunter, it can predict GEIs from both bacteria and archaea through eight features associated with GI: tRNA, Phage, Integrase, Transposase, Highly expressed gene, Gene density, Average intergenic distance, and IVOM (Alien Hunter's Variable Order Interpolated Motifs methodology). The information is based on the study by Langille and collaborators (Evaluation of GEI predictors using a comparative genomics approach—IslandPick) in addition to genome annotations (Che et al. 2014b).

The IslandViewer3 besides being a GEI database is also a web-based prediction software that relies on three methodologies: for genomic comparison analysis it utilizes IslandPick, for gene composition, SIGI-HMM, and search for atypical regions and mobility-related genes, it uses IslandPath-DIMOB. In addition, it uses annotations from other databases to obtain information on pathogenicity, virulence, antibiotic resistance, and homologous genes (Dhillon et al. 2015). Here phylogenetically related genome analyses are needed so other tools such as CVTree and Mauve are used, which build phylogenetic trees of complete genomes and perform multiple genome alignments (Rissman et al. 2009).

Table 3.1 Available bioinformatics resources for the prediction and visualization of Genomic Islands

Tool	Software tool/ database	Genomic signature	References
AlienHunter	SW	ON	Vernikos and Parkhill (2006)
Centroid	SW	GC	Rajan et al. (2007)
Colombo	SW	CU	Waack et al. (2006)
Design-Island	SW	GC + ON	Chatterjee et al. (2008)
EGID	ES	GC + DI + TRI + ON + CU	Che et al. (2011)
GC-Profile	SW	GC	Gao and Zhang (2006)
GEMINI	SW	–	Che et al. (2010)
detector	SW	k-mers + IVOM	Che et al. (2010)
GIHunter	SW	–	Che et al. (2014a)
GI-POP	SW	GC + ON + CU	Lee et al. (2013)
GIPSy	ES	GC + CU	Soares et al. (2016)
GIST	ES	GC + DI + ON + CU	Hasan et al. (2012)
GI-SVM	SW	GC + CU	Lu and Leong (2016a)
HGTector	SW	–	Zhu et al. (2014)
IGIPT	SW	GC + DI + CU	Jain et al. (2011)
INDeGenIUS	SW	ON	Shrivastava et al. (2010)
IslandCompare workflow	ES	GI + DI	Bertelli et al. (2022)
Islander	DB	GC	Hudson et al. (2015)
IslandPath	DB	GC + DI	Hsiao et al. (2003)
IslandPick	SW	–	Langille et al. (2008)
IslandViewer 3	DB+ES	GC + DI + CU	Dhillon et al. (2015)
MJSD	SW	GC	Arvey et al. (2009)
MSGIP	SW	GC	de Brito et al. (2016)
MTGIpick	SW	Tetranucleotide	Yoon et al. (2007)
PAIDB	DB	GC + CU	Yoon et al. (2015))
PAIDB v2.0	DB	GC + DI + CU	Yoon et al. (2015)
PAI-IDA	SW	GC + DI + CU	Tu and Ding (2003)
PIPS	ES	GC + CU	Soares et al. (2012)
Predict Bias	SW	GC + DI + CU	Pundhir et al. (2008)
Pre_GI	DB	GC + ON	Pierneef et al. (2015)
RGPFinder	SW	GC + CU + ON	Ogier et al. (2010)
Sighunt	SW	Tetranucleotide	Jaron et al. (2014)
SIGI-HMM	SW	CU	Waack et al. (2006)
VRprofile	SW	GC + DI + CU	Li et al. (2018)
Zisland Explorer	SW	GC + CU	Wei et al. (2017)
BRIG	SW	CGView + BLAST	Alikhan et al. (2011)
GIV	SW	GIHunter + Circos	Che and Wang (2013)

DB (database); SW (software tool); ES (ensemble software that combines different software tools); GC (G + C content); DI (dinucleotide frequency); TRI (trinucleotide frequency); O (oligonucleotide); CU (codon usage).

3.4 Examples of Bacteria of the Group CMNR in the Context of Genomic Islands

The CMNR group, composed of *Corynebacterium, Mycobacterium, Nocardia,* and *Rhodococcus* species, is characterized by bacteria with an outer membrane composed of diverse lipids such as mycolic acid. This feature is a strong virulent factor which implies a natural antimicrobial resistance (Bansal-Mutalik and Nikaido 2011; Dorella et al. 2006).

3.4.1 Genus *Corynebacterium*

The genus comprises nearly 100 species of Gram-positive bacteria with a large amount of Guanine + Cytosine in the genome. There are extremely pathogenic representatives of this genus to animals and humans, in which the main species for the human health being *Corynebacterium diphtheriae*, and others can be used in industry and food production (Tauch et al. 2016). In the microbiome of the human body, several *Corynebacterium* species have been described as opportunistic bacteria, once they colonizes naturally the gut however, unusually, they have been related to human infections due to contaminations of clinical specimens (Tauch et al. 2016; Mangutov et al. 2021).

Despite the common colonization of the human microbiome, different diseases associated with the respiratory tract, such as pharyngitis, bronchitis, rhinosinusitis, and pneumonia are caused by representatives such as *C. diphtheriae* and *C. ulcerans* (Mangutov et al. 2021). Also, orthopedic infections (such as septic arthritis and osteomyelitis), as well as endocarditis, abscesses, genitourinary tract infections, and various types of infections in both immunologically healthy and immunocompromised patients were described (Kalt et al. 2018).

3.4.1.1 *Corynebacterium pseudotuberculosis*

C. pseudotuberculosis is responsible for triggering chronic infections causing lymphadenitis in addition to other lesions and abscesses in various species of animals. Infections in humans have been documented, normally farm workers and veterinarians who have had contact with contaminated animals, which may culminate in necrotizing lymphadenitis (Join-Lambert et al. 2006).

C. pseudotuberculosis may harbor the gene for the Diphtheria Toxin (DT) (the main cause of *C. diphtheriae* virulence). The gene for this toxin may be acquired via bacteriophages through HGT (Soares et al. 2013). More than 10% of the isolated *C. pseudotuberculosis* microorganisms produce DT, but there are no diphtheria cases associated with the species. However, these strains of *C. pseudotuberculosis* can also produce the dermonecrotic toxin, another relevant virulence factor for this species (Emmerson et al. 1987).

The plasticity level of the *C. pseudotuberculosis* genomes is highly diverse between *ovis* biovar and *equi* biovar strains (Soares et al. 2013). This can be explained because even sharing the same PAIs, genetic deletions are found in different positions in each strain. Some harbor the diphtheria toxin gene and others harbor clusters of pilus genes, where a deletion was noted at the position where the genes should be in the *equi* biovar strains. Also, the clusters of pilus genes show a low similarity between the biovars, because of small deletions, nucleotide substitutions, and frameshift mutations. Finally, some strains harbor the *pld* (Phospholipase D) gene and the *fag* (Fe Acquisition Gene) operon, which are both important for encoding important virulence factors of the species (Soares et al. 2013).

In the strains of biovar *ovis*, besides the elevated genetic similarity rate, a similar deletion pattern was also observed in the same PAIs. In the *equi* biovar strains, it has been witnessed large deletions and a lower level of intra-biovar similarity in GEIs compared to the *ovis* biovar. Furthermore, a clonal-like behavior of the species was indicated when compared to *C. diphtheriae*. Also, the majority of the variable genes in the *ovis* biovar strains were obtained in blocks by horizontal gene transfer and are hugely conserved in the whole biovar. Contrary to this, the *equi* biovar strains showed large variability in island gene content (both inter-and intra-biovar). It's possible to conclude that the acquisition of genes through HGT and the maintenance of the acquired regions along with the plasticity of these regions may be due to the colonization of specific/exclusive hosts (Soares et al. 2013).

Regarding the RIs, the number found was quite versatile among the strains, where the strains with a higher number of RIs have a lower number of isolated resistance genes. Despite this, these genes were grouped in a larger number on islands, concluding that they were more obtained through HGT events and resulting in a greater variability. The number of ARGs in the *Equi* biovar was higher than in the Ovis biovar, including membrane permeases, beta-lactamase, among others (Baraúna et al. 2017).

A total amount of 16 normal and 5 strong PAIs were predicted using GIPSy software in *C. pseudotuberculosis*, where *C. glutamicum* was the non-pathogenic reference species. Also, the software predicted 8, 14, and 16 normal RIs, SIs, and MIs, respectively, whereas 4 RIs, 3 SIs, and 4 MIs were strong.

3.4.2 *Rhodococcus*

There are approximately 57 species of bacteria in the genus *Rhodococcus*. They present a high diversity of metabolic and industrial applications, or potential in bioremediation. However, leastwise six species (*R. equi*, *R. erythropolis*, *R. ruber*, *R. gordoniae*, *R. fascians,* and *R. defluvii*) have been related to animal and plant disorders (Vázquez-Boland et al. 2013; Vázquez-Boland and Meijer 2019).

In *R. equi*, as in the genus *Rhodococcus*, niche specialization is determined by plasmid genetic content. In environmental *Rhodococcus* species, the functions encoded by plasmids are mainly catabolic, while in *R. equi* they promote colonization of the animal host (Vázquez-Boland et al. 2013; von Bargen and Haas 2009). The evolution of the *R. equi* genome is mainly driven by genetic gain/loss mechanisms, with an important support of HGT events. Phages are also numerous in *R. equi* (Petrovski et al. 2013; Salifu et al. 2013; Summer et al. 2011) and likely have an essential place in HGT-driven genome plasticity.

In this species, the conjugative virulence plasmid is an important part of the accessory genome, being a key factor for pathogenesis. Three different plasmids related to the virulence for the host have been identified: the circular variants pVAPA and pVAPB, found in horses and swine, respectively, and the linear pVAPN found in bovine (ruminant) isolates. The pVAP has 80 to 100 kb in size and carries the horizontally acquired vap PAIs, whose products are essential for pathogenesis and survival in macrophages (Vázquez-Boland et al. 2013; Anastasi et al. 2016; Letek et al. 2008).

The different aspects among the three *vap* PAIs are primarily explained by the type of genes. Each *vap* PAI contains a group of homologous virulence-associated genes, in addition to specific non-vap genes present in each of the three PAIs, which tend to be deeply preserved (Anastasi et al. 2016; Letek et al. 2008; Coulson et al. 2010; Valero-Rello et al. 2015). Moreover, they possess several non-vap genes, notably the vir operon harboring two regulators (virR and virS), which activate the expression of vap PAI (Byrne et al. 2007). In addition to vap genes found to this plasmid, several chromosomal genes in *R. equi* appear to have been co-evolved to act under the control of *virR* and *virS presumably* in a *vap* PAI co-regulated network (Coulson et al. 2015; Letek et al. 2010).

Furthermore, these plasmids have specificity for different hosts and can be found in both animal and humans isolates. Therefore, the origin of *R. equi* transmission to human can be inferred from the type of plasmid contained in the variant strains (Takai et al. 2020). The three types of plasmids are found in human isolates (Ocampo-Sosa et al. 2007).

The strict association of certain types of *vap* PAIs with an animal species indicates that the selective pressure is determined by the host. Studies have shown that most strains of *R. equi* in the co-infection by HIV in humans contained type B plasmids (primarily derived from swine). Thus, suggesting that human exposure to this type of animal may be a significant risk factor (Anastasi et al. 2016).

Comparative genomic analyses were accomplished to characterize the discrepancies between the genomes of the recently sequenced *R. equi* WY strain and other previously sequenced *R. equi* genomes. As a result, nine GEIs were identified in *R. equi* WY that showed a variation of 4.3 kb to 29.5 kb in size. They encoded genes involved in virulence, resistance, or niche adaptation, including three unique GEIs in *R. equi* WY. Regarding resistance genes, nine *R. equi* genomes share a 12.2 kb GEI region containing a tunicamycin resistance protein (tmrB), a metallo-β-lactamase, as well as a putative integrase followed by an efflux pump belonging to the major facilitator superfamily (MFS). In addition, such GEIs also have an iupABC operon, and the first iupA gene of this operon, which encodes proteins from an ABC transport system extremely similar to siderophore uptake mechanisms that provide *R. equi* with the capacity to use as a source of iron, the heme and the hemoglobin (Ying et al. 2019).

A recent study performed a broader comparative analysis among *R. equi* and other species genomes belonging to the genus *Rhodococcus*. To accomplish these comparative genomic analyses, 94 *Rhodococcus* complete genome sequences belonging to 22 species were used. The *Rhodococcus* spp. were isolated from different sources, such as plants (*R. fascians*), soil and seawater (*R. erythropolis*), and *R. equi* isolated from both animal hosts (such as equine, swine, and humans) and also from soil (Ying et al. 2019). The GC rate of the 94 genomes varies from 61.67% to 70.67%. Furthermore, the genome sizes of the 22 *Rhodococcus* spp. are different among diverse species (3.89–12.41 Mb). *R. wratislaviensis* had the largest average genome size (9.77 Mb, 9.16–10.38 Mb) and the smallest genome belonged to *R. corynebacterioides* (3.89 Mb), with a difference of 5.8 Mb between them. This enormous diversity of genomes suggests alterability behavior in the *Rhodococcus* genome and may have allowed them to adapt to a wide spectrum of environments, such as soil, plants, water, and animals (Ying et al. 2019).

It is known that GEIs in *R. equi* has consistently been related to niche adaptation, antibiotic resistance, and also virulence and they possibly work together with prophages, therefore contributing to most of the genome expansion (Vázquez-Boland et al. 2013; Petrovski et al. 2013; Summer et al. 2011; Anastasi et al. 2016). Analysis revealed that Two-Component System genes, resistance genes, and virulence factors were substantially enriched in the core genome of *R. equi*. These data suggest the contribution of the core genome to the pathogenicity and niche adaptation of *R. equi* as well. In addition, the comparative genomic analysis demonstrated an identical collinearity relationship shared between *R. equi* genomes, in addition to not showing significant chromosomal rearrangements, and genes located in unaligned regions were acquired mainly in the form of GEIs and prophages (Ying et al. 2019).

In conclusion, as a valued resource for functional genomic researches, comparative analyses have increasingly facilitated a better understanding of the genomic diversity, evolution, and structural variation of the genus *Rhodococcus* and *R. equi*.

3.4.3 Genus *Mycobacterium*

Mycobacterium is a genus belonging to the order *Actinomycetales*, of the family *Mycobacteriaceae*. They are pleomorphic bacilli bacteria with different shapes, ranging from thin filamentous forms to more curved, straight, or slightly curved forms. They are obligate aerobic, non-motile, non-endospore forming, non-encapsulated, and highly pathogenic bacteria, which cause several diseases, including leprosy, tuberculosis, and non-tuberculous mycobacterial infections (Rastogi et al. 2001; Tortoli 2014). This genus contains about 172 species, which are divided into groups, where the main ones are the *M. tuberculosis* Complex, the *M. avium* Complex, and the non-tuberculous mycobacteria (Levy-Frebault and Portaels 1992).

3.4.3.1 *Mycobacterium tuberculosis* Complex

The *Mycobacterium tuberculosis* complex is the group of Mycobacterium species from the genus *Mycobacterium* that cause tuberculosis in humans or other organisms. It includes the *species M. tuberculosis, M. africanum, M. bovis, M. microti, M. canetti, M. caprae, and M. pinnipedii.* The species in this group share a genomic sequence identity of approximately 99.9%, which probably evolved from a single clonal ancestor (Riojas et al. 2018). The two most important species in this complex are *M. tuberculosis* and *M. bovis*. Although *M. bovis* causes tuberculosis in cattle, it may also cause tuberculosis in humans with the same clinic as *M. tuberculosis*. However, this infection is less frequent, due to the control of bovine tuberculosis (Yu et al. 2011).

3.4.3.2 *Mycobacterium tuberculosis*

The main virulence factors of *M. tuberculosis* are the complex lipid wall; the cholesterol catabolism, which is a source of energy and material for the synthesis of lipids of the complex wall; and, proteins and cell envelope lipoproteins important for bacterial adhesion to cells host. These proteins inhibit macrophage antimicrobial responses due to resistance to toxic compounds of the host, mechanisms related to the control of apoptosis and control of the progression and transformation of the phagosome into a phagolysosome, as well as factors that control and regulate gene expression under different conditions of activity of the pathogen (Forrellad et al. 2012).

The active *M. tuberculosis* strains have as one of the head points of their virulent factors, genes acquired by horizontal transfer that are inserted in PAIs (Xie et al. 2014). Genes related to immune evasion and persistent infection of *M. tuberculosis* were found in PAIs, showing the critical importance of those regions for the survival and maintenance of the pathogen (Xie et al. 2014; Arnvig and Young 2012; Marraffini and Sontheimer 2010). Besides that, other GEI regions from this pathogen have shown high sequence variation in their family genes, resulting in proteins with antigenic variability, which may also influence in the immune system evasion (Yu et al. 2011). Another interesting finding is related to the expression rate of the genes of a GEI being different in attenuated strains like *M. bovis* and virulent *M. tuberculosis*, corresponding with different patterns in the maintenance of the infection (Yu et al. 2011).

3.4.4 Genus *Nocardia*

The genus *Nocardia* consists of aerobic and Gram-positive bacteria of the class Actinobacteria, order Actinomycetales, and family Nocardiaceae, many of which are opportunistic pathogens. These bacteria are widely found in plants, gardens, and soil, and to date, 122 species of *Nocardia* have been reported, and more than 50 have been considered clinically relevant and may cause disease in humans (Conville and Witebsky 2010). The most identified species in human diseases are *N. nova*, *N. cyriacigeorgica*, *N. brasiliensis*, *N. abscessus* complex, *N. transvalensis* complex, and *N. farcinica,* the latter being the most common (Wang et al. 2022). The clinical manifestations may be localized or disseminated infections, but sometimes it leads to more serious cases, such as Nocardial osteomyelitis and septicemia. Also, lately there has been an expansion in reports of Nocardia infections worldwide in immunocompetent people (Xu et al. 2021; Martínez-Barricarte 2020; Lu et al. 2020).

The virulence of *Nocardia* is related to the capacity to neutralize phagosomal acidification, inhibit phagosome–lysosome fusion, modulate lysosomal enzymes, and resist toxic oxidative metabolites (Beaman 1994). Also, other virulence characteristic is the important ability to survive in a facultative intracellular way in several human cells, especially in *N. farcinica*, a very virulent species with resistance to several antibiotics (Toyokawa et al. 2021). The combination of Trimethoprim and Sulfamethoxazole (TMP-SMX), for instance, is a drug used for the treatment and as a prophylactic agent for *Nocardia* infections, turning it less effective over time and also demonstrating the ability to acquire mobile elements carrying resistance genes through HGT (Mehta and Shamoo 2020). *Nocardia* is currently a genus of bacteria not deeply studied and few information regarding the prediction of GEIs has been found in the literature.

The virulence of *Nocardia* is associated with immune escape by hindering phagocytosis processes, resistance to toxic oxidative metabolites, in addition to being able to survive in a facultative intracellular way in several human cells. *N. farcinica*, a very virulent species, exhibits resistance to several antibiotics, in addition to demonstrating the ability to acquire mobile elements carrying new resistance mechanisms through horizontal gene transfer (Komaki et al. 2014; Männle et al. 2020; Valdezate et al. 2015; Yasuike et al. 2017). Within this context, a carried-out study showed that the 76 *Nocardia* resistant to TMP-SMX, isolated from patients, belonged to 12 species and 75 of these carried class 1 and/or class 3 integrons, which are mobile elements associated with acquisition of antimicrobial resistance (Valdezate et al. 2015; Gillings 2014). In addition, the strains also carried genes encoding proteins that are involved in the resistance against different antimicrobials as: β-lactams (β-lactamases), aminoglycosides (aminoglycoside-modifying enzymes), macrolides (RNA methylases), tetracyclines (ribosomal protection proteins), as well as efflux pumps (Valdezate et al. 2015). Another study reported in *N. soli* Y48 a total of 17 GEIs composed of 173 genes and 84 hypothetical proteins, including functions such as signal transduction, metabolism (energetic, carbohydrate, nucleotides, and carbon), replication and repair, membrane transport, translation, environmental information processing, and xenobiotic biodegradation (Yang et al. 2019). Furthermore, it was predicted that the transposase genes that may be associated with the possible active HGT in the strain were predominantly related to the EI family and spread throughout the chromosome (Yang et al., 2019). The adaptation of this strain to the soil contaminated by oil and to the degradation of hydrocarbons have been attributed to the presence of these genes, which suggests that there is a great possibility that *N. Soli* Y48 has evolved and acquired GEIs, making the strain more adapted genetically to mineralize, use, or degrade crude oil (Yang et al. 2019).

3.4.5 Artificial Intelligence As an Improving Approach to Genomic Islands Prediction

Small microorganisms, especially bacterial species, are the most plentiful organisms on the earth. Nowadays, bacterial species shows an immense diversity, demonstrating adaptation abilities to the environment over several thousand years. Bacterial species have the capability to acquire genes horizontally from several other ways, comprising other microorganisms such as prokaryotes, viruses, and eukaryotes (Ochman et al. 2000). The genomic sequences of bacteria have suggested that the HGT events in bacteria, and its cluster of genes involvement in HGT play an important role and give the edge to the bacteria and empower them to the adaptation to the environmental habitat (da Silva Filho et al. 2018; Schmidt and Hensel 2004). The notion of GEI was obtained from the pathogenicity island coined by Hacker and his colleagues in uropathogenic *Escherichia coli* to report genomic regions that harbor a group of virulence factors that can be spontaneously deleted (Langille et al. 2008; Hacker et al. 1990).

Programs and software for the prediction of GEIs generally use genomic and sequence comparison techniques. The comparative analysis identifies exclusive regions in genomes of particular pathogens, whereas sequence composition analysis estimates and associates the specific regions with different GEI features in the genome (Lu and Leong 2016b). The comparative genomic analysis predicts variable regions in comparatively close organisms (several genomes), whereas the analysis of sequence composition is performed in one organism (single genome). However, many programs and software are available, but the correctness is inadequate. The use of only one strategy may not be enough to provide adequate results, while multiple different techniques may be a better strategy and can help in GEI identification (da Silva Filho et al. 2018; Lu and Leong 2016b). The majority of bioinformatics programs and software developed for the identification of pathogenicity islands depend on the composition-based methods that handle GEI's specific properties, while other programs try to correlate closely related genomes. Earlier it was revealed that integrating several features of GEIs for the prediction returns better outputs, for which the application of artificial intelligence can be useful in the GEI prediction analysis (Lu and Leong 2016b; Uelze et al. 2020). Many programs and software based on artificial intelligence and machine learning for the identification of GEIs are available that are helping researchers in gene identification that encodes adaptations used by microorganisms (Lu and Leong 2016b).

The first machine learning-based method for structural models of GEIs identification is Relevance Vector Machine (RVM). The datasets were assembled by comparative genomics techniques. Eight different types of genomic features were used to train the model such as IVOM score, insertion point, GI size, gene density, repeats, phage-related protein domains, integrase protein domains, and non-coding RNAs (Lu and Leong 2016b).

The program Genomic Island Hunter is based on a decision tree-based bagging model. Based on the performance metric comparison with other programs, GIHunter showed more accuracy in prediction and has been in use for more than 2000 prokaryotic genomes (Che et al. 2014a). Another program is the GI-SVM, which is formed on a one-class support vector machine (SVM) to identify GEIs in a single genome. It employs composition bias in terms of k-mer content. GI-SVM also allows flexible parameters to identify optimum outcomes for each genome. GI-SVM method is a more sensible and quick approach for researchers in the first steps of detection of GEIs in new sequenced genomes (Lu and Leong 2016a). HGT may account for 1.6%–32.6% of a bacterial genome (Boto 2010). This percentage value suggests that variation throughout the bacterial genome and clades can be featured in GEIs. Most importantly, prediction of GEIs in bacterial species also leads to vaccine and antibiotics development (Coates and Hu 2007) as well as in cancer therapy (Coates and Hu 2007; Bar et al. 2008). For instance, significantly PAIs carry several pathogenicity and virulence genes that help researchers to identify possible vaccine candidates (Moriel et al. 2010; Assaf et al. 2021). Program Shutter Island applies powerful deep neural networks for the identification of GEIs. Shutter Island exhibits a convolutional neural network on visual depictions of the genome for GEIs identification and also exhibits this program based on deep neural networks (Assaf et al. 2021).

An artificial intelligence-based pipeline was developed by Mbulayi Onesime et al., in 2021 for GEIs identification using the chi-square test and random forest algorithm. They used seven different types of sequence features such as the composition of k-spaced nucleic acid pairs, dinucleotide composition, nucleic acid composition, pseudo dinucleotide composition, electron–ion-interaction pseudopotentials of trinucleotide, reverse compliment k-mer, and trinucleotide composition. All used features were filtered by the chi-square test and then the random forest decision tree algorithm was used for GEI prediction. Their experimental outcome demonstrated that the considered method of the pipeline has an improved performance than older methodologies (Onesime et al. 2021).

3.5 Conclusions

The data generated through the high throughput sequencing methods have been remarkably beneficial for biological significance. The comparative genomics study using this data can help us in understanding the molecular machinery of microorganisms. However, this data needs to be efficiently analyzed. Also, phylogenetic and pan-genomics analyses normally highlight the use of conserved regions for several purposes, while the genome plasticity is normally disregarded, even with its profound impact on bacterial adaptation to hosts/environments and bacterial evolution. Here, we discussed the main features of GEIs, their HGT events, the software normally used, some studies in CMNR group, and mostly important, new approaches in Artificial Intelligence that hold the potential for overcoming outdated methods.

References

Akashi H (1994) Synonymous codon usage in Drosophila melanogaster: natural selection and translational accuracy. Genetics 136(3):927–935. [cited 2022 Jun 14]. Available from: https://pubmed.ncbi.nlm.nih.gov/8005445/

Alikhan NF, Petty NK, Ben Zakour NL, Beatson SA (2011) BLAST Ring Image Generator (BRIG): simple prokaryote genome comparisons. BMC Genomics 12:402. [cited 2018 Nov 8]. Available from: http://www.ncbi.nlm.nih.gov/pubmed/21824423

Almpanis A, Swain M, Gatherer D, McEwan N (2018) Correlation between bacterial G+C content, genome size and the G+C content of associated plasmids and bacteriophages. Microb Genom 4(4):e000168. [cited 2022 Jun 14]. Available from: https://www.microbiologyresearch.org/content/journal/mgen/10.1099/mgen.0.000168

Anastasi E, MacArthur I, Scortti M, Alvarez S, Giguére S, Vázquez-Boland JA (2016) Pangenome and phylogenomic analysis of the pathogenic actinobacterium rhodococcus equi. Genome Biol Evol 8(10):3140–3148. [cited 2022 Jun 28]. Available from: https://academic.oup.com/gbe/article/8/10/3140/2939557

Arnvig KB, Young DB (2012) Non-coding RNA and its potential role in Mycobacterium tuberculosis pathogenesis. RNA Biol 9(4):427F. [cited 2022 Jul 29]. Available from: /pmc/articles/PMC3384566/

Arvey AJ, Azad RK, Raval A, Lawrence JG (2009) Detection of genomic islands via segmental genome heterogeneity. Nucleic Acids Res 37(16):5255–5266. [cited 2022 Aug 15]. Available from: https://academic.oup.com/nar/article/37/16/5255/2410410

Assaf R, Xia F, Stevens R (2021) Identifying genomic islands with deep neural networks. BMC Genom 22(3):1–11. [cited 2022 Jun 16]. Available from: https://bmcgenomics.biomedcentral.com/articles/10.1186/s12864-021-07575-5

Azevedo V, Abreu V, Almeida S, Santos A, Soares S, Ali A et al (2011) Whole genome annotation: in silico analysis. Bioinformatics: Trends Methodol. [cited 2022 Jun 14]. Available from: undefined/state.item.id

Baer R, Bankier AT, Biggin MD, Deininger PL, Farrell PJ, Gibson TJ et al (1984) DNA sequence and expression of the B95-8 Epstein—Barr virus genome. Nature 310(5974):207–211. [cited 2022 May 5]. Available from: https://www.nature.com/articles/310207a0

Bansal-Mutalik R, Nikaido H (2011) Quantitative lipid composition of cell envelopes of Corynebacterium glutamicum elucidated through reverse micelle extraction. Proc Natl Acad Sci U S A 108(37):15360–15365. [cited 2022 Jul 26]. Available from: https://www.pnas.org/doi/abs/10.1073/pnas.1112572108

Bar H, Yacoby I, Benhar I (2008) Killing cancer cells by targeted drug-carrying phage nanomedicines. BMC Biotechnol 8(1):1–14. [cited 2022 Jun 16]. Available from: https://bmcbiotechnol.biomedcentral.com/articles/10.1186/1472-6750-8-37

Baraúna RA, Ramos RTJ, Veras AAO, Pinheiro KC, Benevides LJ, Viana MVC et al (2017) Assessing the genotypic differences between strains of Corynebacterium pseudotuberculosis biovar equi through Comparative Genomics. PLoS One 12(1):e0170676. [cited 2022 Jun 16]. Available from: https://journals.plos.org/plosone/article?id=10.1371/journal.pone.0170676

Barcellos FG, Menna P, Batista JSDS, Hungria M (2007) Evidence of horizontal transfer of symbiotic genes from a Bradyrhizobium japonicum inoculant strain to indigenous diazotrophs Sinorhizobium (Ensifer) fredii and Bradyrhizobium elkanii in a Brazilian savannah soil. Appl Environ Microbiol 73(8):2635–2643. [cited 2022 Jun 15].

von Bargen K, Haas A (2009) Molecular and infection biology of the horse pathogen Rhodococcus equi. FEMS Microbiol Rev 33(5):870–891. [cited 2022 Jun 28]. Available from: https://academic.oup.com/femsre/article/33/5/870/562772

Beaman BL (1994) Mechanisms for the virulence of Nocardia. Springer, Dordrecht, pp 561–572. [cited 2022 Jun 16]. Available from: https://link.springer.com/chapter/10.1007/978-94-011-0746-4_39

Bertelli C, Gray KL, Woods N, Lim AC, Tilley KE, Winsor GL et al (2022) Enabling genomic island prediction and comparison in multiple genomes to investigate bacterial evolution and outbreaks. Microbial Genom 8(5):000818. [cited 2022 Aug 15]. Available from: https://www.microbiologyresearch.org/content/journal/mgen/10.1099/mgen.0.000818

Boto L (2010) Horizontal gene transfer in evolution: facts and challenges. Proc Royal Soc B Biol Sci 277(1683):819–827. [cited 2022 Aug 15]. Available from: https://royalsocietypublishing.org/doi/10.1098/rspb.2009.1679

de Brito DM, Maracaja-Coutinho V, de Farias ST, Batista L, v., do Rego TG. (2016) A novel method to predict genomic islands based on mean shift clustering algorithm. PLoS One 11(1): e0146352. [cited 2022 Aug 15]. Available from: https://journals.plos.org/plosone/article?id=10.1371/journal.pone.0146352

vanden Broeck D, Horvath C, de Wolf MJS (2007) Vibrio cholerae: cholera toxin. Int J Biochem Cell Biol 39(10):1771–1775. [cited 2022 Jun 15]. Available from: https://pubmed.ncbi.nlm.nih.gov/17716938/

Bulmer M (1988) Are codon usage patterns in unicellular organisms determined by selection-mutation balance? J Evol Biol 1(1):15–26. [cited 2022 Jun 14]. Available from: https://onlinelibrary.wiley.com/doi/full/10.1046/j.1420-9101.1988.1010015.x

Byrne GA, Russell DA, Chen X, Meijer WG (2007) Transcriptional regulation of the virR operon of the intracellular pathogen Rhodococcus equi. J Bacteriol 189(14):5082–5089. [cited 2022 Jun 28]. Available from: https://journals.asm.org/doi/full/10.1128/JB.00431-07

Callens M, Scornavacca C, Bedhomme S (2021) Evolutionary responses to codon usage of horizontally transferred genes in pseudomonas aeruginosa: gene retention, amelioration and compensatory evolution. Microb Genom 7(6):000587. [cited 2022 Aug 15]. Available from: https://www.microbiologyresearch.org/content/journal/mgen/10.1099/mgen.0.000587

Chaban B, Hughes HV, Beeby M (2015) The flagellum in bacterial pathogens: for motility and a whole lot more. Semin Cell Dev Biol 46:91–103

Chan AP, Choi Y, Clarke TH, Brinkac LM, White RC, Jacobs MR et al (2020) AbGRI4, a novel antibiotic resistance island in multiply antibiotic-resistant Acinetobacter baumannii clinical isolates. J Antimicrob Chemoth 75(10):2760–2768. [cited 2022 Jun 15]. Available from: https://academic.oup.com/jac/article/75/10/2760/5873331

Chatterjee R, Chaudhuri K, Chaudhuri P (2008) On detection and assessment of statistical significance of Genomic Islands. BMC Genom 9(1):1–11. [cited 2022 Aug 15]. Available from: https://bmcgenomics.biomedcentral.com/articles/10.1186/1471-2164-9-150

Che D, Hasan MS, Wang H, Fazekas J, Huang J, Liu Q (2011) EGID: an ensemble algorithm for improved genomic island detection in genomic sequences. Bioinformation 7(6):311–314. [cited 2022 Aug 15]. Available from: https://pubmed.ncbi.nlm.nih.gov/22355228/

Che D, Hockenbury C, Marmelstein R, Rasheed K (2010) Classification of genomic islands using decision trees and their ensemble algorithms. BMC Genomics 11(Suppl. 2):1–9. [cited 2022 Aug 15]. Available from: https://bmcgenomics.biomedcentral.com/articles/10.1186/1471-2164-11-S2-S1

Che D, Wang H (2013) GIV: A Tool for Genomic Islands Visualization. Bioinformation 9(17): 879–882. [cited 2022 Aug 15]. Available from: https://pubmed.ncbi.nlm.nih.gov/24250116/

Che D, Wang H, Fazekas J, Chen B (2014a) An accurate genomic island prediction method for sequenced bacterial and archaeal genomes. J Proteom Bioinform 7(8):1–8. [cited 2022 Aug 15]. Available from: https://www.longdom.org/open-access/an-accurate-genomic-island-prediction-method-for-sequenced-bacterialand-archaeal-genomes-33848.html

Che D, Wang H, Fazekas J, Chen B (2014b) An accurate genomic island prediction method for sequenced bacterial and archaeal genomes. J Proteom Bioinform 7(8):214–221. [cited 2022 Jun 29]. Available from: http://www5.esu.edu/cpsc/bioinfo/dgi

Cho KS, Bang SM, Toh A (2014) Lipids and lipid signaling in drosophila models of neurodegenerative diseases. In: Omega-3 fatty acids in brain and neurological health, vol 1. Academic Press, London, pp 327–336

Coates ARM, Hu Y (2007) Novel approaches to developing new antibiotics for bacterial infections. Br J Pharmacol 152(8):1147–1154. [cited 2022 Jun 29]. Available from: https://onlinelibrary.wiley.com/doi/full/10.1038/sj.bjp.0707432

Colonna Romano N, Fanti L (2022) Transposable elements: major players in shaping genomic and evolutionary patterns. Cells 11(6):1048. [cited 2022 Jun 17]. Available from: /pmc/articles/PMC8947103/

Conville PS, Witebsky FG (2010) The complexity of nocardia taxonomy: implications for the clinical microbiology laboratory. Clin Microbiol Newslett 32(16):119–125

Coulson GB, Agarwal S, Hondalus MK (2010) Characterization of the role of the pathogenicity island and vapG in the virulence of the intracellular actinomycete pathogen Rhodococcus equi. Infect Immun 78(8):3323–3334. [cited 2022 Jun 28]. Available from: https://pubmed.ncbi.nlm.nih.gov/20439471/

Coulson GB, Miranda-CasoLuengo AA, Miranda-CasoLuengo R, Wang X, Oliver J, Willingham-Lane JM et al (2015) Transcriptome reprogramming by plasmid-encoded transcriptional regulators is required for host niche adaption of a macrophage pathogen. Infect Immunity 83(8):3137–3145. [cited 2022 Jun 28]. Available from: https://journals.asm.org/doi/full/10.1128/IAI.00230-15

Craig Venter J, Adams MD, Myers EW, Li PW, Mural RJ, Sutton GG et al (1979) The sequence of the human genome. Science 291(5507):1304–1351. [cited 2022 May 5]. Available from: https://www.science.org/doi/abs/10.1126/science.1058040

Deng W, Burland V, Plunkett G, Boutin A, Mayhew GF, Liss P et al (2002) Genome sequence of Yersinia pestis KIM. J Bacteriol 184(16):4601–4611. [cited 2022 Jun 14]. Available from: https://journals.asm.org/doi/full/10.1128/JB.184.16.4601-4611.2002

Dhillon BK, Laird MR, Shay JA, Winsor GL, Lo R, Nizam F et al (2015) IslandViewer 3: more flexible, interactive genomic island discovery, visualization and analysis. Nucleic Acids Res 43 (W1):W104–W108. [cited 2022 Aug 15]. Available from: https://academic.oup.com/nar/article/43/W1/W104/2467901

Dobrindt U, Janke B, Piechaczek K, Nagy G, Ziebuhr W, Fischer G et al (2000) Toxin genes on pathogenicity islands: impact for microbial evolution. Int J Med Microbiol 290(4–5):307–311. [cited 2022 Jul 29]. Available from: https://pubmed.ncbi.nlm.nih.gov/11111903/

Dorella FA, Pacheco LGC, Oliveira SC, Miyoshi A, Azevedo V (2006) Corynebacterium pseudotuberculosis: microbiology, biochemical properties, pathogenesis and molecular studies of virulence. Veter Res 37(2):201–218

Emmerson A, Hawkey P, Gillespie S (1987) Principles and practice of clinical bacteriology. J Clin Pathol 40(11):1390–1391. [cited 2022 Jun 15]. Available from: https://books.google.com/books/about/Principles_and_Practice_of_Clinical_Bact.html?hl=pt-BR&id=0RN2QgAACAAJ

Forrellad MA, Klepp LI, Gioffré A, García JS, Morbidoni HR, de la Paz Santangelo M et al (2012) Virulence factors of the Mycobacterium tuberculosis complex. Virulence 4(1):3–66. [cited 2022 Jun 16]. Available from: https://www.tandfonline.com/doi/abs/10.4161/viru.22329

Frost LS, Leplae R, Summers AO, Toussaint A (2005) Mobile genetic elements: the agents of open source evolution. Nat Rev Microbiol 3(9):722–732. [cited 2022 Jul 22];3(9). Available from: https://www.nature.com/articles/nrmicro1235

Gao F, Zhang CT (2006) GC-Profile: a web-based tool for visualizing and analyzing the variation of GC content in genomic sequences. Nucleic Acids Res 34(suppl_2):W686–W691. [cited 2022 Aug 15]. Available from: https://academic.oup.com/nar/article/34/suppl_2/W686/2505479

GenBank and WGS Statistics [Internet]. [cited 2022 May 5]. Available from: https://www.ncbi.nlm.nih.gov/genbank/statistics/

Genome List - Genome - NCBI [Internet]. [cited 2022 Jun 14]. Available from: https://www.ncbi.nlm.nih.gov/genome/browse/#!/prokaryotes/

Gillings MR (2014) Integrons: past, present, and future. Microbiol Mol Biol Rev 78(2):257–277. [cited 2022 Jun 28]. Available from: https://journals.asm.org/doi/full/10.1128/MMBR.00056-13

Gogarten JP, Townsend JP (2005) Horizontal gene transfer, genome innovation and evolution. Nat Rev Microbiol 3(9):679–687. [cited 2022 Jun 14]. Available from: https://www.nature.com/articles/nrmicro1204

Gonzalez DL, Giannerini S, Rosa R (2019) On the origin of degeneracy in the genetic code. Interface Focus 9(6):20190038. [cited 2022 Jun 15]. Available from: https://royalsocietypublishing.org/doi/10.1098/rsfs.2019.0038

Griffith F (1928) The significance of pneumococcal types. J Hyg (Lond) 27(2):113–159. [cited 2022 Jul 29]. Available from: https://pubmed.ncbi.nlm.nih.gov/20474956/

Grosjean H, Fiers W (1982) Preferential codon usage in prokaryotic genes: the optimal codon-anticodon interaction energy and the selective codon usage in efficiently expressed genes. Gene 18(3):199–209

Hacker J, Bender L, Ott M, Wingender J, Lund B, Marre R et al (1990) Deletions of chromosomal regions coding for fimbriae and hemolysins occur in vitro and in vivo in various extra intestinal Escherichia coli isolates. Microb Pathog 8(3):213–225

Hacker J, Carniel E (2001) Ecological fitness, genomic islands and bacterial pathogenicity. A Darwinian view of the evolution of microbes. EMBO Rep 2(5):376–381. [cited 2022 Jul 29]. Available from: https://pubmed.ncbi.nlm.nih.gov/11375927/

Hadjilouka A, Paramithiotis S, Drosinos EH (2018) Genetic analysis of the listeria pathogenicity island 1 of listeria monocytogenes 1/2a and 4b isolates. Curr Microbiol 75(7):857–865. [cited 2022 Jun 16];75. Available from: https://link.springer.com/article/10.1007/s00284-018-1458-4

Hasan MS, Liu Q, Wang H, Fazekas J, Chen B, Che D (2012) GIST: genomic island suite of tools for predicting genomic islands in genomic sequences. Bioinformation 8(4):203. [cited 2022 Aug 15]. Available from: /pmc/articles/PMC3302003/

Hickman AB, Dyda F (2015) Mechanisms of DNA transposition. Microbiol Spectr 3(2):MDNA3. [cited 2022 Jun 17]. Available from: /pmc/articles/PMC7422641/

Hoen DR, Hickey G, Bourque G, Casacuberta J, Cordaux R, Feschotte C et al (2015) A call for benchmarking transposable element annotation methods. Mob DNA 6(1):1–9. [cited 2022 Jun 15]. Available from: https://mobilednajournal.biomedcentral.com/articles/10.1186/s13100-015-0044-6

Hsiao W, Wan I, Jones SJ, Brinkman FSL (2003) IslandPath: aiding detection of genomic islands in prokaryotes. Bioinformatics 19(3):418–420. [cited 2022 Aug 15]. Available from: https://academic.oup.com/bioinformatics/article/19/3/418/258350

Hudson CM, Lau BY, Williams KP (2015) Islander: a database of precisely mapped genomic islands in tRNA and tmRNA genes. Nucleic Acids Res 43(D1):D48–D53. [cited 2022 Aug 15]. Available from: https://academic.oup.com/nar/article/43/D1/D48/2437357

Huerta-Cepas J, Forslund K, Coelho LP, Szklarczyk D, Jensen LJ, von Mering C et al (2017) Fast genome-wide functional annotation through orthology assignment by eggNOG-Mapper. Mol Biol Evol 34(8):2115–2122. [cited 2022 Jun 8]. Available from: https://pubmed.ncbi.nlm.nih.gov/28460117/

Jain R, Ramineni S, Parekh N (2011) IGIPT - integrated genomic island prediction tool. Bioinformation 7(6):307–310. [cited 2022 Aug 15]. Available from: https://pubmed.ncbi.nlm.nih.gov/22355227/

Jaron KS, Moravec JC, Martínková N (2014) SigHunt: horizontal gene transfer finder optimized for eukaryotic genomes. Bioinformatics 30(8):1081–1086. [cited 2022 Aug 15]. Available from: https://pubmed.ncbi.nlm.nih.gov/24371153/

Jensen LJ, Julien P, Kuhn M, von Mering C, Muller J, Doerks T et al (2008) eggNOG: automated construction and annotation of orthologous groups of genes. Nucleic Acids Res 36(Database issue):D250. [cited 2022 Jun 8]. Available from: https://pubmed.ncbi.nlm.nih.gov/17942413/

Join-Lambert OF, Ouache M, Canioni D, Beretti JL, Blanche S, Berche P et al (2006) Corynebacterium pseudotuberculosis necrotizing lymphadenitis in a twelve-year-old patient. Pediatr Infect Dis J 25(9):848–851. [cited 2022 Aug 11]. Available from: https://pubmed.ncbi.nlm.nih.gov/16940849/

de Jong NWM, van Kessel KPM, van Strijp JAG (2019) Immune Evasion by Staphylococcus aureus. Microbiol Spectr 7(2) [cited 2022 Jun 15]. Available from: https://pubmed.ncbi.nlm.nih.gov/30927347/

Kalt F, Schulthess B, Sidler F, Herren S, Fucentese SF, Zingg PO et al (2018) Corynebacterium species rarely cause orthopedic infections. J Clin Microbiol 56(12) [cited 2022 Jun 15]. Available from: https://journals.asm.org/doi/full/10.1128/JCM.01200-18

Karlin S, Mrázek J, Campbell AM (1998) Codon usages in different gene classes of the Escherichia coli genome. Mol Microbiol 29(6):1341–1355. [cited 2022 Aug 11]. Available from: https://pubmed.ncbi.nlm.nih.gov/9781873/

Komaki H, Ichikawa N, Hosoyama A, Takahashi-Nakaguchi A, Matsuzawa T, Kichiro S et al (2014) Genome based analysis of type-I polyketide synthase and nonribosomal peptide synthetase gene clusters in seven strains of five representative Nocardia species. BMC Genomics 15(1):1–11. [cited 2022 Jun 28]. Available from: https://bmcgenomics.biomedcentral.com/articles/10.1186/1471-2164-15-323

Koonin E, Makarova KS, Wolf YI (2021) Evolution of microbial genomics: conceptual shifts over a quarter century. Trends Microbiol 29(7):582–592. [cited 2022 Jun 8]. Available from: http://www.cell.com/article/S0966842X2100007X/fulltext

Kresge N, Simoni RD, Hill RL (2005) Chargaff's rules: the work of Erwin Chargaff. J Biol Chem 280(24):172–174. [cited 2022 Jun 14]. Available from: http://www.jbc.org/article/S0021925820615228/fulltext

Krizova L, Nemec A (2010) A 63 kb genomic resistance island found in a multidrug-resistant Acinetobacter baumannii isolate of European clone I from 1977. J Antimicrob Chemother 65(9): 1915–1918. [cited 2022 Jul 29]. Available from: https://pubmed.ncbi.nlm.nih.gov/20558468/

Langille MGI, Hsiao WWL, Brinkman FSL (2008) Evaluation of genomic island predictors using a comparative genomics approach. BMC Bioinform 9(1):1–10. [cited 2022 Jun 14]. Available from: https://bmcbioinformatics.biomedcentral.com/articles/10.1186/1471-2105-9-329

Langille MGI, Hsiao WWL, Brinkman FSL (2010) Detecting genomic islands using bioinformatics approaches. Nat Rev Microbiol 8(5):373–382. [cited 2022 Jun 14]. Available from: https://www.nature.com/articles/nrmicro2350

Lawrence JG, Roth JR (2014) Genomic flux: genome evolution by gene loss and acquisition. In: Organization of the prokaryotic genome. Wiley, Hoboken, pp 263–289. [cited 2022 Jun 14]. Available from: https://onlinelibrary.wiley.com/doi/full/10.1128/9781555818180.ch15

Lee CC, Chen YPP, Yao TJ, Ma CY, Lo WC, Lyu PC et al (2013) GI-POP: a combinational annotation and genomic island prediction pipeline for ongoing microbial genome projects. Gene 518(1):114–123

Letek M, González P, MacArthur I, Rodríguez H, Freeman TC, Valero-Rello A et al (2010) The genome of a pathogenic rhodococcus: cooptive virulence underpinned by key gene acquisitions. PLoS Genet 6(9):e1001145. [cited 2022 Jun 28]. Available from: https://journals.plos.org/plosgenetics/article?id=10.1371/journal.pgen.1001145

Letek M, Ocampo-Sosa AA, Sanders M, Fogarty U, Buckley T, Leadon DP et al (2008) Evolution of the Rhodococcus equi vap pathogenicity island seen through comparison of host-associated vapA and vapB virulence plasmids. J Bacteriol 190(17):5797–5805. [cited 2022 Jun 28]. Available from: https://journals.asm.org/doi/full/10.1128/JB.00468-08

Levy-Frebault V, Portaels F (1992) Proposed minimal standards for the genus Mycobacterium and for description of new slowly growing Mycobacterium species. Int J Syst Bacteriol 42(2): 315–323

Li J, Tai C, Deng Z, Zhong W, He Y, Ou HY (2018) VRprofile: gene-cluster-detection-based profiling of virulence and antibiotic resistance traits encoded within genome sequences of pathogenic bacteria. Brief Bioinform 19(4):566–574. [cited 2022 Aug 15]. Available from: https://academic.oup.com/bib/article/19/4/566/2888442

Lu B, Leong HW (2016a) GI-SVM: a sensitive method for predicting genomic islands based on unannotated sequence of a single genome. J Bioinform Comput Biol 14(1). https://doi.org/10.1142/S0219720016400035
Lu B, Leong HW (2016b) Computational methods for predicting genomic islands in microbial genomes. Comput Struct Biotechnol J 14:200–206
Lu SH, Qian ZW, Mou PP, Xie L (2020) Clinical Nocardia species: Identification, clinical characteristics, and antimicrobial susceptibility in Shandong, China. Bosnian J Basic Med Sci 20(4):531–538. [cited 2022 Jun 16]. Available from: https://www.bjbms.org/ojs/index.php/bjbms/article/view/4764
Lysnyansky I, Calcutt MJ, Ben-Barak I, Ron Y, Levisohn S, Methé BA et al (2009) Molecular characterization of newly identified IS3, IS4 and IS30 insertion sequence-like elements in Mycoplasma bovis and their possible roles in genome plasticity. FEMS Microbiol Lett 294(2):172–182. [cited 2022 Jun 15]. Available from: https://academic.oup.com/femsle/article/294/2/172/512759
Mahillon J, Chandler M (1998) Insertion sequences. Microbiol Mol Biol Rev 62(3):725–774. [cited 2022 Jun 14]. Available from: https://journals.asm.org/doi/full/10.1128/MMBR.62.3.725-774.1998
Mangutov EO, Kharseeva GG, Alutina EL (2021) Corynebacterium spp. - problematic pathogens of the human respiratory tract (review of literature). Klin Lab Diagn 66(8):502–508. [cited 2022 Jun 15]. Available from: https://pubmed.ncbi.nlm.nih.gov/34388322/
Männle D, McKinnie SMK, Mantri SS, Steinke K, Lu Z, Moore BS et al (2020) Comparative genomics and metabolomics in the genus Nocardia. mSystems 5(3) [cited 2022 Jun 28]. Available from: https://journals.asm.org/doi/full/10.1128/mSystems.00125-20
Marraffini LA, Sontheimer EJ (2010) CRISPR interference: RNA-directed adaptive immunity in bacteria and archaea. Nat Rev Genet 11(3):181. [cited 2022 Jul 29]. Available from: /pmc/articles/PMC2928866/
Martínez-Barricarte R (2020) Isolated Nocardiosis, an Unrecognized Primary Immunodeficiency? Fronti Immunol 11:2393
Medini D, Donati C, Tettelin H, Masignani V, Rappuoli R (2005) The microbial pan-genome. Curr Opin Genet Dev 15(6):589–594. [cited 2022 Jul 22]. Available from: https://pubmed.ncbi.nlm.nih.gov/16185861/
Mehta HH, Shamoo Y (2020) Pathogenic Nocardia: a diverse genus of emerging pathogens or just poorly recognized? PLoS Pathogens 16(3):e1008280. [cited 2022 Jun 16]. Available from: https://journals.plos.org/plospathogens/article?id=10.1371/journal.ppat.1008280
Moriel DG, Bertoldi I, Spagnuolo A, Marchi S, Rosini R, Nesta B et al (2010) Identification of protective and broadly conserved vaccine antigens from the genome of extraintestinal pathogenic Escherichia coli. Proc Natl Acad Sci U S A 107(20):9072–9077. [cited 2022 Jun 16]. Available from: www.pnas.org/cgi/doi/10.1073/pnas.0915077107
Nelson MG, Linheiro RS, Bergman CM (2017) McClintock: an integrated pipeline for detecting transposable element insertions in whole-genome shotgun sequencing data. G3 7(8): 2763–2778. [cited 2022 Jun 15]. Available from: https://academic.oup.com/g3journal/article/7/8/2763/6031520
Nichio BTL, Marchaukoski JN, Raittz RT (2017) New tools in orthology analysis: a brief review of promising perspectives. Front Genet 8:165
Nirenberg MW (1963) The genetic code: II. Sci Am 208(3):80–95. Available from: http://www.jstor.org/stable/24936501
Ocampo-Sosa AA, Lewis DA, Navas J, Quigley F, Callejo R, Scortti M et al (2007) Molecular epidemiology of Rhodococcus equi based on traA, vapA, and vapB virulence plasmid markers. J Infect Dis 196(5):763–769. [cited 2022 Jun 28]. Available from: https://pubmed.ncbi.nlm.nih.gov/17674320/
Ochman H, Lawrence JG, Grolsman EA (2000) Lateral gene transfer and the nature of bacterial innovation. Nature 405(6784):299–304. [cited 2022 Jun 16]. Available from: https://pubmed.ncbi.nlm.nih.gov/10830951/

Ogier JC, Calteau A, Forst S, Goodrich-Blair H, Roche D, Rouy Z et al (2010) Units of plasticity in bacterial genomes: new insight from the comparative genomics of two bacteria interacting with invertebrates, Photorhabdus and Xenorhabdus. BMC Genomics 11(1):1–21. [cited 2022 Aug 15]. Available from: https://bmcgenomics.biomedcentral.com/articles/10.1186/1471-2164-11-568

Onesime M, Yang Z, Dai Q (2021) Genomic island prediction via chi-square test and random forest algorithm. Comput Math Methods Med 2021:9969751

Payne SM, Mey AR, Wyckoff EE (2016) Vibrio iron transport: evolutionary adaptation to life in multiple environments. Microbiol Mol Biol Rev 80(1):69–90. [cited 2022 Jun 15]. Available from: https://journals.asm.org/doi/full/10.1128/MMBR.00046-15

Penn K, Jenkins C, Nett M, Udwary DW, Gontang EA, McGlinchey RP et al (2009) Genomic islands link secondary metabolism to functional adaptation in marine Actinobacteria. ISME J 3(10):1193–1203. [cited 2022 Jun 15]. Available from: https://www.nature.com/articles/ismej200958

Petrovski S, Seviour RJ, Tillett D (2013) Genome sequence and characterization of a Rhodococcus equi phage REQ1. Virus Genes 46:588–590. [cited 2022 Jun 28]. Available from: https://link.springer.com/article/10.1007/s11262-013-0887-1

Pierneef R, Cronje L, Bezuidt O, Reva ON (2015) Pre_GI: a global map of ontological links between horizontally transferred genomic islands in bacterial and archaeal genomes. Database 2015:58. [cited 2022 Aug 15]. Available from: https://academic.oup.com/database/article/doi/10.1093/database/bav058/2433191

Preston A, Parkhill J, Maskell DJ (2004) The bordetellae: lessons from genomics. Nat Rev Microbiol 2(5):379–390. [cited 2022 Jun 15]. Available from: https://www.nature.com/articles/nrmicro886

Pundhir S, Vijayvargiya H, Kumar A (2008) PredictBias: a server for the identification of genomic and pathogenicity islands in prokaryotes. Undefined.

Rajan I, Aravamuthan S, Mande SS (2007) Identification of compositionally distinct regions in genomes using the centroid method. Bioinformatics 23(20):2672–2677. [cited 2022 Aug 15]. Available from: https://academic.oup.com/bioinformatics/article/23/20/2672/230015

Rastogi N, Legrand E, Sola C (2001) The mycobacteria: an introduction to nomenclature and pathogenesis. Rev Sci Tech 20(1):21–54. [cited 2022 Jun 29]. Available from: https://pubmed.ncbi.nlm.nih.gov/11288513/

Reichenberger ER, Rosen G, Hershberg U, Hershberg R (2015) Prokaryotic nucleotide composition is shaped by both phylogeny and the environment. Genome Biol Evol 7(5):1380–1389. [cited 2022 Jun 14]. Available from: https://academic.oup.com/gbe/article/7/5/1380/603567

Rice PA, Baker TA (2001) Comparative architecture of transposase and integrase complexes. Nat Struct Biol 8(4):302–307. [cited 2022 Jun 14]. Available from: https://www.nature.com/articles/nsb0401_302

Riojas MA, McGough KJ, Rider-Riojas CJ, Rastogi N, Hazbón MH (2018) Phylogenomic analysis of the species of the Mycobacterium tuberculosis complex demonstrates that Mycobacterium africanum, Mycobacterium bovis, Mycobacterium caprae, Mycobacterium microti and Mycobacterium pinnipedii are later heterotypic synonyms of Mycobacterium tuberculosis. Int J Syst Evol Microbiol 68(1):324–332. [cited 2022 Jun 29]. Available from: https://pubmed.ncbi.nlm.nih.gov/29205127/

Rissman AI, Mau B, Biehl BS, Darling AE, Glasner JD, Perna NT (2009) Reordering contigs of draft genomes using the Mauve Aligner. Bioinformatics 25(16):2071–2073. [cited 2022 Jun 17]. Available from: https://academic.oup.com/bioinformatics/article/25/16/2071/204845

Salifu SP, Valero-Rello A, Campbell SA, Inglis NF, Scortti M, Foley S et al (2013) Genome and proteome analysis of phage E3 infecting the soil-borne actinomycete Rhodococcus equi. Environ Microbiol Rep 5(1):170–178. [cited 2022 Jun 28]. Available from: https://onlinelibrary.wiley.com/doi/full/10.1111/1758-2229.12028

Sanger F, Air GM, Barrell BG, Brown NL, Coulson AR, Fiddes JC et al (1977a) Nucleotide sequence of bacteriophage φX174 DNA. Nature 265:5596. [cited 2022 May 5]. Available from: https://www.nature.com/articles/265687a0

Sanger F, Nicklen S, Coulson AR (1977b) DNA sequencing with chain-terminating inhibitors. Proc Natl Acad Sci U S A 74(12):5463–5467. [cited 2022 May 5]. Available from: https://www.pnas.org

Sanger F, Thompson EO (1953) The amino-acid sequence in the glycyl chain of insulin. 2. The investigation of peptides from enzymic hydrolysates. Biochem J 53(3):366–374. [cited 2022 May 5]. Available from: https://portlandpress.com/biochemj/article/53/3/366/47647/The-amino-acid-sequence-in-the-glycyl-chain-of

Schlüter A, Szczepanowski R, Pühler A, Top EM (2007) Genomics of IncP-1 antibiotic resistance plasmids isolated from wastewater treatment plants provides evidence for a widely accessible drug resistance gene pool. FEMS Microbiol Rev 31(4):449–477. [cited 2022 Jun 15]. Available from: https://academic.oup.com/femsre/article/31/4/449/2399130

Schmidt H, Hensel M (2004) Pathogenicity islands in bacterial pathogenesis. Clin Microbiol Rev 17(1):14–56. [cited 2022 Jun 16]. Available from: https://pubmed.ncbi.nlm.nih.gov/14726454/

Segata N (2018) On the road to strain-resolved comparative metagenomics. mSystems 3(2):e00190. [cited 2022 Jun 14]. Available from: https://journals.asm.org/doi/full/10.1128/mSystems.00190-17

Setubal JC, Almeida NF, Wattam AR (2018) Comparative genomics for prokaryotes. Methods Mol Biol 1704:55–78. [cited 2022 Jun 8]; Available from: https://pubmed.ncbi.nlm.nih.gov/29277863/

Sharp PM, Matassi G (1994) Codon usage and genome evolution. Curr Opin Genet Dev 4(6):851–860

Shrivastava S, Siva Kumar Reddy CV, Mande SS (2010) INDeGenIUS, a new method for high-throughput identification of specialized functional islands in completely sequenced organisms. J Biosci 35(3):351–364. [cited 2022 Aug 15]. Available from: https://link.springer.com/article/10.1007/s12038-010-0040-4

Siguier P, Gourbeyre E, Chandler M (2014) Bacterial insertion sequences: their genomic impact and diversity. FEMS Microbiol Rev 38(5):865–891. [cited 2022 Jun 14]. Available from: https://academic.oup.com/femsre/article/38/5/865/496682

da Silva Filho AC, Raittz RT, Guizelini D, de Pierri CR, Augusto DW, dos Santos-Weiss ICR et al (2018) Comparative analysis of genomic island prediction tools. Front Genet 9:619

Soares SC, AbreuViní VAC, Ramos RTJ, Cerdeira L, Silva A, Baumbach J et al (2012) PIPS: pathogenicity island prediction software. PLoS One 7(2):e30848. [cited 2022 Aug 15]. Available from: https://journals.plos.org/plosone/article?id=10.1371/journal.pone.0030848

Soares SC, Geyik H, Ramos RTJ, de Sá PHCG, Barbosa EGV, Baumbach J et al (2016) GIPSy: genomic island prediction software. J Biotechnol 20(232):2–11

Soares SC, Silva A, Trost E, Blom J, Ramos R, Carneiro A et al (2013) The pan-genome of the animal pathogen Corynebacterium pseudotuberculosis reveals differences in genome plasticity between the biovar ovis and equi strains. PLoS One 8(1):e53818. [cited 2022 Jun 16]. Available from: https://journals.plos.org/plosone/article?id=10.1371/journal.pone.0053818

Sollitto M, Kenny NJ, Greco S, Tucci CF, Calcino AD, Gerdol M (2022) Detecting structural variants and associated gene presence-absence variation phenomena in the genomes of marine organisms. Methods Mol Biol 2498:53–76. [cited 2022 Jul 29]. Available from: https://pubmed.ncbi.nlm.nih.gov/35727540/

Staden R (1979) A strategy of DNA sequencing employing computer programs. Nucleic Acids Res 6(7):2601–2610. [cited 2022 May 5]. Available from: https://academic.oup.com/nar/article/6/7/2601/2384187

Sulaiman S, Yusoff NS, Mun NS, Makmur H, Firdaus-Raih M. Inference of horizontal gene transfer: gaining insights into evolution via lateral acquisition of genetic material. Encyclopedia of bioinformatics and computational biology: ABC of bioinformatics. 2018, Elsevier, San Diego 1;1–3:953–964.

Summer EJ, Liu M, Gill JJ, Grant M, Chan-Cortes TN, Ferguson L et al (2011) Genomic and functional analyses of Rhodococcus equi phages ReqiPepy6, ReqiPoco6, ReqiPine5, and ReqiDocB7. Appl Environ Microbiol 77(2):669–683. [cited 2022 Jun 28]. Available from: https://journals.asm.org/doi/full/10.1128/AEM.01952-10

Swithers KS, Soucy SM, Gogarten JP (2012) The role of reticulate evolution in creating innovation and complexity. Int J Evol Biol 2012:1–10

Takai S, Sawada N, Nakayama Y, Ishizuka S, Nakagawa R, Kawashima G et al (2020) Reinvestigation of the virulence of Rhodococcus equi isolates from patients with and without AIDS. Lett Appl Microbiol 71(6):679–683. [cited 2022 Jun 28]. Available from: https://onlinelibrary.wiley.com/doi/full/10.1111/lam.13386

Tatusov RL, Koonin EV, Lipman DJ (1997) A genomic perspective on protein families. Science 278(5338):631–637. [cited 2022 Jun 8]. Available from: https://pubmed.ncbi.nlm.nih.gov/9381173/

Tauch A, Fernández-Natal I, Soriano F (2016) A microbiological and clinical review on Corynebacterium kroppenstedtii. Int J Infect Dis 48:33–39. [cited 2022 Jun 15]. Available from: https://pubmed.ncbi.nlm.nih.gov/27155209/

Tettelin H, Masignani V, Cieslewicz MJ, Donati C, Medini D, Ward NL et al (2005) Genome analysis of multiple pathogenic isolates of Streptococcus agalactiae: implications for the microbial "pan-genome". Proc Natl Acad Sci U S A 102(39):13950–13955. [cited 2022 Jun 8]. Available from: https://www.pnas.org/doi/10.1073/pnas.0506758102

Teuber M, Meile L, Schwarz F (1999) Acquired antibiotic resistance in lactic acid bacteria from food. In: Lactic acid bacteria: genetics, metabolism and applications. CRC Press, Boca Raton, pp 115–137. [cited 2022 Jun 17]. Available from: https://link.springer.com/chapter/10.1007/978-94-017-2027-4_5

Tortoli E (2014) Microbiological features and clinical relevance of new species of the genus Mycobacterium. Clin Microbiol Rev 27(4):727–752. [cited 2022 Jun 29]. Available from: https://journals.asm.org/doi/full/10.1128/CMR.00035-14

Toyokawa M, Ohana N, Ueda A, Imai M, Tanno D, Honda M et al (2021) Identification and antimicrobial susceptibility profiles of Nocardia species clinically isolated in Japan. Sci Rep 11: 1–11. [cited 2022 Jun 16]. Available from: https://www.nature.com/articles/s41598-021-95870-2

Treangen TJ, Rocha EPC (2011) Horizontal transfer, not duplication, drives the expansion of protein families in prokaryotes. PLoS Genet 7(1):e1001284. [cited 2022 Jun 14]. Available from: https://journals.plos.org/plosgenetics/article?id=10.1371/journal.pgen.1001284

Tu Q, Ding D (2003) Detecting pathogenicity islands and anomalous gene clusters by iterative discriminant analysis. FEMS Microbiol Lett 221(2):269–275. [cited 2022 Aug 15]. Available from: https://academic.oup.com/femsle/article/221/2/269/630498

Tumapa S, Holden MTG, Vesaratchavest M, Wuthiekanun V, Limmathurotsakul D, Chierakul W et al (2008) Burkholderia pseudomallei genome plasticity associated with genomic island variation. BMC Genom 9(1):1–10. [cited 2022 Jul 29]. Available from: https://bmcgenomics.biomedcentral.com/articles/10.1186/1471-2164-9-190

Uelze L, Grützke J, Borowiak M, Hammerl JA, Juraschek K, Deneke C et al (2020) Typing methods based on whole genome sequencing data. One Health Outlook 2:1–19. [cited 2022 Jun 16]. Available from: https://onehealthoutlook.biomedcentral.com/articles/10.1186/s42522-020-0010-1

Valdezate S, Garrido N, Carrasco G, Villalón P, Medina-Pascual MJ, Saéz-Nieto JA (2015) Resistance gene pool to co-trimoxazole in non-susceptible Nocardia strains. Front Microbiol 6 (APR):376

Valero-Rello A, Hapeshi A, Anastasi E, Alvarez S, Scortti M, Meijer WG et al (2015) An invertron-like linear plasmid mediates intracellular survival and virulence in bovine isolates of Rhodococcus equi. Infect Immunity 83(7):2725–2737. [cited 2022 Jun 28]. Available from: https://journals.asm.org/doi/full/10.1128/IAI.00376-15

Vázquez-Boland JA, Giguère S, Hapeshi A, MacArthur I, Anastasi E, Valero-Rello A (2013) Rhodococcus equi: the many facets of a pathogenic actinomycete. Vet Microbiol 167(1–2): 9–33. [cited 2022 Jun 28]. Available from: https://pubmed.ncbi.nlm.nih.gov/23993705/

Vázquez-Boland JA, Meijer WG (2019) The pathogenic actinobacterium Rhodococcus equi: what's in a name? Mol Microbiol 112(1):1–15. [cited 2022 Jun 28]. Available from: https://onlinelibrary.wiley.com/doi/full/10.1111/mmi.14267

Veerachamy S, Yarlagadda T, Manivasagam G, Yarlagadda PK (2014) Bacterial adherence and biofilm formation on medical implants: a review. Proc Inst Mech Eng H 228(10):1083–1099. [cited 2022 Jun 15]. Available from: https://journals.sagepub.com/doi/10.1177/0954411914556137

Vendrell-Mir P, Barteri F, Merenciano M, González J, Casacuberta JM, Castanera R (2019) A benchmark of transposon insertion detection tools using real data. Mob DNA 10(1):1–19. [cited 2022 Jun 15]. Available from: https://mobilednajournal.biomedcentral.com/articles/10.1186/s13100-019-0197-9

Vernikos GS, Parkhill J (2006) Interpolated variable order motifs for identification of horizontally acquired DNA: revisiting the Salmonella pathogenicity islands. Bioinformatics 22(18): 2196–2203. [cited 2022 Aug 15]. Available from: https://academic.oup.com/bioinformatics/article/22/18/2196/316809

Waack S, Keller O, Asper R, Brodag T, Damm C, Fricke WF et al (2006) Score-based prediction of genomic islands in prokaryotic genomes using hidden Markov models. BMC Bioinform 7(1): 1–12. [cited 2022 Aug 15]. Available from: https://bmcbioinformatics.biomedcentral.com/articles/10.1186/1471-2105-7-142

Wan Y, Wick RR, Zobel J, Ingle DJ, Inouye M, Holt KE (2020) GeneMates: an R package for detecting horizontal gene co-transfer between bacteria using gene-gene associations controlled for population structure. BMC Genom 21(1):658. [cited 2022 Jul 22]. Available from: https://bmcgenomics.biomedcentral.com/articles/10.1186/s12864-020-07019-6

Wang H, Zhu Y, Cui Q, Wu W, Li G, Chen D et al (2022) Epidemiology and antimicrobial resistance profiles of the Nocardia species in China, 2009 to 2021. Microbiol Spectrum 10(2): e0156021. [cited 2022 Jun 16]. Available from: https://journals.asm.org/doi/full/10.1128/spectrum.01560-21

Washburne AD, Morton JT, Sanders J, McDonald D, Zhu Q, Oliverio AM et al (2018) Methods for phylogenetic analysis of microbiome data. Nat Microbiol 3:652. [cited 2022 Jun 14]. Available from: https://www.nature.com/articles/s41564-018-0156-0

Watson JD, Crick FHC (1953) Molecular structure of nucleic acids: a structure for deoxyribose nucleic acid. Nature 171:4356. [cited 2022 May 5]. Available from: https://www.nature.com/articles/171737a0

Wei W, Gao F, Du MZ, Hua HL, Wang J, Guo FB (2017) Zisland Explorer: detect genomic islands by combining homogeneity and heterogeneity properties. Brief Bioinform 18(3):357–366. [cited 2022 Aug 15. Available from: https://academic.oup.com/bib/article/18/3/357/2562748

Welch RA, Burland V, Plunkett G, Redford P, Roesch P, Rasko D et al (2002) Extensive mosaic structure revealed by the complete genome sequence of uropathogenic Escherichia coli. Proc Natl Acad Sci U S A 99(26):17020–17024. [cited 2022 Jun 14]. Available from: https://www.pnas.org/doi/10.1073/pnas.252529799

Williams AB (2016) Genome instability in bacteria: causes and consequences. In: Genome stability: from virus to human application, vol 1. Academic Press, London, pp 69–85

Williams D, Fournier GP, Lapierre P, Swithers KS, Green AG, Andam CP et al (2011) A rooted net of life. Biol Direct 6(1):1–20. [cited 2022 Jun 14]. Available from: https://biologydirect.biomedcentral.com/articles/10.1186/1745-6150-6-45

Wu R, Kaiser AD (1968) Structure and base sequence in the cohesive ends of bacteriophage lambda DNA. J Mol Biol 35(3):523–537

Xie J, Zhou F, Xu G, Mai G, Hu J, Wang G et al (2014) Genome-wide screening of pathogenicity islands in Mycobacterium tuberculosis based on the genomic barcode visualization. Mol Biol

Rep 41(9):5883–5889. [cited 2022 Jul 29]. Available from: https://pubmed.ncbi.nlm.nih.gov/25108673/
Xu S, Li Z, Huang Y, Han L, Che Y, Hou X et al (2021) Whole genome sequencing reveals the genomic diversity, taxonomic classification, and evolutionary relationships of the genus Nocardia. PLoS Negl Trop Dis. [cited 2022 Jun 16]. Available from: https://journals.plos.org/plosntds/article?id=10.1371/journal.pntd.0009665
Yang R, Liu G, Chen T, Li S, An L, Zhang G et al (2019) Characterization of the genome of a Nocardia strain isolated from soils in the Qinghai-Tibetan Plateau that specifically degrades crude oil and of this biodegradation. Genomics 111(3):356–366
Yasuike M, Nishiki I, Iwasaki Y, Nakamura Y, Fujiwara A, Shimahara Y et al (2017) Analysis of the complete genome sequence of Nocardia seriolae UTF1, the causative agent of fish nocardiosis: the first reference genome sequence of the fish pathogenic Nocardia species. PLoS One 12(3):e0173198. [cited 2022 Jun 28]. Available from: https://journals.plos.org/plosone/article?id=10.1371/journal.pone.0173198
Ying J, Ye J, Xu T, Wang Q, Bao Q, Li A (2019) Comparative genomic analysis of Rhodococcus equi: an insight into genomic diversity and genome evolution. Int J Genomics 2019:8987436
Yoon SH, Park YK, Kim JF (2015) PAIDB v2.0: exploration and analysis of pathogenicity and resistance islands. Nucleic Acids Res 43(D1):D624–D630. [cited 2022 Aug 15]. Available from: https://academic.oup.com/nar/article/43/D1/D624/2439790
Yoon SH, Park YK, Lee S, Choi D, Oh TK, Hur CG et al (2007) Towards pathogenomics: a web-based resource for pathogenicity islands. Nucleic Acids Res 35(suppl_1):D395–D400. [cited 2022 Aug 15]. Available from: https://academic.oup.com/nar/article/35/suppl_1/D395/1093086
Yu G, Fu X, Jin K, Zhang L, Wu W, Cui Z et al (2011) Integrative analysis of transcriptome and genome indicates two potential genomic islands are associated with pathogenesis of Mycobacterium tuberculosis. Gene 489(1):21–29
Zhou H, Beltrán JF, Brito IL (2021) Functions predict horizontal gene transfer and the emergence of antibiotic resistance. Sci Adv 7(43):eabj5056. [cited 2022 Jun 14]. Available from: https://www.science.org/doi/10.1126/sciadv.abj5056
Zhoua Z, Danga Y, Zhou M, Li L, Yu CH, Fu J et al (2016) Codon usage is an important determinant of gene expression levels largely through its effects on transcription. Proc Natl Acad Sci U S A 113(41):E6117–E6125. [cited 2022 Jun 15]. Available from: www.pnas.org/cgi/doi/10.1073/pnas.1606724113
Zhu Q, Kosoy M, Dittmar K (2014) HGTector: an automated method facilitating genome-wide discovery of putative horizontal gene transfers. BMC Genomics 15(1):1–18. [cited 2022 Aug 15]. Available from: https://bmcgenomics.biomedcentral.com/articles/10.1186/1471-2164-15-717

Microbial Genomic Island Discovery: Visualization and Analysis

4

Ashutosh Kabiraj, Moitri Let, Krishnendu Majhi, Shrabana Sarkar, Bhramar Dutta, Rajendra Kr Roy, Dibyendu Khan, Raju Biswas, Urmi Halder, and Rajib Bandopadhyay

Abstract

Genomic Islands (GIs), the integrative part of the prokaryotic genomes which contain many genes with important biological functions. The islands are one of the main quests of today's concern as they frequently contain genes that are involved in adaptation in diverse environments by providing antimicrobial resistance, virulence, and pathogenicity. The frequency of occurrence of GIs within genome is directly proportional to organism's genomic plasticity and thus the motion of evolution. GIs of prokaryotes can be visualized by using many computational tools. Various databases are spectacularly involved in the analysis of GIs and predictions of their probable functions. Besides pathogenic and antibiotic resistant islands, thermophilic, psychrophilic, acidophilic, halophilic, metal-tolerating prokaryotes, etc., sufficiently harbour GIs within their genomes to adapt to the hectic environments. GIs acquisition through horizontal gene transfer (HGT) or change in frame of genome is supposed to be a driving force of prokaryotic evolution.

Keywords

Genomic Islands · Extremophiles · Databases · Tools · Symbiosis islands

A. Kabiraj · M. Let · K. Majhi · B. Dutta · R. K. Roy · D. Khan · R. Biswas · U. Halder · R. Bandopadhyay (✉)
Microbiology Section, UGC-Centre for Advanced studies, Department of Botany, The University of Burdwan, Golapbag, Bardhaman, West Bengal, India

S. Sarkar
Microbiology Section, UGC-Centre for Advanced studies, Department of Botany, The University of Burdwan, Golapbag, Bardhaman, West Bengal, India

Centro de Investigación de Estudios Avanzados del Maule, Vicerrectoría de Investigación y Posgrado, Universidad Católica del Maule, Talca, Chile

I. Mani et al. (eds.), *Microbial Genomic Islands in Adaptation and Pathogenicity*,
https://doi.org/10.1007/978-981-19-9342-8_4

4.1 Microbial Genomic Islands (GIs): An Overview

For colonization perspective and adaptation within an environment, microbes use to reframe their genetic materials to compensate changing environmental scenario. Horizontal Gene Transfer (HGT) mechanisms like conjugation, transduction, and transformation are the crucial factors provide some clues to face the challenge of critical environmental circumstances (Dobrindt et al. 2004; Bellanger et al. 2014; Assaf et al. 2021). Chromosomal segments acquired through HGT carry some important genes that are responsible for providing adaptation potentialities and fitness to the organism, designated as genomic islands (GIs), and so these appears as a critical member of bacterial mobilome. Larger chromosomal segments coupled with a gene pool, sometimes encode their own transposase or integrase of tyrosine recombinase family, flanking with a few repeated structures and association of tRNA genes are typical characteristics of GIs (Dobrindt et al. 2004; Boyd et al. 2009). GIs containing genes are usually novel or their functions are not elucidated so far, but still they have some roles in adaptation in a specific condition.

It is now well established that GIs are one of the key regulators of bacterial diversity, adaptation within a particular habitat, and of course bacterial evolution (Vale et al. 2022). Its availability varies from species to species, sometimes within the same species inhabiting different habitats. To cope up with ever changing environments, and selection pressure of any contaminants like heavy metals or toxic organic and inorganic materials, antimicrobials, etc., are the crucial factors for acquisition of specific GIs within bacterial chromosomes. According to their functional roles, they could be classified as pathogenic islands, harbouring pathogenic genes; degradation islands, involving in degradation of complex molecules; metabolic islands, related to metabolism of carbon and nitrogenous compounds; and resistant islands, providing resistance capabilities against toxic compounds or antibiotics (van der Meer and Sentchilo 2003; Boyd et al. 2009; Juhas et al. 2009; Langille and Brinkman 2009; Carraro et al. 2014; Bertelli et al. 2019). In the very early 90s, scientists found some pathogenic genes within an *E. coli* strain which was devoid of other strains of *E. coli* (Langille and Brinkman 2009). This discovery may be called as a first step for GIs studies. A distinct difference between prokaryotic and eukaryotic organisms in context to their genetic element is predictable in terms of orthologous sequences which are not only found in the same species but also in the distinctly related eukaryotic species. But the scenario is quite different in prokaryotes where they conserve about 50% of their genome as core genome and genes of rest 50% accessory genome might be unique for the specific strain like., hundred of genome sequences of E. coli comprising of more than 45,000 gene families (Rodriguez-Valera et al. 2016).

For detecting GIs within bacterial genome, there are two basic approaches are available, such as depending on the sequence composition and another is based on comparative genomics. The first technique does not need any reference genome for detection; whereas the latter one needs a reference genome should be selected from the same species. After comparing with the reference genome, this approach finds out GIs from target genome (Langille and Brinkman 2009; Bertelli et al. 2019). There are several programs like SIGI-HMM, PAI-IDA, Centroid, Alien_Hunter, IslandPick, etc., and these programs detect GIs by measuring codon adaptation index, dinucleotide bias, percentage of GC content, etc. (Langille et al. 2008).

After confirming presence of GIs within genome, it is necessary to visualize them for calculating their positions and numbers. So visualization is an important step for studying about GIs. In this chapter, we have discussed comprehensively about genomic island origin, distribution, types, role in adaptation in different hectic conditions in respect to microbial evolution and related tools and finally mentioned some of the relevant databases for studying GIs.

4.2 Origin and Acquisition of GIs

GIs are often considered as the clusters of functionally related genes that are acquired via HGT and have a great impact on the evolutionary lineage of prokaryotes (Jani and Azad 2021). In the late 1980s, the concept of genomic islands (GIs) was first stricken in the mind of J. Hacker and his colleagues as pathogenicity islands (PAIs), who deeply studied the genetic background of the virulence factor of *Escherichia coli* (Hacker et al. 1990). Notably, they observed that the PAIs were unstable regions of chromosomes having variable virulence factors associated with different characteristics and phenotypes (Phillips-Houlbracq et al. 2018). These PAIs islands are of 10 kb–100 kb in size and can be up to 500 kb large. GIs of below 10 kb size are known as genomic islets (Juhas et al. 2009). The evolutionary relationships between different GIs are based on specific sequence and functional homologies. The coding region of genomic islands is not only confined to pathogenicity, but it also comprises other traits like symbiosis aromatic compound and sucrose metabolism (Bertelli et al. 2019), siderophore synthesis (Bertelli et al. 2019), and mercury resistance (Norambuena 2020). This could suggest that GIs evolved for selected adaptive and auxiliary functions.

The acquisition of GIs has happened via horizontal gene transfer. GIs have self-mobility capability; they could excise from their chromosomal region, can transfer itself independently into a different cell, and integrate into the specific target site of host's chromosome (Juhas et al. 2009). GIs are also acquired into the host cell through a well-defined group of the genetic element known as integrative and conjugative elements. It also includes conjugative transposons with multiple integration sites into the host cell (Burrus and Waldor 2004). A group of GIs does not have any self-mobility capability and they transfer via phage packaging, release, and infection (Juhas et al. 2009). In the detailed mechanism of GIs acquisition, it is sometimes inserted into the 30-end of tRNA genes by a phage-like recombinase, named as integrases which acts site specifically (Juhas et al. 2009). Other than GIs, different tRNA could also be inserted by integrases (Williams 2002). Specifically, GI-encoded integrases are related to the lambda, XerD, or P4 families (Juhas et al. 2009). The integrase coding gene, *int* may be located at one extreme part of the island and adjacent to the tRNA gene where they are integrated into GIs. The phage packaging GIs are often observed in staphylococcal pathogenicity islands (SaPIs) (Pantůček et al. 2018; Jin et al. 2021). SaPIs are involved in phage-induced excision, integration, replication, and proliferation resulting in acquisition of GI.

4.2.1 Islands Related to Pollution Degradation

Natural and manmade organic chemicals sometimes cause environmental pollution. Potent bacteria associated with some important adapted genes are responsible for chlorobenzene, nitrobenzene, phenoxyalkanoic acids and atrazine degradation for their own metabolism. Pollutant degrading metabolic genes (referred as 'evolutionary greenhouse') harboured by GIs are intrinsic factors responsible for environmental sustainability (van der Meer and Sentchilo 2003). Tn*4371*-like integrative or conjugative elements harboured in GIs of *Cupriavidus* and *Ralstonia* genera are responsible for toluene degradation (Van Houdt et al. 2012).

4.2.2 Islands Related to Pathogenicity

Genomic islands associated with pathogenicity may be the most studied topic among genomic islands and associated with environmental adaptations. Host organisms usually produce or secrete some chemical compounds that inhibit the growth of pathogens. On the other hand, pathogens adapted to be associated with their own host by producing some toxin degrading proteins. Not only *Yersinia*, a well-known pathogen bears GIs associated pathogenic genes, but also harmless bacteria *E. coli* and *Klebsiella* sometimes acquire pathogenic islands and cause disease (Hacker and Carniel 2001). *E. coli* is a well-known commensal organism present in the intestine of many organisms helping in metabolism, but when it acquires pathogenic islands (PAI), causes diseases. Loss of pathogenic genes from a genome of plasmid sometimes leads to produce non-pathogenic strains. For example, thermophilic strain *Bacillus anthracis* PFAB2 is a novel strain without common virulence genes (Banerjee et al. 2020).

4.3 Islands in Extremophiles

4.3.1 Thermophiles

Horizontal gene transfer mainly provides the advantages to bacteria to adapt in punitive environment (da Silva Filho et al. 2018), one of the example is high temperature. Horizontal gene transfer enlightens the evolutionary explanation on extreme thermophilic bacteria as model for ancient bacteria (Gogarten and Townsend 2005).

The selection of genomic island is controlled by some internal mechanisms which may be a random manner or controlled by specific selection procedure, but it is till date remain mystery. Relative age and movement of genomic island had been studied to analyse the competence level of *Thermus* spp. Several horizontally transferred genomic islands had been studied in genome of *Thermus* spp. by Kumwenda et al. (2014). Genomic island incorporation in chromosome is responsible for DNA amelioration and oligonucleotide distance pattern calculation determining reletive aquisition time (Kumwenda et al. 2014). From the point of evolution it had been found that *Deinococcus* lineage had acquired GI from *Thermus* species. Hence, it can be said that genomic island is most likely to play an evolutionary role in case of *Thermus* species lineage (Kumwenda et al. 2014). On the other hand, GI has also important role in holding the genes responsible for adaptation in extreme environment like high temperature. As per example, in one study done by Mercer et al. (2015)) heat resistance food isolate *Escherichia coli* AW1.7 showed more than 6 min D_{60}-value (highly resistant to heat). A ~14 Kb genomic island consists of many putative heat shock proteins present in 16 open reading frames encoding protease, responsible for highly heat resistance. Hence, this genomic island was converted to locus of heat resistance (LHR) that could help foodborne pathogens to withstand heat (Mercer et al. 2015). An LHR of 15–19 kb comprise of yfdX2, *yfd*X1$_{GI}$, *hde*D$_{GI}$, *orf*11, *kef*B, *trx*$_{GI}$, etc. genes that confers heat resistance in *Enterobacteriaceae* (Mercer et al. 2017). Genomic island PYG1 of 21.4 kb had been identified in the genomic sequence of the hyper-thermophile *Pyrococcus yayanosii*. TheΔPYG1 mutant strain had shown reduced growth at 100 °C compared to the wild one (Li et al. 2016).

4.3.2 Psychrophiles

Microorganisms have evolved diverse cold-adaptation mechanisms to survive and proliferate in the Earth's cold biosphere like cold aquatic and terrestrial ecosystems, or seasonally cold environments. The GIs provided us valuable insight into the unique characteristics of cold-adapted genome that assessed possible HGT events for cold adaptation (Penn et al. 2009; Murray and Grzymski 2007; De Maayer et al. 2014; Bowman 2017). A psychrophilic archaeon, *Methanosarcina burtonii* was reported with 125 genomic islands that represented >50% of the genome. Whereas, the GIs of mesophilic *Methanosarcina* genomes were represented ≤40% of its genome. The high proportion of unclassified genes harboured in genomic island denotes genes from unknown organisms that were perhaps acquired through HGT event. The GIs were overrepresented with different cellular parts such as cell wall and cell membrane (outer part); envelope biogenesis and signal transduction mechanisms (inner metabolism) and these are related to cold adaptation (Allen et al. 2009). Likewise, the GIs of an extreme psychrophilic bacterium *Psychroflexus torquis* harboured most of the genes to dwell in a sea-ice environment. A majority of 44 GIs were represented with insertional elements, addiction modules, and pseudogene. Some GIs were flanked by tRNA genes that are well-known hot-spot for site-specific recombination (Ou et al. 2006; Feng et al. 2014). The GIs of a flavobacterial epiphyte *Psychroflexus torquis* habituating in sea ice algal assemblages harboured genes that encode proteins or enzymes to synthesize polyunsaturated fatty acids, exopolysaccharides, putative antifreeze proteins, and to uptake compatible solutes (Feng et al. 2014; Bowman 2017). The genome of an *Alteromonas* species was represented by 15 specific GIs with genes to provide the ecological fitness in a cold marine environment (Math et al. 2012). An Antarctic deep lake habituating *Halobacterium* species acquired unique gene features like gas vesicle, polyhydroxyalkanoate, bacteriorhodopsin biosynthesis genes to survive in the cold ecosystem (DeMaere et al. 2013).

4.3.3 Halophiles

Halophiles are organisms that grow in saline environment. They can be found in various habitats like hypersaline water, saltern pond crystallizers, salt lakes, saline soil (Dutta and Bandopadhyay 2022). Horizontal acquisitions of various osmoresponsive genes might be a consequence of improving tolerance to salinity stress. Extremely halophilic archaea, *Haloarcula hispanica* possesses insertion element (IS), terminal inverted repeat (TIR), transposase gene (without TIRs) (Woods et al. 1999). 50% of the megaplasmid genes including two prophage regions were reported from GIs of moderately halophilic bacteria, *Pontibacillus*. Most of the chemotaxis genes (*mcp*/*che*) and flagellar motility genes (*fliG*/*M*/*N* and *motA*/*B*) are present in the megaplasmid. These environment sensing genes are harboured in two prophages regions (von Hoyningen-Huene et al. 2021). *Salinicoccus halodurans* genome contained 11 GIs, no CRISPR repeat region. One gene cluster involved in *N*α-acetyl-α-lysine biosynthesis. Genes encoding several hydrolases, stress responsive proteins, i.e., choline and betaine transporters, cold-shock protein, as well as chaperones are also noted (Jiang et al. 2015). Three larger horizontal gene transfer (HGT)-GIs are harboured in *Salinibacter ruber*, flanked by tRNAs and phage-related recombinase, which may participate in HGT events. Metalloresistence island, antibiotic resistance islands are associated with adaptive processes (González-Torres and Gabaldón 2018).

4.3.4 Acidophiles and Alkaliphiles

Acidophiles are organisms that thrive in acidic and sulphur rich environments, acid mine drainage (AMD), and relied on chemoautotrophic production by iron and sulphur oxidation. They are widely distributed in AMD Río Tinto of Spain (Amaral-Zettler et al. 2011), Iron Mountain hot springs in California, USA (Wilmes et al. 2008), and stromatolites (Sriaporn et al. 2020). One of the most studied bacterial genus *Acidithiobacillus* grows at optimum pH <4. Comparative genomics study revealed the genes responsible for survival in the acidic environment, viz., amino acid decarboxylases, deiminase/deaminases group, K^+ transporters, Na^+/H^+ antiporters, modified proton-efflux P-type ATPases (Baker-Austin and Dopson 2007). The genes are mostly present in GIs, are often associated with mobility genes (integrases and transposase), prophage, flanking repeats, plasmid mobilization elements along with atypical GC content (Beard et al. 2021).

Alkaliphiles are organisms that grow efficiently at pH >9. Alkaliphiles are inhabited in various environments like ocean hydrothermal vents, river, soda lakes, and alkaline soils (Grant 2006). GIs of extremely alkaliphilic *Bacillus halodurans* contain transposases, insertion sequences (IS) that facilitate HGT in the course of evolution and also in internal rearrangement of the genome (Takami et al. 2000). *Bacillus pseudofirmus* has phosphoserine aminotransferase, ABC type siderophore transporter, Na + coupled Npt type phosphate transporters, Ktr-type potassium uptake system, and cation/proton antiporters genes whose products contributed adaptations to alkaliphily (Janto et al. 2011).

4.4 Antibiotic Resistance Islands

Advancement in genome sequencing leads to the discovery of involvement of GIs in making antibiotic resistant phenotype within bacterial community. Antibiotic resistant islands often carry more than one antibiotic resistant genes integrated within tRNA gene. These GIs are characterized by terminal integrase or recombinase or insertion sequences, as consequences, GIs are less stable element (Dobrindt et al. 2004). Antibiotic resistant GIs can take part in genetic transfer events like conjugation, transformation, and transduction. GIs that actively participate in conjugation are termed as integrative conjugative elements (ICEs) and thus become the keen interest of modern research (Johnson and Grossman 2015). These ICEs are highly transmissible mobile genetic elements (MGE) and additively self-transmissible due to the presence of insertion sequences. They can exist as an integrated part of nucleoid or may be excised independently, self-replicable extrachromosomal DNA. ICEclc from *Pseudomonas knackmussii*, SXT from *Vibrio cholerae*, pKLC102 from *P. aeruginosa* Tn4371 from *Ralstonia oxalatica* are among the well-defined ICEs (Botelho et al. 2020). For Enterobacteria and other group of bacteria, typical GIs that impart numerous antibiotic resistance features have been reported. Clinically relevant methicillin-resistant *S. aureus* strains have emerged from the so-called SCCmec islands (MRSA). SCCmec islands can range in size from 20 kb to >60 kb, and they could harbour extra resistance features. In certain Proteobacteria, another sort of genomic island provides antibiotic resistance. The SXT island and R391 island of *Vibrio cholerae* and *Providencia rettgeri*, respectively, are the most well-known members of this category. A comparison of these elements indicated a conserved backbone with dedicated areas to the integration, transmission, etc., for these components. However, extra variable regions are also to be noted within these components. SXT-related components have also been discovered in natural settings. The pMERPH element (from *Shewanella putrefaciens*) got from river sediments of UK is an example of antibiotic resistant islands. *V. cholerae* lives a portion of its life cycle in water, suggesting that this group of GIs may have additional, some unidentified features that improve its fitness and/or survival, involved in adaptation and evolution of the species (Dobrindt et al. 2004).

4.5 Catabolic Genomic Islands

Heavy metals are the key toxicant in the environment as they are hazardous, incremental, and tenacious. The heavy metal toxicity effect was shown in every hierarchical level of life including microbes. It could disrupt the cell membrane structure, damage proteins and nucleic acids, and hamper various enzymatic pathways and transcription processes (Chandrangsu et al. 2017) in the microbial cell. To cope with the toxic effect of heavy metals, microbial communities have evolved genetic programs encoding selective function that allows for efflux or sequestration of the heavy metals resulting in reduction of toxic effects. In the efflux system various heavy metal transporters like cation diffusion facilitators (CDF) and PIB-type ATPases were involved, which translocate the metal ions from cytoplasm to periplasmic space (Nies 2016). Notably, the P-type is known as the most relevant heavy metal transporter that uses ATP to efflux heavy metals against their concentration gradients (Nies 2016). In addition, heavy metals are exported from the periplasmic space to the extracellular space across the outer membrane via resistance-nodulation-division (RND)-transport (Greene and Koronakis 2021). The RND transporter is a multi-component system composed of 6 membrane fusion proteins (MFPs), 3 RND transport proteins, and 3 outer membrane factor proteins (Greene and Koronakis 2021). Bacteria have an astonishing potential to confer the heavy metal resistance (HMR) genes within bacterial species through HGT, conjugative plasmids, transposons, and genomic islands (Li et al. 2018b). Genomic islands associated with HMR were described in many bacterial species. For example, the presence of PIB1-ATPase, PIB3-ATPase, PIB4-ATPase, RND-type metal transporter, and metal binding chaperones in *Mucilaginibacter rubeus* and *M. kameinonensis* makes the strain put forward for HMR (Li et al. 2018b). An aquatic ecosystem strain *Listeria welshimeri* harboured a novel LGI2-like genomic island from *L. monocytogenes* that transfers cadmium (Cd) tolerance proteins CadA and makes the strain resistant to Cd (Lee et al. 2021). Different heavy metal related genes are found to be situated within islands of environmental multi-metal resistant strain of *Bordetella petrii*. Arsenic resistant genes like *arsC*, *arsI*, *arsH*, *arsM*, etc., were present in GIs of the strain (Halder et al. 2022). These genes are involved in arsenic tolerance in bacteria by arsenic reduction, methylation, etc. (Kabiraj et al. 2022).

Reactive azo dyes are refractory pollutant containing –N=N– (azo bond) group linked with carbonated skeleton. This reactive azo dye laden textile effluent is being discharged in the aquatic ecosystem with consequent deleterious repercussion (Sarkar et al. 2017).

Genomic Island had important role in harbouring genes related to metabolic process, catabolic expression which have role in environmental adaptation for bacterial isolates. *Shewanella* is known for its high potentiality in dye containing textile effluent bioremediation. As per example, *Shewanella* algae 2NE11, isolated from industrial effluent in Peru, had shown ~97% decolourization against high concentration of anthraquinone dye and ~ 89% decolourization rates for azo dye (Lizárraga et al. 2022). It was also reported to harbour two genomic islands related to horizontal gene transfer showing role in environmental adaptation (Lizárraga et al. 2022). Dye decolourizing genes are associated with this genome, like NADPH-dependent oxidoreductase genes (HU689_04585; HU689_21345; HU689_04700), an FMN-dependent NADH-azoreductase gene (HU689_20695), and heme-dependent Dyp peroxidase gene (HU689_05310) (Lizárraga et al. 2022). In one bacterial consortium SCP (*Stenotrophomonas acidaminiphila* APG1, *Pseudomonas stutzeri* APG2 and *Cellulomonas* sp. APG4) it was found that, APG4 CDS associated category for transport and catabolism could be related with dye (mono-azo dye, Reactive blue 28) degradation (Chen et al. 2020). Maximum number of functional genes had been identified basically in APG2; however, APG1 and APG4 are also associated with it. This scenario further indicates the catabolic reaction related to azo dye degradation (Nanjani et al. 2021). Azo bond breakdown could be noticed by the APG genome due to the presence of a redox mediator. It was found that APG4 contains more number of ORF for NADH:DCIP oxidoreductase which lead to greater functionality (nearly 18 folds higher reductase activity) compared to APG1 and APG2 in azo dye degradation (Nanjani et al. 2021).

4.6 Symbiosis Islands

The role of genomic island is not limited to pathogenicity but have diverse role in symbiosis, aromatic compound metabolism, siderophore synthesis, etc. (Juhas et al. 2009). Many bacteria form symbiotic association with eukaryotic host with the help of symbiotic island. The mosaic structure of island suggests that multiple recombination events have occurred during evolution in a stepwise fashion. Presence of this type of island is also uncertain, i.e. they may not be present in closely related strain of same or different species. This type of island contains nodulation genes, genes related to nitrogen fixation, and other types of genes required for transfer of the island, nodule metabolism, several regulatory genes, etc. Transfer of symbiotic island to a nonsymbiotic mesorhizobia converts their behaviour as symbionts and they could get the ability of nitrogen fixation. A chromosomally integrated element (502 kb), a symbiotic island, from the genome of *Mesorhizobium loti* strain R7A have the ability to transform nonsymbiotic mesorhizobia in the environment to Lotus symbionts (Sullivan et al. 2002). The island also contains several operons that are not required for transfer including operon for vitamin (biotin, thiamine, and nicotinamide) biosynthesis. These operons are not directly linked with symbiosis but they may help bacteria for better competition in rhizospheric environment. Symbiotic nitrogen fixing bacterium *Bradyrhizobium japonicum* harbours a symbiotic island of 681 kb in size that carry cluster of symbiotic genes which are structurally inserted into a val-tRNA gene on the genome (Itakura et al. 2009). Symbiotic nitrogen fixing bacteria, *Mesorhizobium* and *Bradyrhizobium* are involved in root nodule formation with specific types of plants. There are *sym*-genes for regulation of root nodule formation in different stages of plant growth. So, these types of important islands are devoted with nitrogen fixation. More studies also required to know about the stability of these important genes within the bacterial chromosome (Roumiantseva et al. 2018).

4.7 Prediction of GIs

In this Era, robust genome sequencing and emerging interest on role of GIs on bacterial adaptation and evolution drive us to grow interest to assume genomic islands, their position, related genes, etc. Several tools and databases are now available to predict genomic islands. In this section, we will discuss briefly the tools and databases related to GIs identification.

4.7.1 Tools

Bacteria may now be studied by studying their genomic sequences owing to the high sequencing methods. Comparative genome sequence analysis, for example, can identify phenomena like gene gain or loss, or exchange in a genome. Gene gain by horizontal gene transfer makes a bacterium more selective to that particular environment. The study of GIs is crucial for biological and bioinformatics research. So, identifying GIs is one of the most important jobs in genome evolution and gene transfer mechanism research. Nowadays, several tools are available (Table 4.1) for GIs prediction.

4.7.2 Databases

Other than GI prediction tools there are several databases (Table 4.2) available that can be used directly for comparative study. These databases serve as resource to identify integrase site specificity and its evolution (Bertelli et al. 2017).

4.8 Significance of GIs in Prokaryotic Evolution

In 1965, Zuckerkandl and Pauling had found that there is a relationship between nucleotide as well as protein sequences and organism's evolution. Different parameters like codon bias, point mutation, changes in nucleotide sequences, etc., were considered for driving force of evolution. But there are some sudden changes that can be occurred within the genome of prokaryotic cells that cannot be imagined through the grammar of evolution. An example of this sudden change is horizontal gene transfer of genomic islands (Lima et al. 2008).

Table 4.1 List of computational tools used for Genomic Island prediction in Prokaryotes

Tool names	Tool links	Special features	References
IslandPath	http://www.pathogenomics.sfu.ca/islandpath	Aid in detection of genomic islands with genome annotation features	Hsiao et al. (2003)
PAI-IDA	http://compbio.sibsnet.org/projects/pai-ida/	Simple analysis to detect pathogenicity islands and anomalous gene cluster	Tu and Ding (2003)
SIGI (score-based identification of genomic islands)	https://www.uni-goettingen.de/en/research/185810.html	Detect genomic island with high sensitivity	Merkl (2004)
ORFcurator	http://www.genomecurator.org/ORFcurator/	Used for molecular organization of genes and gene clusters	Rosenfeld et al. (2004)
GC-Profile	http://tubic.tju.edu.cn/GC-Profile/	Study for visualization and analysis of the variation of GC percentage and GIs	Gao and Zhang (2006)
MobilomeFINDER	http://mml.sjtu.edu.cn/MobilomeFINDER (Upon request)	High-throughput tools for identification and characterization of island through exploitation of emerging sequence data and PCR-dependent profiling of un-sequenced strains	Ou et al. (2007)
PredictBias	www.davvbiotech.res.in/PredictBias (Upon request)	Genomic and pathogenic islands detection in prokaryotes by analysing sequence composition, insertion elements, and genes associated with virulence factors.	Pundhir et al. (2008)
Design-Island (Detection of Statistically Significant Genomic Island)	http://www.geocities.com/raghuchatterjee/Design-Island.html	This tool does not require any previous data sets and used Monte-Carlo statistical tests	Chatterjee et al. (2008)
INDeGenIUS (Improved N-mer based Detection of Genomic Islands Using Sequence-Clustering)	Upon request	Identification of unique functional islands in complete-sequence of organism	Shrivastava et al. (2010)
EGID (Ensemble Algorithm for Genomic Island Detection)	http://www5.esu.edu/cpsc/bioinfo/software/EGID (Upon request)	Used in horizontal gene transfer and molecular evolution study	Che et al. (2011)
IGIPT (integrated genomic island prediction tool)	http://bioinf.iiit.ac.in/IGIPT/	Allows the users to analyse GIs by simultaneously using	Jain et al. (2011)

(continued)

Table 4.1 (continued)

Tool names	Tool links	Special features	References
		thirteen different measures that give a more precision for prediction	
PIPS (Pathogenicity Island Prediction Software)	http://www.genoma.ufpa.br/lgcm/pips	Using multiple tools for detection of pathogenic island in blended mode	Soares et al. (2016)
GIV (Genomic Island Visualization)	http://www5.esu.edu/cpsc/bioinfo/software/GIV	Detection of the location of GIs in a genome and also gives the supportive features information for GIs	Che and Wang (2013)
GI-POP	http://gipop.life.nthu.edu.tw/	Draft genomes can be submitted in contigs or scaffolds to get highly probable GIs	Lee et al. (2013)
Genomic Island Hunter	http://www.esu.edu/cpsc/che_lab/software/GIHuanter	GI projection method for sequenced bacterial and archaeal genomes. This tool uses sequence composition, mobile gene information, and integrase.	Che et al. (2014)
MSGIP (Mean Shift Genomic Island Predictor)	http://msgip.integrativebioinformatics.me/	Precisely predicted the complete reservoir of GIs in a genome	de Brito et al. (2016)
GI-SVM (Genomic Island-Support Vector Machine)	https://github.com/icelu/GI_Prediction	Highly sensitive tools for prediction of GIs in unannotated sequence of a single genome	Lu and Leong (2016)
GIPSy (genomic island prediction software)	https://www.bioinformatics.org/groups/?group_id=1180	Identify GIs in bacterial genome based on variation of the GC content, tRNA, genomic codons, transposase, etc.	Soares et al. (2016)
Zisland Explorer	http://tubic.tju.edu.cn/Zisland_Explorer/	GIs detection by utilizing both homogeneity and heterogeneity properties	Wei et al. (2017))
Islandviewer4	http://www.pathogenomics.sfu.ca/islandviewer/	Prediction of GIs with their products and features via three integrated tools like IslandPick, SIGI-HMM, and IslandPath-DIMOB	Bertelli et al. (2017)

(continued)

Table 4.1 (continued)

Tool names	Tool links	Special features	References
XenoGI	http://www.cs.hmc.edu/xgiWeb/	Study the history of GIs insertions in a clade of microorganism	Bush et al. (2018)
Vrprofile	http://bioinfo-mml.sjtu.edu.cn/VRprofile	Rapid investigation of antibiotic resistance and virulence gene clusters in pathogenic bacteria	Li et al. (2018b)
MTGIpick	https://github.com/bioinfo0706/MTGIpick	GIs identification from a single, unannotated genome with prior information	Dai et al. (2018)
IslandCafe	https://github.com/mehuljani/IslandCafe	Novel island detection by comparing frequently used tools for GIs analysis	Jani and Azad (2019)
panRGP	https://github.com/labgem/PPanGGOLiN	Prediction of genomic islands diversity via Pangenome analysis	Bazin et al. (2020)
2SigFinder	https://github.com/bioinfo0706/2SigFinder	Statistical analysis for GIs detection from a single genome.	Kong et al. (2020)
SSG-LUGIA	https://nibtehaz.github.io/SSG-LUGIA/	Analysis of new-sequence based genomes and independent of functional annotation of genomes	Ibtehaz et al. (2021)

Secretion systems, encoded by GIs, are not only responsible for release of GIs and GIs encoded products from microbial cells, but also GIs associated chromosomal segments of the host organism. Acquisition of chromosomal segments through GIs is associated with rapid microbial evolution and diversification. A magnificent metabolic change is to be found in recipient microbes (Juhas et al. 2009) as GIs contain huge numbers of genes (Dobrindt et al. 2004). Acquisition of new genes associated with GIs usually counterbalanced by reduction of negative genes which sometimes considered as extra advantage for the organisms. These genetic and metabolic changes ultimately drive organisms to evolve (Juhas et al. 2009). GIs within the recipient microbial chromosome acquire, replace, or disintegrate chromosomal core genes. Gain or loss of genes makes distinct features of a particular strain from rest other strains of the same species. Species like *E. coli,* after acquiring pathogenic islands had become able to cause different diseases in intestine and extraintestine of human beings and other animals (Desvaux et al. 2020). Bacteriophages are sometimes considered as the driving force for bacterial evolution. Bacteria use to develop CRISPR-Cas system to escape itself from viral attack; on the other hand, viruses also modulate their genetic elements which help them to survive against bacterial 'sword'

Table 4.2 Databases related to GIs prediction with some important characteristics

Databases	Feature	Availability	References
Islander	A database associated with integrases and their specific DNA sites in the genomic islands of prokaryotes	http://bioinformatics.sandia.gov/islander (Upon request)	Mantri and Williams (2004)
IslandPick datasets	Investigate the sequence composition-based GIs	http://www.pathogenomics.sfu.ca/islandpick_GI_datasets/	Langille et al. (2008)
PAIDB (Pathogenicity Island Database)	Detection and analysis of antibiotic resistance and pathogenicity islands	http://www.paidb.re.kr/	Yoon et al. (2015)
ICEberg	Study the bacterial mobile genetic elements such as integrative and conjugative elements (ICEs)	http://db-mml.sjtu.edu.cn/ICEberg/	Liu et al. (2019)
VRprofile	Exploration of antibiotic resistance and virulence gene cluster	http://bioinfo-mml.sjtu.edu.cn/VRprofile	Li et al. (2018a)
VCGIDB (*Vibrio cholerae* Genomic Island Database)	Prediction of phylogeny-based upgraded features in a large genome	http://leb.snu.ac.kr/vcgidb	Hur et al. (2019)
DarkHorse	Strong and flexible collection of tools using for prediction of phylogenetically related protein families in both individual HGT in a single genome and large-scale HGT in large genome	http://darkhorse.ucsd.edu/	Podell et al. (2008)
Islandviewer 4	Interactive visualization of GIs in bacterial and archaeal genomes for large-scale datasets	http://www.pathogenomics.sfu.ca/islandviewer/	Bertelli et al. (2017)
DGI	A dataset that comprises GIs derived from 2000 different bacterial genome including PAIs depicted as circular graphical images	http://www5.esu.edu/cpsc/bioinfo/dgi (Upon request)	Che et al. (2014)
GI-POP	Microbial genome annotation dataset including non-coding RNAs, ORF, and GIs. Also, GI-GPS based system is used for genomic islands prediction.	http://gipop.life.nthu.edu.tw/	Lee et al. (2013)
MOSAIC	Study the conserved and diverse segments (i.e., GIs) in the genome	http://genome.jouy.inra.fr/mosaic (Upon request)	Chiapello et al. (2008))

(Vale et al. 2022). However, evolution of GIs itself is quite distinct from the evolutionary lineages of other integrative elements (Boyd et al. 2009). From the analysis of GIs analogous structural and functional characteristics along with their phylogenetic relatedness revealed that GIs may be evolved multiple times.

Bacteriophages usually attack bacteria at different times of bacterial life cycle, integration of viral genome, commonly called as prophage, can acquire more supplementary genes and can be considered as genomic island (Boyd et al. 2009). Cell-wall less small bacterium, *Mycoplasma* is also associated with some genomic islands. Among *Mycoplasma* related bacteriophages, φMFV1 and φMAV1 contain *mem* and *vir* genes, which preferentially encode some membrane anchored surface proteins. *mem* generates a coiled protein, whereas *vir* is responsible for a lipoprotein, eventually acts as putative virulence factor (Citti et al. 2020). Super-integrons are specific types of integron with the ability to stockpile a number of genes that might be associated with antibiotic resistance and can be converted themselves to GIs leading towards development of antibiotics resistant microbes (Dobrindt et al. 2004). After critical review on antibiotics resistance by *Salmonella enterica*, an author reported that this pathogenic bacterium is associated with a class 1 integron. *Salmonella* genomic island 1 (SG1) comprised of 15 kb integron and 27.4 kb backbone with five antibiotic resistant genes (Hall 2010). SGI1 and SGI1-REs were found to be members of large family of integrative genomic elements (IGE) and due to random evolutionary events like insertion, deletion, mutation, etc., their structural shape becomes altered. When members of Gammaproteobacteriaceae catch up these GIs, they distribute GIs easily and more frequently to their related species for adaptation against antimicrobial agents (Cummins et al. 2020). Urease producing bacterium, *Proteus mirabilis* is associated with a novel GI, named as PmGRI1 also responsible for antibiotic resistance (Lei et al. 2020).

Studies on genomic islands of archaea is not as frequent as bacteria, a very few reports are available till date, although archaea are very closely related with bacteria and eukaryotic organisms both genetically and evolutionarily (Makarova et al. 1999).

4.9 Conclusion

Prokaryotes can thrive in all kinds of biomes. In order to adapt in the diverse kinds of environment, they have evolved over time during a variety of events, namely gene rearrangements, mutations, horizontal gene transfer, etc. This evolutionary pattern may contain specific sequence. The horizontal transfer of large gene clusters as genomic islands contains accessory genes for adapting in a specific environmental niche. As evolution is a random and continuous process and GIs have tremendous role in prokaryotic evolution, the question is raised about its stability. Beside developing new tools and databases for genomic islands, now focus is needed on identifying the factors that provide its stability within the bacterial genome.

Acknowledgements The authors are thankful to the Department of Botany, The University Burdwan and DST-FIST; AK is thankful to DHESTBT (WB-DBT), Memo No. [Fc (Sc.)/RS/SF/ BOT./2017-18/22]; ML and RKR are thankful to UGC JRF for financial assistance to conduct this research work, UH is thankful to SRF (State Fund) fellowship [Fc (Sc.)/RS/SF/ BOT./2017-18/22], DK and SS (WBP191579671079) are thankful to Swami Vivekananda Merit cum Means Scholarship. BD is thankful to DST (New Delhi) for PURSE PHASE-II SRF fellowship.

Authorship Contribution Statement RB1, RB2,and UH, conceptualized the idea; AK, ML, KM, SS, BD, DK, RKR, and UH wrote the manuscript. RB2 edited the manuscript. Every author checked and approved the manuscript.

Declaration of Competing Interest Authors are declaring that they have no conflict of interest.

Funding No funding has been received for this study.

References

Allen MA, Lauro FM, Williams TJ et al (2009) The genome sequence of the psychrophilic archaeon, Methanococcoides burtonii: the role of genome evolution in cold adaptation. ISME J 3:1012–1035

Amaral-Zettler LA, Zettler ER, Theroux SM et al (2011) Microbial community structure across the tree of life in the extreme Rio Tinto. ISME J 5:42–50

Assaf R, Xia F, Stevens R (2021) Identifying genomic islands with deep neural networks. BMC Genom 22:1–11

Baker-Austin C, Dopson M (2007) Life in acid: pH homeostasis in acidophiles. Trends Microbiol 15:165–171

Banerjee A, Somani VK, Chakraborty P et al (2020) Molecular and genomic characterization of PFAB2: a non-virulent *Bacillus anthracis* strain isolated from an Indian Hot Spring. Curr Genomics 20:491–507

Bazin A, Gautreau G, Médigue C et al (2020) panRGP: a pangenome-based method to predict genomic islands and explore their diversity. Bioinformatics 36:i651–i658

Beard S, Ossandon FJ, Rawlings DE, Quatrini R (2021) The flexible genome of acidophilic prokaryotes. Curr Issues Mol Biol 40:231–266

Bellanger X, Payot S, Leblond-Bourget N, Guédon G (2014) Conjugative and mobilizable genomic islands in bacteria: evolution and diversity. FEMS Microbiol Rev 38:720–760

Bertelli C, Laird MR, Williams KP et al (2017) IslandViewer 4: expanded prediction of genomic islands for larger-scale datasets. Nucleic Acids Res 45:W30–W35

Bertelli C, Tilley KE, Brinkman FSL (2019) Microbial genomic island discovery, visualization and analysis. Brief Bioinform 20:1685–1698

Botelho J, Mourão J, Roberts AP, Peixe L (2020) Comprehensive genome data analysis establishes a triple whammy of carbapenemases, ICEs and multiple clinically relevant bacteria. Microb Genomics 6. https://doi.org/10.1099/mgen.0.000424

Bowman JP (2017) Genomics of psychrophilic bacteria and archaea. In: Psychrophiles: from biodivers to biotechnology. Springer, Cham, pp 345–387

Boyd EF, Almagro-Moreno S, Parent MA (2009) Genomic islands are dynamic, ancient integrative elements in bacterial evolution. Trends Microbiol 17:47–53

Burrus V, Waldor MK (2004) Shaping bacterial genomes with integrative and conjugative elements. Res Microbiol 155:376–386

Bush EC, Clark AE, DeRanek CA et al (2018) xenoGI: reconstructing the history of genomic island insertions in clades of closely related bacteria. BMC Bioinform 19:1–11

Carraro N, Matteau D, Luo P et al (2014) The master activator of IncA/C conjugative plasmids stimulates genomic islands and multidrug resistance dissemination. PLoS Genet 10:e1004714

Chandrangsu P, Rensing C, Helmann JD (2017) Metal homeostasis and resistance in bacteria. Nat Rev Microbiol 15:338–350

Chatterjee R, Chaudhuri K, Chaudhuri P (2008) On detection and assessment of statistical significance of Genomic Islands. BMC Genom 9:1–11

Che D, Hasan MS, Wang H et al (2011) EGID: an ensemble algorithm for improved genomic island detection in genomic sequences. Bioinformation 7:311

Che D, Wang H (2013) GIV: a tool for genomic islands visualization. Bioinformation 9:879
Che D, Wang H, Fazekas J, Chen B (2014) An accurate genomic island prediction method for sequenced bacterial and archaeal genomes. J Proteomics Bioinform 7:214
Chen G, An X, Feng L et al (2020) Genome and transcriptome analysis of a newly isolated azo dye degrading thermophilic strain Anoxybacillus sp. Ecotoxicol Environ Saf 203:111047
Chiapello H, Gendrault A, Caron C et al (2008) MOSAIC: an online database dedicated to the comparative genomics of bacterial strains at the intra-species level. BMC Bioinform 9:1–9
Citti C, Baranowski E, Dordet-Frisoni E et al (2020) Genomic Islands in mycoplasmas. Genes (Basel) 11:836
Cummins ML, Hamidian M, Djordjevic SP (2020) Salmonella genomic island 1 is broadly disseminated within gammaproteobacteriaceae. Microorganisms 8:161
da Silva Filho AC, Raittz RT, Guizelini D et al (2018) Comparative analysis of genomic island prediction tools. Front Genet 9:619
Dai Q, Bao C, Hai Y et al (2018) MTGIpick allows robust identification of genomic islands from a single genome. Brief Bioinform 19:361–373
de Brito DM, Maracaja-Coutinho V, de Farias ST et al (2016) A novel method to predict genomic islands based on mean shift clustering algorithm. PLoS One 11:e0146352
De Maayer P, Anderson D, Cary C, Cowan DA (2014) Some like it cold: understanding the survival strategies of psychrophiles. EMBO Rep 15:508–517
DeMaere MZ, Williams TJ, Allen MA et al (2013) High level of intergenera gene exchange shapes the evolution of haloarchaea in an isolated Antarctic lake. Proc Natl Acad Sci U S A 110:16939–16944
Desvaux M, Dalmasso G, Beyrouthy R et al (2020) Pathogenicity factors of genomic islands in intestinal and extraintestinal Escherichia coli. Front Microbiol 11:2065
Dobrindt U, Hochhut B, Hentschel U, Hacker J (2004) Genomic islands in pathogenic and environmental microorganisms. Nat Rev Microbiol 2:414–424
Dutta B, Bandopadhyay R (2022) Biotechnological potentials of halophilic microorganisms and their impact on mankind. Beni-Suef Univ J Basic Appl Sci 11(1):1–16
Feng S, Powell SM, Wilson R, Bowman JP (2014) Extensive gene acquisition in the extremely psychrophilic bacterial species Psychroflexus torquis and the link to sea-ice ecosystem specialism. Genome Biol Evol 6:133–148
Gao F, Zhang C-T (2006) GC-Profile: a web-based tool for visualizing and analyzing the variation of GC content in genomic sequences. Nucleic Acids Res 34:W686–W691
Gogarten JP, Townsend JP (2005) Horizontal gene transfer, genome innovation and evolution. Nat Rev Microbiol 3:679–687
González-Torres P, Gabaldón T (2018) Genome variation in the model halophilic bacterium Salinibacter ruber. Front Microbiol 9:1499
Grant WD (2006) Alkaline environments and biodiversity, in extremophilies. Encycl life Support Syst (EOLSS), Dev under Auspices UNESCO Oxford Eolss Publ
Greene NP, Koronakis V (2021) Bacterial metal resistance: coping with copper without cooperativity? MBio 12:e00653–e00621
Hacker J, Bender L, Ott M et al (1990) Deletions of chromosomal regions coding for fimbriae and hemolysins occur in vitro and in vivo in various extra intestinal Escherichia coli isolates. Microb Pathog 8:213–225
Hacker J, Carniel E (2001) Ecological fitness, genomic islands and bacterial pathogenicity. EMBO Rep 2:376–381
Halder U, Biswas R, Kabiraj A et al (2022) Genomic, morphological, and biochemical analyses of a multi-metal resistant but multi-drug susceptible strain of Bordetella petrii from hospital soil. Sci Rep 12:8439. https://doi.org/10.1038/s41598-022-12435-7
Hall RM (2010) *Salmonella* genomic islands and antibiotic resistance in *Salmonella* enterica. Future Microbiol 5:1525–1538
Hsiao W, Wan I, Jones SJ, Brinkman FSL (2003) IslandPath: aiding detection of genomic islands in prokaryotes. Bioinformatics 19:418–420

Hur Y, Chalita M, Ha S et al (2019) VCGIDB: a database and web resource for the genomic islands from Vibrio Cholerae. Pathogens 8:261
Ibtehaz N, Ahmed I, Ahmed MS et al (2021) SSG-LUGIA: single sequence based genome level unsupervised genomic island prediction algorithm. Brief Bioinform 22:bbab116
Itakura M, Saeki K, Omori H et al (2009) Genomic comparison of Bradyrhizobium japonicum strains with different symbiotic nitrogen-fixing capabilities and other Bradyrhizobiaceae members. ISME J 3:326–339
Jain R, Ramineni S, Parekh N (2011) IGIPT-Integrated genomic island prediction tool. Bioinformation 7:307
Jani M, Azad RK (2021) Discovery of mosaic genomic islands in Pseudomonas spp. Arch Microbiol 203:2735–2742
Jani M, Azad RK (2019) IslandCafe: compositional anomaly and feature enrichment assessment for delineation of genomic islands. G3 9:3273–3285
Janto B, Ahmed A, Ito M et al (2011) Genome of alkaliphilic Bacillus pseudofirmus OF4 reveals adaptations that support the ability to grow in an external pH range from 7.5 to 11.4. Environ Microbiol 13:3289–3309
Jiang K, Xue Y, Ma Y (2015) Complete genome sequence of Salinicoccus halodurans H3B36, isolated from the Qaidam Basin in China. Stand Genomic Sci 10:1–7
Jin Y, Zhou W, Yin Z et al (2021) The genetic feature and virulence determinant of highly virulent community-associated MRSA ST338-SCCmec Vb in China. Emerg Microbes Infect 10:1052–1064
Johnson CM, Grossman AD (2015) Integrative and conjugative elements (ICEs): what they do and how they work. Annu Rev Genet 49:577–601
Juhas M, Van Der Meer JR, Gaillard M et al (2009) Genomic islands: tools of bacterial horizontal gene transfer and evolution. FEMS Microbiol Rev 33:376–393
Kabiraj A, Biswas R, Halder U, Bandopadhyay R (2022) Bacterial arsenic metabolism and its role in arsenic bioremediation. Curr Microbiol 79:131
Kong R, Xu X, Liu X et al (2020) 2SigFinder: the combined use of small-scale and large-scale statistical testing for genomic island detection from a single genome. BMC Bioinform 21:1–15
Kumwenda B, Litthauer D, Reva O (2014) Analysis of genomic rearrangements, horizontal gene transfer and role of plasmids in the evolution of industrial important Thermus species. BMC Genom 15:1–13
Langille MGI, Brinkman FSL (2009) IslandViewer: an integrated interface for computational identification and visualization of genomic islands. Bioinformatics 25:664–665
Langille MGI, Hsiao WWL, Brinkman FSL (2008) Evaluation of genomic island predictors using a comparative genomics approach. BMC Bioinform 9:1–10
Lee C-C, Chen Y-PP, Yao T-J et al (2013) GI-POP: a combinational annotation and genomic island prediction pipeline for ongoing microbial genome projects. Gene 518:114–123
Lee S, Parsons C, Chen Y et al (2021) Identification and characterization of a novel genomic island harboring cadmium and arsenic resistance genes in Listeria welshimeri. Biomol Ther 11:560
Lei C-W, Yao T-G, Yan J et al (2020) Identification of Proteus genomic island 2 variants in two clonal Proteus mirabilis isolates with coexistence of a novel genomic resistance island PmGRI1. J Antimicrob Chemother 75:2503–2507
Li J, Tai C, Deng Z et al (2018a) VRprofile: gene-cluster-detection-based profiling of virulence and antibiotic resistance traits encoded within genome sequences of pathogenic bacteria. Brief Bioinform 19:566–574
Li YP, Carraro N, Yang N et al (2018b) Genomic islands confer heavy metal resistance in Mucilaginibacter kameinonensis and Mucilaginibacter rubeus isolated from a gold/copper mine. Genes (Basel) 9:573
Li Z, Li X, Xiao X, Xu J (2016) An integrative genomic island affects the adaptations of the piezophilic hyperthermophilic archaeon Pyrococcus yayanosii to high temperature and high hydrostatic pressure. Front Microbiol 7:1927

Lima WC, Paquola ACM, Varani AM et al (2008) Laterally transferred genomic islands in Xanthomonadales related to pathogenicity and primary metabolism. FEMS Microbiol Lett 281:87–97

Liu M, Li X, Xie Y et al (2019) ICEberg 2.0: an updated database of bacterial integrative and conjugative elements. Nucleic Acids Res 47:D660–D665

Lizárraga WC, Mormontoy CG, Calla H et al (2022) Complete genome sequence of Shewanella algae strain 2NE11, a decolorizing bacterium isolated from industrial effluent in Peru. Biotechnol Rep 33:e00704

Lu B, Leong HW (2016) GI-SVM: a sensitive method for predicting genomic islands based on unannotated sequence of a single genome. J Bioinforma Comput Biol 14:1640003

Makarova KS, Aravind L, Galperin MY et al (1999) Comparative genomics of the Archaea (Euryarchaeota): evolution of conserved protein families, the stable core, and the variable shell. Genome Res 9:608–628

Mantri Y, Williams KP (2004) Islander: a database of integrative islands in prokaryotic genomes, the associated integrases and their DNA site specificities. Nucleic Acids Res 32:D55–D58

Math RK, Jin HM, Kim JM et al (2012) Comparative genomics reveals adaptation by Alteromonas sp. SN2 to marine tidal-flat conditions: cold tolerance and aromatic hydrocarbon metabolism. PLoS One 7:e35784

Mercer R, Nguyen O, Ou Q et al (2017) Functional analysis of genes comprising the locus of heat resistance in Escherichia coli. Appl Environ Microbiol 83:e01400–e01417

Mercer R, Zheng J, Garcia-Hernandez R et al (2015) Genetic determinants of heat resistance in Escherichia coli. Front Microbiol 6:932

Merkl R (2004) SIGI: score-based identification of genomic islands. BMC Bioinform 5:1–14

Murray AE, Grzymski JJ (2007) Diversity and genomics of Antarctic marine micro-organisms. Philos Trans R Soc B Biol Sci 362:2259–2271

Nanjani S, Paul D, Keharia H (2021) Genome analysis to decipher syntrophy in the bacterial consortium 'SCP'for azo dye degradation. BMC Microbiol 21:1–19

Nies DH (2016) The biological chemistry of the transition metal "transportome" of Cupriavidus metallidurans. Metallomics 8:481–507

Norambuena J (2020) Mechanism of resistance focusing on copper, mercury and arsenic in extremophilic organisms, how acidophiles and thermophiles cope with these metals. In: Physiological and biotechnological aspects of extremophiles. Elsevier, San Diego, pp 23–37

Ou H-Y, Chen L-L, Lonnen J et al (2006) A novel strategy for the identification of genomic islands by comparative analysis of the contents and contexts of tRNA sites in closely related bacteria. Nucleic Acids Res 34:e3–e3

Ou H-Y, He X, Harrison EM et al (2007) MobilomeFINDER: web-based tools for in silico and experimental discovery of bacterial genomic islands. Nucleic Acids Res 35:W97–W104

Pantůček R, Sedláček I, Indráková A et al (2018) Staphylococcus edaphicus sp. nov., isolated in Antarctica, harbors the mecC gene and genomic islands with a suspected role in adaptation to extreme environments. Appl Environ Microbiol 84:e01746–e01717

Penn K, Jenkins C, Nett M et al (2009) Genomic islands link secondary metabolism to functional adaptation in marine Actinobacteria. ISME J 3:1193–1203

Phillips-Houlbracq M, Ricard J-D, Foucrier A et al (2018) Pathophysiology of Escherichia coli pneumonia: respective contribution of pathogenicity islands to virulence. Int J Med Microbiol 308:290–296

Podell S, Gaasterland T, Allen EE (2008) A database of phylogenetically atypical genes in archaeal and bacterial genomes, identified using the DarkHorse algorithm. BMC Bioinform 9:1–12

Pundhir S, Vijayvargiya H, Kumar A (2008) PredictBias: a server for the identification of genomic and pathogenicity islands in prokaryotes. In Silico Biol 8:223–234

Rodriguez-Valera F, Martin-Cuadrado A-B, López-Pérez M (2016) Flexible genomic islands as drivers of genome evolution. Curr Opin Microbiol 31:154–160

Rosenfeld JA, Sarkar IN, Planet PJ et al (2004) ORFcurator: molecular curation of genes and gene clusters in prokaryotic organisms. Bioinformatics 20:3462–3465

Roumiantseva ML, Muntyan VS, Cherkasova ME et al (2018) Genomic islands in Sinorhizobium meliloti Rm1021, nitrogen-fixing symbiont of alfalfa. Russ J Genet 54:759–769
Sarkar S, Banerjee A, Halder U et al (2017) Degradation of synthetic azo dyes of textile industry: a sustainable approach using microbial enzymes. Water Conserv Sci Eng 2:121–131
Shrivastava S, Reddy SK, Mande SS (2010) INDeGenIUS, a new method for high-throughput identification of specialized functional islands in completely sequenced organisms. J Biosci 35: 351–364
Soares SC, Geyik H, Ramos RTJ et al (2016) GIPSy: genomic island prediction software. J Biotechnol 232:2–11
Sriaporn C, Campbell KA, Millan M et al (2020) Stromatolitic digitate sinters form under wide-ranging physicochemical conditions with diverse hot spring microbial communities. Geobiology 18:619–640
Sullivan JT, Trzebiatowski JR, Cruickshank RW et al (2002) Comparative sequence analysis of the symbiosis island of Mesorhizobium loti strain R7A. J Bacteriol 184:3086–3095
Takami H, Nakasone K, Takaki Y et al (2000) Complete genome sequence of the alkaliphilic bacterium Bacillus halodurans and genomic sequence comparison with Bacillus subtilis. Nucleic Acids Res 28:4317–4331
Tu Q, Ding D (2003) Detecting pathogenicity islands and anomalous gene clusters by iterative discriminant analysis. FEMS Microbiol Lett 221:269–275
Vale FF, Lehours P, Yamaoka Y (2022) The role of mobile genetic elements in bacterial evolution and their adaptability. Front Microbiol 13:849667
van der Meer JR, Sentchilo V (2003) Genomic islands and the evolution of catabolic pathways in bacteria. Curr Opin Biotechnol 14:248–254
Van Houdt R, Monsieurs P, Mijnendonckx K et al (2012) Variation in genomic islands contribute to genome plasticity in Cupriavidus metallidurans. BMC Genom 13:1–11
von Hoyningen-Huene AJE, Schlotthauer TJ, Schneider D et al (2021) Pontibacillus sp. ALD_SL1 and Psychroflexus sp. ALD_RP9, two novel moderately halophilic bacteria isolated from sediment and water from the Aldabra Atoll, Seychelles. PLoS One 16:e0256639
Wei W, Gao F, Du M-Z et al (2017) Zisland Explorer: detect genomic islands by combining homogeneity and heterogeneity properties. Brief Bioinform 18:357–366
Williams KP (2002) Integration sites for genetic elements in prokaryotic tRNA and tmRNA genes: sublocation preference of integrase subfamilies. Nucleic Acids Res 30:866–875
Wilmes P, Simmons SL, Denef VJ, Banfield JF (2008) The dynamic genetic repertoire of microbial communities. FEMS Microbiol Rev 33:109–132
Woods WG, Ngui K, Dyall-Smith ML (1999) An improved transposon for the halophilic archaeon Haloarcula hispanica. J Bacteriol 181:7140–7142
Yoon SH, Park Y-K, Kim JF (2015) PAIDB v2. 0: exploration and analysis of pathogenicity and resistance islands. Nucleic Acids Res 43:D624–D630

Genomic Islands in Bacterial Genome Evolution and Speciation

5

Ayon Pal, Vivek Roy, Prajesh Dutta, Samarpita Adhikary, Barnan Kr Saha, and Jayanti Saha

Abstract

Bacterial genomes are flexible entities due to the complicated and dynamic makeup of the bacterial chromosomes. The plasticity and evolution of the bacterial genome are facilitated by genome rearrangements, point mutations, and lateral or horizontal gene transfer (HGT). Bacteria benefit from HGT as it facilitates them to adapt to their environment, colonize new niches, and steer evolution in "quantum leaps." Analyses of thousands of bacterial genome sequences have shown that there exist two major divisions in the bacterial genome: the primary core genome and the accessory gene containing flexible genome. Accessory gene acquisition might be mediated by entities designated as genomic islands (GIs), facilitating the process of HGT. GI represents horizontally acquired genes frequently clustered together in bacterial genomes with different GC content, dinucleotide frequencies, codon use, etc., than the neighboring genes. GIs make the genome flexible enough to adapt novel functions over a short life span and encode a diverse range of accessory genes for improved fitness, pathogenicity, resistance potential, metabolic flexibility, ecological adaptability, symbiosis, etc., in the harboring bacteria. Niche dynamicity is the key to having highly flexible genomes in bacteria. The niche exerts selection pressure to retain only helpful information and optimizes genomes based upon the costs and

A. Pal (✉) · V. Roy · P. Dutta · S. Adhikary · B. K. Saha
Microbiology and Computational Biology Laboratory, Department of Botany, Raiganj University, Raiganj, West Bengal, India

J. Saha
Microbiology and Computational Biology Laboratory, Department of Botany, Raiganj University, Raiganj, West Bengal, India

Department of Botany, Acharya Prafulla Chandra Roy Government College, Siliguri, West Bengal, India

I. Mani et al. (eds.), *Microbial Genomic Islands in Adaptation and Pathogenicity*,
https://doi.org/10.1007/978-981-19-9342-8_5

benefits of different strains. This strain-specific propensity to have high island variability contributes to genome plasticity and subsequent genome evolution that might lead to niche specialization of specific bacterial strains. The most significant evolutionary benefit of GI is fostering the genetic flexibility and ability to transmit multiple genes, enabling more effective adaptation and enhancing pertinence in specific ecological niches. The consecutive acquisition and loss of auxiliary genes within the GI and the consequent transmission of chromosomal DNA from the host appear to be the prerequisites for the evolution of bacterial species.

Keywords

Genomic Islands · Bacterial genome evolution · Pathogenicity Islands · Horizontal gene transfer · Genome plasticity

5.1 Genome Plasticity, Horizontal Gene Transfer and Evolution of the Bacterial Genome

Bacteria represent a diverse group of ubiquitous microbial organisms capable of surviving and tolerating a vast range of environments. Bacterial genomes are flexible or plastic, or adaptive entities resulting from the complex and dynamic nature of the bacterial chromosomes (Darmon and Leach 2014). The phenomenon of genome evolution involves different processes through which the content and arrangement of a species' genetic information modify over time. The plasticity and evolution of the bacterial genome are facilitated by various processes, including genetic mutations or genome reshuffling through deletions, duplications, inversions, translocations, and horizontal gene transfer (HGT) (shown in Fig. 5.1). Of these, HGT is unarguably one of the major creative forces that drive bacterial evolution, molding bacterial species' gene repertoires and providing and retaining population diversity. HGT takes place across the bacterial genome and causes the generation of incredibly different adaptations (Ochman et al. 2000). The phenomenon of HGT has cropped up significant ecological differences among closely related bacterial strains belonging to a single "species" taxon (Welch et al. 2002). Bacteria get benefitted from HGT due to their adaptation to various environments through colonization in new niches, thus allowing them to steer evolution in "quantum leaps" (Hacker and Carniel 2001).

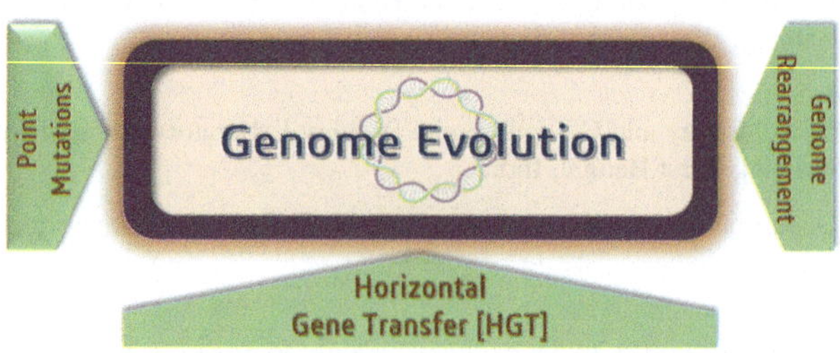

Fig. 5.1 Graphical representation of the major processes facilitating the plasticity and evolution of the bacterial genome like point mutations, genome rearrangements, and horizontal gene transfer

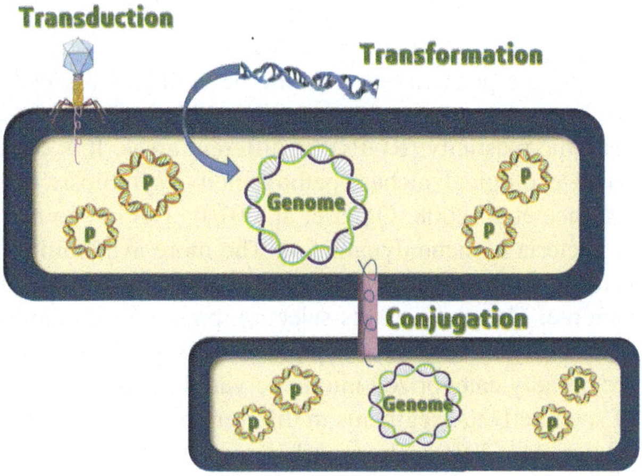

Fig. 5.2 The processes involved in bringing about horizontal gene transfer in bacteria include transformation (cell-free, one-way DNA uptake), transduction (bacteriophage-mediated one-way genetic material transfer from one bacterium to the other), and conjugation (one-way genetic material transfer via physical association between two living bacterial cells)

HGT is generally associated with acquiring DNA fragments that can move from one cell to the other or within a genome, or both (Bellanger et al. 2014). HGT in bacteria is brought about by transformation, transduction, and conjugation (Fig. 5.2), transferring genes across potentially distant bacterial lineages (Bazin et al. 2020). HGT is facilitated by mobile genetic components like conjugative plasmids, transposons, insertion elements, bacteriophages, and genomic islands or GIs (Hacker and Carniel 2001; Milkman 2004). Mobile genetic elements can encode different factors relating to drug resistance, bacterial pathogenicity, production of bacteriocins, and specific metabolic functions associated with the breakdown of xenobiotics chemicals, etc.

The advancement in next-generation shotgun sequencing has expedited the production of large amounts of genome sequence data, revealing insight into genome evolution and speciation. The analyses of thousands of prokaryotic genome sequences have led to the understanding that the bacterial genome is primarily divided into the core genome comprising of the core genes and the flexible genome made up of accessory genes, collectively referred to as "pan-genome," a term first coined by Tettelin and collaborators in 2005 (Tettelin et al. 2005). Generally, core genes encode essential metabolic activities, while accessory genes encode traits that provide fitness to bacteria to thrive under specific growth or environmental conditions. The magnitude of the flexible prokaryotic genome is astonishing. Only around 50% of the genome of any given strain is core. With hundreds of novel flexible genes contributed by each strain, the compounded flexible pool is enormous (Rodriguez-Valera et al. 2016). Acquisition of these accessory genes might be mediated by entities known as GIs, responsible for facilitating HGT. GI represents the cluster of horizontally acquired genes present in bacterial genomes that vary in dinucleotide frequency, GC content, codon usage pattern, etc., compared to the neighboring genes (Busby et al. 2013). While the core genome reflects the evolutionarily conserved character even under severe selection pressure, microbial genome dynamicity is achieved by recurrent gene acquisition and loss. In a microbial species, the plasticity of the genome results from HGT, which facilitates the acquisition of GIs and accelerates the rate of evolution. GIs make the genome flexible enough to adapt novel functions over a short life span.

Genome plasticity or genome flexibility is the phenomenon where the genome has large tracts of strain-specific polymorphic genes, known as the regions of genomic plasticity (RGPs), in different areas. It is critical for microbial survival in novel ecological niches, pathogenicity, symbiosis, and evolution of the genome (Mathee et al. 2008; Ogier et al. 2010). The key to having highly flexible genomes in bacteria is niche dynamicity. The more availability and proximity of exogenous genetic information result in the acquisition of novel information in the bacterial genome. The niche exerts selection pressure to retain only helpful information and optimizes genomes based upon the costs and benefits of different strains. The RGPs are broadly categorized into hypervariable regions that may arise from the deletions of specific DNA segments in different bacterial strains and mobile genetic elements (Ogier et al. 2010). The mobile genetic elements may get transferred from one place to another within the intracellular genome, or bacteria often use various HGT methods for efficient intercellular transfer. The enzymes transposases and recombinases (site-specific) are responsible for the mobility of mobile genetic elements and are coded by genes positioned on either the commonly shared core genome or the mobile genetic elements themselves.

5.2 GIs: Features, Types, Significance, and Plasticity

A GI is a fragment of exotic DNA that has been inserted into the bacterial genome and has clearly defined boundaries (Novick and Ram 2016). These transferable DNA regions are larger than 10–500 kb base-pair sequences (Osborn and Böltner 2002). Hacker et al. (1997) initially described GIs as gene clusters inside a bacterial genome possessing a specified dinucleotide frequency and GC content. GIs contribute significantly to genome flexibility, evolution, and environmental adaptation (Li and Wang 2021) by armoring bacteria with genes that provide antibiotic resistance, virulence traits and even sometime incorporating them with genes encoding enzymes leading to de novo metabolic pathway(s) formation (da Silva Filho et al. 2018).

5.2.1 Features Attributed to a GI

GIs encode a diverse range of accessory genes for improved fitness, secretion, pathogenicity, resistance potential, metabolic flexibility, ecological adaptability, symbiosis, etc., in the harboring bacteria (Darmon and Leach 2014). GIs have many other distinguishing features which delineate them from the other genomic regions. The specialized components of GIs are the presence of flexible sequences that differ from the core genome, the occurrence of genes for self-mobilization (viz. insertion sequences or ISs, integrases, and transposases), direct repeats (DRs) for flanking, and particular integration sites (Juhas et al. 2009; Schmidt and Hensel

Fig. 5.3 A graphical representation of the generalized organization of genomic island integrated within the bacterial chromosome (GI). DR = direct repeats, IS = insertion sequence and *int* = gene coding for integrase. The functional gene content of GIs are used to classify them into some subtypes, such as (i) pathogenicity island (PAI) that encode genes essential for bacterial pathogenicity/virulence, (ii) resistance island (RI) that encode antimicrobial resistance genes, (iii) symbiosis island (SI), and (iv) metabolic island (MI) that encode for adaptive metabolic abilities

2004). The recombination or transposition events during the integration process of GIs into the chromosome may generate the DRs. A generalized graphical representation of GI is given in Fig. 5.3. The evolution of GIs occurs through genome exchange, gene reduction, or further acquisition of some transposable elements (i.e., mobile genetic elements). The gene content between related strains of a species may vary considerably in each subset of islands. The G + C content (25–75%) of GIs is specific; they differ from the core chromosomal regions of bacteria (Schmidt and Hensel 2004). GIs are mostly inserted at 3′ end of genes coding for tRNA, and the region is often known as a hotspot for GI insertion (Lu and Leong 2016; Williams 2002). However, structurally every GI is almost the same in having a recombination module (integrase/excisionase module), two attachment sites (*attR* and *attL*, at the right and left end, respectively), and sometimes a recombination directionality factor (RDF). The tyrosine/serine recombinase enzyme family is involved in the exchange process during GIs integration (Boyd et al. 2009; Desvaux et al. 2020).

5.2.2 Types of GIs

The flexible gene content of GIs was classified and named based on the types of functions they encipher (Rainey and Oren 2011). These are (i) pathogenicity islands (PAIs) encoding genes relating to bacterial pathogenicity or virulence, (ii) symbiosis islands possessing similar structural attributes to pathogenicity islands though encoding proteins facilitating the mutual relationship of bacteria and multicellular organisms (Rainey and Oren 2011), (iii) antibiotic resistance islands that encode antimicrobial resistance genes, and (iv) metabolic/catabolic genomic islands enabling bacteria for adaptive metabolic abilities in degrading xenobiotic chemicals (Bertelli et al. 2019). Several kinds of mobile transposable elements including integrons, integrative and conjugative elements (ICEs), prophages, etc., are included within this wide dimensions of GI. GIs are mainly distinguished depending on the acquisition mechanism (i.e., transformation, conjugation, or transduction) and accompanying mobile elements (like transposases, ISs, and integrases) facilitating GI mobilization and transmission (Bertelli et al. 2019; Juhas et al. 2009; Langille et al. 2010; Soucy et al. 2015). According to Boyd et al. (2009), the GIs refer to a discrete class of evolutionarily ancient integrative components that are not "degenerate relics of prophages, episomes, integrons or ICEs."

New GI insertion is followed by subsequent changes in the cellular or colonial morphology, function, or even the lifestyle of the accepting organism. Some GIs are PAIs of the strain *Salmonella* SPI1 and *Listeria monocytogenes*, symbiosis islands (SIs) of *Bradyrhizobium, Mesorhizobium loti* strain R7A, acid fitness islands (AFIs) of *Escherichia coli*, defense island (DIs) of *Microcystis aeruginosa, Shewanella* sp. strain ANA-3, resistance islands (RIs) of *Haemophilus influenzae, Acinetobacter baumannii*, saprophytic islands of some *Escherichia coli* strains, a phenol degrading ecological island of *Pseudomonas putida*, xanthan gum production island (a metabolic island) of *Xanthomonas* (Arashida et al. 2022; Chan et al. 2015; Hadjilouka et al. 2018; Lerminiaux et al. 2020; Lima et al. 2008; Makarova et al. 2011; Mates et al. 2007; Van Elsas et al. 2011; Xiang et al. 2021). Different ecological niches exhibit diverse selection pressure under which bacteria with similar GIs play different functions. A bacterium may contain a variety of GIs within its genome to perform various functions.

5.2.3 Significance of GIs

The evolutionary benefit of GI is that many genes (like the whole operon dealing with novel traits) can be transferred horizontally to the recipient's genome, instigating significant modifications in the recipient's characteristics. According to the theory of "selfish operon," genes concerned with a particular function are grouped to promote their HGT (Lawrence and Roth 1996). GI can improve adaptability and competitiveness within the niche, thus providing selective benefits under certain growth conditions. The most significant evolutionary benefit of GI is fostering the genetic capacity and flexibility to transmit multiple genes, enabling more effective adaptation and enhancing fitness in particular ecological niches (Dobrindt et al. 2004).

5.2.4 Plasticity in GIs

Bacteria harbor IS elements, plasmids, prophages, transposons, GIs, ICEs as MGEs. The ISs are the simplest MGEs ($<$ 2.5 kb in size), which transmit no genes except those that encode machinery essential for their insertion at various DNA sites (Siguier et al. 2014). Plasmids have self-replication ability, and their intercellular transfer involves conjugation in prokaryotic cells (Smillie et al. 2010). Prophages integrate with bacterial chromosomes, and their mobility depends on transduction (Brüssow et al. 2004). Transposons may transfer from one intra-genomic region to another intra-genomic region and do not undergo any HGT. GIs are larger

(10–500 kb) MGEs with genes for self-mobility and other essential genes for strain-specific functions. The staphylococcal pathogenicity islands (SaPIs), GTAs, and ICEs are the three most common types of GIs. The first two are presumably descended from prophage forebears and have retained crucial prophage architectural traits. The third group likely originated from conjugative plasmids, which acquired additional characteristics and transformed into mosaics. While GTAs and ICEs independently influence HGT, the SaPIs depend on certain bacteriophages. The ICEs principally transmit their own DNA, whereas the GTAs solely convey the unlinked host DNA, but the SaPIs are a blend of both ICE and GTA. It is assumed that immobile GIs are variations of mobile ones (Novick and Ram 2016).

Not all GIs have genes for autonomous transfer. For example, SaPIs need helper phages for mobility (Lindsay et al. 1998). GIs share discrete segments of DNA between similar strains. However, their formation and ability to acquire accessory genes in the syntenic block lead to bacterial adaptation, genome diversification, and evolution. Different molecular events like recombination, deletion, duplication, inversion, etc., make GIs flexible.

The SaPIs of Gram-positive bacteria have phage integrase and excisionase homologs. The functional relatedness of SaPIs with phage makes them highly mobile with a unique lifestyle. The helper phage-mediated transfer of SaPIbov5 involves a prophage *cos* site in it. Due to phage interference SaPIs increase the transfer of chromosomal adaptive genes for host virulence. Genomic data from *Staphylococcus* implies SaPIs have typical phage-related genome organization but are sharply different from their progenitor prophage. Phage-related elements of streptococci and lactococci have orthology patterns similar to the SaPIs. Genome-based analysis also suggests that the widespread and diversified nature of SaPIs all over the bacterial genomic world successfully builds evolutionary strategy (Chen et al. 2015; Dokland 2019; Novick et al. 2010; Novick and Ram 2016).

The ICEs catalyze self-excision through site-specific recombination regardless of conjugation and integration. ICEs have diverse modular structures, i.e., gene clusters with different functions like conjugation, integration or excision, and adaptation. Conjugation modules have two types: a single-stranded DNA-based MOB/MPF module where MOB is the relaxase protein family, and MPF is a mating pair formation protein family. The second is a double-stranded DNA-based module that encodes SpoIIIE/FtsK protein family DNA translocator (Tra proteins) (Álvarez-Rodríguez et al. 2020; Besprozvannaya et al. 2013; Guglielmini et al. 2011). A double-stranded conjugation module containing ICE, regarded as actinomycete ICE (AICE), is found in actinomycetes (Johnson and Grossman 2015; Te Poele et al. 2008). The MOB acts on the 5′ end of ICEs and initiates DNA transfer through the MPF into the recipient bacterial cell. On the other hand, Tra proteins form channels and act on the cis-acting locus of circular AICE to transfer. The integration/excision modules encode enzymes for recombination events, generally, tyrosine/serine recombinase (or DDE transposase) those identify repetitive flanking sequences of ICEs. ICEs containing such integration/excision modules are usually found at the 3′ end of genes coding for tRNA. In case of different housekeeping genes, they can be found either at the 3′ or 5′ end. ICEs Ecoc54N from *Escherichia coli*, Tn*5397* from *Clostridium difficile*, Tn*1806* from *Streptococcus pneumoniae*, Tn*6012* and ICE*6013* from *Staphylococcus aureus*, and TnGBS elements from *Streptococcus agalactiae* have integration/excision modules with either tyrosine/ serine recombinase or DDE transposase (Antonenka et al. 2006; Bellanger et al. 2014; Camilli et al. 2011; Guérillot et al. 2013; Mingoia et al. 2016; Sansevere et al. 2017; Ternan et al. 2012; Wang et al. 2006).

The fundamental mechanisms behind GIs plasticity and evolution are acquiring, exchanging, or deleting different modules. Comparisons of sequences between ICEs provide information about the occurrence of enormous exchanges within different modules. It is suggested that the host specificity of ICEs evolved through the site-specific recombination between conjugation and integration modules of various ICEs (Burrus et al. 2002). Module exchanges also occur between other GIs. Sequence comparisons also reveal many GIs developed by deletion mutations in mobility modules. In some GIs, for example, ICE_*2603_tRNA*^{Lys} from *S. agalactiae*, genes (like *orf*D) associated with intercellular or intracellular mobility are deleted by incorporating the copy of insertion elements (IS*1193*). The combination of a GI within another GI, followed by subsequent restructuring, resulted in the acquisition of novel modules. Insertion of non-mobilizable ISs, induce deletion or inversion of neighbor sequences besides inactivation. In the high-pathogenicity island (HPI) of *Y. pestis* KIM, ISs are connected with conjugation module deletion (Bellanger et al. 2014; Chen et al. 2010; Puymège et al. 2013).

5.2.4.1 Driving Forces for GIs Plasticity

The GIs vary between individual strains and help adapt to new ecological niches, host-cell interaction, and virulence. This strain-specific propensity to have high island variability contributes to genome plasticity and subsequent genome evolution that might lead to niche specialization of specific bacterial strains. The variable strain-specific genomic DNA segments may evolve through rearrangement events like recombination, deletion, insertion, duplication, amplification, inversion, or tandem accretion.

Recombination

This is a crucial mechanism to keep genome flexibility and fitness. Homologous and non-homologous recombination are genetic information exchange procedures between DNA sequences with higher or lower identity, respectively (Didelot et al. 2012). In homologous recombination, the recombination rate reduces if the sequences are poorly identical. Non-homologous recombination occurs during either DNA synthesis or strand breakage and causes the addition of new genetic material through HGT and site-directed recombination. The insertions of new DNA segments often lead to deletions and subsequent hairpin formation due to strand slippage. The new DNA segments introduce genome diversity and synteny loss in bacteria. In homologous recombination, the involvement of Rec enzymes results in the single-strand or double-strand DNA gap repair; therefore, the recombination rates increase. The Rec enzymes, on the other hand, negatively regulate non-homologous recombination, lowering the recombination rate. Site-directed recombination involves

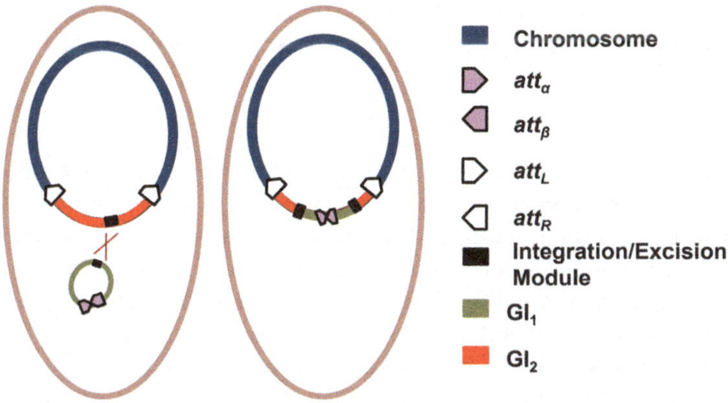

Fig. 5.4 GI plasticity via recombination. The plasticity of GIs may come through the recombination process between two separate GIs. Here, the red cross mark in the left diagram depicts the recombination between the integration/excision modules of GI_1 and GI_2. This event leads to a new GI formation, as shown on the right side of the figure

tyrosine/serine recombinase or DDE transposase to recognize the flanking repeats causing the exchange of DNA segments between integration/excision modules of different GIs. Both types of recombination play indispensable roles in bacterial genome evolution. The plasticity of GIs may come through the recombination process between two separate GIs, as shown in Fig. 5.4. A transposon Tn*6022* mediated recombination is indicated as the source of GI formation in many *Acinetobacter baumannii* strains (Patel 2016; Peters et al. 2014). From the ecological perspective, the overlapping niches of strains bring more opportunities to exchange genetic information between GI modules than lineages living in distant niches.

Deletion

Inactivation or deletion of part of the genome in bacteria facilitates genome evolution. While acquiring novel genes enhances the bacterial colonization potential, gene loss enables niche specialization. The "Streamlining" and "Black Queen" hypotheses assume that the loss of superfluous genes confers the survival ability or the fitness cost in bacteria lowering the metabolic burden (Giovannoni et al. 2005; Morris et al. 2012). The "Streamlining" theory is dependent on the fact that the challenging environmental circumstances favor robust selection to minimize cell complexity. On the other hand, the "Black Queen" hypothesis specifies that an organism should stop performing its costly function under certain conditions. Both hypotheses support reductive genomic evolution. These trends are visualized when host-specialized, highly virulent bacterial pathogens evolve from a vast range of hosts through massive gene loss. In endosymbionts, genome shrinkage is essential, which occurs possibly by deletion. Many new flagellar genes have been identified in the flagellar operon (*fli* operon) of *Salmonella enterica* var. *typhimurium* LT2. Deletions in the 2.07Mbp region in *fli* operon resulted in increased fitness (Frye et al. 2006; Koskiniemi et al. 2012). The plasticity and evolution of GIs also depend on the deletion of different modules. GIs are often sensitive to ISs inclusion leading to the inactivation or loss of superfluous genes, thus lowering the energy/mass expenditure.

Insertion

Insertion is the process of new gene acquisition that induces adaptive alterations in genome architectures and confers bacteria the ability to survive in new ecological niches. The transposition ability of ISs helps them to integrate anywhere in the genome. Their lodging is accidental in the genome, and several mechanisms control their disrupting activity resulting in genome innovation. The introduction of ISs in the genome creates an opportunity to evolve and accommodate better adaptive ability to the stressors. In the genome of nosocomial isolates of *Staphylococcus epidermidis*, insertion element IS*256* facilitates the tolerance to drugs making them multi-drug resistant. The insertion occurs in the GIs, consequently acquiring novel gene clusters (Dengler Haunreiter et al. 2019; Espadinha et al. 2019; Otto 2009). Some GIs may integrate within other GIs, followed by subsequent reorganization. Several mechanisms are responsible for the evolution of such mobilizable GIs. The non-mobilizable GIs that do not encode self-conjugative machinery can insert within ICEs. In the genome of *Mesorhizobium loti* USDA110, about 64 ISs are found within the 680 kb ICE*MlSymR7A* (Kaneko et al. 2002; Sullivan et al. 2013). Varieties of transposable elements have also been recognized in many ICEs or related GIs. Some ICEs also bear prophages. For example, Tn*6164* from *C. difficile* bears a complete prophage (Hargreaves et al. 2016). A graphical representation of deletion and insertion in GI is shown in Fig. 5.5.

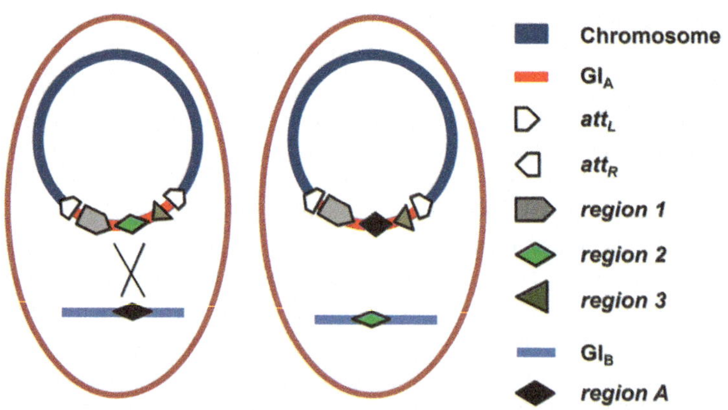

Fig. 5.5 Schematic presentation of deletion and insertion in GI. Insertion of *region 2* from GI_A into the GI_B and region A from GI_B into the GI_A leads to the loss and gain of regions islands (Melnyk et al. 2019)

Duplication and Inversion

Duplications are dynamic forces that aid in surviving in an unfriendly environment. Multiple amplification of the genomic regions results in tandem repetitive sequences. Inversion is another process when a DNA fragment is excised and reconnected in the opposite direction elsewhere in the genome. In general, the inversion sequences are bordered by inverted repetitions and are occasionally found inside a coding region. This reversal modifies the gene expression profile and alters the phenotypic characteristics of bacterial species. The evolution of distinct bacterial lineages confers adaptability by coupled gene duplication-amplification in response to drugs. It has been found that multistep adaptive development is preceded by gene amplification. As a result, mutation occurs in the additional copies while stabilizing the other copies of essential genes, enhancing fitness. Both duplication and inversion frequently target GIs, leading to plasticity. A schematic presentation of deletion and insertion in GI is depicted in Fig. 5.6. *E. coli* ST58 contains two GIs, PAI-1 and PAI-2, both islands sharing a great deal of genetic information. A duplication event at two distinct tRNA–Phe–GAA sites led to the development of these two progenitor islands from a single island. This duplication event was followed by inversion, and PAI-1 and PAI-2 acquired separate sets of genes throughout time (Wyrsch et al. 2020). IS-mediated duplication has been found in the symbiosis island of WN105 mutant of *Bradyrhizobium diazoefficiens* USDA110 (Arashida et al. 2022).

Tandem Accretion

GIs also originate from site-specific recombination and subsequent tandem accretion-deletion of CIMEs and ICEs (Fig. 5.7). The site-specific accretion resulting from gene gain and loss is a key tool for GI flexibility and evolution. The composite structure of ICE*St1* and related GIs from *Streptococcus thermophilus* demonstrated these components developed via site-specific recombinations and deletions. At the 3′ end of the *fda* locus in seven distinct strains of

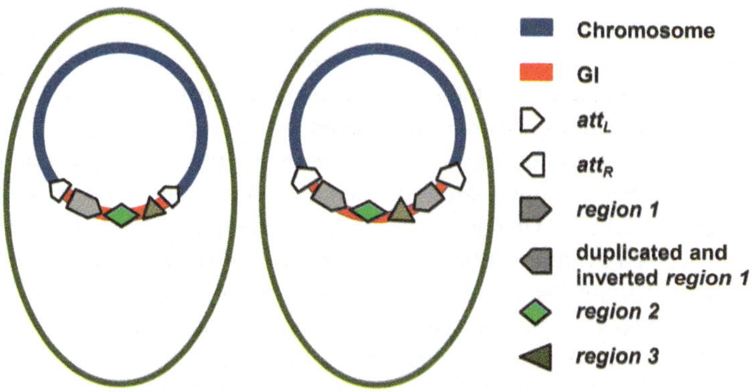

Fig. 5.6 Duplication and inversion in GI. Left diagram shows normal GI, whereas right shows GI with duplicated and inverted *region 1* (Wyrsch et al. 2020)

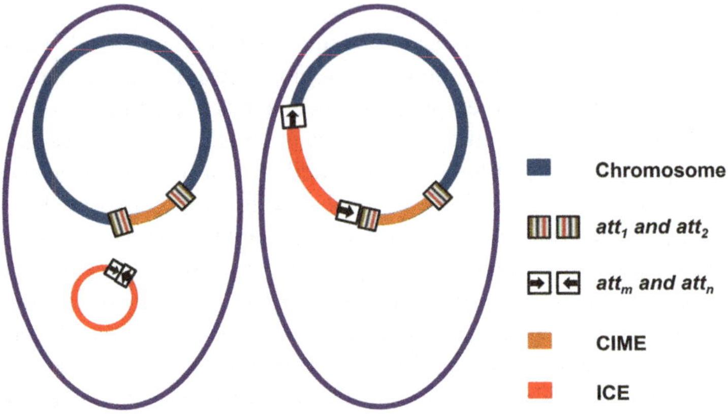

Fig. 5.7 Tandem accretion of GIs. The diagram shows the integration of ICE adjacent to the att_1 of the resident CIME by site-specific recombination, causing composite GI formation (Bellanger et al. 2014)

S. thermophilus, four forms of ICE*St1*-related ICEs with comparable conjugation and recombination modules were identified. These elements are flanked by different site-specific attachment sites (*att*) that are strongly connected to attachment sites of two distinct *cis*-mobilizable elements (CIMEs) CIME19258 and CIME302. This results in site-specific recombination, which results in the excision of ICEs and the subsequent incorporation of CIMEs at the 3′ end of *fda*. Moreover, genome analysis showed identically shortened sequences at the *att* regions of these ICEs and CIMEs (Bellanger et al. 2014; Pavlovic et al. 2004).

5.3 GIs and Bacterial Evolution

The idea of prokaryotic species is complicated, and it is commonly assumed that such species are formed because of continuous processes combining gene loss and gain facilitated by HGT (Lawrence 2001). The acquisition and loss of auxiliary genes within the GI and the probable transmission of chromosomal DNA from the host may be a precondition for bacterial species evolution. Since GIs may integrate themselves into the host chromosomes by excision, conjugated-mediated self-transfer into a new host, and reintegration, they can transmit a piece of the host genome into the recipient bacteria. This process of receiving donor (foreign) DNA may open the way for bacterial evolution when donor DNA is integrated into the genome of the host via transformation. The unique self-transfer type IV secretion system (T4SS) of *Neisseria gonorrhoeae* (encoded by a horizontally acquired huge gonococcal genetic island or GGI), enables both secretion and spread of the host chromosomal DNA. Later, when taken up by the transformation process, the secreted chromosomal DNA (via GGI-encoded type T4SS) can undergo recombination along with the host's chromosome, adding to antigenic variation and drug resistance. Thus, GIs appear to influence the evolution of the host bacteria. This change occurs in several Gram-negative or Gram-positive, environmental or pathogenic bacteria like *N. gonorrhoeae*, *Acinetobacter* sp. ADP1, *Haemophilus influenza*, *Pseudomonas stutzeri*, *Bacillus subtilis*, *Streptococcus pneumoniae*, and *Ralstonia solanacearum*. In addition to transformation, the transmission of GIs across bacterial species can also be mediated through conjugation and bacteriophages. The SaPIs are the most well-known example of this phenomenon. There have been several reports of such bacteriophage-mediated transfers of GIs, that includes the HPI and the GIs of *Yersinia pseudotuberculosis* and *Prochlorococcus* sp. (marine cyanobacteria), respectively (Juhas et al. 2009).

5.3.1 GIs in Bacterial Genome Evolution and Shapiro's Geographical Metaphors

In his article "How clonal are bacteria over time?," Shapiro (2016) has convincingly employed different geographical metaphors to explain how GIs are potentially associated with genome evolution. According to Shapiro, horizontal transfer (recombination) rates fluctuate significantly across the genome so that an entire population, except for a few loci, can be clonal. These loci are referred to as GIs. The term "peninsula" provides a simile that might better depict the connection of the islands to the microbial genomes. An island remains evolutionarily independent compared to the mainland genome, but their fates may become associated. For instance, a bacterium may obtain any gene from the enormous microbial gene pool. The acquired gene enables the bacteria to colonize into a new ecological niche, initiating a clonal expansion where the acquired gene's fate and its new host genome are inextricably intertwined, at least during the clonal expansion period. Certain microbial genomes may include an inordinate number of islands to the point where there is no mainland but a large archipelago. Archipelagos are not always stable throughout time, and they can occasionally combine into continents. When the ecological conditions are favorable, any genome from the panmictic gene pool can break free from the "gravitational pull" of recombination and embark on clonal growth (Shapiro 2016).

With the help of bioinformatics, it became apparent that novel genes with unknown functions are present within GIs. These novel genes lack orthologs in other bacterial species and might furnish the host bacterium with various adaptive functions. Different accessory functions like additional metabolic pathways, resistance toward antibiotics and harmful drugs, pathogenesis, symbiosis, or traits involved in increasing microbial fitness, are encoded by GIs.

Although plasticity (Dobrindt et al. 2004) within GI alters bacterial lifestyle or behavior, it is not unfamiliar that bacterial "quantum leaps" in context with evolution is mediated by GI (Juhas et al. 2009), as it has the intrinsic property to transfer a large set of genes and integrate as a whole into the recipient genome, thereby promoting bacterial diversity and speciation (Guo et al. 2012). Some of the GI-mediated features that can offer a significant selective benefit to the host bacteria and thus leap a step toward evolution and speciation are discussed briefly below.

5.3.2 GI and Heavy Metal Tolerance

GIs may provide an advantage for living in harsh toxic environments leading to the evolution of resistant strains. One of the most critical toxic environments includes acid mine drainage (AMD) which contains high concentrations of toxic heavy metals (including arsenic). A comparative genomic study on different isolates of *Thiomonas* bacteria obtained from AMD of Carnoulès, France, revealed the presence of more than 20 GIs. The findings also indicated that arsenic-associated GIs had evolved differently in two closely related *Thiomonas* strains, resulting in varied survival capabilities in As-rich environments (Freel et al. 2015).

GIs contributing to heavy metal tolerance in *Mucilaginibacter* spp. have been observed in isolates obtained from gold/copper mines. Clusters of genes that may be connected with mobile genetic elements were discovered by analyzing the location of heavy metal resistance determinants. These loci contained genes for tyrosine recombinases (integrases) and subunits of T4SS, letting integration/excision and conjugative transfer of many GIs, respectively. The supposed presence of many CTnDOT-related GIs in the genomes of *Mucilaginibacter* may have a crucial task in genome evolution and subsequent adaptation (Vásquez-Ponce et al. 2018).

In *Cupriavidus metallidurans*, GI variation has contributed to the variable plasticity of the genome. *C. metallidurans* represents a versatile multi-metal resistant bacteria. Comparative genomic hybridization of sixteen *C. metallidurans* strains revealed that the broad arsenal of heavy metal resistance factors was well preserved across all strains of *C. metallidurans*. Contrarily, the transposable elements found in strain CH34 were not observed in the other strains but displayed an indirect pattern associated with a specific geographical location or biotope. One set of strains had nearly all transposable components, whereas the second group had a substantially lower proportion. This was also manifested in their capacity to break down toluene and thrive on carbon dioxide and hydrogen gas in an autotrophic manner. Both of these are connected to distinct GIs of the Tn4371 family (Van Houdt et al. 2012).

5.3.3 GIs in Secondary Metabolism, Pathway Evolution and Xenobiotic Degradation

GIs have been found in a distinct environmental (marine) bacteria *Salinispora*, belonging to the phylum Actinobacteria. Genomic comparison of *S. tropica* with *S. arenicola* displayed the distribution of three-quarters of species-specific genes within 21 GIs associated with the production of secondary metabolites, also establishing a connection between secondary metabolism and functional adaptation. All species-specific biosynthetic pathways are found in GIs, most of which are found in *S. arenicola,* contributing to its worldwide distribution in different habitats. Gene duplication and acquisition dominate genome evolution, which provides rapid chances for generating novel bioactive compounds in the case of secondary metabolism. The horizontal sharing of secondary metabolic pathways performs a major functional role in acquiring natural product biosynthetic gene clusters, which also serve as the driving force for maintaining bacterial diversity (Penn et al. 2009). GIs have been identified as hotspots for biosynthetic gene cluster acquisition in Salinispora (Letzel et al. 2017). Using something apparently like a plug-and-play paradigm of evolution, clusters of acquired biosynthetic genes are targeted to certain GI and can replace each other (Letzel et al. 2017).

GIs seem to have a significant role in developing novel pathways through "patchwork assembly" (a novel combination of previously existing pathways) (Dobrindt et al. 2004; Guzman and Harris 2015; Mingoia et al. 2016). Studies reveal that plasmids, transposons, and GIs include catabolic genes that encode functionalities with the capacity to digest xenobiotic substances. For instance, *Ralstonia oxalatica* Tn437, a GI, might break down chlorobiphenyl. First discovered in *Pseudomonas knackmussii* B13, the *clc* element could utilize chloroaromatic chemicals as a carbon source. In the field of biodegradation, ICE*clc* is the best-known ICE. It bears selective genes for the ortho-cleavage of chlorocatechols and aminophenol metabolism, and these are *clc* and *amn* genes respectively). This component is capable of metabolizing 3-chlorocatechol, 3-chlorobenzoate, 4-chlorocatechol, as well as aminophenol. Due to its self-transfer capabilities, it can insert itself into the genomes of different Proteobacteria based on environmental circumstances (Klockgether et al. 2006). Based on amino acid homology searches, similar clc-like elements from bacterial genomes have already been isolated around the world, including in *Xylella fastidiosa* 9a5c (a plant pathogen), *Pseudomonas aeruginosa* C (a clinical isolate), *P. aeruginosa* S17GM (an environmental isolate) and *Xanthomonas campestris* (Lacour et al. 2006). *Pseudomonas*-like clc components have also been identified in a *Ralstonia* sp. JS705 isolate (reported from contaminated groundwater). This clc element codes for chlorobenzene to chlorocatechol metabolizing enzymes with 85–100% nucleotide similarity in the conserved area (Klockgether et al. 2006). These discoveries demonstrate that the clc element can spread to new environments and obtain new functionalities within its current location (van der Meer and Sentchilo 2003). GIs conferring the capacity to degrade xenobiotics and toxic compounds biologically has been detected in many bacterial populations. Some examples of GI with biodegradation functions are listed in Table 5.1.

5.3.4 GIs and Siderophore Expressing Bacteria

Bacteria are well known for expressing iron uptake systems, known as siderophores. Siderophores represent low molecular weight secondary metabolites synthesized and then released into their environment, where they chelate ferric iron to combat iron deficiency (Neilands 1995; Thode et al. 2018). This is an adaptation to survive in an iron-restricted environment and is associated with virulence. Genes encoding for

Table 5.1 List of GIs with biodegradation potential for different xenobiotics

Sl. No.	GI	Biodegradation potential for	References
1	100 kb clc element	Chlorocatechols, aminophenols	Gaillard et al. (2006))
2	90 kb bph–sal element	Biphenyl	Nishi et al. (2000)
3	55 kb biphenyl catabolic transposon Tn4371	Biphenyl	Toussaint et al. (2003)
4	100 kb Tn3-like Alteromonas sp. SN2 transposon	Naphthalene	Jin et al. (2011)
5	232 kb phn island in *Delftia* sp. Cs1–4	Polynuclear aromatic hydrocarbons	Hickey et al. (2012)
6	(per)chlorate reduction-associated genomic island (PRI)	Perchlorate reduction	Melnyk et al. (2011)

siderophores are distributed in several pathogenic and non-pathogenic bacterial species harboring GIs. Examples include HPI in *Yersinia* sp.; SHI-2, SRL, and SHI-3 in various species of the genus *Shigella;* and PPI-1 in *S. pneumoniae* (Dobrindt et al. 2004). Such GIs can serve as fitness islands in environmental bacteria or PAI in pathogenic bacteria.

5.3.5 GIs and Bacterial Secretion Systems

The evolution of pathogenicity in bacteria through acquiring GIs (virulence-carrying genes) is a well-established phenomenon. During evolution, bacteria may have gained new genes by HGT, or their current genes may have acquired mutation. One fine example of the successful host-pathogen interaction is represented by several classes of protein secretory systems encoded by GIs (Martínez 2013). The type III secretion systems (T3SS) or "contact-dependent" secretion systems are complex multiprotein machinery (Scherer and Miller 2001). T3SS is generally expressed by pathogenic bacteria infecting plants and animals, including genera like *Yersinia, Shigella, Salmonella, Pseudomonas,* enteropathogenic *E. coli* (EPEC), *Erwinia* and *Rhizobium.* In some exceptional cases, two T3SSs are expressed within a single pathogen, each necessary at a different infection stage. In *S. enterica*, out of the two T3SS (harbored by SPI-1 and 2), one is essential for the initial interaction and penetration into the eukaryotic target cell (intestinal epithelium cells). At the same time, the other is essential for systemic infection (Juhas et al. 2009).

GIs of many bacterial pathogens encode T4SSs, translocating bacterial effector proteins through the bacterial membrane and plasma membrane into eukaryotic host cells. T4SSs, in turn, mediate HGT, which contributes to plasticity of the genome, development of infectious diseases, and the spread of drug resistance and other attributes related to virulence. The architecture of the genetic determinants of T4SS is diverse and comprises numerous genes grouped as a single functional unit (Juhas et al. 2007). The T4SS has been extensively studied in *Agrobacterium tumefaciens*. Unlike the T3SS system, this complex system is unique as it delivers nucleoprotein complexes and effector proteins into plant cells, contributing to pathogenicity directly (Dobrindt et al. 2004).

5.3.6 GIs and Antimicrobial Resistance

One of the most significant routes for acquiring drug resistance is GIs. The advent of methicillin-resistant *Staphylococcus aureus* (MRSA) has largely been attributed to the so-called staphylococcal cassette chromosome methicillin-resistant (SCC*mec*) islands present in the genome of *S. aureus*. This island can also integrate with other MGEs and might confer resistance against additional antibiotics, thus representing a hotspot with variable size (20 kb to ≥60 kb) (Dobrindt et al. 2004). MRSA is resistant to various antibiotics like methicillin, penicillins, kanamycin, tobramycin, bleomycin, tetracycline, macrolide, lincosamide, streptogramin, vancomycin and also to heavy metals (Juhas et al. 2009). Origin of the SSC*mec* island in *S. aureus* is yet to be established; however, comparative bioinformatic studies have proposed that it could have originated from other staphylococcal species via HGT, such as *S. sciuri*, *S. fleuretti*, *S. epidermidis*, or *S. haemolyticus*. Reports have shown the existence of SSC*mec* in *S. epidermidis* well before its discovery from *S. aureus*. Thus, SCC*mec* in *S. epidermidis* might act like a pool of resistance genes contributing to the evolution of multi-drug-resistant *S. aureus*. In another case, *mecA* was naturally found to be present in the chromosome of *S. fleuretti* and therefore was supposed to be the original source of the *mecA* in the SCC*mec*. Many dynamic SCC*mec* islands have been discovered from *S. haemolyticus* genome, thus indicating *S. haemolyticus* to be a potent carrier for methicillin-resistant genes (Juhas 2019).

The genus *Enterococcus* has become a chief cause of nosocomial infections, and the prime player in the swift expansion of such enterococcal infection comprises those of drug-resistant strains. Besides genomic modification and HGT, GIs also play a crucial role in the acquisition of drug resistance. In studies where the whole genome sequences of some *E. faecium* and *E. faecalis* (carrying many resistance genes) were screened to analyze the correlation between antibiotic resistance genes (ARGs) and GI transmission, two observations became distinct, firstly the prevalent nature of GIs in *Enterococcus*, and secondly, antibiotic-resistant genomic islands (ARGIs) contributing significantly to the dissemination of some ARGs. The above study has clearly shown the existence of 119 GIs in 37 strains, with an average value of 3.2 in each strain (universal presence of GI in *Enterococcus*). GIs in these strains was found to harbor variant ARGs, including aminoglycosides, chloramphenicol, glycopeptides/peptides, lincosamides, streptomycin and multi-resistant efflux pumps. The ARGs identified in the enterococcal ARGIs are, *mdtG* (encodes an efflux pump providing resistance against with fosfomycin), *tetM* (tetracycline resistance), *dfrG* (diaminopyrimidine antibiotic resistance), *lnuG* (lincosamide resistance), *fexA*, (an efflux pump providing chloramphenicol resistance). Besides encoding for drug-resistant, some of the GIs have been credited with mobility-related elements, like genes for conjugation, transposase or excisionase. Credible relationships among enterococcal strains and GIs were found to indicate frequent genetic exchanges within and between *Enterococcus* strains. Regular genetic exchanges among all the strains (comprising *E. faecium* and *E. faecalis*) mediated by GIs were not unusual (Li and Wang 2021). The high plasticity of *Enterococcus* genome has been allocated to the conjoint action of HGT, and either gain or loss of genetic information. According to Darwin, for the evolution of an organism, environmental selection pressure must have served as the driving force. This hypothesis fits well in the case of the MDR *Enterococcus*, isolated from complex ecological niches, like hospitals, medical clinics, farmlands, contaminated water, stools, humans, and pigs, and where harsh environmental factors always exist (such as antimicrobial compounds, organic and inorganic biocides, and heavy metals). The GIs harboring numerous novel genes due to their self-mobility or non-mobility might have integrated into the bacterial genome. Consequently, the recipient organism develops a new metabolic potential to enhance fitness or adaptability (Li and Wang 2021).

Many *Salmonella enterica* serovars responsible for gastrointestinal sickness are resistant to antibiotics due to GIs bearing a class 1 integron that contains the resistance genes. Studies have suggested that *Salmonella* genomic island 1 (SGI-1) retains a complex multi-drug resistance segment, imparting resistance against many antibiotics, including tetracycline, ampicillin, chloramphenicol/florfenicol, sulfamethoxazole and streptomycin/spectinomycin. The SGI1-associated MDR region comprises a complex integron harboring the *aadA2*, *floR*, bla_{PSE}, *tetR*, and *tetG* genes (Vo et al. 2010). Many *Salmonella* serovars and *Proteus mirablis* possess SGI1 or similar islands harboring diverse resistance gene sets. SGI1 is a mobilizable integrative element transferable experimentally into *E. coli* (Hall 2010).

Cholera, caused by *Vibrio cholerae*, is a dreadful disease. Reports of multidrug-resistant *V. cholerae* strains have been frequent over the past few decades. The spread of determinants related to resistance is primarily due to mobile genetic elements such as the SXT / R391 integrated conjugate element, IncC plasmid, and GI. Transmission of the IncC plasmid is activated by the master activator AcaCD (Rivard et al. 2020). The regulatory network of AcaCD extends to the chromosomally integrated GIs. A discrete and novel mobile genomic island (MGI) MGI*Vch*Hai6 integrated into the chromosome of a multidrug-resistant *V. cholerae* HC-36A1 isolate (Carraro et al. 2016) contains an integron In104-like multi-drug resistance element and a mercury resistance transposon, analogous to SGI1. Acquisition of MGI*Vch*Hai6 plays a vital role in resistance against β-lactams, chloramphenicol, trimethoprim, tetracycline, sulfamethoxazole, and streptomycin/spectinomycin (Carraro et al. 2016).

Pseudomonas aeruginosa is a severe threat to burn patients and the immune-compromised. Different high-risk clonal strains, like ST111, ST175, and ST235 carry genes that give resistance to β-lactam antibiotics (Roy Chowdhury et al. 2016). GIs play a determining role in the spread of resistance to a wide variety of effective antibiotics, such as metallo-β-lactams and extended-spectrum β-lactams. Strains of

P. aeruginosa (ST) 235 carry Tn6162 and Tn6163 in GI1 and GI2, respectively. The class 1 integron coupled with Tn6163 in GI2 carries a *blaGES-5–aacA4–gcuE15–aphA15* cassette range that confers resistance to aminoglycosides, including carbapenems. Studies suggest that the evolution of GI2 could have occurred from a novel ICE. GI2 is winged by a repeat motif region (direct) of 12 nucleotide bases and codes for integration, conjugative transfer and ICE-specific proteins (Roy Chowdhury et al. 2016).

Another Gram-negative opportunistic pathogen, *Acinetobacter baumannii*, is a nosocomial pathogen that causes serious health hazards to immunocompromised patients. Nowadays, *A. baumannii* has been garnering considerable attention widely due to its rapid capacity to build up multi-drug resistance. The sequences of many *A. baumannii* genomes have divulged a vast collection of ARGs, several of which are connected with transposable elements and ISs, and it might be found in GIs, known as AbaR (Leal et al. 2020; Liu et al. 2014). Different AbaR islands have been found that vary in size, and are dynamically reshaped primarily because of recombinases, transposases, and integrases (Leal et al. 2020). Few resistance genes are present within plasmid, which can be intra- and interspecies exchanged, even by prophages (Leal et al. 2020). Studies have detected the presence of a novel GI (GIBJ4) in the drug-sensitive strain BJ4 possessing metal resistance genes inserted into the position where AbaR-like RIs commonly reside in other strains of *A. baumannii* (Liu et al. 2014). In *A. baumannii* several antibiotic resistance determinants are also present outside the RIs, such as integrons, chromosomal intrinsic antibiotic resistance genes, and the blaOXA-23-containing transposon Tn2009 (Liu et al. 2014).

5.3.7 In Planta GI Mediated Bacterial Evolution

GIs can transmit across bacteria in vitro but not during the infection process in the host. Lovell et al. (2009) have demonstrated that horizontal transmission of a GI (PPHGI-1) occurs in planta between strains of the plant pathogen *Pseudomonas syringae* pv. Phaseolicola (Pph). This study reveals that the transfer of PPHGI-1 across Pph strains by transformation involves four unique steps: (i) excision of the GI from the bacterial chromosome, (ii) release of the circular episome from the bacterium, (iii) relocation into competent bacterial cells, and (iv) integration at a particular *att* site. Transformation, the simplest method of DNA exchange, may thus accomplish the evolution of bacterial pathogens via HGT (Lovell et al. 2009).

5.3.8 GIs in Evolution of Pathogenic Bacteria

Many bacteria incorporate PAIs, a subset of GI, in their chromosomes. PAIs are specialized islands comprising arrays of genes whose expression leads to pathogenicity (virulence) and disease. These PAIs provide fitness to the PAI-positive bacteria directly or indirectly by increasing their chances of survival in vivo and/or transmission to new hosts and contributing to genome evolution. This PAI-mediated fitness can be well observed during the onset of clinical symptoms, which is directly linked to the pathogenicity or lesions triggered by the pathogenic or virulent bacteria (Hacker and Carniel 2001). PAIs are the most well-known GI which has been most exhaustively studied. The various determinants of pathogenic bacteria responsible for the pathogenesis or disease are embedded within these PAIs and also on various mobile elements like extra-chromosomal plasmids, phages, insertion elements, and transposons (Schmidt and Hensel 2004). PAIs are ubiquitous in both Gram-positive and Gram-negative pathogens. Some examples of Gram-negative bacteria harboring PAIs include *Salmonella* spp., *Neisseria* spp., *Shigella* spp., *Yersinia* spp., *Helicobacter pylori, E. coli, Pseudomonas* spp., *Vibrio cholerae, Porphyromonas gingivalis,* and *Francisella* spp. Examples of Gram-positive pathogenic bacteria having PAIs include *Staphylococcus aureus, Streptococcus* spp., *Listeria* spp., *Clostridium* spp., and *Enterococcus* spp. The presence of a certain PAI, on the whole, is specific to a pathogenic bacterium or a specific strain of bacteria. A particular bacteria can also have more than one PAI in its genome (Gal-Mor and Finlay 2006). Bacterial species or strains equipped with PAI have an inherent advantage over their non-PAI-bearing counterparts when it comes to pathogenicity. Most virulence determinants for the typical *Salmonella enterica* are present both in the chromosome and within PAI, termed as *Salmonella* pathogenicity islands (SPIs). Both SPI-1 and SPI-2 are essential in determining virulence. SPI-1 encoded T3SS proteins (Shea et al. 1996) build up complex machinery for the translocation of various effector proteins from the extracellular *S. enterica* into the host (eukaryotes) cells (Schmidt and Hensel 2004). SPI-1 also codes for various regulators, some acting as transcriptional activators and others as inhibitors or repressors of SPI-1 genes. The most crucial are HilA, HilE, and LeuO; these gene products and many more work in a complex way to tightly regulate SPI-1 genes (Lou et al. 2019). HPI, a discrete PAI, is naturally found in all the virulent serotypes of *Yersinia* sp., namely *Y. pestis, Y. enterocolitica and Y. pseudotuberculosis* but is completely lacking in serotypes with low virulence. Due to the high instability of HPI, it has also been established in various other members of enterobacteria (Carniel et al. 1996). *Clostridium difficile,* an aerobic bacterium, produces various toxins leading to diarrhea and pseudo-membranous colitis. However, toxin production is restricted only to the virulent (toxigenic) variant but is lacking in the non-virulent type. Comparing both variants revealed that the genes coding for the toxins are integrated into a PAI termed PaLoc (pathogenicity locus) (Braun et al. 1996). The first case of vancomycin resistance in bacterial pathogen was described from a clinical isolate of *E. faecalis.* Genes related to virulence determinants are incorporated in a 154 kb PAI. It has been found that commensal *E. faecalis* got transformed into virulent ones by acquiring GI-encoding virulence factors from the virulent *E. faecalis* (Juhas et al. 2009; Shankar et al. 2002). Members of the genus *Pseudomonas* have been discovered from multiple habitats, and some species are even considered opportunistic pathogens. Genomic analysis has shown the presence of different types of PAI in strains of *P. aeruginosa,* termed PAGI (viz., PAGI-1, 2, 3), thereby conferring various adaptive traits (Battle et al. 2009). PAGI-1 (from strain PAO1 and patients with urinary infection) was found to contain genes coding for numerous dehydrogenases (with unknown potentials) and proteins able to sense redox-cycling agents, thus, indicating a protective role of this island against reactive oxygen species (ROS) damage. The latter two, PAGI-2 and PAGI-3, were discovered from the type C strain of *P. aeruginosa* (isolated from cystic fibrosis patients) and strain SG17M (aquatic strain), respectively. The island PAGI-2 contains numerous genes encoding transporters, regulators, and proteins needed for biosynthetic pathways. The crucial one seems to be the proteins involved in the biogenesis of cytochrome C, thus providing a selective advantage to the bacteria to thrive in an environment with oxidative stress and deprived of iron. Cytochrome C-mediated iron uptake and inactivation of free radicals appeared to be the player behind the scenes. Furthermore, PAGI-3 was found to be a metabolic island without any virulence factors (Schmidt and Hensel 2004). Six more novel islands were identified through a subtractive hybridization approach from *P. aeruginosa* clinical isolates *(*Battle et al. 2009*)*.

In conclusion, it may be stated that GIs are still an enigma. The evolution of prokaryotes, particularly that of eubacteria, is primarily impacted by forces like HGT, and the GIs contribute vastly to this direction. With the availability of a deluge of whole genome sequence information due to next-generation sequence techniques, the puzzle of bacterial evolution and speciation has started to unwrap, albeit slowly. With more than 400,000 prokaryotic whole genome sequences in global databases at present, biologists have an enormous amount of data to analyze and decipher newer paradigms in bacterial evolution.

References

Álvarez-Rodríguez I, Arana L, Ugarte-Uribe B, Gómez-Rubio E, Martín-Santamaría S, Garbisu C, Alkorta I (2020) Type IV coupling proteins as potential targets to control the dissemination of antibiotic resistance. Front Mol Biosci 7:201

Antonenka U, Nölting C, Heesemann J, Rakin A (2006) Independent acquisition of site-specific recombination factors by asn tRNA gene-targeting genomic islands. Int J Med Microbiol 296: 341–352

Arashida H, Odake H, Sugawara M, Noda R, Kakizaki K, Ohkubo S, Mitsui H, Sato S, Minamisawa K (2022) Evolution of rhizobial symbiosis islands through insertion sequence-mediated deletion and duplication. ISME J 16:112–121

Battle SE, Rello J, Hauser AR (2009) Genomic islands of Pseudomonas aeruginosa. FEMS Microbiol Lett 290:70–78. https://doi.org/10.1111/j.1574-6968.2008.01406.x

Bazin A, Gautreau G, Médigue C, Vallenet D, Calteau A (2020) panRGP: a pangenome-based method to predict genomic islands and explore their diversity. Bioinformatics 36:i651–i658. https://doi.org/10.1093/bioinformatics/btaa792

Bellanger X, Payot S, Leblond-Bourget N, Guédon G (2014) Conjugative and mobilizable genomic islands in bacteria: evolution and diversity. FEMS Microbiol Rev 38:720–760. https://doi.org/10.1111/1574-6976.12058

Bertelli C, Tilley KE, Brinkman FSL (2019) Microbial genomic island discovery, visualization and analysis. Brief Bioinform 20:1685–1698. https://doi.org/10.1093/bib/bby042

Besprozvannaya M, Pivorunas VL, Feldman Z, Burton BM (2013) SpoIIIE protein achieves directional DNA translocation through allosteric regulation of ATPase activity by an accessory domain. J Biol Chem 288:28962–28974

Boyd EF, Almagro-Moreno S, Parent MA (2009) Genomic islands are dynamic, ancient integrative elements in bacterial evolution. Trends Microbiol 17:47–53. https://doi.org/10.1016/j.tim.2008.11.003

Braun V, Hundsberger T, Leukel P, Sauerborn M, Von Eichel-Streiber C (1996) Definition of the single integration site of the pathogenicity locus in Clostridium difficile. Gene 181:29–38. https://doi.org/10.1016/S0378-1119(96)00398-8

BrüSsow H, Canchaya C, Hardt W-D (2004) Phages and the evolution of bacterial pathogens: from genomic rearrangements to lysogenic conversion. Microbiol Mol Biol Rev 68:560–602

Burrus V, Pavlovic G, Decaris B, Guédon G (2002) Conjugative transposons: the tip of the iceberg. Mol Microbiol 46:601–610

Busby B, Kristensen DM, Koonin EV (2013) Contribution of phage-derived genomic islands to the virulence of facultative bacterial pathogens. Environ Microbiol 15:307–312. https://doi.org/10.1111/j.1462-2920.2012.02886.x

Camilli R, Bonnal RJ, Del Grosso M, Iacono M, Corti G, Rizzi E, Marchetti M, Mulas L, Iannelli F, Superti F (2011) Complete genome sequence of a serotype 11A, ST62 Streptococcus pneumoniae invasive isolate. BMC Microbiol 11:1–13

Carniel E, Guilvout I, Prentice M (1996) Characterization of a large chromosomal "high-pathogenicity island" in biotype 1B Yersinia enterocolitica. J Bacteriol 178:6743–6751. https://doi.org/10.1128/jb.178.23.6743-6751.1996

Carraro N, Rivard N, Ceccarelli D, Colwell RR, Burrus V (2016) IncA/C conjugative plasmids mobilize a new family of multidrug resistance islands in clinical vibrio cholerae non-O1/non-O139 isolates from Haiti. MBio 7:e00509–e00516

Chan AP, Sutton G, Depew J, Krishnakumar R, Choi Y, Huang X-Z, Beck E, Harkins DM, Kim M, Lesho EP (2015) A novel method of consensus pan-chromosome assembly and large-scale comparative analysis reveal the highly flexible pan-genome of Acinetobacter baumannii. Genome Biol 16:1–28

Chen J, Ram G, Penadés JR, Brown S, Novick RP (2015) Pathogenicity island-directed transfer of unlinked chromosomal virulence genes. Mol Cell 57:138–149

Chen PE, Cook C, Stewart AC, Nagarajan N, Sommer DD, Pop M, Thomason B, Thomason MPK, Lentz S, Nolan N (2010) Genomic characterization of the Yersinia genus. Genome Biol 11:1–18

Da Silva Filho AC, Raittz RT, Guizelini D, De Pierri CR, Augusto DW, Dos Santos-Weiss ICR, Marchaukoski JN (2018) Comparative analysis of Genomic Island prediction tools. Front Genet 9:619. https://doi.org/10.3389/fgene.2018.00619

Darmon E, Leach DRF (2014) Bacterial genome instability. Microbiol Mol Biol Rev 78:1–39. https://doi.org/10.1128/MMBR.00035-13

Dengler Haunreiter V, Boumasmoud M, Häffner N, Wipfli D, Leimer N, Rachmühl C, Kühnert D, Achermann Y, Zbinden R, Benussi S (2019) In-host evolution of Staphylococcus epidermidis in a pacemaker-associated endocarditis resulting in increased antibiotic tolerance. Nat Commun 10:1–14

Desvaux M, Dalmasso G, Beyrouthy R, Barnich N, Delmas J, Bonnet R (2020) Pathogenicity factors of genomic islands in intestinal and extraintestinal Escherichia coli. Front Microbiol 11:2065

Didelot X, Méric G, Falush D, Darling AE (2012) Impact of homologous and non-homologous recombination in the genomic evolution of Escherichia coli. BMC Genomics 13:1–15

Dobrindt U, Hochhut B, Hentschel U, Hacker J (2004) Genomic islands in pathogenic and environmental microorganisms. Nat Rev Microbiol 2:414–424

Dokland T (2019) Molecular piracy: redirection of bacteriophage capsid assembly by mobile genetic elements. Viruses 11:1003

Espadinha D, Sobral RG, Mendes CI, Méric G, Sheppard SK, Carriço JA, De Lencastre H, Miragaia M (2019) Distinct phenotypic and genomic signatures underlie contrasting pathogenic potential of Staphylococcus epidermidis clonal lineages. Front Microbiol 10:1971

Freel KC, Krueger MC, Farasin J, Brochier-Armanet C, Barbe V, Andrès J, Cholley P-E, Dillies M-A, Jagla B, Koechler S, Leva Y, Magdelenat G, Plewniak F, Proux C, Coppée J-Y, Bertin PN, Heipieper HJ, Arsène-Ploetze F (2015) Adaptation in toxic environments: arsenic Genomic Islands in the bacterial genus Thiomonas. PLoS One 10:e0139011. https://doi.org/10.1371/journal.pone.0139011

Frye J, Karlinsey JE, Felise HR, Marzolf B, Dowidar N, Mcclelland M, Hughes KT (2006) Identification of new flagellar genes of Salmonella enterica serovar Typhimurium. J Bacteriol 188:2233–2243

Gaillard M, Vallaeys T, Vorhölter FJ, Minoia M, Werlen C, Sentchilo V, Pühler A, Van Der Meer JR (2006) The clc element of Pseudomonas sp. strain B13, a genomic island with various catabolic properties. J Bacteriol 188:1999–2013. https://doi.org/10.1128/JB.188.5.1999-2013.2006

Gal-Mor O, Finlay BB (2006) Pathogenicity islands: a molecular toolbox for bacterial virulence. Cell Microbiol 8:1707–1719. https://doi.org/10.1111/j.1462-5822.2006.00794.x

Giovannoni SJ, Tripp HJ, Givan S, Podar M, Vergin KL, Baptista D, Bibbs L, Eads J, Richardson TH, Noordewier M (2005) Genome streamlining in a cosmopolitan oceanic bacterium. Science 309:1242–1245

Guérillot R, Da Cunha V, Sauvage E, Bouchier C, Glaser P (2013) Modular evolution of Tn GBS s, a new family of integrative and conjugative elements associating insertion sequence transposition, plasmid replication, and conjugation for their spreading. J Bacteriol 195:1979–1990

Guglielmini J, Quintais L, Garcillán-Barcia MP, De La Cruz F, Rocha EP (2011) The repertoire of ICE in prokaryotes underscores the unity, diversity, and ubiquity of conjugation. PLoS Genet 7: e1002222

Guo FB, Wei W, Wang XL, Lin H, Ding H, Huang J, Rao N (2012) Co-evolution of genomic islands and their bacterial hosts revealed through phylogenetic analyses of 17 groups of homologous genomic islands. Genet Mol Res 11:3735–3743. https://doi.org/10.4238/2012.October.15.5

Guzman MG, Harris E (2015) Dengue. Lancet 385:453–465. https://doi.org/10.1016/s0140-6736(14)60572-9

Hacker J, Blum-Oehler G, Mühldorfer I, Tschäpe H (1997) Pathogenicity islands of virulent bacteria: structure, function and impact on microbial evolution. Mol Microbiol 23:1089–1097

Hacker J, Carniel E (2001) Ecological fitness, genomic islands and bacterial pathogenicity. A Darwinian view of the evolution of microbes. EMBO Rep 2:376–381. https://doi.org/10.1093/embo-reports/kve097

Hadjilouka A, Paramithiotis S, Drosinos EH (2018) Genetic analysis of the Listeria Pathogenicity Island 1 of Listeria monocytogenes 1/2a and 4b isolates. Curr Microbiol 75:857–865

Hall RM (2010) Salmonella genomic islands and antibiotic resistance in Salmonella enterica. Future Microbiol 5:1525–1538. https://doi.org/10.2217/fmb.10.122

Hargreaves KR, Thanki AM, Jose BR, Oggioni MR, Clokie MR (2016) Use of single molecule sequencing for comparative genomics of an environmental and a clinical isolate of Clostridium difficile ribotype 078. BMC Genomics 17:1–12

Hickey W, Chen S, Zhao J (2012) The phn island: a new Genomic Island encoding catabolism of Polynuclear aromatic hydrocarbons. Front Microbiol 3:125. https://doi.org/10.3389/fmicb.2012.00125

Jin HM, Jeong H, Moon EJ, Math RK, Lee K, Kim HJ, Jeon CO, Oh TK, Kim JF (2011) Complete genome sequence of the polycyclic aromatic hydrocarbon-degrading bacterium Alteromonas sp. strain SN2. J Bacteriol 193:4292–4293. https://doi.org/10.1128/jb.05252-11

Johnson CM, Grossman AD (2015) Integrative and conjugative elements (ICEs): what they do and how they work. Annu Rev Genet 49:577–601. https://doi.org/10.1146/annurev-genet-112414-055018

Juhas M (2019) Genomic Islands and the evolution of multidrug-resistant bacteria. In: Villa TG, Viñas M (eds) Horizontal gene transfer breaking borders between living kingdoms. Springer, Cham, pp 143–153. https://doi.org/10.1007/978-3-030-21862-1_4

Juhas M, Crook DW, Dimopoulou ID, Lunter G, Harding RM, Ferguson DJP, Hood DW (2007) Novel type IV secretion system involved in propagation of genomic islands. J Bacteriol 189: 761–771. https://doi.org/10.1128/JB.01327-06

Juhas M, Van Der Meer JR, Gaillard M, Harding RM, Hood DW, Crook DW (2009) Genomic islands: tools of bacterial horizontal gene transfer and evolution. FEMS Microbiol Rev 33:376–393. https://doi.org/10.1111/j.1574-6976.2008.00136.x

Kaneko T, Nakamura Y, Sato S, Minamisawa K, Uchiumi T, Sasamoto S, Watanabe A, Idesawa K, Iriguchi M, Kawashima K (2002) Complete genomic sequence of nitrogen-fixing symbiotic bacterium Bradyrhizobium japonicum USDA110. DNA Res 9:189–197

Klockgether J, Reva O, Tümmler B (2006) Spread of genomic islands between clinical and environmental strains. Paper presented at the Prokaryotic Diversity: Mechanisms and Significance: Published for the Society for General Microbiology, held at The University of Warwick April 2006, 01/01

Koskiniemi S, Sun S, Berg OG, Andersson DI (2012) Selection-driven gene loss in bacteria. PLoS Genet 8:e1002787

Lacour S, Gaillard M, Meer J (2006) Genomic islands and evolution of catabolic pathways. In: Logan NA, Lappin-Scott HM, Oyston PCF (eds) Prokaryotic diversity: mechanisms and significance. Cambridge University Press, Cambridge, pp 255–273. https://doi.org/10.1017/CBO9780511754913.016

Langille MG, Hsiao WW, Brinkman FS (2010) Detecting genomic islands using bioinformatics approaches. Nat Rev Microbiol 8:373–382

Lawrence JG (2001) Catalyzing bacterial speciation: correlating lateral transfer with genetic headroom. Syst Biol 50:479–496. https://doi.org/10.1080/10635150120286

Lawrence JG, Roth JR (1996) Selfish operons: horizontal transfer may drive the evolution of gene clusters. Genetics 143:1843–1860

Leal NC, Campos TL, Rezende AM, Docena C, Mendes-Marques CL, De Sá Cavalcanti FL, Wallau GL, Rocha IV, Cavalcanti CL, Veras DL (2020) Comparative genomics of Acinetobacter baumannii clinical strains from Brazil reveals polyclonal dissemination and selective exchange of mobile genetic elements associated with resistance genes. Front Microbiol 11:1176

Lerminiaux NA, Mackenzie KD, Cameron AD (2020) Salmonella pathogenicity island 1 (SPI-1): the evolution and stabilization of a core genomic type three secretion system. Microorganisms 8: 576

Letzel A-C, Li J, Amos GCA, Millán-Aguiñaga N, Ginigini J, Abdelmohsen UR, Gaudêncio SP, Ziemert N, Moore BS, Jensen PR (2017) Genomic insights into specialized metabolism in the marine actinomycete Salinispora. Environ Microbiol 19:3660–3673. https://doi.org/10.1111/1462-2920.13867

Li W, Wang A (2021) Genomic islands mediate environmental adaptation and the spread of antibiotic resistance in multiresistant Enterococci - evidence from genomic sequences. BMC Microbiol 21:55. https://doi.org/10.1186/s12866-021-02114-4

Lima WC, Paquola AC, Varani AM, Van Sluys M-A, Menck CF (2008) Laterally transferred genomic islands in Xanthomonadales related to pathogenicity and primary metabolism. FEMS Microbiol Lett 281:87–97

Lindsay JA, Ruzin A, Ross HF, Kurepina N, Novick RP (1998) The gene for toxic shock toxin is carried by a family of mobile pathogenicity islands in Staphylococcus aureus. Mol Microbiol 29:527–543

Liu F, Zhu Y, Yi Y, Lu N, Zhu B, Hu Y (2014) Comparative genomic analysis of Acinetobacter baumannii clinical isolates reveals extensive genomic variation and diverse antibiotic resistance determinants. BMC Genomics 15:1–14

Lou L, Zhang P, Piao R, Wang Y (2019) Salmonella pathogenicity island 1 (SPI-1) and its complex regulatory network. Front Cell Infect Microbiol 9:270. https://doi.org/10.3389/fcimb.2019.00270

Lovell HC, Mansfield JW, Godfrey SA, Jackson RW, Hancock JT, Arnold DL (2009) Bacterial evolution by genomic island transfer occurs via DNA transformation in planta. Curr Biol 19:1586–1590. https://doi.org/10.1016/j.cub.2009.08.018

Lu B, Leong HW (2016) Computational methods for predicting genomic islands in microbial genomes. Comput Struct Biotechnol J 14:200–206

Makarova KS, Wolf YI, Snir S, Koonin EV (2011) Defense islands in bacterial and archaeal genomes and prediction of novel defense systems. J Bacteriol 193:6039–6056

Martínez JL (2013) Bacterial pathogens: from natural ecosystems to human hosts. Environ Microbiol 15:325–333. https://doi.org/10.1111/j.1462-2920.2012.02837.x

Mates AK, Sayed AK, Foster JW (2007) Products of the Escherichia coli acid fitness island attenuate metabolite stress at extremely low pH and mediate a cell density-dependent acid resistance. J Bacteriol 189:2759–2768

Mathee K, Narasimhan G, Valdes C, Qiu X, Matewish JM, Koehrsen M, Rokas A, Yandava CN, Engels R, Zeng E (2008) Dynamics of Pseudomonas aeruginosa genome evolution. Proc Natl Acad Sci U S A 105:3100–3105

Melnyk RA, Engelbrektson A, Clark IC, Carlson HK, Byrne-Bailey K, Coates JD (2011) Identification of a perchlorate reduction genomic island with novel regulatory and metabolic genes. Appl Environ Microbiol 77:7401–7404. https://doi.org/10.1128/AEM.05758-11

Melnyk RA, Hossain SS, Haney CH (2019) Convergent gain and loss of genomic islands drive lifestyle changes in plant-associated Pseudomonas. ISME J 13:1575–1588. https://doi.org/10.1038/s41396-019-0372-5

Milkman R (2004) Horizontal transfer, genomic diversity, and genomic differentiation. In: Microbial evolution. Wiley, Hoboken, pp 295–318

Mingoia M, Morici E, Marini E, Brenciani A, Giovanetti E, Varaldo PE (2016) Macrolide resistance gene erm (TR) and erm (TR)-carrying genetic elements in Streptococcus agalactiae: characterization of ICE Sag TR7, a new composite element containing IME Sp 2907. J Antimicrob Chemother 71:593–600

Morris JJ, Lenski RE, Zinser ER (2012) The Black Queen Hypothesis: evolution of dependencies through adaptive gene loss. MBio 3:e00036–e00012

Neilands JB (1995) Siderophores: structure and function of microbial iron transport compounds. J Biol Chem 270:26723–26726. https://doi.org/10.1074/jbc.270.45.26723

Nishi A, Tominaga K, Furukawa K (2000) A 90-kilobase conjugative chromosomal element coding for biphenyl and salicylate catabolism in Pseudomonas putida KF715. J Bacteriol 182:1949–1955

Novick RP, Christie GE, Penadés JR (2010) The phage-related chromosomal islands of Gram-positive bacteria. Nat Rev Microbiol 8:541–551

Novick RP, Ram G (2016) The floating (pathogenicity) island: a genomic dessert. Trends Genet 32:114–126. https://doi.org/10.1016/j.tig.2015.11.005

Ochman H, Lawrence JG, Groisman EA (2000) Lateral gene transfer and the nature of bacterial innovation. Nature 405:299–304. https://doi.org/10.1038/35012500

Ogier J-C, Calteau A, Forst S, Goodrich-Blair H, Roche D, Rouy Z, Suen G, Zumbihl R, Givaudan A, Tailliez P (2010) Units of plasticity in bacterial genomes: new insight from the comparative genomics of two bacteria interacting with invertebrates, Photorhabdus and Xenorhabdus. BMC Genomics 11:1–21

Osborn AM, Böltner D (2002) When phage, plasmids, and transposons collide: genomic islands, and conjugative- and mobilizable-transposons as a mosaic continuum. Plasmid 48(3):202–212. https://doi.org/10.1016/s0147-619x(02)00117-8

Otto M (2009) Staphylococcus epidermidis—the 'accidental' pathogen. Nat Rev Microbiol 7:555–567

Patel S (2016) Drivers of bacterial genomes plasticity and roles they play in pathogen virulence, persistence and drug resistance. Infect Genet Evol 45:151–164

Pavlovic G, Burrus V, Gintz B, Decaris B, Guedon G (2004) Evolution of genomic islands by deletion and tandem accretion by site-specific recombination: ICESt1-related elements from Streptococcus thermophilus. Microbiology 150:759–774

Penn K, Jenkins C, Nett M, Udwary DW, Gontang EA, Mcglinchey RP, Foster B, Lapidus A, Podell S, Allen EE, Moore BS, Jensen PR (2009) Genomic islands link secondary metabolism to functional adaptation in marine Actinobacteria. ISME J 3:1193–1203. https://doi.org/10.1038/ismej.2009.58

Peters JE, Fricker AD, Kapili BJ, Petassi MT (2014) Heteromeric transposase elements: generators of genomic islands across diverse bacteria. Mol Microbiol 93:1084–1092

Puymège A, Bertin S, Chuzeville S, Guédon G, Payot S (2013) Conjugative transfer and cis-mobilization of a genomic island by an integrative and conjugative element of Streptococcus agalactiae. J Bacteriol 195:1142–1151

Rainey FA, Oren A (2011) 1 - Taxonomy of prokaryotes – introduction. In: Rainey F, Oren A (eds) Methods in microbiology, vol 38. Academic Press, Cambridge, pp 1–5. https://doi.org/10.1016/B978-0-12-387730-7.00001-2

Rivard N, Colwell RR, Burrus V (2020) Antibiotic resistance in vibrio cholerae: mechanistic insights from IncC plasmid-mediated dissemination of a novel family of genomic islands inserted at trmE. mSphere 5:e00748–e00720

Rodriguez-Valera F, Martin-Cuadrado AB, López-Pérez M (2016) Flexible genomic islands as drivers of genome evolution. Curr Opin Microbiol 31:154–160. https://doi.org/10.1016/j.mib.2016.03.014

Roy Chowdhury P, Scott M, Worden P, Huntington P, Hudson B, Karagiannis T, Charles IG, Djordjevic SP (2016) Genomic islands 1 and 2 play key roles in the evolution of extensively drug-resistant ST235 isolates of Pseudomonas aeruginosa. Open Biol 6:150175

Sansevere EA, Luo X, Park JY, Yoon S, Seo KS, Robinson DA (2017) Transposase-mediated excision, conjugative transfer, and diversity of ICE 6013 elements in Staphylococcus aureus. J Bacteriol 199:e00629–e00616

Scherer CA, Miller SI (2001) Chapter 7 - Molecular pathogenesis of salmonellae. In: Groisman EA (ed) Principles of bacterial pathogenesis. Academic Press, San Diego, pp 265–333. https://doi.org/10.1016/B978-012304220-0/50008-X

Schmidt H, Hensel M (2004) Pathogenicity islands in bacterial pathogenesis. Clin Microbiol Rev 17:14–56

Shankar N, Baghdayan AS, Gilmore MS (2002) Modulation of virulence within a pathogenicity island in vancomycin-resistant Enterococcus faecalis. Nature 417:746–750. https://doi.org/10.1038/nature00802

Shapiro BJ (2016) How clonal are bacteria over time? Curr Opin Microbiol 31:116–123. https://doi.org/10.1016/j.mib.2016.03.013

Shea JE, Hensel M, Gleeson C, Holden DW (1996) Identification of a virulence locus encoding a second type III secretion system in Salmonella typhimurium. Proc Natl Acad Sci U S A 93:2593–2597. https://doi.org/10.1073/pnas.93.6.2593

Siguier P, Gourbeyre E, Chandler M (2014) Bacterial insertion sequences: their genomic impact and diversity. FEMS Microbiol Rev 38:865–891

Smillie C, Garcillán-Barcia MP, Francia MV, Rocha EP, De La Cruz F (2010) Mobility of plasmids. Microbiol Mol Biol Rev 74:434–452

Soucy SM, Huang J, Gogarten JP (2015) Horizontal gene transfer: building the web of life. Nat Rev Genet 16:472–482

Sullivan JT, Brown SD, Ronson CW (2013) The NifA-RpoN regulon of Mesorhizobium loti strain R7A and its symbiotic activation by a novel LacI/GalR-family regulator. PLoS One 8:e53762

Te Poele EM, Bolhuis H, Dijkhuizen L (2008) Actinomycete integrative and conjugative elements. Antonie Van Leeuwenhoek 94:127–143

Ternan NG, Jain S, Srivastava M, Mcmullan G (2012) Comparative transcriptional analysis of clinically relevant heat stress response in Clostridium difficile strain 630. PLoS One 7:e42410

Tettelin H, Masignani V, Cieslewicz MJ, Donati C, Medini D, Ward NL, Angiuoli SV, Crabtree J, Jones AL, Durkin AS (2005) Genome analysis of multiple pathogenic isolates of Streptococcus agalactiae: implications for the microbial "pan-genome". Proc Natl Acad Sci U S A 102:13950–13955

Thode SK, Rojek E, Kozlowski M, Ahmad R, Haugen P (2018) Distribution of siderophore gene systems on a Vibrionaceae phylogeny: database searches, phylogenetic analyses and evolutionary perspectives. PLoS One 13:e0191860. https://doi.org/10.1371/journal.pone.0191860

Toussaint A, Merlin C, Monchy S, Benotmane MA, Leplae R, Mergeay M, Springael D (2003) The biphenyl- and 4-chlorobiphenyl-catabolic transposon Tn4371, a member of a new family of genomic islands related to IncP and Ti plasmids. Appl Environ Microbiol 69:4837–4845. https://doi.org/10.1128/AEM.69.8.4837-4845.2003

Van Der Meer JR, Sentchilo V (2003) Genomic islands and the evolution of catabolic pathways in bacteria. Curr Opin Biotechnol 14:248–254. https://doi.org/10.1016/s0958-1669(03)00058-2

Van Elsas JD, Semenov AV, Costa R, Trevors JT (2011) Survival of Escherichia coli in the environment: fundamental and public health aspects. ISME J 5:173–183

Van Houdt R, Monsieurs P, Mijnendonckx K, Provoost A, Janssen A, Mergeay M, Leys N (2012) Variation in genomic islands contribute to genome plasticity in Cupriavidus metallidurans. BMC Genomics 13:111. https://doi.org/10.1186/1471-2164-13-111

Vásquez-Ponce F, Higuera-Llantén S, Pavlov MS, Marshall SH, Olivares-Pacheco J (2018) Phylogenetic MLSA and phenotypic analysis identification of three probable novel Pseudomonas species isolated on King George Island, South Shetland, Antarctica. Braz J Microbiol 49:695–702. https://doi.org/10.1016/j.bjm.2018.02.005

Vo ATT, Van Duijkeren E, Gaastra W, Fluit AC (2010) Antimicrobial resistance, class 1 integrons, and genomic island 1 in salmonella isolates from Vietnam. PLoS One 5:e9440. https://doi.org/10.1371/journal.pone.0009440

Wang H, Smith MC, Mullany P (2006) The conjugative transposon Tn 5397 has a strong preference for integration into its Clostridium difficile target site. J Bacteriol 188:4871–4878

Welch RA, Burland V, Plunkett G, Redford P, Roesch P, Rasko D, Buckles EL, Liou S-R, Boutin A, Hackett J, Stroud D, Mayhew GF, Rose DJ, Zhou S, Schwartz DC, Perna NT, Mobley HLT, Donnenberg MS, Blattner FR (2002) Extensive mosaic structure revealed by the complete genome sequence of uropathogenic *Escherichia coli*. Proc Natl Acad Sci 99:17020–17024. https://doi.org/10.1073/pnas.252529799

Williams KP (2002) Integration sites for genetic elements in prokaryotic tRNA and tmRNA genes: sublocation preference of integrase subfamilies. Nucleic Acids Res 30:866–875

Wyrsch ER, Chowdhury PR, Jarocki VM, Brandis KJ, Djordjevic SP (2020) Duplication and diversification of a unique chromosomal virulence island hosting the subtilase cytotoxin in Escherichia coli ST58. Microb Genom 6:e000387

Xiang L, Li G, Wen L, Su C, Liu Y, Tang H, Dai J (2021) Biodegradation of aromatic pollutants meets synthetic biology. Synth Syst Biotechnol 6:153–162

Genomic Islands in the Gut Microbiome: Current Knowledge and the Application in the Probiotics Field

6

Duy-Ha Nguyen, Nguyen Thai Son, and Dinh Toi Chu

Abstract

Intestinal microorganisms play a significant role in human health, they are considered part of the human being. The number of functional genes in intestinal microorganisms far exceeds those in human functional genes and contributes to many metabolic processes related to the state of health and pathology of the body. The number and degree of diversity of intestinal microorganisms varies with a number of factors such as age, psychology, eating habits, drug consumption, place of residence in the digestive tract, and medical condition of the body. All bacteria, viruses, their genomes and environmental factors found in the human colon make up the intestinal microbiome. Microorganisms are closely connected to the hosts and are distinguished by genomic islands (GIs), which play an important role in microbe evolution and genetics. At this time, with the development of genome sequencing tools, the relationships between bacteria and hosts are increasingly well established and the role of gut microflora is confirmed. In this chapter, we find out about the genomic islands present in the intestinal microflora and focus on pathogenicity islands (PAI), symbiotic islands, antibiotic resistance genetic (ARG), and mobile genomic elements (MGEs). The connection between GIs and human health and its role in the way microorganisms evolve. The role of genomic islands is linked to human health in relation to probiotics, especially probiotics of current interest.

D.-H. Nguyen · N. T. Son
Department of Microbiology, Vietnam Military Medical University, Hanoi, Vietnam

D. T. Chu (✉)
Center for Biomedicine and Community Health, International School, Vietnam National University, Hanoi, Vietnam

Faculty of Applied Sciences, International School, Vietnam National University, Hanoi, Vietnam

I. Mani et al. (eds.), *Microbial Genomic Islands in Adaptation and Pathogenicity*,
https://doi.org/10.1007/978-981-19-9342-8_6

Keywords

Genomic islands · Gut microbiome · Current knowledge · Probiotics field

6.1 Introduction

The human intestinal microbiota is believed to contain more than 10^{14} microorganisms and has a genome that is 100 times larger than the human genome (Gill et al. 2006). The whole length of the digestive tract is covered in bacteria, which play a crucial role in controlling gut physiology (Natividad and Verdu 2013), improves the integrity of the intestinal epithelium, contributes to the digestion of food (den Besten et al. 2013), protects the body from pathogens and strengthens the body's immune system (Bäumler and Sperandio 2016; Gensollen et al. 2016). However, the microbial components in the intestinal tract are unstable, which can be changed by external factors such as individual, family, race, gender, age, eating habits, region, and lifestyle, which affect the health of the host. Mutations, horizontal gene transfer, and gene rearrangement are all examples of how microorganisms evolve and adapt. Secondary genes are unusual portions of the bacterial genome called genotypes that play a crucial role in evolution and adaptation. They are derived by horizontal gene transfer (HGT), and microorganism diversity (Juhas et al. 2009). There are numerous types of GIs based on the biological function of genes on the island. Aside from symbiotic islands and pathogenicity islands, antibiotic-resistant islands contain genes that encode proteins linked to antibiotic resistance (Schmidt and Hensel 2004). GIs has a size of 10–200 kb with a specific content of %GC and a frequency of dinucleotides (Juhas et al. 2009). The genes of GIs are often collected to fulfill specific functions and are beneficial for bacterial growth (Hacker and Carniel 2001). As a result, they provide selective advantages to microorganisms that have genomic islands within the population. Pathogenicity islands, for example, can induce significant changes in bacterial phenotype and are predominantly studied in genetic islands. In the intestinal tract, some pathogenicity islands are associated with the virulence of bacterial pathogens and clinical symptoms of the disease. Both *V. cholerae* and *E. coli* bacteria that produce intestinal toxins stimulate bowel movement and release water into the intestines of infected people result in the spread of bacteria directly through feces. Microbial transmission is often a result of coded pathogen factors such as adhesives and toxins. Antibiotic-resistant gene islands have also been studied extensively in GIs-related issues. The level of antibiotic resistance causes a great deal of difficulty in treating infectious diseases. Increased bacterial pathogenicity and antibiotic resistance factors are linked to GIs, showing their role in microorganism evolution (Juhas et al. 2009). A factor implicated in the genetic genetics of antibiotic resistance is the horizontal transfer of HGT genes, which promotes the development of superbacteria containing antibiotic-resistant genes (Lerminiaux and Cameron 2019). The primary cause of the spread of antibiotic-resistant genes in humans and animals is horizontal gene transfer via plasmids (Liu et al. 2020). Variability, synthesis, and gene transfer are all pathways that lead to the transfer of antibiotic-resistant genes. Some substances facilitate the transfer of ARG genes among certain bacterial strains such as *E.coli*, *S. typhirumium* (Zhang et al. 2017). HGT is mediated by moving genetic factors which are DNA fragments that encode proteins that help with intracellular motion (inside cells) and between cells (cell mobility). The mobile genetic elements is copious and consist of phages, plasmids and gene islands (Flores-Ríos et al. 2019). MGEs are essentially genomic DNA sections that can migrate from one genomic region to another or between genomes. Bacterial genome are made of MGEs as 11% of the *Clostridium difficile* genome consists of mobile genetic element, providing the bacterium with a remarkable genetic character (Sebaihia et al. 2006). MGEs have a component including plasmid, insertion sequence (IS) elements, intergrated elements (IE). It has been demonstrated that GIs can move from one cell to the next step by step (1) achieved through horizontal transfer, (2) corresponds to the suitable site, incorporated into the host chromosomal, (3) generated as a result of gene rearrangement, loss, and acquisition, (4) take out the chromosome (5) send to a different person (Juhas et al. 2009).

6.2 Common and Specific Features of Gut Microbiome

A group of microorganisms that dwell in the human gastrointestinal system is known as gut microflora, in the human intestine, including bacteria, fungus, viruses, their genomes, and environmental circumstances (Marchesi and Ravel 2015). In the digestive tract, microorganisms have a metabolic relationship and interdependency, which are prevalent in the human gut and play a role in the intestinal endothelium balance, constituting the natural environment for most symbiotic bacteria (Guarner 2015). The intestinal microflora and the host have a symbiotic relationship in which the host provides the environment and nutrients for bacteria to survive and grow, while the microflora aids metabolism, digestion, and intestinal immune system strength by providing beneficial nutrients such as vitamins and short-chain fatty acids (SCFAs) (Kau et al. 2011; McDermott and Huffnagle 2014). Firmicutes, Bacteroidetes, Proteobacteria, and Actinobacteria make up the majority of bacteria in healthy humans microbiomes (Wright et al. 2015). The lack of diversity in the intestinal microbiome has been linked to a variety of illnesses, including obesity and inflammatory bowel disease (Turnbaugh et al. 2008; Qin et al. 2010), the great diversity of bacterial vaginosis. Currently, the most commonly used method for determining the classification and assessment of the diversity of species of organisms is gene sequencing, which encodes the small subunit of ARN ribosome (16S rRNA). For phylogenetic investigations of microbial communities and providing classification names to bacteria, 16S rRNAs are considered the gold standard (Huse et al. 2012). Another research strategy is to sequence the entire genome, which will reveal all of the genes present in the sample. Genome sequencing also allows for the research of functional and metabolic networks, as well as the identification of genes from non-bacterial species, such as viruses, yeasts, and protozoa (Li et al. 2014). Differences exist between the microorganisms that live in the intestinal tract and the intestinal mucosa in the same individual. Moreover, the species of bacteria found in the gut vary from the colon to the rectum. For instance, in the small intestine, where there are plenty of nutrients and oxygen, antibacterial peptides with higher pH so Lactobacillaceae and Enterobacteriae dominate. Slow transit periods and the lack of simple sugars in the colon encourage the proliferation of anaerobic bacteria such as Bacteroidaceae, Prevotellaceae, Rikenellaceae, Lachnospiraceae, and Ruminococcaceae, which break down polysaccharides (Donaldson et al. 2016). In areas far away from the colon, species that degrade mucus and proteins are common. At the genus and phylum level, bacteria linked with the mucous membranes of the rectal endile are more stable, but plaques are not homogeneous in the same location of the gut. Most strains of intestinal microorganisms have been resident for decades, though their performance changes over time for a given individual... However, factors such as high-carb diet lead to increase in the ratio of Firmicutes and Bacteroidetes (Maier et al. 2017), lifestyle (smoking, walking, physical activity), intestinal diseases which impact the microbial composition of the host intestine (Gilbert et al. 2018; Allaband et al. 2019). Intestinal microflora also varies with age, growing through childhood to adulthood and decreasing when old. The intestinal microflora of adults has a steady maintenance structure over time. Microflora in infants is less stable, from a sterile condition to a densely populated area of microorganisms and at a stable stage of development in adulthood. However, the composition of microorganisms in newborn also depends heavily on the reproductive form of the child (Palmer et al. 2007). For example, babies frequently receive the same vaginal microbiota as their mothers', primarily species like Lactobacillus, Prevotella, Sneathia spp. The intestinal microbiota of babies born through cesarean section is comparable to that of the mother's skin, with Staphylococcus, Corynebacterium, and Propionibacterium spp. (Dominguez-Bello et al. 2010). These characteristics of nourishment also affect intestinal microorganisms. A study showed that in breastfed babies, the microbiome is quickly dominated by Bifidobacterial bacteria, and a more diverse microbiome than babies fed with formula (Favier et al. 2002). Food-induced alterations in the microbiome have played a part in the varied rates of age-related health deterioration in senior persons, according to a link between indications of weakness, nutrition, and markers of inflammation. Most bacterial groups' species diversity is reduced as people get older, making the microbiome less responsive to external stressors (Salazar et al. 2013). The change of prevailing species in certain groups of bacteria, the decline of useful bacteria, the increase of harmful bacteria, genders, ethnicity, and geographic location have an effect on the taxonomic composition of microbial biomes (Gaulke and Sharpton 2018).

6.3 The Gut Microbiome Flexible Gene Pool

In bacterial genetics, horizontal gene transfer is crucial. They are usually 10–500 kb in size and can be integrated into bacterial chromosomes to provide favorable aspects to bacterial cells. Bacterial growth in the new environment is aided by factors, such as metabolic capacity, antibiotics resistance, and toxic substances (Soucy et al. 2015; Dobrindt et al. 2004). The original formation of genome islands is a horizontal genetic factor that produces bacterial genetic characteristics. Mobile genomic islands are tiny genomic islands (around 20 kb) that can travel from one cell to next due to a common aberrant mechanism (Daccord et al. 2010). Within the bacterial genome, they always have a number of flexible genes. They can move through the bacterial genome or between bacteria through mechanisms like feeding, loading, gene transfer. In bacterial genome analysis, we consider GIs to be part of the flexible genetic capital (Dobrindt et al. 2004), which is associated with bacterial genetics. Genetic islets may be present in pathogenic or non-pathogenic bacteria. These are clusters of genes that cover certain Kbp's that are reflected in GC content, that may differ from other parts of the central genome. GIs usually code for virulence or adaptive traits and are often associated with the tRNA or the integrated gene at one end of the genetic islet (Dobrindt et al. 2004). Most GIs derived from genetic factors have lost the mobile gene through evolution, but some retain the ability to propagate from one cell to another through fusion or loading.

They are called integrative and conjugative elements (ICEs) and prophages respectively. ICEs are elements that, depending on their location, encode their own elimination by recombination. Their transfer through integration and integration, regardless of integration and convergence method. For example, self-conjugated and integrated GIs have been found in *Streptococcus agalactiae*, which does not rely on site-specific recombination but on transposase DDE (Brochet et al. 2009; Guérillot et al. 2013). As a result, the ICEs definition should also include conjugated MGEs that circulate and integrate through a DDE translocation enzyme. ICE is available in sizes varying from 11 kb (pSAM2 of *S. ambofaciens*) to 674 kb (PAISt from *Streptomyces turgidiscabies*) (Pernodet et al. 1984; Huguet-Tapia et al. 2011; Kers et al. 2005). Many genome islands are encircled by repetitive structures and contain remnants of various subcellular and cellular genetic components including phages, plasmids, and sequential insertion elements. These transport genes differ considerably even among ICEs carriers of closely related conjugated and recombinant modules.

Antibiotics, heavy metals, or antimicrobial, sacaroza catabolism, bacterial synthesis, pathogenesis, or symbiosis are examples of functions unrelated to their ability to move in and out of cells that can provide hosts a major selection advantage or even transform their lifestyle (Juhas et al. 2009; Dobrindt et al. 2004; Wozniak and Waldor 2010). The genomes of intestinal bacteria are highly variable, partly because of the transverse transfer of the mobile genomic elements, including plasmid, bacteria, and conjugated translocations (CTns) (Burrus and Waldor 2004; Ochman et al. 2000). Phages and CTns play an important role in functional gene transfer, including antibiotic resistance, excretion systems, and secretions, as well as other pathogenic features among bacterial hosts of the intestinal microbiome (Rodríguez-Blanco et al. 2012).

6.4 Genomic Islands of the Gut Microbiome

6.4.1 Pathogenicity Islands

Pathogenic islands are important factors for bacterial virulence. It is another form of bacterial gene factor, described for the potentially pathogenic strain of *Escherichia coli (*Blum et al. 1994*)*. Various genes play a part in adaptation, intracellular proliferation, bacterial survival and spread, and these genes also determine the degree of bacterial virulent activity that is known as the genes situated in the region called "islands are associated with pathogenesis" in the chromosomes of pathogenic bacteria which code different products and determine the virulence of bacteria. When these gene regions, which are found only in pathogenic species, are transferred to non-pathogenic species through cross-functional gene transfer. They also deliver complex virulence factors to the bacteria to which they are transferred. Pathogen islands are probably mostly carriers of functional genes or encode mobile elements such as integration, translocation, and insertion sequencing factors (Schmidt and Hensel 2004). The structure of the PAIs may contain one or more virulent genes that are present only in bacterial pathogens but in the same closely related bacterial species. The size of the PAIs (10–200 Kb)constitutes the major part of the genome size, which is different from the basic genome in terms of GC composition and G C content. It is present within the genome, which is proof of foreign origin and reception of horizontal genes. PAIs are typically located close to tRNA genes, which are often combined with genetic factors such as phage, plasmid, insertion factor or transposon (Hacker et al. 1997). These cellular components are involved in the recombination of gene segments, which results in rearrangement, insertion, paragraph loss, and, as a result, pai variation. Pathogenic pathways in Salmonella use two T3SS-1 coded by Pathogen Island 1 (SPI-1) and Pathogen Island 2 (SPI-2) (Dobrindt et al. 2010). T3SS-1 is encoded by SPI-1, a cluster gene with a 40-kb domain that controls the expression of numerous pathogenic genes (Lou et al. 2019). The SPI-1 gene's islands play an important function in the entrance of bacteria into the intestine and stimulate neutrophil growth. This process involves transcription kits with the expression of invF and hilA as these are SPI-1 gene transcription activators. PAI-coding genes can be used by *E.coli* bacteria to manufacture a variety of toxins, including α-haemolysin, CNF-1 and deforming toxins and colibactin. In inflammatory bowel disease (IBD), *E.coli* PAI increases adherence to intestinal epithelial cells, crossing the mucosal layer, improvement antibacterial properties, entrance in the epithelium, permeability and enhance the immune system (Palmela et al. 2018). The pathogenic island *Vibrio cholerae* identified as VPI contains the *aldA* and *tagA* genes that are involved in bacterial pathogenic mechanisms. VPI is about 40 kb in size and contains a G + C content of 35%, whereas the average G + C concentration of the *V. cholera* genome is between 47% and 49%. VPI is also inserted next to *ssrA*, a gene similar to tRNA that is surrounded by att locations. The toxin-coding VPI works as a necessary colony factor in contributing to *V. cholera* adhesion in the epithelial area (Palmela et al. 2018; Karaolis et al. 1998). Cytolysin toxin (*Cyl*) and the surficial protein *Esp* are two virulence factors produced by *E. faecalis*. The virulence factors contribute by the genes in PAI that are 154-kb in size and contain *cyl, esp* operon and some unknown functional genes. This zone has a G + C content of 32.2%, which is below the amount of *E. faecalis* 's core genome and implies that the bacteria genome is genetically unstable. Genes within the PAIs have sequential similarities to those of intestinal plasmids, possibly due to the chromosomal integration of a plasmid (Dobrindt et al. 2010).

6.4.2 Symbiosis Islands

The symbiotic genes of bacteria are generally carried on symbiotic islands or plasmids which can be exchanged horizontally between different bacterial species. The symbiotic genes involved in transverse transfer have different phylogenes for their host's main genome. The symbiotic island is larger than any studied island so far, comprising 10% of the host genome and its ability to transmit has been proven (Sullivan and Ronson 1998).

6.4.3 Antibiotic Resistance Genes

Antibiotic resistance is rising as a result of misuse in clinical settings. Many kinds of ultra-resistant bacteria exist nowadays, making the treatment of infectious diseases difficult. Horizontal gene transfer is one of the most well-known resistance mechanisms and the primary cause of the creation of multi-antibiotic-resistant bacteria. Horizontal gene transfer occurs when bacteria with no paternal tie to their offspring exchange genes (Soucy et al. 2015). The main mechanisms associated with HGT are the synthesis process, the power variable and the load. Combining several genetic factors, such as transopon and plasmid, is the most effective (Redondo-Salvo et al. 2020). Sequencing the whole genome of six drug-resistant diseases in the intestine has resulted in numerous genetic factors and horizontal gene transfers (Kumar et al. 2017). In particular, new mechanisms of antibiotic resistance are constantly emerging that some species of intestinal microflora, such as Campylobacter, appear to be multi-resistant genetic islets. Recently it has been discovered to contain a number of genes associated with resistance to aminoglycosides and macrolides. In *Enterobacter sp.* and *P. aeruginosa*, excessive occurrence of *blaAmpC* is caused by regulatory genetic mutations (Ruppé et al. 2015). In the presence of *ampC* cephalosporinase, this overexpressed gene disrupts the enzyme-muscle ratio, thus causing resistance to penicillin and broad-spectrum cephalosporin. Other methods of resistance to penicilin in Pneumococci include genetic alterations that bind to the protein 2b (pbp2b) in penicilin. Mutations in the *rpsL* gene, which codes for the ribosomal protein, cause structural alterations. Changes in the target drug binding, resulting in resistance to various aminoglycoside antibiotics. The transfer of medication-resistant genesis is mediated by bacteria to new bacterial species such as those of methicillin-resistant *Staphylococcus aureus* (MRSA)

(Hasan et al. 2021) that have acquired the drug-resistant *mecA* gene from other bacterial species by transferring the feed (Hasan et al. 2021). The natural transformation occurs when antibiotic-resistant genes from the DNA of dead bacteria can be acquired and embedded into the chromosomes of other bacteria through homogeneous recombination. Natural transformation has occurred in numerous clinically pathogenic bacterial species such as *Streptococcus pneumoniae* that have acquired penicillin-binding proteins (PBP2Bs) from the bacterium *Streptococcus mitis*; N*eisseria gonorrhoeae* acquire the ceftriaxone-resistant *penA* gene by natural transformation (Unemo et al. 2012). In addition, pharmacoresistant genes are also inherited by MGE such as different types of plasmids and transposons (Frost et al. 2005; Alekshun and Levy 2007). Most drug-resistant genes for gram-negative bacteria which cause clinical transfer of gene diseases by plasmid. For example, carbapenemase enzymes of enterobacteria and other gram-negative bacteria are resistant to antibiotic-resistant genes. It contains plasmid which can be rapidly spread to other sensitive bacteria by conjugation (Paterson and Bonomo 2005). Human health is jeopardized by drug resistance to conjugated plasmidic-mediated medicines. Because the enzyme carbapenemases—the final antibacterial weapon against Gram-negative bacteria. ...as well as other beta-lactams have been found on a variety of plasmids (Kumarasamy et al. 2010; Yong et al. 2009). Genetic transmission by integrated chromosome factors is also achieved by mediating the synthesis process. This mechanism of drug-resistant spread occurs mostly in Gram-negative bacteria and also in some gram-positive bacterial species such as *Streptococci spp*. (San 2018). Many drug resistances have been spread by plasmids such as the conjugate plasmid carrying the SGI1 gene of *Salmonella spp*. and carbapenem resistance in *Acinetobacter baumannii* that is associated with a pABTJ1 conjugated plasmid (Vo et al. 2010; Huang et al. 2012). During the evolution, the conjugate was metabolized between *Klebsiella pneumonia* and *E. coli* in terms of drug-resistant carbapenems in hospitalized patients (Mulvey et al. 2011). Plasmid clearly leads to antibiotic resistance in both the environment and clinical illnesses.

6.4.4 Chromosomal Mobile Genetic Elements

Many creatures, including plants, animals, and people, have evolved and genetically varied as a result of GMEs (Bock 2010; Iyer et al. 2004). The human genome contains approximately 200 genes that have similarities to those that code bacterial proteins, possibly due to the horizontal transfer of bacteria during human evolution. In the intestinal bacterial genome, genes from starchy fecal bacteria were found to be associated with lateral transgenic with intestinal microflora (Arias et al. 2012). Some intestinal bacteria have gained genes important for bacterial evolution and development through conjugation, such as bacteroids that have acquired new beneficial genes from microorganisms in the environment (Hehemann et al. 2010). Bacteroids are the most common bacteria in the human intestinal microbiome, and they use the outer membrane protein complex to harvest huge amounts of food and host glycoprotein. MGEs is composed of members such as phage, plasmid, transposon, and insertion sequence (Leplae et al. 2010). They play an important role in mobilizing and reorganizing genes, whether in a genome or between bacterial cells. Plasmid's function in gut bacteria includes cell membrane biosynthesis, homeostasis, nutrient absorption, and bacteriocin production (Claesson et al. 2006; Jones 2010). Plasmids have been found to contain functions that help bacteria to thrive, such as epithelial cell adhesion, virulence factors, and antibiotic resistance. In rats with *Salmonella spp.*-induced inflammatory bowel illness, the presence of *E.coli* in the intestinal tract increased, resulting in the transfer of the colicinplasmid gene p2 from *Salmonella spp.* to *E. coli (*Stecher et al. 2012*)*.. Plasmids play a key role in the spread of antibiotic-resistant genes among pathogens, resulting in the establishment of multidrug-resistant bacteria strains. Transposons are often associated with clinical resistance like in enterobacteria, the gene *blaNDM-1* in pathogens is a synthetic transposon inside and between bacterial species (Nordmann et al. 2012). Antibiotic resistance genes in *E. coli* are linked to genetic elements such transposons and gene bands (Poirel et al. 2018). In *Clostridium difficile* bacteria, genetic factors constitute the virulence in their genomes (Sebaihia et al. 2006). The main virulence factors of bacteria transferred to genetic islets through integrated CT. Therefore, MGEs have a profound effect on the pathogenic ability of bacteria. MGEs also have an impact on bacterial biology, such as the dissemination of antibiotic resistance and other phenotypes that help bacteria thrive in harsh settings.

6.5 The Application in the Probiotics Field

Probiotics are described as "a living microorganism that confers a benefit to the host when provided in appropriate amounts." (Probiotics in food: health and nutritional properties and guidelines for evaluation: report of a joint FAO/WHO expert consultation on evaluation of health and nutritional properties of probiotics in food including powder milk with live lactic acid bacteria, Cordoba, Argentina, 1—4 October 2001 [and] Report of a joint FAO/WHO working group on drafting guidelines for the evaluation of probiotics in food, London, Ontario, Canada, 30 April–1 May 2002. Rome [Italy]: Food and Agriculture Organization of the United Nations, World Health Organization 2006) *Lactobacilli* and *Bifidobacteria* are the most common bacteria employed. The role of probiotics is demonstrated by interacting with intestinal microflora, improving beneficial microflora and limiting harmful microflora, and regulating the body's immune system. The activation of the *pepQ* gene by *CcpA* in the investigation of genomic islands of lactic acid bacteria involved in carbon and nitrogen metabolism (Zomer et al. 2007). *GlnR* and *CodY* are two nitrogen control genes found in lactic acid bacteria. *GlnR* and *CodY* are two nitrogen-control genes found in lactic acid bacteria (Kormelink et al. 2012). *Lactococcus lactis* MG 136 has CodY control over more than 30 genes involved in amino acid metabolism (Guédon et al. 2005). Furthermore, the *Lactococcus lactis* IL1403 bacterium strain possesses pili-producing genes that may be expressed and engaged in biofilm formation (Oxaran et al. 2012). The presence of pili-producing genes confers the bacterium a competitive edge in different environments. Furthermore, certain beneficial bacteria used as probiotics have the ability to inhibit intestinal pathogens like *Bifidobacterium thermophilum* RBL67, which regulates the expression of genes linked to SPI-1 and SPI-2 resulting in excessive energy expenditure and protective activity against *Salmonella* infection (Tanner et al. 2016). In lactic bacteria, CTns are known to be resistant to tetracycline, vancomycin, and erythromycin, which increase the tolerance of bacteria in the intestine to undesired factors followed by the potential use of laboratory bacteria for probiotic production (Broaders et al. 2013). Genome analysis between two *L. rhamnosus* strains that revealed the presence of genomic islands can transport and metabolize protease-dependent sugars encoded by *spaCBA (*Kankainen et al. 2009*)*. The spaC gene has been found to promote *L. rhamnosus* adhesion to human intestinal mucus. It also demonstrates the role of the *spaC* gene in bacterial survival in the digestive tube. Analyzing the mutation of the *Tad* gene group in *Bifidobacterium breve* UCC200 is an important factor for bacterial invasion and uptake in the intestinal parenchyma of mice (O'Connell Motherway et al. 2011).

This research reveals that a variety of cell surface components have a role in probiotic attachment to the human gut epithelium. The *glgBCDAP - amy - pgm* gene, found on an operon in *Lactobacillus acidophilus*, contributes to glycogen metabolism and is involved in the bacteria's energy production, carbohydrate, and amino acid metabolism (Eydallin et al. 2010). Probiotic activity and *L. acidophilus* retention in the human intestinal environment are aided by glycogen metabolism (Goh and Klaenhammer 2013).

6.6 Conclusion

The intestinal microflora is a part of the human body that participates in the metabolism of the body and is implicated in the appearance of various pathologies. Discovering and analyzing the genetic information of the intestinal microbiota is as important as analyzing the human genome. Genetic information on the intestinal microbiota, genes codifying the main microbial functions, antibiotic resistance mechanisms and bacterial virulence genes. In particular, islets of pathogenic genes and gene transfer mechanisms of gut bacteria and bacterial genetic factors shared by all humans (base microbiota). This gives insight into the biological roles of gut microorganisms. The human body's physiology, as well as the genetic information from the patient's gut microbiota, can be used to diagnose and cure disease. The transfer of antibiotic genes between bacterial species has become increasingly resistant to antibiotics. The gut environment is conducive to genetic transfer through plasmids and transmissible factors. The main cause of several ailments in the human body is an imbalance in the composition of the gut microflora. Probiotics have been demonstrated to address abnormalities in the intestinal microbiota composition by increasing the quantity of beneficial bacteria, enhancing intestinal epithelial function, and boosting the host's immune system. Certain species with probiotic potential have genetic expressions that increase the competitiveness and tolerance of bacteria in the gastro-intestinal environment.

Acknowledgments We would like to thank Hue Vu Thi (Center for Biomedicine and Community Health, International School, Vietnam National University, Hanoi, Vietnam) for critical reading and checking to improve the manuscript.

List of Abbreviations

GIs	Genomic Islands
PAI	Pathogenicity Islands.
ARG	Antibiotic Resistance Genetic.
MGEs	Mobile Genomic Elements.
HGT	Horizontal Gene Transfer.
ICEs	Integrative and Conjugative Elements.
IS	Insertion Sequence.
IE	Integrated Elements.

References

Gill SR, Pop M, Deboy RT et al (2006) Metagenomic analysis of the human distal gut microbiome. Science 312(5778):1355–1359

Natividad JM, Verdu EF (2013) Modulation of intestinal barrier by intestinal microbiota: pathological and therapeutic implications. Pharmacol Res 69(1):42–51

den Besten G, van Eunen K, Groen AK, Venema K, Reijngoud DJ, Bakker BM (2013) The role of short-chain fatty acids in the interplay between diet, gut microbiota, and host energy metabolism. J Lipid Res 54(9):2325–2340

Bäumler AJ, Sperandio V (2016) Interactions between the microbiota and pathogenic bacteria in the gut. Nature 535(7610):85–93

Gensollen T, Iyer SS, Kasper DL, Blumberg RS (2016) How colonization by microbiota in early life shapes the immune system. Science 352(6285):539–544

Juhas M, van der Meer JR, Gaillard M, Harding RM, Hood DW, Crook DW (2009) Genomic islands: tools of bacterial horizontal gene transfer and evolution. FEMS Microbiol Rev 33(2): 376–393

Schmidt H, Hensel M (2004) Pathogenicity islands in bacterial pathogenesis. Clin Microbiol Rev 17(1):14–56

Hacker J, Carniel E (2001) Ecological fitness, genomic islands and bacterial pathogenicity. A Darwinian view of the evolution of microbes. EMBO Rep 2(5):376–381

Lerminiaux NA, Cameron ADS (2019) Horizontal transfer of antibiotic resistance genes in clinical environments. Can J Microbiol 65(1):34–44

Liu Y, Tong Z, Shi J, Jia Y, Yang K, Wang Z (2020) Correlation between exogenous compounds and the horizontal transfer of plasmid-borne antibiotic resistance genes. Microorganisms 8(8): 1211

Zhang Y, Gu AZ, He M, Li D, Chen J (2017) Subinhibitory concentrations of disinfectants promote the horizontal transfer of multidrug resistance genes within and across genera. Environ Sci Technol 51(1):570–580

Flores-Ríos R, Quatrini R, Loyola A (2019) Endogenous and foreign nucleoid-associated proteins of bacteria: occurrence, interactions and effects on mobile genetic elements and Host's biology. Comput Struct Biotechnol J 17:746–756

Sebaihia M, Wren BW, Mullany P et al (2006) The multidrug-resistant human pathogen Clostridium difficile has a highly mobile, mosaic genome. Nat Genet 38(7):779–786

Marchesi JR, Ravel J (2015) The vocabulary of microbiome research: a proposal. Microbiome 3(1): 31

Guarner F (2015) The gut microbiome: what do we know? Clin Liver Dis (Hoboken) 5(4):86–90

Kau AL, Ahern PP, Griffin NW, Goodman AL, Gordon JI (2011) Human nutrition, the gut microbiome and the immune system. Nature 474(7351):327–336

McDermott AJ, Huffnagle GB (2014) The microbiome and regulation of mucosal immunity. Immunology 142(1):24–31

Wright EK, Kamm MA, Teo SM, Inouye M, Wagner J, Kirkwood CD (2015) Recent advances in characterizing the gastrointestinal microbiome in Crohn's disease: a systematic review. Inflamm Bowel Dis 21(6):1219–1228

Turnbaugh P, Hamady M, Yatsunenko T et al (2008) A core gut microbiome in obese and lean twins. Nature 457:480–484

Qin J, Li R, Raes J et al (2010) A human gut microbial gene catalogue established by metagenomic sequencing. Nature 464(7285):59–65

Huse SM, Ye Y, Zhou Y, Fodor AA (2012) A core human microbiome as viewed through 16S rRNA sequence clusters. PLoS One 7(6):e34242

Li J, Jia H, Cai X et al (2014) An integrated catalog of reference genes in the human gut microbiome. Nat Biotechnol 32(8):834–841

Donaldson GP, Lee SM, Mazmanian SK (2016) Gut biogeography of the bacterial microbiota. Nat Rev Microbiol 14(1):20–32

Maier TV, Lucio M, Lee LH et al (2017) Impact of dietary resistant starch on the human gut microbiome, metaproteome, and metabolome. mBio 8(5):e01343–e01317

Gilbert JA, Blaser MJ, Caporaso JG, Jansson JK, Lynch SV, Knight R (2018) Current understanding of the human microbiome. Nat Med 24(4):392–400

Allaband C, McDonald D, Vázquez-Baeza Y et al (2019) Microbiome 101: studying, analyzing, and interpreting gut microbiome data for clinicians. Clin Gastroenterol Hepatol 17(2):218–230

Palmer C, Bik EM, DiGiulio DB, Relman DA, Brown PO (2007) Development of the human infant intestinal microbiota. PLoS Biol 5(7):e177

Dominguez-Bello MG, Costello EK, Contreras M et al (2010) Delivery mode shapes the acquisition and structure of the initial microbiota across multiple body habitats in newborns. Proc Natl Acad Sci U S A 107(26):11971–11975

Favier CF, Vaughan EE, De Vos WM, Akkermans AD (2002) Molecular monitoring of succession of bacterial communities in human neonates. Appl Environ Microbiol 68(1):219–226

Salazar N, López P, Valdés L et al (2013) Microbial targets for the development of functional foods accordingly with nutritional and immune parameters altered in the elderly. J Am Coll Nutr 32(6):399–406

Gaulke CA, Sharpton TJ (2018) The influence of ethnicity and geography on human gut microbiome composition. Nat Med 24(10):1495–1496

Soucy SM, Huang J, Gogarten JP (2015) Horizontal gene transfer: building the web of life. Nat Rev Genet 16(8):472–482

Dobrindt U, Hochhut B, Hentschel U, Hacker J (2004) Genomic islands in pathogenic and environmental microorganisms. Nat Rev Microbiol 2(5):414–424

Daccord A, Ceccarelli D, Burrus V (2010) Integrating conjugative elements of the SXT/R391 family trigger the excision and drive the mobilization of a new class of Vibrio genomic islands. Mol Microbiol 78(3):576–588

Brochet M, Da Cunha V, Couvé E, Rusniok C, Trieu-Cuot P, Glaser P (2009) Atypical association of DDE transposition with conjugation specifies a new family of mobile elements. Mol Microbiol 71(4):948–959

Guérillot R, Da Cunha V, Sauvage E, Bouchier C, Glaser P (2013) Modular evolution of TnGBSs, a new family of integrative and conjugative elements associating insertion sequence transposition, plasmid replication, and conjugation for their spreading. J Bacteriol 195(9):1979–1990

Pernodet JL, Simonet JM, Guérineau M (1984) Plasmids in different strains of Streptomyces ambofaciens: free and integrated form of plasmid pSAM2. Mol Gen Genet 198(2):35–41

Huguet-Tapia JC, Badger JH, Loria R, Pettis GS (2011) Streptomyces turgidiscabies Car8 contains a modular pathogenicity island that shares virulence genes with other actinobacterial plant pathogens. Plasmid 65(2):118–124

Kers JA, Cameron KD, Joshi MV et al (2005) A large, mobile pathogenicity island confers plant pathogenicity on Streptomyces species. Mol Microbiol 55(4):1025–1033

Wozniak RA, Waldor MK (2010) Integrative and conjugative elements: mosaic mobile genetic elements enabling dynamic lateral gene flow. Nat Rev Microbiol 8(8):552–563

Burrus V, Waldor MK (2004) Shaping bacterial genomes with integrative and conjugative elements. Res Microbiol 155(5):376–386

Ochman H, Lawrence JG, Groisman EA (2000) Lateral gene transfer and the nature of bacterial innovation. Nature 405(6784):299–304

Rodríguez-Blanco A, Lemos ML, Osorio CR (2012) Integrating conjugative elements as vectors of antibiotic, mercury, and quaternary ammonium compound resistance in marine aquaculture environments. Antimicrob Agents Chemother 56(5):2619–2626

Blum G, Ott M, Lischewski A et al (1994) Excision of large DNA regions termed pathogenicity islands from tRNA-specific loci in the chromosome of an Escherichia coli wild-type pathogen. Infect Immun 62(2):606–614

Hacker J, Blum-Oehler G, Mühldorfer I, Tschäpe H (1997) Pathogenicity islands of virulent bacteria: structure, function and impact on microbial evolution. Mol Microbiol 23(6): 1089–1097

Dobrindt U, Zdziarski J, Salvador E, Hacker J (2010) Bacterial genome plasticity and its impact on adaptation during persistent infection. Int J Med Microbiol 300(6):363–366

Lou L, Zhang P, Piao R, Wang Y (2019) Salmonella Pathogenicity Island 1 (SPI-1) and its complex regulatory network. Front Cell Infect Microbiol 9:270

Palmela C, Chevarin C, Xu Z et al (2018) Adherent-invasive Escherichia coli in inflammatory bowel disease. Gut 67(3):574–587

Karaolis DKR, Johnson JA, Bailey CC, Boedeker EC, Kaper JB, Reeves PR (1998) A *Vibrio cholerae* pathogenicity island associated with epidemic and pandemic strains. Proc Natl Acad Sci U S A 95(6):3134–3139

Sullivan JT, Ronson CW (1998) Evolution of rhizobia by acquisition of a 500-kb symbiosis island that integrates into a phe-tRNA gene. Proc Natl Acad Sci U S A 95(9):5145–5149

Redondo-Salvo S, Fernández-López R, Ruiz R et al (2020) Pathways for horizontal gene transfer in bacteria revealed by a global map of their plasmids. Nat Commun 11(1):3602

Kumar P, Bag S, Ghosh TS et al (2017) Molecular insights into antimicrobial resistance traits of multidrug resistant enteric pathogens isolated from India. Sci Rep 7(1):14468

Ruppé É, Woerther P-L, Barbier F (2015) Mechanisms of antimicrobial resistance in gram-negative bacilli. Ann Intensive Care 5(1):61–61

Hasan CM, Dutta D, Nguyen ANT (2021) Revisiting antibiotic resistance: mechanistic foundations to evolutionary outlook. Antibiotics 11(1):40

Unemo M, Golparian D, Nicholas R, Ohnishi M, Gallay A, Sednaoui P (2012) High-level cefixime- and ceftriaxone-resistant Neisseria gonorrhoeae in France: novel penA mosaic allele in a successful international clone causes treatment failure. Antimicrob Agents Chemother 56(3): 1273–1280

Frost LS, Leplae R, Summers AO, Toussaint A (2005) Mobile genetic elements: the agents of open source evolution. Nat Rev Microbiol 3(9):722–732

Alekshun MN, Levy SB (2007) Molecular mechanisms of antibacterial multidrug resistance. Cell 128(6):1037–1050

Paterson DL, Bonomo RA (2005) Extended-spectrum beta-lactamases: a clinical update. Clin Microbiol Rev 18(4):657–686

Kumarasamy KK, Toleman MA, Walsh TR et al (2010) Emergence of a new antibiotic resistance mechanism in India, Pakistan, and the UK: a molecular, biological, and epidemiological study. Lancet Infect Dis 10(9):597–602

Yong D, Toleman MA, Giske CG et al (2009) Characterization of a new metallo-beta-lactamase gene, bla(NDM-1), and a novel erythromycin esterase gene carried on a unique genetic structure in Klebsiella pneumoniae sequence type 14 from India. Antimicrob Agents Chemother 53(12): 5046–5054

San MA (2018) Evolution of plasmid-mediated antibiotic resistance in the clinical context. Trends Microbiol 26(12):978–985

Vo AT, van Duijkeren E, Gaastra W, Fluit AC (2010) Antimicrobial resistance, class 1 integrons, and genomic island 1 in Salmonella isolates from Vietnam. PLoS One 5(2):e9440

Huang H, Yang ZL, Wu XM et al (2012) Complete genome sequence of Acinetobacter baumannii MDR-TJ and insights into its mechanism of antibiotic resistance. J Antimicrob Chemother 67(12):2825–2832

Mulvey MR, Grant JM, Plewes K, Roscoe D, Boyd DA (2011) New Delhi metallo-β-lactamase in Klebsiella pneumoniae and Escherichia coli, Canada. Emerg Infect Dis 17(1):103–106

Bock R (2010) The give-and-take of DNA: horizontal gene transfer in plants. Trends Plant Sci 15(1):11–22

Iyer LM, Aravind L, Coon SL, Klein DC, Koonin EV (2004) Evolution of cell-cell signaling in animals: did late horizontal gene transfer from bacteria have a role? Trends Genet 20(7): 292–299

Arias MC, Danchin EGJ, Coutinho P, Henrissat B, Ball S (2012) Eukaryote to gut bacteria transfer of a glycoside hydrolase gene essential for starch breakdown in plants. Mob Genet Elements 2(2):81–87

Hehemann J-H, Correc G, Barbeyron T, Helbert W, Czjzek M, Michel G (2010) Transfer of carbohydrate-active enzymes from marine bacteria to Japanese gut microbiota. Nature 464(7290):908–912

Leplae R, Lima-Mendez G, Toussaint A (2010) ACLAME: a classification of mobile genetic elements, update 2010. Nucleic Acids Res 38(Database issue):D57–D61

Claesson MJ, Li Y, Leahy S et al (2006) Multireplicon genome architecture of Lactobacillus salivarius. Proc Natl Acad Sci U S A 103(17):6718–6723

Jones BV (2010) The human gut mobile metagenome: a metazoan perspective. Gut Microbes 1(6): 415–431

Stecher B, Denzler R, Maier L et al (2012) Gut inflammation can boost horizontal gene transfer between pathogenic and commensal Enterobacteriaceae. Proc Natl Acad Sci U S A 109(4): 1269–1274

Nordmann P, Dortet L, Poirel L (2012) Carbapenem resistance in Enterobacteriaceae: here is the storm! Trends Mol Med 18(5):263–272

Poirel L, Madec JY, Lupo A et al (2018) Antimicrobial resistance in Escherichia coli. Microbiol Spectr 6(4). https://doi.org/10.1128/microbiolspec.ARBA-0026-2017

Probiotics in food: health and nutritional properties and guidelines for evaluation: report of a joint FAO/WHO expert consultation on evaluation of health and nutritional properties of probiotics in food including powder milk with live lactic acid bacteria, Cordoba, Argentina, 1–4 October 2001 [and] Report of a joint FAO/WHO working group on drafting guidelines for the evaluation of probiotics in food, London, Ontario, Canada, 30 April–1 May 2002. Rome [Italy]: Food and Agriculture Organization of the United Nations, World Health Organization; 2006

Zomer AL, Buist G, Larsen R, Kok J, Kuipers OP (2007) Time-resolved determination of the CcpA regulon of Lactococcus lactis subsp. cremoris MG1363. J Bacteriol 189(4):1366–1381

Kormelink TG, Koenders E, Hagemeijer Y et al (2012) Comparative genome analysis of central nitrogen metabolism and its control by GlnR in the class Bacilli. BMC Genomics 13(1):191

Guédon E, Sperandio B, Pons N, Ehrlich SD, Renault P (2005) Overall control of nitrogen metabolism in Lactococcus lactis by CodY, and possible models for CodY regulation in Firmicutes. Microbiology 151(Pt 12):3895–3909

Oxaran V, Ledue-Clier F, Dieye Y et al (2012) Pilus biogenesis in Lactococcus lactis: molecular characterization and role in aggregation and biofilm formation. PLoS One 7(12):e50989

Tanner SA, Chassard C, Rigozzi E, Lacroix C, Stevens MJA (2016) Bifidobacterium thermophilum RBL67 impacts on growth and virulence gene expression of Salmonella enterica subsp. enterica serovar Typhimurium. BMC Microbiol 16(1):46

Broaders E, Gahan CG, Marchesi JR (2013) Mobile genetic elements of the human gastrointestinal tract: potential for spread of antibiotic resistance genes. Gut Microbes 4(4):271–280

Kankainen M, Paulin L, Tynkkynen S et al (2009) Comparative genomic analysis of Lactobacillus rhamnosus GG reveals pili containing a human- mucus binding protein. Proc Natl Acad Sci U S A 106(40):17193–17198

O'Connell Motherway M, Zomer A, Leahy SC et al (2011) Functional genome analysis of Bifidobacterium breve UCC2003 reveals type IVb tight adherence (Tad) pili as an essential and conserved host-colonization factor. Proc Natl Acad Sci U S A 108(27):11217–11222

Eydallin G, Montero M, Almagro G et al (2010) Genome-wide screening of genes whose enhanced expression affects glycogen accumulation in Escherichia coli. DNA Res 17(2):61–71

Goh YJ, Klaenhammer TR (2013) A functional glycogen biosynthesis pathway in Lactobacillus acidophilus: expression and analysis of the glg operon. Mol Microbiol 89(6):1187–1200

Genomic Islands in Nutritional Fitness and Adaptation

7

Praveenya Tirunagari, Megha Choudhary, and Sushanta Deb

Abstract

Genomic islands have attracted wide attention and are extensively used to study the evolution and adaptation strategies of drug-resistant pathogenic and environmental non-pathogenic strains. Horizontal genes, nucleotide substitution, and DNA recombination are few among the several underlying dynamic genetic determinant that shapes the genomic islands in microbial genomes. Unlike pathogenic bacteria, genomic islands in nonpathogenic/environmental bacteria contain genes that are linked to the functions of secondary metabolites, isoprenoids, and metabolic enzymes which promotes bacterial adaptability to changing environments. Here an attempt has been made to emphasize on the importance of genomic islands in commensal, symbiotic, and environmental bacteria. Additionally, the recent lessons learned from pathogenicity islands in pathogenic microbes have also been discussed.

Keywords

Genomic islands · Resistance islands · HGT · Microorganism and Adaptation

P. Tirunagari
Department of Biotechnology, Indian Institute of Technology, Kharagpur, India

M. Choudhary
CSIR-Institute of Microbial Technology, Chandigarh, India

All India Institute of Medical Sciences (AIIMS), New Delhi, India

S. Deb (✉)
All India Institute of Medical Sciences (AIIMS), New Delhi, India

Department of Molecular Biology & Bioinformatics, Tripura University, Suryamaninagar, Tripura, India

I. Mani et al. (eds.), *Microbial Genomic Islands in Adaptation and Pathogenicity*,
https://doi.org/10.1007/978-981-19-9342-8_7

7.1 Introduction

Horizontal gene transfer (HGT) widens the extent of variable genetic content in microbial genome that encodes genes beneficial for their adaptation in changing environmental conditions. These genes lie on contiguous stretch of DNA, frequently found in variable genetic regions, and are often considered as genomic islands (GEIs). Bacterial and archeal genome evolve through a diverse genetic mechanism, which involves mutations, gene rearrangements, and the acquisition of genes from other microbes, phages, and environments (Dobrindt et al. 2004; Aminov 2011). Genes located in genomic islands perform a variety of functions in clinical and environmental strains, the majority of which are involved in pathogenicity and adaptation. Based on GEIs mobility it can be categorized into mobile GEI, where the movement is supported by conjugative elements (ICEs) while nonmobile GEIs remain tightly integrated into the host bacteria (Juhas 2019). The highly dynamic nature of GEIs is the key factor contributing to substantial genetic variation in pathogenic strains. Recently, studies have illustrate the role of GEIs in the pathogenicity of outbreak and non-outbreak strains (Ingle et al. 2016; Winstanley et al. 2009). Additionally, several other phenomena such as symbiosis, secondary metabolites synthesis, toxic compounds neutralization and thermal adaptation have been found to be associated with the microbial genomic islands. A prolonged symbiotic relationship between animals and microbes substantially influences genomic evolution and modifies the metabolic features of associated bacteria. For instance, in the Acetobacteraceae family, persistent symbiotic associations between insects and their gut-associated microbes provide evidence of horizontal acquisition of genes associated with amino acid export, coenzyme transport, energy metabolism, and defense mechanisms (Brown and Wernegreen 2019). GEI analysis has become a widely accepted approach to track disease outbreaks and the evolution of accessory genetic material of pathogens. The growing availability of genomes and rapid decrease in sequencing cost gives us easy access to dedicated databases, genome analysis tools, and sequence analysis algorithms (Gardy and Loman 2018; Bertelli and Greub 2013). In the past few decades, several algorithms and bioinformatics software were developed to identify GIs in bacterial genomes, however, most of them are standalone software that requires basic programming skills making it unintuitive for biologists (Bertelli and Greub 2013). Tools with more user-friendly features like graphical user interfaces or web servers recently have come forward to predict and interactively visualize the genomic islands (Bertelli, Laird and Williams, K.P. 2017).

7.2 Genomic Islands and Pathogens

Genomic islands are the bacterial gene clusters that seem to have been obtained through horizontal gene transfer, where Conjugation (Fig. 7.1), transformation and transduction (Fig. 7.2), are the most common modes of transmission (Jain et al. 2002; Chen et al. 2005) (Håvarstein 1998). Integrative conjugative elements (ICEs),

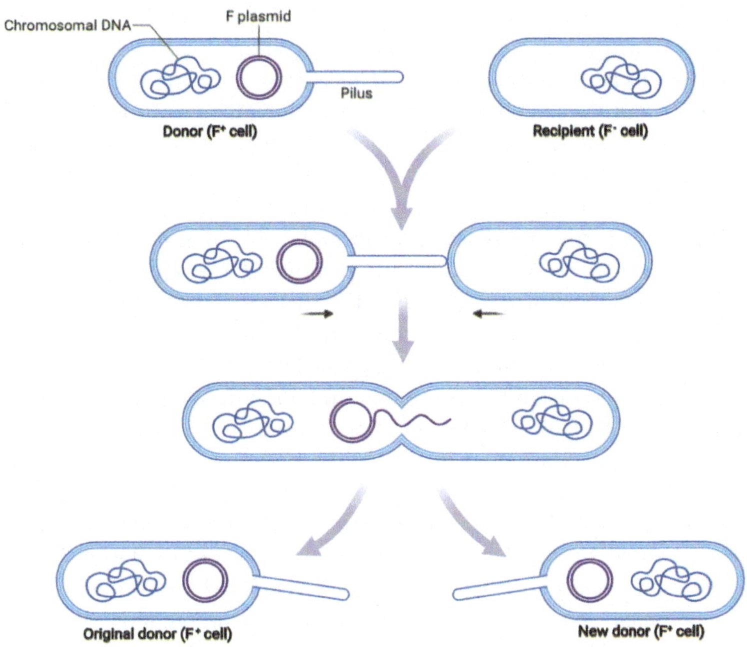

Fig. 7.1 Schematic Representation of Bacterial Conjugation process

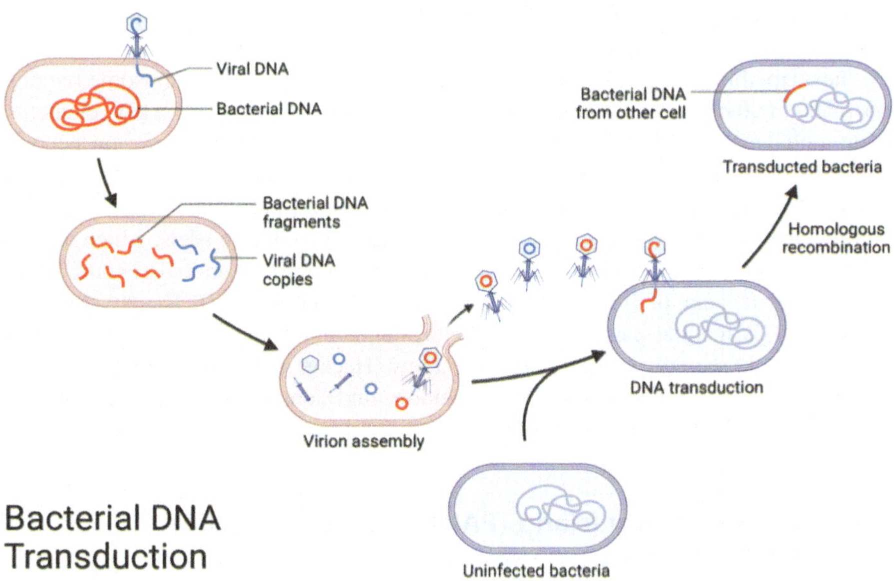

Fig. 7.2 Steps and Mechanism of Bacterial DNA Transduction Process

which are GIs transmitted by conjugation mode, are among the most well-studied types of GIs (Partridge et al. 2018; Botelho et al. 2020; Boyd et al. 2016).The capacity of GIs to be transferred between species is an intriguing trait (Juhas et al. 2009). Type II and IV secretion systems are frequently linked to GI changes that occur naturally. Because many of the encoded genes of Genomic islands (GEIs) synthesize toxins or other pathogenicity factors since they were also first discovered in pathogenic bacteria they were given the name pathogenicity islands (Sulaiman et al. 2019). PAIs were first discovered in pathogenic E.coli (Blum et al. 1994). There is mounting evidence that non-pathogenic species also include the components of pathogenicity islands. The islands comprise transfer genes, integrases, and IS elements in addition to sequences obtained from phages and/or plasmids. These potentially unstable DNA building blocks are commonly inserted into tRNA genes. The instability is brought on by the flanking direct repeats, which commonly resemble phage attachment sites and encourage bacterial genome integration and excision. (Hacker et al. 1997; Buchrieser et al. 1998). Pathogenicity islands (PAIs), which are mobile genetic components, have been linked to the rapid evolution of bacterial diseases (Schaechter et al. 2004). Identification of the mechanisms behind bacterial genomic diversification and evolution has been the focus of genomic research over the past 10 years. After evaluating expanding bacterial genomics sequences for Homologous Recombination and Horizontal Gene Transfer (HGT), the conventional idea that clonal divergence and occasional selection is the route of prokaryotic evolution has been widened to include genome exchange and loss. Apart from the core genes that code for essential functions, bacteria's genome also contains a large number of "fexibe" genes that provide other features that may be useful in specific circumstances.

The adaptable gene pool includes mobile, auxiliary, variable chromosome regions, plasmids, pathogenicity islands, integrons, insertion sequence (IS) elements, bacteriophages, and transposons. The impact of horizontal gene transfer on genome plasticity is seen in the diversification and adaption of microorganisms. Genomic islands (GEIs), which ranging from a few kb to 500 kb, permit a significant degree of horizontal gene transfer. Hacker et al. first described genomic islets, which are GIs with a size of less than 10 kb and are collections of genes in a bacterial genome with a particular percentage of GC and dinucleotides (Hacker et al. 1997). The physiological actions of GIs include pathogenicity, phenol degradation, antibiotic resistance, iron absorption, and secretory activity, to name a few (Hacker and Carniel 2001). They are also necessary for genetic plasticity, evolution, and adaptation to changing habitat (Dobrindt et al. 2004; Juhas et al. 2009).

7.3 Pathogenicity Island (PAIs) and Its Correlation with Virulence

PAIs are a type of genomic island that has the same overall structure and content as other genomic islands and can vary in size ranging from 10 to 200 kbp (Dobrindt et al. 2004;Tormo et al. 2008). The bulk of PAIs discovered (75 percent) had tRNA flanking sequences. According to the evidence, PAIs appear to be acquired horizontally by one or more lateral transfer events. Within some PAIs, there is evidence of a single large transfer event, whereas others are more "mosaic-like." The "mosaic-like" makeup of some PAIs is caused by many, distinct lateral transfer events (Hacker et al. 1997; Schmidt and Hensel 2004). Pathogenicity is connected with the expression of disease-related factors in pathogenic bacteria that are absent in non-pathogenic bacteria. Unlike the conserved "core" genome of bacteria, the "flexible" pool encodes for certain virulence factors beneficial in some environments. If the donor and recipient bacteria are closely related, the PAI GC content may not differ from that of the core genome (Hacker et al. 1997). Essential traits including antibiotic resistance, symbiosis, fitness, and adaptation in general are influenced by pathogenic and genomic islands. Genome plasticity, or the gaining or losing of genetic information, is essential for the adaptive evolution of pathogenic bacteria because it permits the inheritance of intricate disease-related traits in a single step due to the simultaneous acquisition of numerous genes in HGT (Ochman et al. 2000). On mobile or formerly mobile genomic elements, such as PAIs, virulence genes are frequently found (Juhas 2019). (Jain et al. 2002),while DNA has been transported by plasmid, phage, virulence genes being found on mobile or formerly mobile genetic elements that include PAIs is a recent finding (Boyd and Brüssow 2002; Feigel et al. 2002). In-depth genomic regions (PAIs) that are less frequently observed in closely related non-pathogenic bacteria were discovered to have originated from lysogenic bacteriophages and plasmids and are present in pathogenic variations. They are usually flanked by repeat sequences, contain one or more virulence genes, differ from the rest of the chromosome in terms of G + C composition, and are frequently linked to tRNA genes. In addition, PAIs frequently contain genes that code for integrases or transposons and are inherently unstable (Hacker et al. 1997).

According to the crucial functions they encode, virulence genes found on PAIs in a broad range of organisms such as Gram-negative, Gram-positive bacteria as well as in humans and may be categorized into different groups: (Table 7.1).

7.4 Resistance Islands (REIs) for Pathogen Adaptation

Despite conferring pathogenic and drug-resistant traits to the bacterium, genomic Islands are also well known for their role in phenol degradation, iron intake, and other secretory activity of microbes. The accommodation and maintenance of the antibiotic resistance genes in the bacterial genomes depends heavily on these genomic islands. A no. of dynamic genetic factors such as horizontal gene transfer and mutations contribute greatly to increase genomic plasticity, which facilitates the acquisition of drug/antibiotic-resistant genes in pathogenic bacterial species (Cattoir and Giard 2014) (Hacker and Carniel 2001). In the terms of pathogen adaptation, it is well known that pathogenicity islands (PAIs) and antimicrobial resistance islands (REIs) are two major subsets of the genomic island. Pathogenicity island acquires the genes mainly encoded for virulence factor, which includes enterotoxin,

Table 7.1 Virulence genes characters and their functions

Adherence elements	Aid bacterial attachment to host surfaces
Siderophores	Assure adequate Fe^{3+} ion solubilization and uptake
Capsules	Resist phagocytosis and provide defence against other immune system components of the host
Endotoxin (LPS) (belonging to Gram-negative organisms)	Induce inflammation and the host complement pathway
Exotoxins	Affect function of eukaryotic cells by interfering with signal transduction and structures within cells.
Invasins	Facilitate the entrance of microorganisms into eukaryotic cells
Secretion systems, type III and IV	Toxins or modulins that are essential for their targeted delivery into eukaryotic cells and that change how the host interacts with them. Example: T3SSs and T4SSs Interfere with apoptotic and signaling cascades while facilitating entry into non-phagocytic cells in the host

colonization factor, iron uptake transporter, and type III secretion system (Yoon et al. 2015), while genes acquired by antimicrobial resistance islands were confined to the factors that involved in the antibiotic resistance mechanism. Antimicrobial resistance islands confer the resistance benefit against the antibiotic and drug molecule to escape from therapeutics, eventually giving rise to multidrug-resistant pathogenic strains. For instance, interspecies recombination events in mixed microbial population results in the acquisition of large Carbapenem resistance islands by *Acinetobacter baumannii* from other pathogenic *Acinetobacter* species (Godeux et al. 2022). A number of databases dedicated to antibiotic resistance genes were reported earlier named ARDB, CARD, and BacMet, however, databases related to resistance islands are yet to be developed (Liu and Pop 2009; McArthur 2009). The genes that constitute the resistance Islands are indulged in different antibiotic resistance mechanisms in bacteria such as antibiotic degradation, target modification, efflux pumps, and many more (Fig. 7.3) (McArthur et al. 2013). The underlying factors that involved in acquisition of antibiotic resistance island still remain unexplored due to lack of experimental validations supporting the genomic observations.

7.5 Pathogen Fitness

The characteristics that assist an organism in surviving, multiplying, and/or transmitting within a specific biological niche are referred to as its pathogen fitness (Preston et al. 1998). Positively chosen genomic islands with activities that promote bacterial fitness, either directly or indirectly, are referred to as 'fitness islands'. Insight into the resistance gene transfer mechanism underlying the development of multidrug-resistant isolates is important and requires knowing the GIs, particularly the resistant GIs.

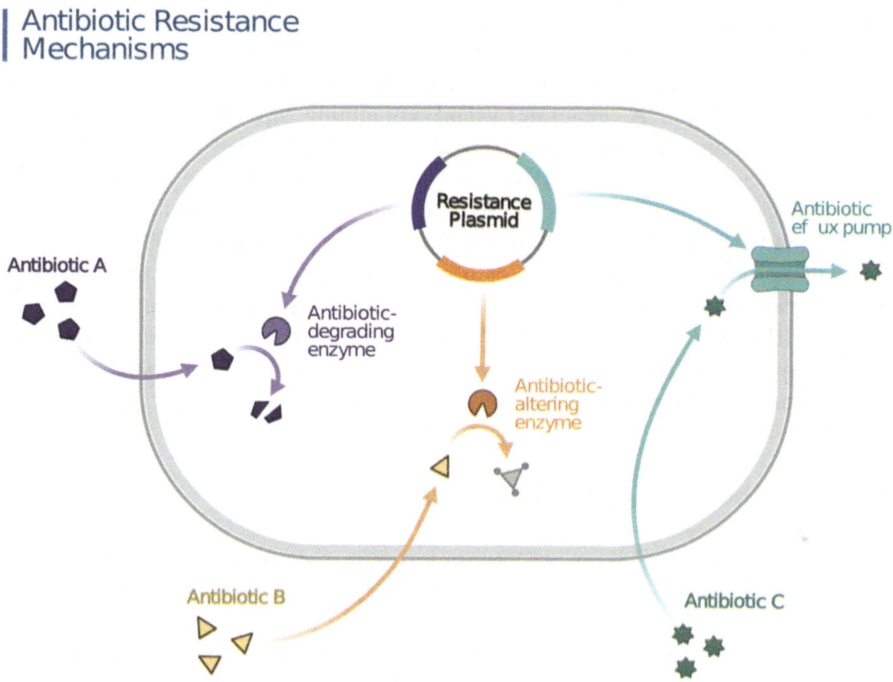

Fig. 7.3 Strategies adopted by bacteria to develop antibiotic resistance phenotypes

The results of an artificial intelligence and in silico study suggest that evolutionary relationships control the fitness and functional compatibility of horizontally transferred genes in new hosts. The findings also revealed that codon use, GC content, and mRNA-folding energy all have a minimal effect on heterologous gene transfer (Hacker and Carniel 2001; Emamalipour et al. 2020). The Darwinian rules ("survival of the fittest") apply to both eukaryotes and prokaryotes in their evolution. Under specific environmental conditions, possessing a genomic island may give a selective advantage to develop some beneficial traits in microbes that improve their transmission, survival, and colonization within a niche (stress, in vivo settings, antibacterial drug exposure). As Preston et al. have already suggested, genomic islands promoting the fitness of microorganisms are deemed as "fitness islands" from a functional standpoint (Arber 2000). In these conditions, genetic fitness islands bestow novel traits that increase the bacterial host's ability for adaptation. Genetic fitness islands in these conditions confer unique features that increase the bacterial host's capacity for adaptation. A true pathogenicity island aids directly or indirectly in the growth of true pathogens, whereas a symbiosis island can be referred to as a fitness island mainly facilitating the bacterial interaction with their host (Ochman and Moran 2001).

Adhesins and toxins, which are encoded by PAIs, phages, or plasmids and frequently function as a direct result of their activity, are pathogenicity factors (Waldor and Mekalanos 1996; Karaolis et al. 1998). These factors' effects on pathogenicity appear to be a direct result of pressures on evolution. This is true for intestinal infections as well as respiratory pathogens, where pathogenicity factors make it easier for these pathogens to spread and hence have a favorable impact on microbial evolution. A genomic island containing the genes that code for yersiniabactin, an iron-uptake mechanism, was first recorded from a pathogenic *Yersinia* species (Carniel et al. 1996). In *E. coli*, certain adhesion factors (such P-, S-, and F1C-fimbriae) are produced by commensal strains prevalent in the human gut flora and are encoded by genomic islands (Hacker and Carniel 2001). PAIs have genes that produce type III or IV secretion systems. Again, these secretion systems might be termed PAIs if they convey proteins implicated in the infectious process. The type III system includes strains of *Salmonella*, *Shigella*, and *Yersinia*, while the type IV system includes *Helicobacter pylori* and *Legionella pneumophila* (Galan and Collmer 1999; Parsot and Sansonetti 1999; Censini et al. 1996).

7.6 Pathogenicity of *Clostridium* Species

The *Clostridium* genus is an endospore-forming, strictly anaerobic bacteria and numerous species that are significant for human diseases. *Clostridium* cluster XIVa and IV species colonize, the main bacteria in the gut, accounts for 10–40% of the total microorganisms in the gut. Around 250 species of *Clostridium* exist, including ordinary free-living bacteria (Koukou 2021) among them the main disease-causing species in humans are:

- *Clostridium botulinum*
- *Clostridium tetani*
- Clostridium *perfringens*
- *Clostridium Sordellii*

Antibiotic-associated diarrhea (AAD) imbalances the gut microbiota caused by antibiotic usage, with multiple pathways implicated in the illness process (McFarland 1998). Antibiotics cause C. difficile to produce strong toxins, which are the main virulence factors. *Clostridium difficile* infection (CDI) is one kind of AAD that can cause serious gastrointestinal problems. *Clostridium difficile* (formerly *Clostridioides difficile*) is a Gram-positive anaerobic bacillus that forms spores. *C. difficile* spores may persist for extended durations on inanimate things (resisting heat, acid, and antibiotics), which is one of the reasons why this bacterium can cause so many difficulties in hospitals. *Clostridium difficile* causes illness in humans by producing two protein exotoxins (toxin A and toxin B), which are cytotoxic to colonic epithelial cells and spread via the fecal-oral pathway (Rupnik et al. 2009). When comparing dangerous and benign strains of *C. difficile*, a unique locus encoding the toxin genes tcdA and tcdB was found (Braun et al. 1993). A 19.6-kb insertion between genes that are adjacent in nontoxigenic *C. difficile* strains makes up this PaLoc (short for "pathogenicity locus"), which has five ORFs. There are no genes associated with genetic mobility or instability in PaLoc, and it is unclear which genetic mechanisms give rise to hazardous and harmless strains (Cohen et al. 2000). The 19-kilobyte Pai gene, which is found in the genome of virulent strains of *C. difficile* that cause antibiotic-associated diarrhea or its fatal form, pseudomembranous colitis, encodes two high-molecular-weight toxins named TcdA and TcdB (a cytotoxin). It is intriguing that a 115-bp segment that creates a 20-bp hairpin loop only shows up in non-pathogenic variants (not in pathogenic strains). It might be where this pathogenicity island's chromosomes integrate (Braun et al. 1996). The presence of a 115-bp stretch that results in a 20-bp hairpin loop only in non-pathogenic forms is intriguing (not in pathogenic strains).

7.7 Pathogenicity Locus of *Clostridium difficile*

The production of toxins by *C. difficile* is only possible when PaLoc is present (Cohen et al. 2000). It was interestingly found that PaLoc insertions and deletions can lead to toxin loss when particularly virulent *C. difficile* isolates were compared to other toxigenic isolates. According to studies on the expression of the PaLoc gene, TcdC and TcdD, which are encoded by the PaLoc gene, respectively function as negative and positive regulators of PaLoc gene expression (Burks et al. 1997; Spigaglia and Mastrantonio 2002). While there was no evidence of PAI in the genome sequencing research for Clostridium tetani and Clostridium perfringens (Shimizu et al. 2002; Brüggemann et al. 2003).

7.8 Pathogenicity Islands of *Citrobacter* Species

The Enterobacteriaceae family includes the *Citrobacter* genus, a group of aerobic, Gram-negative, rod-shaped bacteria that use citrate as their primary carbon source. The locus for enterocyte effacement (LEE) has also been found in *C. rodentium*, the organism that causes nursing mice to develop transmissible murine colonic hyperplasia. Mice with *C. rodentium* infections had ruffled coats, mild diarrhea, and delayed growth. Mice can experience rectal prolapse in extreme circumstances, which can result in low to high mortality rates (Luperchio et al. 2000; Schiavo and van der Goot 2001). In the LEE of *Citrobacter rodentium*, horizontal transfer between attaching and effacing pathogens has been observed (Deng et al. 2001). Although the *C. rodentium* LEE shares 41 ORF with the EPEC and EHEC LEE, it differs due to the location of the RORF1 and RORF/espG genes and the presence of numerous insertion sequences. In selC, the LEE of *C. rodentium* is absent. It is surrounded by sequences that are homologous to the *Shigella* plasmid R100 and the EHEC plasmid pO157 on one side and an operon that contains an ABC transport system and an IS element on the other. The animal model mentioned above may help with the examination of the LEE in *C. rodentium* (Deng et al. 2001).

7.9 Genomic Islands in Environmental Microorganisms

Due to the recent advent of cost-effective next-generation sequencing technology, a myriad of whole-genome sequencing data become easily accessible to the public database which promotes the extensive genomic study of non-pathogenic/ environmental bacteria from different ecological niches. Pathogenicity is not the sole factor that constitutes the Genomic Islands, other phenomena such as commensalism, environmental fitness, and metabolic adaptability have emerged as major factors responsible for the extension of the genetic island in microbial genomes (Fig. 7.4) (Hacker and Carniel 2001). The association of the integrase gene with the extrachromosomal genomic island provides flexibility to organisms to evolve more rapidly through genetic rearrangements and gene acquisition processes (Dobrindt et al. 2004). Horizontal gene acquisition and Genomic island formation are likely to occur in densely populated niches with large number of diverse bacterial species.

7.10 GEIs for Adaptation in Toxic Environments

Environmental factors can act as a driving force shaping the evolution of bacterial genomic islands. Thus, heavy metal toxic habitats that arise naturally or through anthropogenic activity provide environmental stress to the dwelling species of that environment, which makes them competent to survive in unfavorable toxic conditions. It has been reported that *Thiomonas* species isolated from toxic acid mine drainage contain a specific genomic island with the capability to oxidize arsenite that confer resistance against extremely toxic habitats (Freel et al. 2015). Studies have demonstrated a close association between antibiotic-resistant and toxic compound-resistant phenotypes that follow the principle of cross-resistance

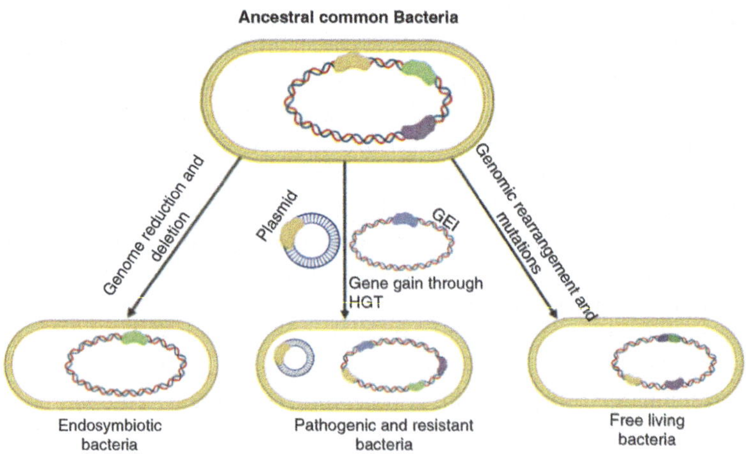

Fig. 7.4 Horizontal gene transfer and genomic island Integration in various groups of Bacteria

mechanisms (Arsène-Ploetze et al. 2018). It has also been observed that two *Thiomonas* strains containing GEI from toxic metal-contaminated habitats facilitate the development of antibiotic resistance and toxic compound resistance phenotype (Arsène-Ploetze et al. 2018). Furthermore, studies have reported that beta-carbonic anhydrase (β-CA) genes were horizontally transferred from prokaryotes to eukaryotes (Zolfaghari Emameh et al. 2016). Through the evolutionary process, β-CA gene from the prokaryotic genomic island has integrated into the eukaryotic genome (Zolfaghari Emameh et al. 2018). Hence, the above findings support that genomic islands promote metabolic adaptation and resistance to hazardous environmental factors.

7.11 Gene Acquisitions and Adaptation of Marine Bacterial Species

Anciently, it was believed that marine bacterial adaption required sodium-ion to maintain osmotic balance, but later it has been observed that several genetic determinants are responsible for bacterial adaptation in the marine environment. In marine *Salinispora*, it has been observed that gene acquisition and gene loss influence the marine adaptation capability, acquisition of transporter genes from other marine bacteria promotes adaptation of *Salinispora* species while gene loss renders this genus incapable to survive outside the ocean (Penn and Jensen 2012). Furthermore, genomic insight on streptomycete revealed that horizontal gene transfer cause to the acquisition of gene clusters that encodes for hybrid isoprenoids (HIs) in the marine *Streptomyces* genus, acquired isoprenoids gene cluster contributed to develope adaptive ability to survive in marine environment (Gallagher and Jensen 2015). *Bacillus methylotrophicus* a plant growth promoter produces lanthipeptides with antibacterial activity, genes related to lanthipeptide synthesis were found to reside in the genomic island which is probably associated with the functional adaptation of that species (Dias et al. 2015). Genomic investigation on marine actinobacteria disclosed that genes involved in secondary metabolite biosynthesis located in 21 genomic islands of two *Salinispora* species correlate to the functional adaptation of those bacteria (Penn et al. 2009). *Arthrobacter* genomic islands of diverse ecotypes provide insights into multiple genetic factors that exhibit specialized phenotype, which includes antibiotic resistance, multidrug efflux, and carbon metabolism facilitating its adaptability to different environmental niches (Gushgari-Doyle et al. 2022).

7.12 Extremophiles and Horizontal Gene Transfer

An established fact is that hyperthermophilic species show high diversity in their genetic content (Dobrindt et al. 2004). Studies have described the existence of conjugative plasmids in thermophilic archaeal species (Schleper et al. 1995) and demonstrated its capability to integrate with genomes of other species (Peng et al.

2000). Using the Genome-based Subtraction Hybridization method the dissimilar genomic regions between two *Thermotoga maritima* species were investigated. An additional genomic region acquired by HGT that is absent in other *Thermotoga* species encodes for genes involved in arabinosidase expression, rhamnose biosynthesis, and alcohol dehydrogenase and is believed to influence the species adaptability to the environment (Nesbø and Doolittle 2003). Extremophiles are prone to participate in DNA exchange between the same species, to escape the detrimental effect of DNA damage due to higher temperatures in an environment (van Wolferen et al. 2013). In acidophiles, horizontal gene transfer and gene loss events equipped them with an increased capacity to adapt to changing environments (Zhang et al. 2017).

7.13 Gene Exchange Between Archaea and Bacteria

Despite the importance of horizontal gene transfer in genome evolution, it was not fully exploited to study molecular evolution for decades (Cohan 1994). The increasing trend of genomic analysis has enlightened the role of HGT events in genomic evolution and its influence to promote the adaptive capability of microbes. In the present time, it has become more evident that throughout the evolution, bacteria and archaea have exchanged genes to better adapt to the dynamic environmental condition (Guglielmini et al. 2013). Thermophilic and anaerobic are the two bacterial groups, most frequently shared a considerable number of genes with archaea. Geothermal springs, oil wells, and marine sediments are major sites where genes were frequently horizontally transferred between archaea and bacteria (Fuchsman et al. 2017). Bioinformatics analysis mainly investigates codon usage and distinct base composition to identify the horizontally transferred genes in genomes (Nakamura et al. 2004). Three major strategies have been adopted by archaea and bacteria for genetic exchange, which include transduction, conjugation, and natural transformation. Natural transformation and conjugation are two phenomena opposite to each other based on their invasive mechanism, during natural transformation recipient cell has control over transferred DNA in contrast conjugation mainly follows the command from the donor cell. In comparison to mesophilic bacteria, hyperthermophiles acquired a relatively higher number of genes from archaea through the HGT mechanism (van Wolferen et al. 2013).

7.14 Conclusion

Recent advancements in NGS technologies and bioinformatics tools have prompted comprehensive investigation of the genomic content, organization, and potential functions of microbial genomes, which helps to characterize intrinsic as well as horizontally acquired genes of microbes. Genomic analyses have provided an in-depth understanding of the evolution, virulence, and adaptation of microbial species. Advancement in genomic techniques has made it possible to characterize the Genomic island by studying the dynamics and complexity of the microbial genomes. Genomic islands have been used as a target to examine the emergence of novel functional capabilities in pathogenic as well as nonpathogenic microbes. A number of studies have established the fact that other than pathogenic microbes, GEIs played a significant role to develop adaptive capabilities also in microbes from the geothermal and marine environment. The available information on the genomic island has improved our understanding greatly on microbial adaptation in diverse ecological habitats.

References

Aminov RI (2011) Horizontal gene exchange in environmental microbiota. Front Microbiol 2:158

Arber W (2000) Genetic variation: molecular mechanisms and impact on microbial evolution. FEMS Microbiol Rev 24(1):1–7

Arsène-Ploetze F, Chiboub O, Lièvremont D, Farasin J, Freel KC, Fouteau S, Barbe V (2018) Adaptation in toxic environments: comparative genomics of loci carrying antibiotic resistance genes derived from acid mine drainage waters. Environ Sci Pollut Res 25(2):1470–1483

Bertelli C, Greub G (2013) Rapid bacterial genome sequencing: methods and applications in clinical microbiology. Clin Microbiol Infect 19(9):803–813

Bertelli C, Laird MR, Williams KP, Simon Fraser University Research Computing Group, Lau BY, Hoad G, Winsor GL, Brinkman FSL (2017) IslandViewer 4: expanded prediction of genomic islands for larger-scale datasets. Nucleic Acids Res 45(W1):30–35

Blum G, Ott M, Lischewski A, Ritter A, Imrich H, Tschäpe H, Hacker J (1994) Excision of large DNA regions termed pathogenicity islands from tRNA-specific loci in the chromosome of an Escherichia coli wild-type pathogen. Infect Immun 62(2):606–614

Botelho J, Mourão J, Roberts AP, Peixe L (2020) Comprehensive genome data analysis establishes a triple whammy of carbapenemases, ICEs and multiple clinically relevant bacteria. Microb Genom 6(10):mgen000424

Boyd DA, Lalancette C, Lévesque S, Golding GR (2016) Characterization of a genomic island harbouring a new vanD allele from Enterococcus faecium N15-508 isolated in Canada. J Antimicrob Chemother 71(7):2052–2054

Boyd EF, Brüssow H (2002) Common themes among bacteriophage-encoded virulence factors and diversity among the bacteriophages involved. Trends Microbiol 10(11):521–529

Braun V, Hundsberger T, Leukel P, Sauerborn M, von Eichel-Streiber C (1996) Definition of the single integration site of the pathogenicity locus in Clostridium difficile. Gene 181(1–2):29–38

Braun V, Schönherr R, Hobbie S (1993) Enterobacterial hemolysins: activation, secretion and pore formation. Trends Microbiol 1(6):211–216

Brown BP, Wernegreen JJ (2019) Genomic erosion and extensive horizontal gene transfer in gut-associated Acetobacteraceae. BMC Genomics 20(1):1–15

Brüggemann H, Bäumer S, Fricke WF, Wiezer A, Liesegang H, Decker I, Herzberg C, Martinez-Arias R, Merkl R, Henne A, Gottschalk G (2003) The genome sequence of Clostridium tetani, the causative agent of tetanus disease. Proc Natl Acad Sci 100(3):1316–1321

Buchrieser C, Brosch R, Bach S, Guiyoule A, Carniel E (1998) The high-pathogenicity island of Yersinia pseudotuberculosis can be inserted into any of the three chromosomal asn tRNA genes. Mol Microbiol 30(5):965–978

Burks AW, Shin D, Cockrell G, Stanley JS, Helm RM, Bannon GA (1997) Mapping and mutational analysis of the IgE-binding epitopes on Ara h 1, a legume vicilin protein and a major allergen in peanut hypersensitivity. Eur J Biochem 245(2):334–339

Carniel E, Guilvout I, Prentice M (1996) Characterization of a large chromosomal "high-pathogenicity island" in biotype 1B Yersinia enterocolitica. J Bacteriol 178(23):6743–6751. https://doi.org/10.1128/jb.178.23.6743-6751.1996. PMID: 8955291; PMCID: PMC178570
Cattoir V, Giard JC (2014) Antibiotic resistance in Enterococcus faecium clinical isolates. Expert Rev Anti-Infect Ther 12(2):239–248
Censini S, Lange C, Xiang Z, Crabtree JE, Ghiara P, Borodovsky M, Rappuoli R, Covacci A (1996) cag, a pathogenicity island of Helicobacter pylori, encodes type I-specific and disease-associated virulence factors. Proc Natl Acad Sci 93(25):14648–14653
Chen I, Christie PJ, Dubnau D (2005) The ins and outs of DNA transfer in bacteria. Science 310(5753):1456–1460
Cohan FM (1994) The effects of rare but promiscuous genetic exchange on evolutionary divergence in prokaryotes. Am Nat 143(6):965–986
Cohen SH, Tang YJ, Silva J Jr (2000) Analysis of the pathogenicity locus in Clostridium difficile strains. J Infect Dis 181(2):659–663
Deng W, Li Y, Vallance BA, Finlay BB (2001) Locus of enterocyte effacement from Citrobacter rodentium: sequence analysis and evidence for horizontal transfer among attaching and effacing pathogens. Infect Immun 69(10):6323–6335
Dias L, Caetano T, Pinheiro M, Mendo S (2015) The lanthipeptides of Bacillus methylotrophicus and their association with genomic islands. Syst Appl Microbiol 38(8):525–533
Dobrindt U, Hochhut B, Hentschel U, Hacker J (2004) Genomic islands in pathogenic and environmental microorganisms. Nat Rev Microbiol 2(5):414–424
Emamalipour M, Seidi K, Zununi Vahed S, Jahanban-Esfahlan A, Jaymand M, Majdi H, Amoozgar Z, Chitkushev LT, Javaheri T, Jahanban-Esfahlan R, Zare P (2020) Horizontal gene transfer: from evolutionary flexibility to disease progression. Front Cell Develop Biol 8: 229
Feigel A, Kotler Z, Sfez B (2002) Scalable interference lithography alignment for fabrication of three-dimensional photonic crystals. Opt Lett 27(9):746–748
Freel KC, Krueger MC, Farasin J, Brochier-Armanet C, Barbe V, Andrès J, Cholley PE, Dillies MA, Jagla B, Koechler S, Leva Y (2015) Adaptation in toxic environments: arsenic genomic islands in the bacterial genus Thiomonas. PLoS One 10(9):e0139011
Fuchsman CA, Collins RE, Rocap G, Brazelton WJ (2017) Effect of the environment on horizontal gene transfer between bacteria and archaea. PeerJ 5:e3865
Galan JE, Collmer A (1999) Type III secretion machines: bacterial devices for protein delivery into host cells. Science 284:1322–1328
Gallagher KA, Jensen PR (2015) Genomic insights into the evolution of hybrid isoprenoid biosynthetic gene clusters in the MAR4 marine streptomycete clade. BMC Genomics 16(1): 1–13
Gardy JL, Loman NJ (2018) Towards a genomics-informed, real-time, global pathogen surveillance system. Nat Rev Genet 19(1):9–20
Godeux AS, Svedholm E, Barreto S, Potron A, Venner S, Charpentier X, Laaberki MH (2022) Interbacterial transfer of carbapenem resistance and large antibiotic resistance islands by natural transformation in pathogenic Acinetobacter. MBio 13(1):e02631–e02621
Guglielmini J, de la Cruz F, Rocha EP (2013) Evolution of conjugation and type IV secretion systems. Mol Biol Evol 30(2):315–331
Gushgari-Doyle S, Lui LM, Nielsen TN, Wu X, Malana RG, Hendrickson AJ, Carion H, Poole FL, Adams MW, Arkin AP, Chakraborty R (2022) Genotype to ecotype in niche environments: adaptation of Arthrobacter to carbon availability and environmental conditions. ISME Commun 2(1):1–10
Hacker J, Blum-Oehler G, Mühldorfer I, Tschäpe H (1997) Pathogenicity islands of virulent bacteria: structure, function and impact on microbial evolution. Mol Microbiol 23(6): 1089–1097
Hacker J, Carniel E (2001) Ecological fitness, genomic islands and bacterial pathogenicity. EMBO Rep 2:376–381

Håvarstein LS (1998) Bacterial gene transfer by natural genetic transformation. APMIS 106(S84): 43–46

Ingle DJ, Tauschek M, Edwards DJ, Hocking DM, Pickard DJ, Azzopardi KI, Amarasena T, Bennett-Wood V, Pearson JS, Tamboura B, Antonio M (2016) Evolution of atypical enteropathogenic E. coli by repeated acquisition of LEE pathogenicity island variants. Nat Microbiol 1(2):1–9

Jain R, Rivera MC, Moore JE, Lake JA (2002) Horizontal gene transfer in microbial genome evolution. Theor Popul Biol 61(4):489–495

Juhas M (2019) Genomic Islands and the evolution of multidrug-resistant bacteria. In: Horizontal gene transfer. Springer, Cham, pp 143–153

Juhas M, Van Der Meer JR, Gaillard M, Harding RM, Hood DW, Crook DW (2009) Genomic islands: tools of bacterial horizontal gene transfer and evolution. FEMS Microbiol Rev 33(2): 376–393

Karaolis DK, Johnson JA, Bailey CC, Boedeker EC, Kaper JB, Reeves PR (1998) A Vibrio cholerae pathogenicity island associated with epidemic and pandemic strains. Proc Natl Acad Sci 95(6):3134–3139

Koukou I (2021) Management of Clostridium botulinum in food: A predictive food microbiology approach. DTU Food

Liu B, Pop M (2009) ARDB—antibiotic resistance genes database. Nucleic Acids Res 37(suppl_1): D443–D447

Luperchio SA, Newman JV, Dangler CA, Schrenzel MD, Brenner DJ, Steigerwalt AG, Schauer DB (2000) Citrobacter rodentium, the causative agent of transmissible murine colonic hyperplasia, exhibits clonality: synonymy of *C. rodentium* and mouse-pathogenic Escherichia coli. J Clin Microbiol 38(12):4343–4350

McArthur AG, Waglechner N, Nizam F, Yan A, Azad MA, Baylay AJ, Bhullar K, Canova MJ, De Pascale G, Ejim L, Kalan L (2013) The comprehensive antibiotic resistance database. Antimicrob Agents Chemother 57(7):3348–3357

McFarland LV (1998) Epidemiology, risk factors and treatments for antibiotic-associated diarrhea. Dig Dis 16(5):292–307

Nakamura Y, Itoh T, Matsuda H, Gojobori T (2004) Biased biological functions of horizontally transferred genes in prokaryotic genomes. Nat Genet 36(7):760–766

Nesbø CL, Doolittle WF (2003) Targeting clusters of transferred genes in Thermotoga maritima. Environ Microbiol 5(11):1144–1154

Ochman H, Moran NA (2001) Genes lost and genes found: evolution of bacterial pathogenesis and symbiosis. Science 292(5519):1096–1099

Ochman H, Lawrence JG, Groisman EA (2000) Lateral gene transfer and the nature of bacterial innovation. Nature 405(6784):299–304

Parsot C, Sansonetti PJ (1999) The virulence plasmid of Shigellae: an archipelago of pathogenicity islands? In: Pathogenicity islands and other mobile virulence elements. ASM Press, Washington, pp 151–165

Partridge SR, Kwong SM, Firth N, Jensen SO (2018) Mobile genetic elements associated with antimicrobial resistance. Clin Microbiol Rev 31(4):e00088–e00017

Peng X, Holz I, Zillig W, Garrett RA, She Q (2000) Evolution of the family of pRN plasmids and their integrase-mediated insertion into the chromosome of the crenarchaeon Sulfolobus solfataricus. J Mol Biol 303(4):449–454

Penn K, Jensen PR (2012) Comparative genomics reveals evidence of marine adaptation in Salinispora species. BMC Genomics 13(1):1–12

Penn K, Jenkins C, Nett M, Udwary DW, Gontang EA, McGlinchey RP, Foster B, Lapidus A, Podell S, Allen EE, Moore BS (2009) Genomic islands link secondary metabolism to functional adaptation in marine Actinobacteria. ISME J 3(10):1193–1203

Preston GM, Haubold B, Rainey PB (1998) Bacterial genomics and adaptation to life on plants: implications for the evolution of pathogenicity and symbiosis. Curr Opin Microbiol 1(5): 589–597

Rupnik M, Wilcox MH, Gerding DN (2009) Clostridium difficile infection: new developments in epidemiology and pathogenesis. Nat Rev Microbiol 7(7):526–536

Schaechter M, View From Here Group (2004) Escherichia coli and Salmonella 2000: the view from here. EcoSal Plus 65(1):119

Schiavo G, van der Goot FG (2001) The bacterial toxin toolkit. Nat Rev Mol Cell Biol 2(7): 530–537

Schleper C, Holz I, Janekovic D, Murphy J, Zillig W (1995) A multicopy plasmid of the extremely thermophilic archaeon Sulfolobus effects its transfer to recipients by mating. J Bacteriol 177(15):4417–4426

Schmidt H, Hensel M (2004) Pathogenicity islands in bacterial pathogenesis. Clin Microbiol Rev 17(1):14–56

Shimizu T, Ohtani K, Hirakawa H, Ohshima K, Yamashita A, Shiba T, Ogasawara N, Hattori M, Kuhara S, Hayashi H (2002) Complete genome sequence of Clostridium perfringens, an anaerobic flesh-eater. Proc Natl Acad Sci 99(2):996–1001

Spigaglia P, Mastrantonio P (2002) Molecular analysis of the pathogenicity locus and polymorphism in the putative negative regulator of toxin production (TcdC) among Clostridium difficile clinical isolates. J Clin Microbiol 40(9):3470–3475

Sulaiman S, Yusoff NS, Mun NS, Makmur H, Firdaus-Raih M (2019) Inference of horizontal gene transfer: gaining insights into evolution via lateral acquisition of genetic material. In: Reference module in life sciences. Elsevier, Amsterdam, pp 953–964

Tormo MA, Ferrer MD, Maiques E, Ubeda C, Selva L, Lasa Í, Calvete JJ, Novick RP, Penadés JR (2008) Staphylococcus aureus pathogenicity island DNA is packaged in particles composed of phage proteins. J Bacteriol 190(7):2434–2440

van Wolferen M, Ajon M, Driessen AJ, Albers SV (2013) How hyperthermophiles adapt to change their lives: DNA exchange in extreme conditions. Extremophiles 17(4):545–563

Waldor MK, Mekalanos JJ (1996) Lysogenic conversion by a filamentous phage encoding cholera toxin. Science 272(5270):1910–1914

Winstanley C, Langille MG, Fothergill JL, Kukavica-Ibrulj I, Paradis-Bleau C, Sanschagrin F, Thomson NR, Winsor GL, Quail MA, Lennard N, Bignell A (2009) Newly introduced genomic prophage islands are critical determinants of in vivo competitiveness in the Liverpool Epidemic Strain of Pseudomonas aeruginosa. Genome Res 19(1):12–23

Yoon SH, Park YK, Kim JF (2015) PAIDB v2. 0: exploration and analysis of pathogenicity and resistance islands. Nucleic Acids Res 43(D1):D624–D630

Zhang X, Liu X, Liang Y, Guo X, Xiao Y, Ma L, Miao B, Liu H, Peng D, Huang W, Zhang Y (2017) Adaptive evolution of extreme acidophile Sulfobacillus thermosulfidooxidans potentially driven by horizontal gene transfer and gene loss. Appl Environ Microbiol 83(7):e03098–e03016

Zolfaghari Emameh R, Barker HR, Hytönen VP, Parkkila S (2018) Involvement of β-carbonic anhydrase genes in bacterial genomic islands and their horizontal transfer to protists. Appl Environ Microbiol 84(15):e00771–e00718

Zolfaghari Emameh R, Barker HR, Tolvanen ME, Parkkila S, Hytönen VP (2016) Horizontal transfer of β-carbonic anhydrase genes from prokaryotes to protozoans, insects, and nematodes. Parasit Vectors 9(1):1–12

Genomic Islands Involved in Iron Uptake

8

Stanzin Choksket, Harshvardhan, Ashish Jain, Suresh Korpole, and Vishakha Grover

Abstract

Iron is an important element for all life forms. In microbial life, it plays a significant bearing either as an important growth factor and/or cofactor for various metabolic processes in case of environmental bacteria or as a virulence determinant for many pathogenic microorganisms to affect their disease-causing ability. Microorganisms have developed a variety of modes to acquire iron from local environment. In iron scarcity conditions, many bacteria adopt specific strategies to fulfill their iron requisite and survive. Distinct genetic machinery targeted for iron uptake and utilization have been documented and have been extensively studied. Different microorganisms harbor distinct genomic islands specifically intended to accomplish the iron uptake and few have been described in detail to provide insights into this important area. The current chapter provides an update on the various microbial mechanisms of iron uptake, general aspects of bacterial genomic islands and the details of the genomic islands involved in the microbial iron uptake mechanisms.

Keywords

Iron uptake · Genomic islands · Ferric ion · Siderophores · Heme · Yersiniabactin

S. Choksket · Harshvardhan · S. Korpole
CSIR-Institute of Microbial Technology, Chandigarh, India
e-mail: stanzin@imtech.res.in; harshvardhan@imtech.res.in; suresh@imtech.res.in

A. Jain
Dental Institute, Rajendra Institute of Medical Sciences, Ranchi, Jharkhand, India

V. Grover (✉)
Dr. HS Judge Institute of Dental Sciences and Hospital, Panjab University, Chandigarh, India

I. Mani et al. (eds.), *Microbial Genomic Islands in Adaptation and Pathogenicity*,
https://doi.org/10.1007/978-981-19-9342-8_8

8.1 Introduction

Iron is a vital component for all life forms including humans and microbes. It has a significant bearing on day-to-day life processes such as the interactions with bacteria, defensive ability, and many other vital metabolic activities considered crucial for life processes (Wessling-Resnick 2010; Nairz et al. 2010; Markel et al. 2007). The majority of living things have evolved to obtain iron from their nearby environment using the conserved evolutionary approach. Iron plays a vital role as a cofactor in bacterial replication and growth requirements, synthesis of DNA, energy metabolism, and protection against oxidative species. Thus, the deficiency of iron may hamper many essential bacterial metabolic processes (Sheftel et al. 2012). In such conditions, generally, ferric uptake is initiated by the derepression or activity of several iron acquisition gene programs. On the other hand, Ferric uptake regulator (FUR) proteins act as a regulator guided by iron concentration to stop iron intake in the presence of adequate amount of iron present in the local environment. Siderophores formation, cell surface enzymes that reduce iron to be used by the microbe and biochemical mediators, e.g., cytotoxins stimulating release of iron from host contribute toward the microbial modes of iron acquisition (Yilmaz and Li 2018).

It is important to understand iron metabolism to have better insights into microbial survival processes including the vital housekeeping as well as advanced mechanisms of adaptation (Andrews 2011; Utley 1990; Hederstedt et al. 2020; Mey et al. 2005). Some bacteria even utilize iron oxidation as a resource for energy production. Since iron is such a pivotal nutrient that almost all prokaryotes need this vital element for survival except the lactic acid bacteria, which can manage to survive without it. The versatility of iron may be gathered from the fact that it has a range of −300 to +700 mV as redox potential based activities, depending on the nature of the ligands and the local environment. However, microbes utilize the compound best in the lower oxidizing form of iron ion (+3-valence) at neutral pH. The compound also plays an essential role in modulating the expression of some virulence factors (Symeonidis and Marangos 2012; Andrews et al. 2003a, 2003b; Golonka et al. 2019; Kramer et al. 2020; Grenier and Tanabe 2011; Zhang et al. 2012). Thus, iron is a crucial element for the majority forms of gut bacteria except for *Bacteroides* spp. (Rocha et al. 1991; Otto et al. 1990) and *Enterobacteriaceae* (Carpenter and Payne 2014). Another exceptional strain is *Lactobacilli* species, which can grow without iron (Imbert and Blondeau 1998; Weinberg 1997). Some probiotic bacteria like *Lactobacillus plantarum 299v*, help in the increased absorption in host organisms (Botta et al. 2021).

Although relatively low iron concentrations are consistent with satisfactory growth yields of microbes, still many times low-level iron around the bacteria owing to the presence of iron-binding proteins, e.g., ferritin, lactoferrin, and transferring and must exercise a huge deal of adaptation to maintain the needful concentrations of the vital element. In the oral cavity, salivary lactoferrin serves as a crucial element in the maintenance of iron levels and metabolism in the context of oral microorganisms (van der Hoeven et al. 1984; Weinberg 2001; Lönnerdal and

Iyer 1995). The resident oral bacteria have devised a large variety of mechanisms to meet their iron requirements, e.g., reducing Fe^{3+} to the soluble Fe^{2+} via a membrane-associated ferric reductase activity (Evans et al. 1986), siderophores as high-affinity ferric chelators (Ge et al. 2009; Moelling et al. 2007), directly acquiring human transferring (Duchesne et al. 1999), or importing heme via heme binding protein (Liu et al. 2006). The essential requirement coupled with the limited bioavailability, and microbial iron regulation mechanisms make iron an important deterministic factor for microbial composition in the oral cavity (Wang et al. 2012).

8.2 Iron Uptake in Bacteria

There are many acquiring systems in the microbes that can work toward maintaining a balanced concentration of iron, for example, siderophore production, with its associated receptors and transport protein, etc. is one prime acquisition system that ensures the availability of iron in iron-deficient conditions to microbes. Similarly in very extreme conditions, there are microbial species known as extremophiles which tend to possess characteristics from diverse groups of bacteria, as their adaptive measures to sustain in that extremely harsh environment. One of these bacterial groups halophiles has its siderophores in distinctive structural characterization different from the usual structure seen in rest of the microbes.

Iron assimilation in bacteria occurs by diverse pathways, as several different resources of iron may be utilized to uptake environmental iron. Gram-negative bacteria, accomplish iron transport by complexing iron to a carrier which with the help of receptor protein in the outer membrane shall transport across based on iron concentration. A particular receptor protein is usually specific to bind a unique iron-carrier complex. The receptor proteins viz. TonB, ExbB, and ExbD generally work as a functional unit to act as gated pores utilizing the electrochemical potential of cell membranes and allowing the passage of iron or iron chelates across the cellular environment. Most of the bacteria have a single set of receptor export proteins that can interact further with multiple compounds; however, two microbial species, i.e., *Vibrio cholerae* and *Pseudomonas aeruginosa* have two sets of TonB systems which are spanning into the intracellular environment. There are observed subtle differences in the Gram-positive and Gram-negative microbes in harboring these receptor proteins. In Gram-positive cells, generally, the lipoprotein attached to cell membrane is hanging into the peptidoglycan region, whereas Gram-negative bacteria require a fully-fledged system of outer membrane receptor protein, Ton B system, and an ABC transporter out of which Ton B and ABC transporters are known to have broader specificity.

8.2.1 Siderophore-Mediated Iron Transport Systems

Siderophores are highly specific compounds which have low molecular mass (61,000 Da) and a huge affinity toward the ferric form of iron (Andrews et al.

2003a, 2003b). These are secreted by many lower life forms including bacteria and fungi and can reach huge concentrations up to 200 mg in response to iron-deficient conditions. So far, approximately 500 kinds of these compounds have been known and are classified according to the functional group involved in the binding of iron (Byers and Arceneaux 1998; Ratledge and Dover 2000). These are one of the prime mechanisms to obtain iron by microbial species, and sustain life in iron-deficient conditions. Two broad categories which are most well studied are catecholate or o hydroxamate types Siderophores from hexadentate octahedral complexes with ferric iron typically employ hydroxamates, which serve as much efficacious Fe^{3+} ligands (Winkelmann 1991; Winkelmann 2002). The siderophores compounds so formed or their modifications, for example, by the addition of some glucose moieties or the breakdown products, etc. may be synthesized in an iron-deficient environment and once these compounds are released from cells they are internalized by the microbes present using the available transport systems. Extensive studies into the siderophore biosynthesis mechanisms have revealed a new group of enzymes with phosphopantetheine transferases, the enzymatic reaction mediated by this particular class of enzyme enables the binding and aggregation of the activated precursor compounds to produce the final product by donating the P-pant to the peptide synthetases to make iron available for membrane receptor proteins in conjunction with the protein-dependent ABC transporters. Citrate is another important mediator compound which when present in an environment above the concentration of 0.1 Milli mole per liter serves as a useful resource iron transport system as it acts similar to dihydroxy benzoic acid and dihydroxy benzoyl serine which are the metabolic precursor and a breakdown product of siderophore enterobactin. In comparison to biosynthesis, the release of iron from these compounds is not elucidated much. Once these compounds are within the cellular environment, the mechanism responsible for release is usually thought to be either the lesser affinity of these compounds to ferrous form or the reducing environment of the cytoplasm where enzymatic reduction of iron is considered to release the free siderophores. There have been many in vitro studies to demonstrate the presence of such reduction activities employing ferri-siderophore reductase activities. Usually, these enzymes have a broad range of substrate specificity and are not affected by the presence or absence of iron in the environment. Many new variants of these enzymes are being discovered, e.g., Fhu F from *E. coli* which is specific for three ferric hydroxamate siderophores. In terms of transport across membranes, ABC transporters are the most widely used set of ATP-utilizing transporters. There are other transporter families such as MFS and RND which are utilized by specific species of the genera *Vibrio, Bordetella, Legionella* and *Pseudomonas*. The efflux pumps employed in iron transport have workability to many substrates, for example, *E.coli* utilizes two proteins, i.e., ENTs and Tol C for its enterobactin secretion. Tol C is an outer membrane protein that is vital for efflux pump functioning in many bacteria. In specific situations such as in the case of a high acidic environment, iron is oxidized for energy production by bacteria and different mechanism of acquisition of the element from environment has been described. Considerate research effort has been focused in the recent past on bioinformatics approach to study iron transport proteins. It was found that

Acidithiobacillus ferrooxidans, *Acidithiobacillus thiooxidans*, and *Acidithiobacillus caldus* possess the corresponding genetic information required for the oxidation of ferrous iron (Andrews et al. 2003a, 2003b).

Acidithiobacillus ferrooxidans, *Acidithiobacillus thiooxidans*, and *Acidithiobacillus caldus* all possess the corresponding genetic information that is responsible for the oxidation of ferrous iron. These microorganisms have helped study some facets of the mechanism and suggested that *A. ferrooxidans* is capable of complementing the functions of Dfur *E. coli* (Quatrini et al. 2005). Other bioinformatic studies revealed that genetic machinery for iron metabolism was upregulated in *Ferroplasma acidarmanus* (Potrykus et al. 2011) particularly in iron-poor conditions. Osorio et al. 2008; Potrykus et al. 2011 have shared some of the computational analysis-based genetic information in this regard. *A. ferrooxidans* and *A. thiooxidans* are known to have genetically an efficient setupfor iron metabolism including citrate (a siderophore) production, a citrate efflux pump and a Ton B-dependent Fe(III)-dicitrate transport systems (Osorio et al. 2008). Also, Potrykus et al. (2011) reported the production of a siderophore system in *F. acidarmanus*, without much elucidation on the structural details. A number of phylogenetic studies have been specifically carried out in acidic soil environments on secondary metabolite production from rhizobia, yet not much information on the structure could be obtained (Yadav et al. 2011). The isolated siderophores include *Pseudomonas putida* (7), *P. fluorescens* (2), *Rhizobium radiobacter* (1), *P. syringae* (1), and *Bacillus atrophaeus* (1) Karagoz et al. (2012). Another report from Verma et al. also measured the production activity of siderophore from a microbe viz. *P. fluorescens, yet hardly any information on the structural aspects was obtained* (Verma et al. 2007). Only a compound pyoverdine from *Pseudomonas* species was considered as a potential candidate (Kalinowski et al. 2006).

8.2.2 Iron Acquisition Through Ferrous and Ferric Ions

Utilizing the outer membrane, the iron can enter in the form of Fe. Microbes have multiple systems in place to enable iron absorption from the extracellular environment. Cell cytoplasmic membrane transporters generally have an affinity for the transition metals at large, yet show some limited affinity for ferrous form. Microbial iron (II) transport system, known as Feo, has been well studied in *E. coli* and serves an important role in ensuring iron supply under anaerobic conditions. FeoB is a huge constitutional membrane protein, which spans approximately to involve 700–800 amino acid residues and functions as the mainstay of this transport system. FeoB functions optimally with two other regulated small proteins, FeoA and FeoC, and is ATP/GTP dependent for transport.

Streptococcus mutans, being an aerotolerant bacteria has a different system to uptake iron. They utilize reductase enzyme on the outer surface of the cell to convert surface-bound iron into ferrous form so that it may be taken inside the cellular cytoplasm. Gram-negative flora, e.g., *Serratia* has an acquisition system for iron which is a kind of ABC transporter and has been described for the first time in this bacteria. The same system named Sfu-type transport is responsible for iron acquisition in *Haemophilus*, *Yersinia*, *Actinobacillus*, and *Neisseria*. Ferric iron transporters along with uptake systems in outer membrane also facilitate iron transport in many microbial organisms (Andrews et al. 2003a, 2003b).

8.2.3 Iron Acquisition from Heme

Animals typically have bonded iron because it is found in cells as heme and acts as a prosthetic group for many animal proteins, most notably hemoglobin. Other, heme-containing proteins are myoglobin, cytochromes, etc. and to obtain free iron from these substances, it has to be released by the hemolysis of red blood corpuscles. The extracellular heme further binds to multiple plasma proteins and glycoproteins and thus, becomes available as a possible source of iron. For most bacterial species such as *Shigella, vibrio,* and *Yersinia* the entire heme molecule may enter the cell. On the other hand, all strains of *E. coli* are not able to obtain iron from heme but certain strains having an ABC transporter for dipeptides may complex with hemoglobin transporter and serve as additional roles in iron transport from heme. Several other microbial species, for example, *Nesseria* have a designated transporter for heme which is a TonB-dependent transport system. Another variant of extracellular heme binding protein is HasA which can uptake free or bound hemoglobin and transfer iron to receptors at the outer cellular membrane. *E. coli* having a specific plasmid pqv also secretes hemoglobin binding protein. All these transfer proteins require functional TonB-dependent outer membrane systems. Specifically, hemophilus strains can utilize heme associated with hemopexin, which also requires a functional TonB protein (Andrews et al. 2003a, 2003b).

8.2.4 Iron Uptake via Transferrin and Lactoferrin

TF and LF are the vital components involved in the extracellular iron transport system, particularly in vertebrate forms of life. These are two structurally related glycoproteins, which harbor an ability to bind two ferric ions each. Many of the bacterial species such as *Neisseria* and *Haemophilus* absorb iron with the help of an iron receptor transferrin and do not make siderophores. This bipartite receptor contains two distinct proteins, Tbpa and Tbpb, which are present on the surface and iron regulated. Protein b is a lipoprotein that discriminates between appo and holo transferrin and its expression is not mandatory for transferrin-based iron uptake. The uptake of ferric siderophore and vitamin B12 via bacterial cell membranes is carried out by TbpA, a member of TonB-dependent transporters, via a protein gradient. This protein differs from other mechanisms where the whole complex is internalized, but in this case, iron is removed from the transferrin at the cell surface (Cornelissen 2003). Another protein of the same family with 60% amino acids similar to transferrin is Lactoferrin. It was first discovered from bovine milk, though the protein shares similar composition, secondary and tertiary structures to transferrin, it differs in their biological activities from transferring. There are three different isoforms to LF with Alpha form being the only form binding iron whereas beta and gamma both have ribonuclease activity only. LF helps the utilization of iron by pathogenic bacteria as it may directly accept ferric iron from other proteins and serves as an important player by preventing bacteria to sequester iron, which is an essential requirement for their well-being and virulence (Andrews et al. 2003a, 2003b).

8.2.5 Low-Affinity Iron Uptake Systems

The low-affinity iron transport systems comes to the mainstream when the concentration of iron becomes limited in extraneous environments and are generally observed in more than 5–10 mmol l–1 iron present in the vicinity. The prime requisite for these systems to work is the presence of iron in the Fe form. Typically, the feo system proteins or a cell membrane-associated transporter system that has broad selectivity for interacting molecules like divalent cations, such as the CorA protein of *E. coli* and *Salmonella typhimurium*, are used to ingest the Fe form (Earhart 2009).

8.2.6 Iron Uptake Mechanisms in Pathogenic Bacteria

Most of the iron in higher forms like mammals exists usually in a bound state to iron binding proteins and maintains a reduced level of the free extracellular iron, which is unable to support the growth of microbial species, hence infections. Apart from this, the host also produces many oxygen transport and storage proteins such as hemoglobin and myoglobin, which carry heme as an Iron containing central molecule. These molecules further reduce the amount of iron for pathogenic bacteria (Litwin and Calderwood 1993). Pathogenic flora employs several mechanisms to cope with these systems and develop many sophisticated mechanisms to sequester iron in such low iron conditions. For instance, iron deficiency triggers the Shiga-like toxin I of enterohemorrhagic *E. coli* (Calderwood and Mekalanos 1987). Some others can combat the host-driven iron restriction via siderophores. Siderophores act as competitive analogs to host iron-binding proteins, whereas some siderophore-based transport systems serve as portals for ejective host colonization (Williams and Warner 1980; Winkelmann 1991). However, many pathogens may directly utilize the host iron complexes via receptor-mediated transport systems, instead of free iron or releasing the iron from its bound form (Andrews et al. 2003a, 2003b).

8.3 Bacterial Iron Homeostasis

Iron is an important biomolecule involved in multiple biological processes. The functionality of iron in microbial life depends on many factors such as if it is in free or bound form with proteins, exists as a mono or binuclear species and whether it forms a member of some iron-based complex such as iron-sulfur clusters or heme groups. Iron molecules existing within a protein complex, appear a little bit more manageable in the local environment as the presence of protein allows a controlled interface for the diverse functionality of the iron assuming the necessary redox potential, geometry, and spin state. Oxygen availability and iron metabolism are very closely linked as the bacteria thrive in varying conditions of both these compounds over a great range. The most common shift of the diverse form of iron was a transit from a soluble form, i.e., ferrous state to a quiet stable insoluble form of ferric ion, which makes the element much less available. Also, this particular form seems to have a potential for toxicity, with oxygen being available in the vicinity. Iron mediates many important biochemical reactions specifically related to oxidation mechanisms such as superoxide formation, hydroxyl free radicals production, and Fenton reactions in the cellular environment and prevents the biological damage incident on cellular DNA. The regulation of iron sulfur cluster is primarily accomplished by an autoregulated protein termed as ferric uptake regulator(Fur), which serves as a repressor in the regulatory mechanism for most iron uptake promoters (Vogt et al. 2021; Delany et al. 2003). Thus, to ensure adequate iron supply for normal cell cycle and cellular activities, the microbes have to develop unique strategies for iron uptake from environment depending upon the amount of available iron concentrations in the local milieu. The most significant five mechanisms used by bacteria to manage iron are High-affinity transport enabling iron scavenging, intracellular storage for periods of iron shortage, redox stress resistance systems enabling the inducible release of radicals, downregulation of certain iron-containing proteins, and a highly orchestrated iron responsive regulatory system which mediates all the associate machinery in coherence according to the prevailing environment, ecological niche, and phylogeny (Andrews et al. 2003a, 2003b).

8.4 Genetic Aspects of Iron Uptake Mechanisms in Bacteria

The bacterial genomes keep evolving with diverse processes, for example, mutations, genetic rearrangements, and even by horizontal gene transfers across multiple species. Thus, many accessory genes are acquired by horizontal transport from syntenic blocks which have been recently recognized as the emergence of genomic islands. These islands have led to the belief that they contribute to the diversification and adaptation of microbial species. These alterations impact the genetic characteristics of the species in a big way as significant features such as antibiotic susceptibility, virulence, and many catabolic processes are carried via genomic islands (Schmidt and Hensel 2004).

GEIs are frequently identified as distinct DNA stretches that separate closely related strains. These are considered important portals of genome plasticity and evolution, as these tend to allow huge flexibility and adaptation in characteristics as needed for the survival of the organisms. With the advent of novel analysis methods and advances in bioinformatic technologies, a great deal of information is emerging to understand these genetic structural entities in a better and clearer way. This information further paves the path for a variety of biotechnological applications in diverse fields of life (Juhas et al. 2009).

8.4.1 Genomic Islands

Genomic Islands are also known popularly as "pathogenicity" Islands. The term came in last some years to denote the genomic islands encoding pathogenicity features and has attracted a great deal of interest from researchers (Hacker and Carniel 2001). These have been first and most studied in the case of *E. coli* but subsequently, many other pathogenic microorganisms also showed the presence of these specific entities, which were related to the pathogenicity of the microorganism. Genome sequence studies showed a much wider prevalence of these genetic islands and demonstrated that actually these represented a pool of genes for adaptation and flexibility. Genomic Islands usually range between 10 and 100 kilobases in length (Hacker and Kaper 2000). They typically contain genetic sequences acquired from plasmids and/or phages, especially integrases or IS elements. These most frequently emerge as tRNA genes' unstable component parts. Instability stems from the flanking direct repeats, which appear similar to phage attachment sites and alterations for transit across the genomes by disintegration and assimilation into the new genome (Hacker et al. 1997). Usually, they carry distinct function-related gene clusters. These islands appear distinct from the core genome depending on their G + C composition and codon use. The advantage of genomic islands over smaller genetic changes is that these help the transfer of multiple genes in one go and its incorporation en bloc into the recipient genome. The gene responsible for encoding the siderophore yersiniabactin-mediated iron absorption mechanism is called HPI and has a 36–43 kb GEI. Numerous bacterial GEIs contain type III and type IV secretion systems (T3SS and T4SS), which directly influence pathogenicity and horizontal gene transfer by transferring proteins or nucleoprotein complexes. *P. aeruginosa* PA14 also encodes many biomolecules involved in iron metabolism. *Magnetospirillum gryphiswaldense* is known to carry a 130-kb unstable region of probable ancient GEI origin, lately has been documented as a region containing multiple IS causing huge amounts of genetic alterations and also mediating an interesting phenomenon of magnetosome biomineralization. The anaerobic bacterium *Geobacter sulfurreducens* also exhibited the presence of ancient GEIs with deteriorated integrases with no detection of transfer or excision (Butler et al. 2007). There were grouped genes in this 300 kb region that were involved in the anaerobic metabolism of organic compounds.

8.4.2 GEIs—General Features

GEIs are distinct DNA segments that vary among mobile, closely related bacterial strains. They represent an expanded class of DNA elements with a huge range of size and diversity beyond PAIs. These entities reveal differential coding capacity for several important functional aspects such as ability to coexist, metal assimilations, and toxicities in addition to pathogenicity functions (Hacker et al. 1990; Groisman and Ochman 1996; Sullivan et al. 2002; Gaillard et al. 2006; Larbig et al. 2002). These observations state their strong distinctive ability to provide the organism with a selective advantage of many adaptive, flexible attributes, and auxiliary functions.

The extremes of variability in the genetic makeup and the associated functional profiles sometimes make it difficult to suggest a typical definition and landscape of work for GEIs. They usually appear as a unique entity, with salient differences from the rest of the chromosome. GEIs appear as a component part of tRNA genes, often as ICEs. GEIs are frequently followed by 16–20 bp of perfectly or almost perfectly repeated direct sequences (Dobrindt et al. 2004). Different evolutionary-related and ancient GEI families have been recognized based on prediction of sequence and functional similarities (Burrus et al. 2002; Juhas et al. 2007; Vernikos and Parkhill 2008).

Direct repeats serve as identifying sequences for enzymatic activity required for the excision of sequence and often appear depending upon site-specific integration of the island in the target site (Schmidt and Hensel 2004). These regions frequently contain functional or hidden genes that code for integrases, plasmid conjugation systems, phages carrying insertion elements, or transposons, which are involved in transferring genetic material to or erasing DNA from the element (Buchrieser et al. 1998; Gal-Mor and Finlay 2006). GEIs are often associated with the transfer of genetic machinery specifically implicated to provide a selective advantage for the bacteria harboring GEIs (Dobrindt et al. 2004; Schmidt and Hensel 2004).

8.4.3 Genomic Island Identification Methods

Microbial isolation and culturing may or may not be required as part of the approaches for studying genomic islands. However, more thorough analyses, like phenotypic screening of (meta)genomic libraries for traits suggestive of mobile and transferable DNA elements, do require bacterial cultivation when specific and more concrete information is required, such as with comparative genomic analysis using suppressive subtractive hybridization, more frequently for strain-specific DNA regions, or DNA–DNA hybridization for mapping known genes. To detect any horizontally acquired DNA, flanking sequences alterations in tRNA genes and repeat structures, e.g., insertion sequences (IS), the microbes are cultured by standard methods first and further analysis of the overall base composition, G + C content / skew, and delta-differences are characterized. Another unique Island probing strategy that uses counter-selectable markers and other techniques to examine the stability of a DNA region is likewise characterized as a technique involving the isolation and growth of bacteria (Dobrindt et al. 2004).

8.4.4 Types of Genomic Islands—Contributions to Ecological Adaptation and the Pathogenesis

It is well documented that genomic islands are critical in transferring a part of the host chromosome into other related or unrelated bacterial genomes (Hochhut et al. 2000). These GEIs undergo genetic recombination and can modify the characteristics of bacteria. Given that many crucial clinical or fitness features are transmitted via genomic islands, these changes may eventually have an impact on the evolution of the host bacteria (Hamilton et al. 2005). The categorization of these fitness islands into different kinds is not merely based upon their genetic constitution, but more often are related to the specific niche and the organism, into which they are present. Thus, the functionality of the genomic island may be diverse in different environments (Groisman and Ochman 1996; Vernikos and Parkhill 2008), for example, genomic islands contributing toward iron uptake may act as a pathogenic factor in *yersiniabactin,* however, the same genome island is harmless in other nonpathogenic species and constitutes constitutive machinery contributing to the general well-being of the bacteria. These are also termed as an ecological islands, with some role assigned in the cellular metabolism. If such an island is present in bacteria residing in a host with virulence features, it shall be termed a pathogenicity island. On the contrary, integration of the island into the genome of a nonvirulent microbe may constitute saprophytic islands.

In this context, fitness is defined as characteristics that help an organism survive, spread, and/or transmit within a particular ecological niche. Because it increases microbial transmission, survival, or colonization within a niche, having a genomic island may confer a selective advantage under particular environmental conditions. From a functional perspective, genomic islands that increase the fitness of the recipient microorganisms should be called "fitness islands," by Preston et al. In contrast, a genuine pathogenicity island assists directly or indirectly in developing lesions. As a result, a "symbiosis island" is a form of Fitness Island that aids bacteria in favorably interacting with their hosts. In contrast, a genuine pathogenicity island assists directly or indirectly in developing lesions (Juhas et al. 2009).

8.5 Genomic Islands Implicated in Iron Acquisition

Iron is vital for both pathogenic and nonpathogenic bacteria to propagate bacterial populations. Regarding harmful bacteria, generally, availability of iron is much limited in higher forms as most of it remains strongly bound to specific binding proteins like transferrin and lactoferrin. However, there are some species capable of adhering directly to these carrier molecules and obtaining iron, e.g., *H. influenzae* (Finlay and Falkow 1997). Siderophores, in pathogenic *Yersinia* also serve as the means for acquiring iron from environment as they bind to iron more strongly as compared to the host's proteins (Hacker and Carniel 2001). A yersiniabactin coding "high pathogenicity island" (HPI) has been identified in *Klebsiella* also (Hacker and Carniel 2001).

Except for a few exceptions like *Borrelia burgdorferi* strains and *lactobacilli*, most bacteria need iron as a growth factor. As siderophores are the prime mode of acquisition of iron by most pathogenic bacteria, thus islands encoding siderophore systems are considered as PAIs and as "fitness islands" when they enhance the niche adaptive characteristics, in case of environmental bacteria. "High pathogenicity island" (HPI), initially described in virulent *Yersinia* spp. is one such example and has been documented in many disease causing and other bacteria. A progenitor island to HPI has been observed in an *E. coli* isolate related to functional T4SS, active DNA-processing region, and genes necessary for mobilization. Thus, it has been suggested to simulate an IC, which can disseminate effectively across genomes. Many other siderophore systems are located on plasmids, PAIs, and GEIs of diverse microbiota indicating the enhanced capacity of the organisms for maintaining good iron levels by virtue of intaking extraneous DNA (Dobrindt et al. 2004).

Though the presence of very similar HPI-borne ORFs among different species suggests that probably such acquisitions are not very old in terms of evolutionary changes of these bacteria; however, the underlying mechanisms responsible for this kind of mediating the horizontal transfer yet need exploration (Lesic and Carniel 2005; Juhas et al. 2009).

Three human pathogenic strains of *Yersinia, Y. pestis, Y. pseudotuberculosis, and Y. enterocolitica*, have been used as suitable model organisms to research siderophores and their effects on bacterial fitness and metabolism (Rakin et al. 2012). In fact the species delineation into different pathogenic abilities has been based on their ability to deal with iron uptake and expression of a siderophore yersiniabactin (Ybt), which remains a major determinant for murine pathogenicity (Carniel et al. 1987; Heesemann 1987; Rakin et al. 1994). Contrary to this, other two lesser pathogenic strains, i.e., *Y. pseudotuberculosis and Y. enterocolitica* (low and apathogenic) lack Ybt which led to the belief that Ybt solely exists as an endogenous iron regulatory system for highly pathogenic species of *Yersinia* only.

8.5.1 Yersiniabactin

This *Yersinia* species' siderophore system has undergone the most in-depth research (Perry and Fetherston 2011). It was discovered that Ybt shares a lot of structural similarities with two additional chemicals, pyochelin and anquibactin, which are made by the bacteria *Pseudomonas aeruginosa and Vibrio anguillarum*, respectively. Phenolate, thiazoline, and thiazolidine rings were observed in the structure of the molecule. Non-ribosomal peptide synthetase (NRPS) and polyketide synthetase (PKS) pathways were employed in the synthesis from salicylate, three cysteine molecules and three methyl groups (Gehring et al. 1998). *Six genes, notably irp1—irp5 and irp9 in Y. enterocolitica and irp1–2, ybtU, T, E, and S in Y.* pestis and *Y. pseudotuberculosis* are found in the Ybt cluster. Chorismate is immediately transformed into salicylate, the YbtS precursor, via Irp9 (YbtS) (Pelludat et al. 2003). The activated salicylate is transferred to HMWP2 via the Irp5 (YbtE) salicyl-AMP ligase (encoded by irp2). Irp8 (YbtX in *Y. pestis*), which was believed to be in charge of exporting Ybt (Fetherston et al. 1999); was documented to have some alternate pathways also working as YbtX—mutants could also secrete the molecule. Fe-Ybt assimilation occurs through both outer and inner membrane proteins such as FyuA and Irp6-7, respectively (Fetherston et al. 1999; Brem et al. 2001; Perry and Fetherston 2011). There is limited information on the periplasmic ferric Ybt transport protein. The major synthetic and transport genetic machinery of Ybt are contained in the mobile genomic element showing recombination activity to a particular site as seen in P4-like integrase (Rakin et al. 1999; Carniel 2001). It is widely distributed in the family Enterobacteriaceae owing to its locus present on the mobile element accounts (Schubert et al. 2004; Antonenka et al. 2005). Multiple factors (environmental and host) control the synthesis and activity of YbT system. Iron-loaded ferric uptake regulator (Fur) protein is involved in the repression of the biosynthetic and transport genes (Fetherston et al. 1999; Anisimov et al. 2005). Apart from this, there is an activation controlled after the transcription by P-pant transferase considered essential for phosphopantetheinylation NRP/PK synthetases (Bobrov et al. 2002). The loss of activity of the Ybt system results in a significant loss of the pathogenicity and virulence of the bacteria (de Almeida et al. 1993; Heesemann et al. 1993; Rakin et al. 1994; Bearden et al. 1997; Pelludat et al. 1998; Brem et al. 2001). *Y. enterocolitica 1B* harbors a single siderophore, i.e., Ybt. The organism loses its siderophore activity and virulence on its inactivation (Schwyn and Neilands 1987). These findings have come from a comparative analysis of the ybt genes expression in LB media and in vivo growth conditions. The ybt genes were upregulated after *Y. pestis* inoculation through various routes such as subcutaneous (Sebbane et al. 2006) and intranasal (Lathem et al. 2005; Liu et al. 2009) in different animal experimental studies. These findings were however, not recapitulated in the spleen and liver in mice (Liu et al. 2009) and in the flea (Vadyvaloo et al. 2007; Vadyvaloo et al. 2010). On the contrary, a strong evidence for its role in pathogenicity was gathered from the mammalian hosts. This Ybt siderophore system not only contributes by iron uptake only, rather it imparts the bacterial survival, particularly to some species of *E. coli* implicated in urinary infections by enabling the microbial species to tackle the excess and toxicity of copper in local environment (Chaturvedi et al. 2012).

8.5.2 Pseudochelin

Another closely related genetic locus to ybt gene cluster, for siderophore system, has been observed in *Y. pestis* and *Y. pseudotuberculosis* strains, which is very potent, however, no such locus has been seen in *Y. enterocolitica*. These are accordingly named as Yersinia non-ribosomal peptide (ynp) (Perry and Fetherston 2004; Forman et al. 2010). It was initially referred to as the second HPI in *Y. pestis* CO92 (Parkhill et al. 2001), however, due to the absence of mobile elements such as integrase-encoding genes or recombination sites, it was later altered to refer to a separate system. In *Y. pseudotuberculosis IP32953*, the ynp locus (YPTB3290–3298) contains putative NRPS and PKS siderophore assembly genes, transport genes, and outer membrane receptor coding genes. This locus does not contain a salicylate synthase (ybtS), suggesting an alternative biosynthetic precursor. Enzymatic analysis methods have revealed that the YPTB3296 and YPTB3297 genes exhibited three thiazoline rings and molecular mass of 404.4/460.37 m/z with and without Fe^{3+}, respectively (Forman et al. 2010). Ybt and Ynp clusters have diverse distribution patterns in *Y. pseudotuberculosis O1*. In *Y. pestis CO92*, the ynp cluster has two chromosomal loci where YPO1011–1012 codes for the receptor and N-terminal part while YPO0770–0778 have the rest of the genes. Such observations over a period clarified the specific functional status of the ynp locus in *Y. pestis* (Forman et al. 2010) and it was found to be specific to the sporadic *Y. pseudotuberculosis* strains and named as pseudochelin (Pch) coding locus. The expression of the ynp genes was supported by multiple investigations including in vivo and in vitro studies using microarray and quantitative real-time PCR. It was thought to be another Fur-dependent locus that got upregulated in iron scarcity conditions (Han et al. 2007; Gao et al. 2008). A lot of evidence mounted on the upregulation of the putative ynp receptor gene (YPTB3298) in *Y. pestis and Y. pseudotuberculosis in low iron conditions from different researchers* (Han et al. 2007; Gao et al. 2008; Rosso et al. 2008). Further, it was observed that growth conditions also altered the upregulation of the cluster as human plasma could witness greater upregulation as compared to LB media (Chauvaux et al. 2007). Studies using the murine pneumonic model revealed that pseudochelin played a function in the early stages of infection (Lathem et al. 2005). Similar findings in the rat bubonic infection model supported the observed expression level (YPTB3298 and YPTB3296) in vivo as opposed to the expression level in LB media (Sebbane et al. 2006; Vadyvaloo et al. 2010). When cells were cultured in PMH2 medium, putative Ynp receptor was found in higher concentrations in the outer membrane extracts of *Y. pestis* KIM6 at 26C compared to 37C (Pieper et al. 2009). The YbtD siderophore assembly tool could make up for transferase components (Forman et al. 2010). Although the production of aberrant Ybt like molecules by the ynp operon has not yet been proven in laboratory trials, this data has provided valuable insight into the potential connections with other siderophore systems that already exist (Miller et al. 2010).

8.5.3 Yersiniachelin

Another siderophore system which is somewhat different from the rest is a purified fractional part of the autoagglutination factor (AF) of the Ybt-negative strain *Y. pestis* EV76 (Podladchikova and Rykova 2006). It is a 17,485-kDa complex protein. The AF protein was discovered to be the Hcp (hemolysin coregulated protein)-like component of an extracellular apparatus of the type six secretion system (T6SS) of Gram-negative bacteria (Podladchikova et al. 2011). Yersiniachelin (Ych) revealed hydroxamate groups in chemical analysis. All *Y. pestis* and *Y. pseudotuberculosis* strains have shown the presence of this cluster except *Y. enterocolitica*. Removal of three genes responsible for the synthesis of the putative hydroxamate siderophore (ysu IHG) in a knockout experiment for *Y. pestis* EV76 resulted in the absence of the production of Ych (Podladchikova et al. 2012) support the accountability of the ysu locus for production and assimilation of the siderophore. *Bordetella* spp. has shown a very similar system NRPS independent alc cluster for another hydroxamate siderophore alcaligin (Alc) primarily mediating the growth, differentiation, and the disease-causing ability of the bacteria (Brickman et al. 2008, 2011; Brickman and Armstrong 2009). Diamine precursors are assembled by an enzyme mechanism that involves multiple steps and is controlled by the alc cluster in *Bordetella* species. The four siderophore synthesis-related genes (ysuGHIJE and odc, alcABC), outer membrane receptors (ysuR and fauA), ferric iron reductase genes, and Ysu and Alc clusters have commonalities (ysuF and alcD). The ysu cluster and *Y. pestis* genomes, however, do not contain the homolog of the alcE biosynthetic gene, which codes for an iron-sulfur protein involved in the C-hydroxylation of precursor molecules during Alc biosynthesis. This is where they diverge from one another. The macrocyclic hydroxamate siderophore has several conceivable structural modifications (Forman et al. 2010). Second, unlike *Bordetella* cells, the ysu cluster lacks a homolog of the alcS gene (Brickman and Armstrong 2005). The same's export mechanisms are scarcely known. Third, the alc cluster lacks the ysuABCD genes that code for cellular membrane trafficking. *Bordetella* completes this specific stage through a different mechanism (Brickman et al. 2011). Multiple structurally unique siderophores carried by the FbpA ABC transporter move iron from the periplasm across the cytoplasmic membrane. Fourth, *Bordetella's* synthesis and transport-related genes lack a homolog in the ysu cluster for the AraC-type transcriptional regulator AlcR (Brickman and Armstrong 2009). From numerous study types, such as microarray in vitro, ex vivo growth in human immune cells, and in vivo animal studies, there is growing evidence that *Y. pestis* expresses the ysu genes.

Previous observations regarding *Y. pestis* strains in laboratory studies reveal that the temperature does not affect the transcription of ysu genes (Han et al. 2004; Motin et al. 2004), however, presence of iron tends to downregulate and vice versa (Zhou et al. 2006; Gao et al. 2008). In a comparative analysis (Han et al. 2004; Han et al. 2007), it was found that high salinity and hyperosmotic stress Omp R-dependent signals also have an effect on the cluster, with OmpR potentially serving the same function as AlcR in the expression of the alc genes. Plasma upregulation of ysu genes for *Y. pestis* and *Y. pseudotuberculosis* were noted in ex vivo studies (Chauvaux et al. 2007; Rosso et al. 2008). No change was observed in expression inside J774.1 macrophage-like cells in Ybt-negative *Y. pestis strains* (Fukuto et al. 2010). Within the cells, many genes controlling iron levels were downregulated indicating that the organism is able to maintain its iron homeostasis with adequate levels of iron. The ysu genes failed to express in flea vector in vivo (Vadyvaloo et al. 2010), whereas did not differ in expression in case of mice lungs, spleen, and liver compared to in vitro LB medium (Lathem et al. 2005; Liu et al. 2009). The upregulation of these in the bubo of s.c. infected rats (Sebbane et al. 2006) actually hinted at their role in the early infective phase. Proteomic analysis unraveled the encoding process of ysu genes (Pieper et al. 2009). The biosynthetic protein and siderophore receptor were deciphered to be the part of membrane proteome and YsuR protein and YsuG (AlcC homolog) expression increased in the outer membrane and periplasm during iron starvation (Pieper et al. 2010). It was anticipated that probably the final steps of Ych synthesis are occurring in the periplasm. Thus, the third siderophore system encoded by the ysu locus is expressed in extremely pathogenic *Yersinia*. Several studies, including mutagenesis investigations, transcriptome, and proteomic analyses, as well as the separation of the Ych siderophore from the strain, have confirmed this finding. However, the exposure of Ych linked to the Hcp-like protein YPO502 in iron-rich environments and the upregulation of its production in response to stress implies that Ych has another function in the physiology of *Y. pestis* (Rakin et al. 2012).

The *Yersinia* HPI exhibits pathognomonic characteristics of a pathogenicity island since it is a significant chromosomal region that contains crucial virulence genes and has a tRNA gene nearby. The area has significant variation in terms of the insertion sequences, G + C content, and an integrase gene, from chromosome in general (Koczura and Kaznowski 2003; Carniel 2002).

The HPI is the most widely distributed and well-studied PAI among Enterobacteriaceae and has been used to understand the horizontal gene transfer in this process (Schubert et al. 2009). This system establishes the coding of the siderophore yersiniabactin, as a very efficient and significant mechanism for the maintenance of iron homeostasis. The genetic sequence of the island is pretty conserved in different species of microbes with primarily two kinds of variants witnessed in ExPECs: A minimal proportion (approx. 1%) of HPI-positive *E. coli* strains has a completely self-transmissible integrative conjugative element island. While the majority of *E. coli*, or around 99% of them, exhibit a non-self-transmissible island combined with a deletion of about 30 kb, containing genes necessary for the island's mobilization (Schubert and Norenberg 2010; Messerer et al. 2017). The SHI-2 pathogenicity island of *Shigella flexneri*, which contains the aerobactin genes, is a second significant PAI (Vokes et al. 1999; Koczura and Kaznowski 2003).

Koczura and Kaznowski (2003) studied the genetic makeup and its diversity of *Yersinia* HPI in clinical *K. pneumoniae* isolates. Its location was almost the same way as observed in *E. coli* (Karch et al. 1999) and *Y. enterocolitis* (Carniel et al. 1996) strains, i.e., close to asn T gene. However, in the *Y. pseudotuberculosis and Y. pestis*, its location was variable in relation to any of the three or two asn tRNA genes, respectively (Buchrieser et al. 1998). A 30-bp deletion inside the gene was inferred based on the short version of PCR output for fyuA gene (strains RK 74 and RK 75). The same observations have been documented in some earlier investigations also for some of the disease-causing strains of *E. coli* (Schubert et al. 1998). Even the integrase gene has seen deletions. As was the case with Yersinia bacteria (Bach et al. 2000), IS100 was not constraining the left boundary in *Klebsiella* spp. strains and some *E. coli* isolates (Karch et al. 1999). According to the scientists, the primers were complimentary to any deletions or lack of conservation in DNA sequences found throughout the experiment. This information has previously been published for another strain of *Klebsiella oxytoca*. The arrangement of the HPI genes in the *K. pneumoniae* strains was compared to that of *Yersinia* spp. and *E. coli* and suggested that these species may have undergone horizontal gene transfer (Koczura and Kaznowski 2003).

8.5.4 Specific PAIs Involved in Iron Uptake

Numerous bacteria have created capture systems that can absorb the iron complexed with different molecules like citrate ions (FEC system) or heme that is coupled to iron (chu system). The bacteria utilize Sit systems or generate siderophores, which are PK-NRP chemicals (Braun 2003). Some siderophores, such as enterobactin (*E. coli*), are broadly dispersed and are encoded by the organism's core DNA. There are compounds like Lipocalin-2, which can prevent the iron uptake released from some siderophores such as enterobactin via the FepA receptor of bacteria (Fischbach et al. 2006). To circumvent this host defense response, some bacteria develop iron uptake systems that encode the PAI. The three siderophores that are most frequently observed are salmochelin, yersiniabactin, and aerobactin (Braun 2003). Salmochelin is a well-known glycosylated enterobactin that human Lipocalin-2 cannot inhibit (Fischbach et al. 2006). Urinary tract infections are brought on by extraintestinal pathogenic *E. coli* (ExPEC), which has a high prevalence of salmochelin and yersiniabactin (Henderson et al. 2009). Salmochelin tends to maintain a high level of bacteremia, and permits the persistent passage of bacteria via the blood–brain barrier in a mice meningitis model. The characterization of ChuA and Hma, two outer membrane heme receptors, ensures the fitness of the bacteria in a UTI mouse model (Hagan and Mobley 2009). The PAI-encoded iron regulation also mediates the transport of manganese and ferrous iron, which renders the organism resistant to oxidative stress. In research models using rats and humans, these pathways help explain urofitness (Snyder et al. 2004).

The locus of adhesion and autoaggregation (LAA) is a composite 86-kb PAI composed of four modules (Montero et al. 2017). Module III of this system has a role in iron regulation as it encodes an adhesin Iha which is a homolog to an iron regulatory gene. Other components show great similarity to the integrase encoded in SHI-2 (Shigella pathogenicity island 2) PAI.

The relics of the island's numerous stepwise assemblies, including a large number of incomplete transposases and IS elements (Moss et al. 1999; Vokes et al. 1999). SHI-2 encodes the iuc ABCD operon that produces the hydroxamate siderophore known as aerobactin (Lawlor and Payne 1984; Vokes et al. 1999; Braun 2003), as well as the iut A gene that codes for the outer membrane receptor for aerobactin complexed to iron (Crosa 1989).

This island has a unique ABCD locus that encodes a ferrous iron and manganese uptake system from the ABC transporter family, which is quite similar to what was seen for *Salmonella*. This system was first described in *S. flexneri* (Kehres et al. 2002; Runyen-Janecky et al. 2006; Fisher et al. 2009). This pathogenicity islands have been documented in many members of *E. coli*, including both commensals and pathogenic strains, in four clustered insertion sites at diverse locations in the chromosome (Fisher et al. 2009).

8.6 Model for Horizontal Transfer of PAIs

Genomic islands, conserved in *E. coli* chromosomal backbone are crucial elements of horizontal transfer. However, the transfer of these across diverse strains within the *E. coli* species has much-limited understanding so far. HPI has been utilized to understand horizontal transfer of genetic content. To comprehend the transfer of DNA islands within the *E. coli* species, Schubert et al. 2009 used DNA transfer studies and sequence-based approaches (Multi Locus Sequence Typing). The scientists presented evidence that homologous DNA recombination and conjugative transfer are key roles in the horizontal transfer of genetic information, including the pathogenicity island within *E. coli.*

The asnT tRNA locus is where HPI is introduced in *E. coli*. Of the four asn tRNA gene copies found in *E. coli*, the HPI is positioned at the asnT gene in the majority of the ECOR strain; however, the asnV gene is only linked to the island in the ECOR31 strain. This particular strain harbors a huge genetic cluster as HPI, which tend to resemble an arrangement like conjugative plasmids. Since the HPI is fixed in the *E. coli* genome, thus the plausibility of its transfer exists either by a clonal distribution or by multiple independent insertions into *E. coli* from other microbes. Further, a passive transfer mechanism by horizontal transfer cannot be ruled out which actually may follow the transfer with homologous recombinations and integrations in the recipient's genome. In fact, the passive transfer was demonstrated via experimental evidence also. It was established that the presence of the rec A gene was necessary for the subsequent integration. The researchers defined the transmission and spread of PAIs in general and the HPI in specific. Even though there are some minor variations among the different *E. coli* strains, almost all *E. coli* HPIs appear to be recent descendants of a single progenitor. The rapid proliferation of the species suggests a significant selective pressure among its members, which is highly congruent high distribution of the HPI among all extraintestinal pathogenic *E. coli* (Schubert et al. 2009).

8.7 Siderophores and Colibactin Genotoxin Biosynthetic Pathway Interactions in *Escherichia coli*

A series of enzymes known as phosphopantetheine transferases (PPTases) are involved in the manufacture of the genotoxin colibactin and low-molecular weight iron chelators (siderophores). EntD (a component of the core genome) and ClbA are the only two PPTases that have been clearly discovered in *E. coli* (on the pks pathogenicity island, coding for colibactin).

The Pks Island is observed to be physically connected with the high pathogenicity island (HPI) coding for yersiniabactin production in a subset of highly pathogenic strains of *E. coli* I, except a gene encoding its homologous PPTase. The authors demonstrated that function interchangeability for ClbA as it could help in siderophores synthesis. Only the destruction or Inactivation of both entD and clbA components completely stopped the disease causing potential of the bacteria in a mouse sepsis model. Whereas any of these being present in a functional form could sustain the survival of ExPEC in vivo. The research revealed for the first time a potential interconnection between various phosphopantetheinyl-dependent pathways that could result in the creation of functionally different secondary metabolites for particular bacteria. The authors speculated that such kind of association is possible owing to the promiscuity of the ClbA PPTase and points toward the intricacy of the virulence regulation mechanisms in bacteria (Martin et al. 2013).

Another pertinent example of complex bacterial interactions in the context of iron regulation comes from the *B. cereus* group, which forms bacillibactin (BB), a 2,3-dihydroxybenzoyl-Gly-Thr trilactone siderophore that shares a huge similarity to the enterobactin (Ent) of Gram-negative bacteria. Two members, i.e., *B. anthracis* and *B. cereus*, however, produce petrobactin (PB) as a prime compound (Zawadzka et al. 2009; Koppisch et al. 2005; Wilson et al. 2006). PB is also produced by *B. cereus* and *B. thuringiensis* isolates (Koppisch et al. 2008), and is known to impact the general growth and virulence mechanisms of the bacteria. Siderocalin, binds to several siderophores and hinders its usage by pathogenic bacteria, however, the unbound siderophores get assimilated to promote the virulence of the bacteria (Abergel et al. 2006;Goetz et al. 2002; Hoette et al. 2008). PB is a unique compound containing two 3,4-catecholate moieties and a citrate-based backbone, whose origin was traced in the marine bacterium *Marinobacter hydrocarbonoclasticus* (Barbeau et al. 2002). Ferric PB undergoes photolysis and results in an iron reduction, decarboxylation, and oxidation of the ligand with the production of a 3-ketoglutarate residue at the previous position of the citryl moiety (Barbeau et al. 2002). The PB photoproduct (PB^{ν}) acts to significant removal of the iron from transferrin (Abergel et al. 2008). The photoreactivity of ferric complexes with citrate siderophores, like PB and aerobactin, was thought to mediate the formation of ferrous iron available in the environment (Barbeau 2006). Ferric complex of PB and its photoproduct help both in the growth and in the survival of *Bacillus subtilis* under poor availability of iron (Garner et al. 2004).

In Gram-positive bacteria, the iron–siderophore complexes are extracted from the local environment by specific transport proteins, i.e., ABC-type transporters (Brown and Holden 2004). *B. subtilis* the prototype of the group has five such membrane anchors that identify and transport the cell formed BB and a variety of xenosiderophores (Moore and Helmann 2005; Ollinger et al. 2006). These substrate-binding proteins (SBPs), which are similar to the periplasmic binding proteins (bi-lobed structures) of Gram-negative bacteria but have higher selectivity and affinities toward the ligands, scavenge iron-loaded siderophores. A binding protein known as FeuA can distinguish between BB and the similar triscatecholate Ent (Miethke et al. 2006). Different SBPs identify ferric citrate, ferrichromes, ferrioxamines, and photoreactive citrate-based hydroxamates such as schizokinen (Sch) and arthrobactin (YfmC, FhuD, YxeB, and YfiY, respectively). Certain *B. subtilis* SBPs have been found to utilize the same ABC transporter, FhuBGC, as *Staphylococcus aureus* (Sebulsky and Heinrichs 2001). It has been determined that the *B. cereus* group only contains one such mechanism, a ferric dicitrate transporter, which is necessary for *B. cereus* 569 to be fully virulent in a lepidopteran infection model (Harvie and Ellar 2005).

8.8 Conclusion

Genomic islands make an important portal for horizontal gene transfer and mediate a huge variety of adaptive functions for microbes. Conceptually they are different from plasmids due to their integration ability into the host genetic makeup, though do not require a high replication cost than plasmids. They tend to distribute in a far wide variety of unrelated bacterial species and strongly depend on the ecological positioning and lifestyle of the host species that are environmental or pathogenic. Examination of various genes impacting traits such as resistance to antimicrobials, their ability to survive in close community, and general adaptation can be mediated through genomic Islands. Iron uptake systems are significant metabolic adaptation which is mediated by genomic islands in many bacterial species true for *Enterobacter* species of bacteria and members of the *Pseudomonas* group. *Yersenia* spp. is the model organism to understand the genomic islands involved in iron uptake and has been extensively studied. Increasing the knowledge about the genomic islands involved in the iron uptake may help identify suitable targets to modulate the bacterial resources both in the environmental and virulence aspects.

References

Abergel RJ, Wilson MK, Arceneaux JEL (2006) The anthrax pathogen evades the mammalian immune system through stealth siderophore production. Proc Natl Acad Sci U S A 103:18499–18503. https://doi.org/10.1073/pnas.0607055l0

Abergel RJ, Zawadzka AM, Raymond KN (2008) Petrobactin-mediated iron transport in pathogenic bacteria: coordination chemistry of an unusual 3,4-catecholate/citrate siderophore. J Am Chem Soc 130:2124–2125. https://doi.org/10.1021/ja077202g

de Almeida AM, Guiyoule A, Guilvout I et al (1993) Chromosomal irp2 gene in *Yersinia*: distribution, expression, deletion and impact on virulence. Microb Pathog 14:9–21. https://doi.org/10.1006/mpat.1993.1002

Andrews SC (2011) Making DNA without iron - induction of a manganese-dependent ribonucleotide reductase in response to iron starvation. Mol Microbiol 80:286–289. https://doi.org/10.1111/j.1365-2958.2011.07594.x

Andrews SC, Robinson AK, Rodríguez-Quiñones F (2003a) Bacterial iron homeostasis. FEMS Microbiol Rev 2-3:215–237. https://doi.org/10.1016/S0168-6445(03)00055-X

Andrews SC, Robinson AK, Rodríguez-Quiñones F (2003b) Bacterial iron homeostasis. FEMS Microbiol Rev 27:215–237. https://doi.org/10.1016/S0168-6445(03)00055-X

Anisimov R, Brem D, Heesemann J (2005) Transcriptional regulation of high pathogenicity island iron uptake genes by YbtA. Int J Med Microbiol 295:19–28. https://doi.org/10.1016/j.ijmm.2004.11.007

Antonenka U, Nölting C, Heesemann J et al (2005) Horizontal transfer of *Yersinia* high-pathogenicity island by the conjugative RP4attB target presenting shuttle plasmid. Mol Microbiol 57:727–734. https://doi.org/10.1111/j.1365-2958.2005.04722.x

Bach S, de Almeida A, Carniel E (2000) The Yersinia high-pathogenicity island is present in different members of the family Enterobacteriaceae. FEMS Microbiol Lett 183:289–294. https://doi.org/10.1111/j.1574-6968.2000.tb08973.x

Barbeau K (2006) Photochemistry of organic iron(III) complexing ligands in oceanic systems. Photochem Photobiol 82:1505–1516. https://doi.org/10.1111/j.1751-1097.2006.tb09806.x

Barbeau K, Zhang G, Live DH, Butler A (2002) Petrobactin, a photoreactive siderophore produced by the oil-degrading marine bacterium Marinobacter hydrocarbonoclasticus. J Am Chem Soc 124:378–379. https://doi.org/10.1021/ja0119088

Bearden SW, Fetherston JD, Perry RD (1997) Genetic organization of the yersiniabactin biosynthetic region and construction of avirulent mutants in *Yersinia pestis*. Infect Immun 65:1659–1668. https://doi.org/10.1128/iai.65.5.1659-1668.1997

Bobrov AG, Geoffroy VA, Perry RD (2002) Yersiniabactin production requires the thioesterase domain of HMWP2 and YbtD, a putative phosphopantetheinylate transferase. Infect Immun 70: 4204–4214. https://doi.org/10.1128/IAI.70.8.4204-4214.2002

Botta A, Barra NG, Lam NH et al (2021) Iron reshapes the gut microbiome and host metabolism. J Lipid Atheroscl 10:160–183. https://doi.org/10.12997/jla.2021.10.2.160

Braun V (2003) Iron uptake by *Escherichia coli*. Front Biosci 8:s1409–s1421. https://doi.org/10.2741/1232

Brem D, Pelludat C, Rakin A (2001) Functional analysis of yersiniabactin transport genes of *Yersinia enterocolitica*. Microbiology 147:1115–1127. https://doi.org/10.1099/00221287-147-5-1115

Brickman TJ, Armstrong SK (2005) *Bordetella* AlcS transporter functions in alcaligin siderophore export and is central to inducer sensing in positive regulation of alcaligin system gene expression. J Bacteriol 187:3650–3661. https://doi.org/10.1128/JB.187.11.3650-3661.2005

Brickman TJ, Armstrong SK (2009) Temporal signaling and differential expression of *Bordetella* iron transport systems: the role of ferrimones and positive regulators. Biometals 22:33–41. https://doi.org/10.1007/s10534-008-9189-9

Brickman TJ, Cummings CA, Liew SY et al (2011) Transcriptional profiling of the iron starvation response in *Bordetella pertussis* provides new insights into siderophore utilization and virulence gene expression. J Bacteriol 193:4798–4812. https://doi.org/10.1128/JB.05136-11

Brickman TJ, Hanawa T, Anderson MT et al (2008) Differential expression of *Bordetella pertussis* iron transport system genes during infection. Mol Microbiol 70:314. https://doi.org/10.1111/j.1365-2958.2008.06333.x

Brown JS, Holden DW (2004) Iron acquisition by gram-positive bacterial pathogens. Microbes Infect 4:1149–1156. https://doi.org/10.1016/S1286-4579(02)01640-4

Buchrieser C, Prentice M, Carniel E (1998) The 102-kilobase unstable region of *Yersinia pestis* comprises a high pathogenicity island linked to a pigmentation segment which undergoes internal rearrangement. J Bacteriol 180:2321–2329. https://doi.org/10.1128/JB.180.9.2321-2329.1998

Burrus V, Pavlovic G, Decaris B (2002) Conjugative transposons: the tip of the iceberg. Mol Microbiol 46:601–610. https://doi.org/10.1046/j.1365-2958.2002.03191.x

Butler JE, He Q, Nevin K (2007) Genomic and microarray analysis of aromatics degradation in *Geobactermetallireducen*s and comparison to a *Geobacter* isolate from a contaminated field site. BMC Genomics 8. https://doi.org/10.1186/1471-2164-8-180

Byers BR, Arceneaux JEL (1998) Microbial iron transport: iron acquisition by pathogenic microorganisms. In: Sigel A, Sigel H (eds) Metal ions in biological systems, vol 35. Marcel Dekker, New York, pp 37–66

Calderwood SB, Mekalanos JJ (1987) Iron regulation of Shiga-like toxin expression in *Escherichia coli* is mediated by the fur locus. J Bacteriol 169:4759–4764. https://doi.org/10.1128/jb.169.10.4759-4764.1987

Carniel E (2001) The *Yersinia* high-pathogenicity island: an iron-uptake island. Microbes Infect 3: 561–569. https://doi.org/10.1016/S1286-4579(01)01412-5

Carniel E (2002) Plasmids and pathogenicity islands of *Yersinia*. Curr Top Microbiol Immunol 264: 89–108. https://doi.org/10.1007/978-3-642-56031-6_6

Carniel E, Guilvout I, Prentice M (1996) Characterization of a large chromosomal high-pathogenicity island in biotype 1B *Yersinia enterocolitica*. J Bacteriol 178:674351. https://doi.org/10.1128/jb.178.23.6743-6751.1996

Carniel E, Mazigh D, Mollaret HH (1987) Expression of iron-regulated proteins in *Yersinia* species and their relation to virulence. Infect Immun 55:277–280. https://doi.org/10.1128/iai.55.1.277-280.1987

Carpenter C, Payne SM (2014) Regulation of iron transport systems in *Enterobacteriaceae* in response to oxygen and iron availability. J Inorg Biochem 133:110–117. https://doi.org/10.1016/j.jinorgbio.2014.01.007

Chaturvedi KS, Hung CS, Crowley JR et al (2012) The siderophore yersiniabactin binds copper to protect pathogens during infection. Nat Chem Biol 8:731–736. https://doi.org/10.1038/nchembio.1020

Chauvaux S, Rosso ML, Frangeul L et al (2007) Transcriptome analysis of *Yersinia pestis* in human plasma: an approach for discovering bacterial genes involved in septicaemic plague. Microbiology 153:31123123. https://doi.org/10.1099/mic.0.2007/006213-0

Cornelissen CN (2003) Transferrin-iron uptake by gram-negative bacteria. Front Biosci 1(8):d836–d847. https://doi.org/10.2741/1076

Crosa JH (1989) Genetics and molecular biology of siderophore-mediated iron transport in bacteria. Microbiol Rev 53:517–530. https://doi.org/10.1128/mr.53.4.517-530.1989

Delany I, Spohn G, Rappuoli R (2003) An anti-repression Fur operator upstream of the promoter is required for iron-mediated transcriptional autoregulation in *Helicobacter pylori*. Mol Microbiol 50:1329–1338. https://doi.org/10.1046/j.1365-2958.2003.03757.x

Dobrindt U, Hochhut B, Hentschel U (2004) Genomic islands in pathogenic and environmental microorganisms. Nat Rev Microbiol 2:414–424. https://doi.org/10.1038/nrmicro884

Duchesne P, Grenier D, Mayrand D (1999) Binding and utilization of human transferrin by *Prevotella nigrescens*. Infect Immun 67:576–580. https://doi.org/10.1128/IAI.67.2.576-580.1999

Earhart CF (2009) Iron metabolism. In: Schaechter M (ed) Encyclopedia of microbiology, 3rd edn. Academic Press, Oxford, pp 210–218

Evans SL, Arceneaux JE, Byers BR, Martin ME, Aranha H (1986) Ferrous iron transport in *Streptococcus mutans*. J Bacteriol 168:1096–1099. https://doi.org/10.1128/jb.168.3.1096-1099.1986

Fetherston JD, Bertolino VJ, Perry RD (1999) YbtP and YbtQ: two ABC transporters required for iron uptake in *Yersinia pestis*. Mol Microbiol 32:289–229. https://doi.org/10.1046/j.1365-2958.1999.01348.x

Finlay BB, Falkow S (1997) Common themes in microbial pathogenicity revisited. Microbiol Mol Biol Rev 61(2):136–169. https://doi.org/10.1128/mmbr.61.2.136-169.1997

Fischbach MA, Lin H, Zhou L et al (2006) The pathogen-associated iroA gene cluster mediates bacterial evasion of lipocalin 2. Proc Natl Acad Sci U S A 103:16502–16507. https://doi.org/10.1073/pnas.0604636103

Fisher CR, Davies NMLL, Wyckoff EE et al (2009) Genetics and virulence association of the *Shigella flexneri* sit iron transport system. Infect Immun 77:1992–1999. https://doi.org/10.1128/IAI.00064-09

Forman S, Paulley JT, Fetherston JD et al (2010) Yersinia ironomics: comparison of iron transporters among Yersinia pestis biotypes and its nearest neighbor, *Yersinia pseudotuberculosis*. Biometals 2:275–294. https://doi.org/10.1007/s10534-009-9286-4

Fukuto HS, Svetlanov A, Palmer LE et al (2010) Global gene expression profiling of *Yersinia pestis* replicating inside macrophages reveals the roles of a putative stress-induced operon in regulating type III secretion and intracellular cell division. Infect Immun 78:3700–3715. https://doi.org/10.1128/IAI.00062-10

Gaillard M, Vallaeys T, Vorholter FJ et al (2006) The clc element of *Pseudomonas* sp. strain B13, a genomic island with various catabolic properties. J Bacteriol 188:1999–2013. https://doi.org/10.1128/JB.188.5.1999-2013.2006

Gal-Mor O, Finlay BB (2006) Pathogenicity islands: a molecular toolbox for bacterial virulence. Cell Microbiol 8:1707–1719. https://doi.org/10.1111/j.1462-5822.2006.00794.x

Gao H, Zhou D, Li Y et al (2008) The iron-responsive fur regulon in *Yersinia pestis*. J Bacteriol 190:3063–3075. https://doi.org/10.1128/JB.01910-07

Garner BL, Arceneaux JEL, Byers BR (2004) Temperature control of a 3,4-dihydroxybenzoate (protocatechuate)-based siderophore in *Bacillus anthracis*. Curr Microbiol 49:89–94. https://doi.org/10.1007/s00284-004-4286-7

Ge R, Sun X, He Q (2009) Iron acquisition by *Streptococcus* species: an updated review. Front Biol China 4:392–401. https://doi.org/10.1007/s11515-009-0035-4

Gehring AM, DeMoll E, Fetherston JD et al (1998) Iron acquisition in plague: modular logic in enzymatic biogenesis of yersiniabactin by *Yersinia pestis*. Chem Biol 5:573–586. https://doi.org/10.1016/S1074-5521(98)90115-6

Goetz DH, Holmes MA, Borregaard N (2002) The neutrophil lipocalin NGAL is a bacteriostatic agent that interferes with siderophore-mediated iron acquisition. Mol Cell 10:1033–1043. https://doi.org/10.1016/S1097-2765(02)00708-6

Golonka R, Yeoh BS, Vijay-Kumar M (2019) The iron tug-of-war between bacterial siderophores and innate immunity. J Innate Immun 11:249–262. https://doi.org/10.1159/000494627

Grenier D, Tanabe S (2011) Transferrin as a source of iron for *Campylobacter rectus*. J Oral Microbiol 3:3. https://doi.org/10.3402/jom.v3i0.5660

Groisman EA, Ochman H (1996) Pathogenicity islands: bacterial evolution in quantum leaps. Cell 87:791–794. https://doi.org/10.1016/S0092-8674(00)81985-6

Hacker J, Bender L, Ott M et al (1990) Deletions of chromosomal regions coding for fimbriae and hemolysins occur in vitro and in vivo in various extraintestinal Escherichia coli isolates. Microb Pathog 8:213–225. https://doi.org/10.1016/0882-4010(90)90048-U

Hacker J, Blum-Oehler G, Mühldorfer I et al (1997) Pathogenicity island of virulent bacteria: structure, function and impact on microbial evolution. Mol Microbiol 23:1089–1097. https://doi.org/10.1046/j.1365-2958.1997.3101672.x

Hacker J, Carniel E (2001) Ecology, fitness, genomic islands and bacterial pathogenicity. A Darwinian view of the evolution of microbes. EMBO Rep 2:376–381. https://doi.org/10.1093/embo-reports/kve097

Hacker J, Kaper JB (2000) Pathogenicity islands and the evolution of microbes. Annu Rev Microbiol 54:641–679

Hagan EC, Mobley HL (2009) Haem acquisition is facilitated by a novel receptor Hma and required by uropathogenic *Escherichia coli* for kidney infection. Mol Microbiol 71:79–91. https://doi.org/10.1111/j.1365-2958.2008.06509.x

Hamilton HL, Dominguez NM, Schwartz KJ, Hackett KT, Dillard JP (2005) *Neisseria gonorrhoeae* secretes chromosomal DNA via a novel type IV secretion system. Mol Microbiol 55:1704–1721. https://doi.org/10.1111/j.1365-2958.2005.04521.x

Han Y, Qiu J, Guo Z et al (2007) Comparative transcriptomics in *Yersinia pestis*: a global view of environmental modulation of gene expression. BMC Microbiol 7:96. https://doi.org/10.1186/1471-2180-7-96

Han Y, Zhou D, Pang X et al (2004) Microarray analysis of temperature-induced transcriptome of *Yersinia pestis*. Microbiol Immunol 48:791–805. https://doi.org/10.1111/j.1348-0421.2004.tb03605.x

Harvie DR, Ellar DJ (2005) A ferric dicitrate uptake system is required for the full virulence of *Bacillus cereus*. Curr Microbiol 50:246–250. https://doi.org/10.1007/s00284-004-4442-0

Hederstedt L, Gorton L, Pankratova G (2020) Two routes for extracellular electron transfer in *Enterococcus faecalis*. J Bacteriol 202:e00725-19. https://doi.org/10.1128/JB.00725-19

Heesemann J (1987) Chromosomal-encoded siderophores are required for mouse virulence of enteropathogenic *Yersinia* species. FEMS Microbiol Lett 48:229–233. https://doi.org/10.1111/j.1574-6968.1987.tb02547.x

Heesemann J, Hantke K, Vocke T et al (1993) Virulence of *Yersinia enterocolitica* is closely associated with siderophore production, expression of an iron-repressible outer membrane polypeptide of 65,000 Da and pesticin sensitivity. Mol Microbiol 8:397–408. https://doi.org/10.1111/j.1365-2958.1993.tb01583.x

Henderson JP, Crowley JR, Pinkner JS et al (2009) Quantitative metabolomics reveals an epigenetic blueprint for iron acquisition in uropathogenic *Escherichia coli*. PLoS Pathog 5:e1000305. https://doi.org/10.1371/journal.ppat.1000305

Hochhut B, Marrero J, Waldor MK (2000) Mobilization of plasmids and chromosomal DNA mediated by the SXT element, a constin found in *Vibrio cholerae* O139. J Bacteriol 182:2043–2047. https://doi.org/10.1128/JB.182.7.2043-2047.2000

Hoette TM, Abergel RJ, Xu J (2008) The role of electrostatics in siderophore recognition by the immunoprotein siderocalin. J Am Chem Soc 130:17584–17592. https://doi.org/10.1021/ja8074665

van der Hoeven JS, de Jong M, Rogers AH, Camp P (1984) A conceptual model for the co-existence of *Streptococcus* spp. and *Actinomyces* spp. in dental plaque. J Dent Res 63:389–392. https://doi.org/10.1177/00220345840630030601

Imbert M, Blondeau R (1998) On the iron requirement of lactobacilli grown in chemically defined medium. Curr Microbiol 37:64–66. https://doi.org/10.1007/s002849900339

Juhas M, Power PM, Harding RM et al (2007) Sequence and functional analyses of *Haemophilus* spp. genomic islands. Genome Biol 8:R23. https://doi.org/10.1186/gb-2007-8-11-r237

Juhas M, van der Meer JR, Gaillard M et al (2009) Genomic islands: tools of bacterial horizontal gene transfer and evolution. FEMS Microbiol Rev 33:376–393. https://doi.org/10.1111/j.1574-6976.2008.00136

Kalinowski BE, Johnsson A, Arlinger J et al (2006) Microbial mobilization of uranium from shale mine waste. Geomicrobiol J 23:157–164. https://doi.org/10.1080/01490450600599197

Karagoz K, Ates F, Karagoz H et al (2012) Characterization of plant growth-promoting traits of bacteria isolated from the rhizosphere of grapevine grown in alkaline and acidic soils. Eur J Soil Biol 50:144–150. https://doi.org/10.1016/j.ejsobi.2012.01.007

Karch H, Schubert S, Zhang D (1999) A genomic island, termed high-pathogenicity island, is present in certain non-O157 Shiga toxin-producing *Escherichia coli* clonal lineages. Infect Immun 67:5094–6001. https://doi.org/10.1128/IAI.67.11.5994-6001.1999

Kehres DG, Janakiraman A, Slauch JM et al (2002) SitABCD is the alkaline Mn(2+) transporter of *Salmonella enterica* serovar Typhimurium. J Bacteriol 184:3159–3166. https://doi.org/10.1128/jb.184.12.3159-3166.2002

Koczura R, Kaznowski A (2003) Occurrence of the Yersinia high-pathogenicity island and iron uptake systems in clinical isolates of *Klebsiella pneumoniae*. Microb Pathog 35:197–202. https://doi.org/10.1016/s0882-4010(03)00125-6

Koppisch AT, Browder CC, Moe AL (2005) Petrobactin is a primary siderophore synthesized by *Bacillus anthracis* str. Sterne under conditions of iron starvation. Biometals 18:577–585. https://doi.org/10.1007/s10534-005-1782-6

Koppisch AT, Dhungana S, Hill KK (2008) Petrobactin is produced by both pathogenic and non-pathogenic isolates of the *Bacillus cereus* group of bacteria. Biometals 21:581–589. https://doi.org/10.1007/s10534-008-9144-9

Kramer J, Özkaya Ö, Kümmerli R (2020) Bacterial siderophores in community and host interactions. Nat Rev Microbiol 18:152–163. https://doi.org/10.1038/s41579-019-0284-4

Larbig KD, Christmann A, Johann A et al (2002) Gene islands integrated into tRNA(Gly) genes confer genome diversity on a *Pseudomonas aeruginosa* clone. J Bacteriol 184:6665–6680. https://doi.org/10.1128/JB.184.23.6665-6680.2002

Lathem WW, Crosby SD, Miller VL et al (2005) Progression of primary pneumonic plague: a mouse model of infection, pathology, and bacterial transcriptional activity. Proc Natl Acad Sci U S A 102:17786–17791. https://doi.org/10.1073/pnas.0506840102

Lawlor KM, Payne SM (1984) Aerobactin genes in *Shigella* spp. J Bacteriol 160:266–272. https://doi.org/10.1128/jb.160.1.266-272.1984

Lesic B, Carniel E (2005) Horizontal transfer of the highpathogenicity island of *Yersinia pseudotuberculosis*. J Bacteriol 187:3352–3358. https://doi.org/10.1128/JB.187.10.3352-3358.2005

Litwin CM, Calderwood SB (1993) Role of iron in regulation of virulence genes. Clin Microbiol Rev 6:137–149. https://doi.org/10.1128/CMR.6.2.137

Liu H, Wang H, Qiu J et al (2009) Transcriptional profiling of a mice plague model: insights into interaction between *Yersinia pestis* and its host. J Basic Microbiol 49:92–99. https://doi.org/10.1002/jobm.200800027

Liu X, Olczak T, Guo H-C (2006) Identification of amino acid residues involved in heme binding and hemoprotein utilization in the *Porphyromonas gingivalis* heme receptor HmuR. Infect Immun 74:1222–1232. https://doi.org/10.1128/IAI.74.2.1222-1232.2006

Lönnerdal B, Iyer S (1995) Lactoferrin: molecular structure and biological function. Annu Rev Nutr 15:93–11. https://doi.org/10.1177/0022034584063003060

Markel TA, Crisostomo PR, Wang M et al (2007) The struggle for iron: gastrointestinal microbes modulate the host immune response during infection. J Leukoc Biol 81:393–400. https://doi.org/10.1189/jlb.0906579

Martin P, Marcq I, Magistro G et al (2013) Interplay between siderophores and colibactin genotoxin biosynthetic pathways in Escherichia coli. PLoS Pathog 9(7):e1003437. https://doi.org/10.1371/journal.ppat.1003437

Messerer M, Fischer W, Schubert S (2017) Investigation of horizontal gene transfer of pathogenicity islands in *Escherichia coli* using next generation sequencing. PLoS One 12(7):e0179880. https://doi.org/10.1371/journal.pone.0179880

Mey AR, Wyckoff EE, Kanukurthy V (2005) Iron and fur regulation in *Vibrio cholerae* and the role of fur in virulence. Infect Immun 73:8167–8178. https://doi.org/10.1128/IAI.73.12.8167-8178.2005

Miethke M, Klotz O, Linne U (2006) Ferri-bacillibactin uptake and hydrolysis in *Bacillus subtilis*. Mol Microbiol 61:1413–1427. https://doi.org/10.1111/j.1365-2958.2006.05321.x

Miller MC, Fetherston JD, Pickett CL et al (2010) Reduced synthesis of the Ybt siderophore or production of aberrant YBT-like molecules activates transcription of yersiniabactin genes in *Yersinia pestis*. Microbiology 156:2226–2238. https://doi.org/10.1099/mic.0.037945-0

Moelling C, Oberschlacke R, Ward P et al (2007) Metal-dependent repression of siderophore and biofilm formation in *Actinomyces naeslundii*. FEMS Microbiol Lett 275:214–220. https://doi.org/10.1111/j.1574-6968.2007.00888.x

Montero DA, Velasco J, Fetal DC (2017) Locus of adhesion and autoaggregation (LAA), a pathogenicity island present in emerging Shiga Toxin-producing *Escherichia coli* strains. Sci Rep 7:7011. https://doi.org/10.1038/s41598-017-06999-y

Moore CM, Helmann JD (2005) Metal homeostasis in *Bacillus subtilis*. Curr Opin Microbiol 8: 188–195. https://doi.org/10.1016/j.mib.2005.02.007

Moss JE, Cardozo TJ, Zychlinsky A (1999) The selC-associated SHI-2 pathogenicity island of *Shigella flexneri*. Mol Microbiol 33:74–83. https://doi.org/10.1046/j.1365-2958.1999.01449.x

Motin VL, Georgescu AM, Fitch JP et al (2004) Temporal global changes in gene expression during temperature transition in *Yersinia pestis*. J Bacteriol 186:6298–6305

Nairz M, Schroll A, Sonnweber T et al (2010) The struggle for iron—a metal at the host-pathogen interface. Cell Microbiol 12:1691–1702. https://doi.org/10.1111/j.1462-5822.2010.01529.x

Ollinger J, Song KB, Antelmann H, Hecker M, Helmann JD (2006) Role of the Fur regulon in iron transport in *Bacillus subtilis*. J Bacteriol 188:3664–3673. https://doi.org/10.1128/JB.188.10.3664-3673.2006

Osorio H, Martinez V, Nieto PA et al (2008) Microbial iron management mechanisms in extremely acidic environments: comparative genomics evidence for diversity and versatility. BMC Microbiol 8:1. https://doi.org/10.1186/1471-2180-8-203

Otto BR, Sparrius M, Verweij-van Vught AM, MacLaren DM (1990) Iron-regulated outer membrane protein of *Bacteroides fragilis* involved in heme uptake. Infect Immun 58:3954–3958. https://doi.org/10.1128/iai.58.12.3954-3958.1990

Parkhill J, Wren BW, Thomson NR et al (2001) Genome sequence of *Yersinia pestis*, the causative agent of plague. Nature 413:523–527. https://doi.org/10.1038/35097083

Pelludat C, Brem D, Heesemann J (2003) Irp9, encoded by the high-pathogenicity island of *Yersinia enterocolitica*, is able to convert chorismate into salicylate, the precursor of the siderophore yersiniabactin. J Bacteriol 185:5648–5653. https://doi.org/10.1128/JB.185.18.5648-5653.2003

Pelludat C, Rakin A, Jacobi CA et al (1998) The yersiniabactin biosynthetic gene cluster of *Yersinia enterocolitica*: organization and siderophore-dependent regulation. J Bacteriol 180:538–546. https://doi.org/10.1128/JB.180.3.538-546.1998

Perry RD, Fetherston J (2004) Iron and heme uptake systems. In: Carniel E, Hinnebusch BJ (eds) *Yersinia* molecular and cellular biology. Horizon Bioscience, Norfolk, pp 257–283

Perry RD, Fetherston JD (2011) *Yersinia*bactin iron uptake: mechanisms and role in Yersinia pestis pathogenesis. Microbes Infect 13:808–817. https://doi.org/10.1016/j.micinf.2011.04.008

Pieper R, Huang ST, Clark DJ et al (2009) Integral and peripheral association of proteins and protein complexes with *Yersinia pestis* inner and outer membranes. Proteome Sci 7:5. https://doi.org/10.1186/1477-5956-7-5

Pieper R, Huang ST, Parmar PP et al (2010) Proteomic analysis of iron acquisition, metabolic and regulatory responses of *Yersinia pestis* to iron starvation. BMC Microbiol 10:30. https://doi.org/10.1186/1471-2180-10-30

Podladchikova O, Antonenka U, Heesemann J, Rakin A (2011) *Yersinia pestis* autoagglutination factor is a component of the type six secretion system. Int J Med Microbiol 301:562–569. https://doi.org/10.1016/j.ijmm.2011.03.004

Podladchikova O, Rykova V, Antonenka U et al (2012) *Yersiniapestis* autoagglutination is mediated by HCP-like protein and siderophore Yersiniachelin (Ych). Adv Exp Med Biol 954: 289–292. https://doi.org/10.1007/978-1-4614-3561-7_36

Podladchikova ON, Rykova VA (2006) Isolation and characterization of autoagglutination factor of *Yersinia pestis* Hms$^-$ cells. Biochemistry (Mosc) 71:1192–1199. https://doi.org/10.1134/S0006297906110046

Potrykus J, Jonna VR, Dopson M (2011) Iron homeostasis and responses to iron limitation in extreme acidophiles from the *Ferroplasma* genus. Proteomics 11:52–63. https://doi.org/10.1002/pmic.201000193

Quatrini R, Jedlicki E, Holmes DS (2005) Genomic insights into the iron uptake mechanisms of the biomining microorganism *Acidithiobacillus ferrooxidans*. J Ind Microbiol Biotechnol 1(11–12): 606–614. https://doi.org/10.1007/s10295-005-0233-2

Rakin A, Saken E, Harmsen D et al (1994) The pesticin receptor of Yersinia enterocolitica: a novel virulence factor with dual function. Mol Microbiol 13:253–263. https://doi.org/10.1111/j.1365-2958.1994.tb00420.x

Rakin A, Schneider L, Podladchikova O (2012) Hunger for iron: the alternative siderophore iron scavenging systems in highly virulent Yersinia. Front Cell Infect Microbiol 2:151. https://doi.org/10.3389/fcimb.2012.00151

Rakin A, Schubert S, Pelludat C (1999) The high-pathogenicity island of yersiniae. In: Kaper JB, Hacker J (eds) Pathogenicity Islands and other mobile virulence elements. ASM Press, Washington, DC, pp 77–90. https://doi.org/10.1128/9781555818173.ch5

Ratledge C, Dover LG (2000) Iron metabolism in pathogenic bacteria. Annu Rev Microbiol 54: 881–941

Rocha ER, de Uzeda M, Brock JH (1991) Effect of ferric and ferrous iron chelators on growth of *Bacteroides fragilis* under anaerobic conditions. FEMS Microbiol Lett 68:45–50. https://doi.org/10.1111/j.1574-6968.1991.tb04567.x

Rosso ML, Chauvaux S, Dessein R et al (2008) Growth of Yersinia pseudotuberculosis in human plasma: impacts on virulence and metabolic gene expression. BMC Microbiol 8:211. https://doi.org/10.1186/1471-2180-8-211

Runyen-Janecky L, Dazenski E, Hawkins S et al (2006) Role and regulation of the *Shigella flexneri* sit and MntH systems. Infect Immun 74:4666–4672. https://doi.org/10.1128/IAI.00562-06

Schmidt H, Hensel M (2004) Pathogenicity islands in bacterial pathogenesis. Clin Microbiol Rev 17:14–56. https://doi.org/10.1128/CMR.17.1.14-56.2004

Schubert S, Darlu P, Clermont O et al (2009) Role of intraspecies recombination in the spread of pathogenicity islands within the *Escherichiacoli* species. PLoS Pathog 5:e1000257. https://doi.org/10.1371/journal.ppat.1000257

Schubert S, Norenberg D, Clermont O et al (2010) Prevalence and phylogenetic history of the TcpC virulence determinant in *Escherichiacoli*. Int J Med Microbiol 300:429–434. https://doi.org/10.1016/j.ijmm.2010.02.006

Schubert S, Rakin A, Heesemann J (2004) The *Yersinia* high-pathogenicity island (HPI): evolutionary and functional aspects. Int J Med Microbiol 294:83–94. https://doi.org/10.1016/j.ijmm.2004.06.026

Schubert S, Rakin A, Karch H et al (1998) Prevalence of the 'high-pathogenicity island' of Yersinia species among *Escherichiacoli* strains that are pathogenic to human. Infect Immun 66:480–485. https://doi.org/10.1128/IAI.66.2.480-485.1998

Schwyn B, Neilands JB (1987) Universal chemical assay for the detection and determination of siderophores. Anal Biochem 160:47–56. https://doi.org/10.1016/0003-2697(87)90612-9

Sebbane F, Lemaitre N, Sturdevant DE et al (2006) Adaptive response of *Yersinia pestis* to extracellular effectors of innate immunity during bubonic plague. Proc Natl Acad Sci U S A 103:11766–11771. https://doi.org/10.1073/pnas.0601182103

Sebulsky MT, Heinrichs DE (2001) Identification and characterization of *fhuD1* and *fhuD2*, two genes involved in iron-hydroxamate uptake in *Staphylococcus aureus*. J Bacteriol 183:4994–5000. https://doi.org/10.1128/JB.183.17.4994-5000.2001

Sheftel AD, Mason AB, Ponka P (2012) The long history of iron in the Universe and in health and disease. Biochim Biophys Acta 1820:161–187. https://doi.org/10.1016/j.bbagen.2011.08.002

Snyder JA, Haugen BJ, Buckles EL et al (2004) Transcriptome of uropathogenic *Escherichia coli* during urinary tract infection. Infect Immun 72:6373–6381. https://doi.org/10.1128/IAI.72.11.6373-6381.2004

Sullivan JT, Trzebiatowski JR, Cruickshank RW et al (2002) Comparative sequence analysis of the symbiosis island of *Mesorhizobium loti* strain R7A. J Bacteriol 184:3086–3095. https://doi.org/10.1128/JB.184.11.3086-3095.2002

Symeonidis A, Marangos M (2012) Iron and microbial growth. In: Priti R (ed) Insight and control of infectious disease in global scenario. IntechOpen, London, pp 289–330

Utley JR (1990) Pathophysiology of cardiopulmonary bypass: current issues. J Card Surg 5:177–189. https://doi.org/10.1111/j.1540-8191.1990.tb01036.x

Vadyvaloo R et al (2007) island. Microbes Infect 3:561–569. https://doi.org/10.1016/s1286-4579(01)01412-5

Vadyvaloo V, Jarrett C, Sturdevant DE (2010) Transit through the flea vector induces a pretransmission innate immunity resistance phenotype in *Yersinia pestis*. PLoS Pathog 6:e1000783. https://doi.org/10.1371/journal.ppat.1000783

Verma R, Naosekpam AS, Kumar S et al (2007) Influence of soil reaction on diversity and antifungal activity of fluorescent pseudomonads in crop rhizospheres. Bioresour Technol 98:1346–1352. https://doi.org/10.1016/j.biortech.2006.05.030

Vernikos GS, Parkhill J (2008) Resolving the structural features of genomic islands: a machine learning approach. Genome Res 18:331–342. https://doi.org/10.1101/gr.7004508

Vogt AS, Arsiwala T, Mohsen M (2021) On iron metabolism and its regulation. Int J Mol Sci 22: 4591. https://doi.org/10.3390/ijms22094591

Vokes SA, Reeves SA, Torres AG, Payne SM (1999) The aerobactin iron transport system genes in *Shigella flexneri* are present within a pathogenicity island. Mol Microbiol 33:63–73. https://doi.org/10.1046/j.1365-2958.1999.01448.x

Wang R, Kaplan A, Guo L (2012) The influence of iron availability on human salivary microbial community composition. Microb Ecol 64:152–161. https://doi.org/10.1007/s00248-012-0013-2

Weinberg ED (1997) The *Lactobacillus* anomaly: total iron abstinence. Perspect Biol Med 40:578–583. https://doi.org/10.1353/pbm.1997.0072

Weinberg ED (2001) Human lactoferrin: a novel therapeutic with broad spectrum potential. J Pharm Pharmacol 53:1303–1310. https://doi.org/10.1211/0022357011777792

Wessling-Resnick M (2010) Iron homeostasis and the inflammatory response. Annu Rev Nutr 30: 105–122. https://doi.org/10.1146/annurev.nutr.012809.104804

Williams PH, Warner PJ (1980) ColV-plasmid mediated, colicin V-independent iron uptake system of invasive strains of *Escherichia coli*. Infect Immun 29:411–416. https://doi.org/10.1128/iai.29.2.411-416.1980

Wilson MK, Abergel RJ, Raymond KN (2006) Siderophores of *Bacillus anthracis*, *Bacillus cereus*, and *Bacillus thuringiensis*. Biochem Biophys Res Commun 348:320–325. https://doi.org/10.1016/j.bbrc.2006.07.055

Winkelmann G (1991) Structural and sterochemical aspects of iron transport in fungi. Biotechnol Adv 8:207–223. https://doi.org/10.1016/0734-9750(90)90013-2

Winkelmann G (2002) Microbial siderophore-mediated transport. Biochem Soc Trans 30:691–696

Yadav S, Kaushik R, Saxena AK (2011) Diversity and phylogeny of plant growth-promoting bacilli from moderately acidic soil. J Basic Microbiol 51:98–106. https://doi.org/10.1002/jobm.201000098

Yilmaz B, Li H (2018) Gut microbiota and iron: the crucial actors in health and disease. Pharmaceuticals 11:98. https://doi.org/10.3390/ph11040098

Zawadzka A, Aberge R, Nichiporuk R (2009) Siderophore-mediated iron acquisition systems in Bacillus cereus: identification f receptors for anthrax virulence-associated petrobactin. Biochemistry 48:3645–3657. https://doi.org/10.1021/bi8018674

Zhang XH, Austin B, Symeonidis A, Marangos M (2012) Iron and microbial growth. In: Priti R (ed) Insight and control of infectious disease in global scenario. IntechOpen, London, pp 289–330

Zhou D, Qin L, Han Y et al (2006) Global analysis of iron assimilation and fur regulation in *Yersinia pestis*. FEMS Microbiol Lett 258:917. https://doi.org/10.1111/j.1574-6968.2006.00208.x

Genomic Islands in Uropathogenic *Escherichia coli*

9

Shriparna Mukherjee and Ranadhir Chakraborty

Abstract

Discrete DNA segments known as genomic islands are responsible for establishing horizontally transmitted genes in a population. This region might be excised, integrated into the host's chromosome, or transferred en route to different bacteria via all three modes of gene transfer. Genomic islands may aid in adaptability, can code for a variety of roles, and may be involved in disease or symbiosis. A genomic island can be identified by certain sequence features, including direct repeats, insertion sequence elements, tRNA genes, and mobility genes (including transposases and integrases). More than one virulence genes, like that coding for adhesins, toxins, or invasins, are carried by pathogenicity islands. They might be contained in a plasmid or found on a bacterial chromosome. Large genomic islands can be found in *Escherichia coli* strains that are uropathogenic (UPECs). UPECs are the root cause of at least 80% of all urinary system infections that the general population contracts. There is an expanding list of virulence factors found in UPECs. Several virulence genes are linked among UPEC isolates. Virulence factor-encoding genes are linked and localized to a region in the UPECs' genome that resembles a pathogenicity island. The chapter provided a thorough analysis of the genomic islands' potential use in UPECs.

Shriparna Mukherjee and Ranadhir Chakraborty contributed equally to this work

S. Mukherjee
Department of Botany, Prasannadeb Women's College, Jalpaiguri, India

R. Chakraborty (✉)
OMICS Laboratory, Department of Biotechnology, University of North Bengal, Siliguri, India

I. Mani et al. (eds.), *Microbial Genomic Islands in Adaptation and Pathogenicity*,
https://doi.org/10.1007/978-981-19-9342-8_9

Keywords

Urinary tract infection · Pathogenicity Island · Uropathogenic *Escherichia coli* · Virulence factors · Omics technology · Mobile DNA element · tRNA-encoding gene

9.1 General Introduction

Escherichia coli is an "aero-anaerobic" Gram-negative bacillus. *E. coli*'s evolutionary substructure is significantly influenced by phylogenetic groups, A, B1, B2, C, D, E, F, and G. Each of these phylogroups is associated with the diverse strains' way of life (Milkman 1973; Selander et al. 1987; Escobar-Páramo et al. 2004; Clermont et al. 2019). *E. coli*'s major habitat is the intestines of vertebrates, where it interacts symbiotically with the host. However, if these commensal strains pick up virulence factors, they can become lethal, resulting in life-threatening illnesses (Kaper et al. 2004; Le Gall et al. 2007; Tenaillon et al. 2010). One of the following three groups can be used to categorize *E. coli*:

***Group I*: Commensal (nonpathogenic) *E. coli*.** They have a symbiotic relationship with the host and do not cause disease.

***Group II*: Intestinal pathogenic *E. coli* (InPEC)** They bring on colitis or gastroenteritis [often called diarrhea-causing (=diarrheagenic) *Escherichia coli* (DEC)]. DECs are distributed into six distinctly characterized pathotypes. These are enteropathogenic (EHEC), enterohemorrhagic (EHEC), enterotoxigenic (ETEC), enteroaggregative (EAEC), enteroinvasive (EIEC), and diffusely adherent (DAEC) *E. coli*.

***Group III*: Extraintestinal pathogenic *E. coli* (ExPEC).**

In order to include *E. coli* strains causing illness in organs external to the intestinal system, the acronym ExPEC was proposed in the year 2000. ExPEC isolates are facultative pathogens that can harmlessly colonize human gastrointestinal region, but they can also spread to organs of different systems like blood-vascular, urinary, and central nervous systems inducing pathogenesis (Wiles et al. 2008; Kohler and Dobrindt 2011). ExPECs are differentiated from those exhibiting commensalism or InPECs with the aid of genotyping techniques or by using advanced phylogenetic tools including multilocus sequence typing (MLST) or Kmer analysis of whole genome sequences. Genetically, ExPECs are distinguished via a wide variety of virulence genes concerned with high-affinity iron acquisition system, invasion strategies, and evasion of host immune defense system. ExPEC strains must possess at least two of the virulence genes *pap*A and/or *papc*, *sfa*/*foc*, *afa*/*dra*BC, *kps*M II, and/or *iut*A (Sarowska et al. 2019). ExPEC strains were formerly divided into four pathotypes based on the place of isolation [urinary tract

(UPEC) or tissues covering the brain and spinal cord of the newborn (NMEC) or birds (APEC)] and disease associations (sepsis: SEPEC) (Ewers et al. 2007). Since *E. coli* strains from several pathotypes were found to spread disease in non-specific locations, it is currently thought that the classification, as stated above, is no longer accurate (Dale and Woodford 2015) plus these strains have common genealogical characteristics (Jorgensen et al. 2019).

9.2 Genomic Islands in *E. coli*

According to *E. coli* phylogeny based on whole genome sequences, commensal *E. coli* strains were originally virulence free. The current shape of the *E. coli* genome has really been altered through horizontal gene transfer comprising transformation, transduction, and conjugation events. Although the genomes of *E. coli* and *Salmonella enterica* display 70% identity, they have separated due to the acquisition of large DNA pieces known as genomic islands (GEIs). According to bioinformatic investigations, GEIs typically contain more new genes than the rest of the genome, or genes that have no orthologs in other species (Hsiao et al. 2005). Symbiosis, metabolism of sucrose and aromatic compounds, resistance to mercury, siderophore production, and pathogenicity functions are all included in the coding capacity of GEIs. Pathogenicity islands, symbiosis islands, metabolic islands, fitness islands, and resistance islands are all terms that can be used to describe GEIs indicating that genomic islands have been actively chosen for auxiliary and adaptable roles.

The chromosomal DNAs of the *E. coli* strains incapable of causing disease that are similar or somewhat interrelated sometimes lack single or bunch of virulence genes. Those missing virulence genes are encoded by a set of large (>10 kb) integrative elements known as pathogenicity islands. PAIs are unable to self-replicate and also, they cannot mobilize themselves. In this respect they do differ from those of other extrachromosomal conjugative mobile genetic element or integrative and conjugative elements (ICEs). Pathogenicity islands usually carry portions of mobile genetic elements like bacteriophages, plasmids, and insertion sequence (IS) elements. They are flanked by repetitive structures and frequently coupled with either the genes that encode tRNA or *att* sites specific for temperate bacteriophages. PAIs have unique genomic traits, which is compelling proof of their horizontal acquisition and foreign origin. When compared to the host organisms it is found that frequency of occurrence of dinucleotides, codon usage pattern, also the G + C contents of PAIs differ markedly from that of host organism. With time, composition of the sequence or the codon usage pattern of PAI can improve, becoming more comparable to that of the core region. Therefore, the difference in GC content could represent a new acquirement or the result of evolution.

Virulence factors (≥1), found missing in the chromosomes of nonpathogenic strains, similar or allied, are encoded by pathogenic islands, a large (>10 kb) set of integrative factors. Pathogenicity islands are not self-replicating or self-mobilizable, contrary to other extrachromosomal DNA elements and ICEs. They frequently contain fragments of other mobile DNA elements, such as plasmids, bacteriophages, and insertion sequence (IS) elements. PAIs commonly link with genes coding for tRNA or temperate bacteriophage-specific *att* sites, and are flanked by repetitive structures. Comparing the DNA sequences of PAIs with the rest of the chromosomal DNA, revealing pathogenicity islands' uniqueness in genomes, stands as solid proof in favor of lateral gene transfer. When compared to the host organisms, PAIs frequently have different G + C contents, dinucleotide frequencies or Kmers, and codon usage patterns. The nucleotide sequence or codon usage pattern of the PAI region can improve through time, becoming more comparable to that of the core region. Divergence in GC content could thus be the result of evolution or the possession of new genes.

9.3 Characteristics of Pathogenic Genomic Islands in *E. Coli*

Hacker and Kaper (2000) discovered GEIs in UPEC that had genes related with virulence, and they were given the name pathogenicity islands (PAIs). Studies employing UPEC strain 536 [O6:K15:H31] provided the first insight into the occurrence of PAIs in bacterial pathogens. The pathogenesis of ExPEC and evolutionary development of bacterial pathogens were studied using one of the ExPEC *E. coli* model strains. Five distinct PAI characteristics were discovered after examining nucleotide sequences of PAI I_{536} to PAI III_{536} including left and right flanks (spanning approximately a 270-kb DNA region) of a uropathogenic *E. coli* strain 536 (O6:K15:H31). They were: (i) a connection with genes coding for tRNAs; (ii) a difference in G + C content (compared to the G + C content of the major genomic regions); (iii) the presence of repetitive DNA sequences on either flanks of a PAI; (iv) the presence of abundant coding sequences of genes, with identified or unidentified properties, forming a mishmash structure; and (v) occurrence of several sections of itinerant DNA sequences. The PAIs (PAI I_{536} to PAI III_{536}) had molecular sizes ranging from 68 to 102 kb, indicating the presence of ORFs coding for unknown proteins as well as multiple ORFs coding for virulence factors homologous to putative virulence proteins explained in the case of PAIs from other extraintestinal pathogenic *E. coli* (ExPEC) isolates. Genes encoding a siderophore system, found in PAI IV_{536}, and first discovered in pathogenic *Yersinia*, form the foundation of a high-pathogenicity island. Generally speaking, PAI I_{536} to PAI III_{536} are mishmash structures made up of several DNA pieces that have the most nucleotide resemblance to genomic sections of newly discovered EPECs, including those of *Shigella flexneri* and the O157:H7 strains EDL933 and Sakai (*she* and SHI-2 PAIs). The different plasmids belonging to three members of the order Enterobacterales (pColV, pB171, pO157, and pAPEC-1 of *E. coli*; pWR100 and pWR501 of *Shigella* spp.; and pMT1 and pYVe227 of *Yersinia* spp.) have virulence genes highly identical to many segments of PAI I_{536} to PAI III_{536}. Other fragments are yet-to-be-identified DNA sequences that lack DNA-level similarity. These various areas are interleaved with each other and contain putative ORFs that have not yet been found but may have unknown roles, as well as functional and nonfunctional or truncated ORFs that are previously known. MGEs (mobile genetic elements) including bacteriophages, plasmids, and *IS* elements make up a sizable portion of the ORFs found on these PAIs. It is interesting to note that PAI IV_{536}, a broad-host-range PAI found in numerous enterobacteria, only has functional ORFs.

9.3.1 Molecular Characteristics of PAI I_{536}

The tRNA-encoding gene *sel*C and PAI I_{536} are genetically connected. The island PAI I_{536} is 76,843 bp in length, with 16 bp straight repetitions on either side of it, and has 46% G + C composition. The inactive integrase gene of bacterial virus origin ($ORF1_{I\text{-}536}$) located immediately after *sel*C, found homologous to several intP4-like genes identified for pathogenicity islands and to the phage φR73 gene positioned at gene coding for tRNA. In addition to previously documented virulence-linked genes discovered on PAI I_{536}, for example, the genes coding for alpha-hemolysin, and identifying functional elements along the sequence of PAI I_{536} disclosed only a pair of unnamed potential adhesin genes without any sequence homology. Accordingly, translated protein sequences as well as their genetic organization, the putative ORF $15_{I\text{-}536}$ to ORF $18_{I\text{-}536}$ and ORF $37_{I\text{-}536}$ to ORF $42_{I\text{-}536}$ correspond to genes encoding F17- and CS12-like adhesins. However, ORF $18_{I\text{-}536}$ translated peptide exhibits homology to the F17a fimbrial subunit as well as the *P. mirabilis* fimbrial protein, UcaA, to aid in adhesion to epithelial cells bridging the urinary space and underlying tissue, raising the possibility that fimbriae holding this subunit is conceivably connected to UTIs.

Putative ORF $47_{I\text{-}536}$ encoding adhesin-like protein go before two presumptive ATP-binding cassette transporter-coding ORFs (ORF $45_{I\text{-}536}$ and ORF $46_{I\text{-}536}$). Although the amino acid sequences of these ORFs are comparable to those of three neighboring genes (NMB0586, NMB0587, and NMB0588) in the *Neisseria meningitidis* strain MC58, they do not share any DNA homology. Putative ORFs $2_{I\text{-}536}$ and $3_{I\text{-}536}$ are intriguing ORFs that have not yet been described. These related ORFs (whose sizes are alike) share 43 and 39 percent identity in their translated peptide sequences with *N. gonorrhoeae*'s 374 amino acid long modification methylase, NgoFVII. However, the first 89 nucleotides of each ORFs' DNA exhibit just 78 percent identity, ruling out the likelihood of gene duplication.

9.3.2 Molecular Characteristics of PAI II_{536}

A 102-Kb long PAI II_{536} is linked with the presence of tRNA gene *leu*X. PAI II_{536}, with direct 18 base pair direct repeats on its either side, has G + C content of 46 percent. The ORF $1_{II\text{-}536}$ (located just after *leu*X) encodes an active bacteriophage P4 integrase protein. ORFs encoding P-associated fimbrial adhesins (ORF $6_{II\text{-}536}$ to ORF $17_{II\text{-}536}$) and a bunch of genes coding for α-hemolysin have been identified as virulence factors located on PAI II_{536}. Other likely virulence-linked genetic sequences found on this genomic island include ORF which encodes *E. coli* Hek adhesin (ac. no. AY040859), and two presumed coding sequence of genes (ORF

$40_{II\text{-}536}$ and ORF $41_{II\text{-}536}$) that lack DNA sequence homology. The protein coded by ORF $40_{II\text{-}536}$ shares similarities with filamentous protein (FHA) that provides a principal attachment factor for adherence (filamentous hemagglutinin-like adhesions) made by *Yersinia pestis*, *Pseudomonas aeruginosa*, and *Bordetella pertussis*. The ORF located immediately before the gene that codes for hemagglutinin and is required for adhesin secretion is homologous to ORF $41_{II\text{-}536}$. Another unnamed ORF is ORF $35_{II\text{-}536}$, the putative protein of which displays homology to a portion of the *Herpetosiphon aurantiacus*'*s* modification methylase, HgiDII (EMBL accession no. P25265). An additional DNA section of 4 kb that is missing from *E. coli* K-12 strain MG1655 immediately follows the right-hand direct repeat structure instead of the unchanged sequence reads in the genome unique to *E. coli* K-12. There is just one supposed ORF encoding unknown function protein in this genomic area of *E. coli* O157:H7 strain Z5892. The typical *E. coli* K-12 chromosomal sequence, which starts with the ORF coding for the YjhS protein, is preceded by the aforementioned 4 kb DNA section.

9.3.3 Molecular Characteristics of PAI III_{536}

The gene *thr*W, which codes for tRNA, is related to PAI III_{536}. The GC composition and length of PAI III_{536} is 47% and 68,124 bp, respectively. The active integrase gene (ORF $1_{III\text{-}536}$) located just after *thr*W bears the most resemblance to SfX phage's *int* gene (which identifies *thr*W during insertion into the chromosome). 47 bp direct repeats surround this PAI on either side. There is evidence that the Tsh hemoglobin protease, present in some disease-causing *E. coli* strains in birds, contributes to the pathogenicity of some bacteria (Dozois et al. 2000). DNA sequences with similarity to *int* gene segments, present in various bacteriophages infecting *S. flexneri* and *E. coli*, follow segments of homologous Iso *IS*1's DNA sequences coding for *ins*B and *ins*A at intersection point of PAI (Dobrindt et al. 2002). These findings show that the DNA sequence between two genes, *thr*W and *yag*U, is composed of horizontally acquired sequences, including PAI III_{536}, and could be thought of as being flanked by *att*P sites of *Shigella* bacteriophage sequences responsible for integration after *thr*W. Shorter tRNA-encoding genes have also been discovered for PAI II from pathogenic *E. coli* CFT073 (Rasko et al. 2001), highlighting the significance of these epicenter where alien DNA could integrate into the chromosome. It is believed that lateral gene transfer is how these DNA regions were obtained.

There are more known genes encoding for *E. coli* virulence factors in the PAI III_{536} area, according to a sequence study of the complete region. Other virulence factors' expressing genes, ORF $17_{III\text{-}536}$ to ORF $25_{III\text{-}536}$, and ORF $27_{III\text{-}536}$ to ORF $31_{III\text{-}536}$, borne on PAI III_{536}, were the homologous genes encoding adhesin (*sfa*) and siderophore (*iro*) in *S. flexineri*. Moreover, *S. flexneri* 2a's *she* PAI (EMBL accession no. AF200692) contains the homolog of *sap*, the autotransporter-adhesin coding ORF $52_{III\text{-}536}$. The ORF $36_{III\text{-}536}$ putatively encodes a heme-binding receptor protein similar to *Y. pestis*'s HmuR. Smaller portions of the putative proteins lysine decarboxylase and cadaverine antiporter, which were translated from ORFs $47_{\mathrm{III\text{-}536}}$ and $48_{\mathrm{III\text{-}536}}$, respectively, show sequence homology with related genes *cad*A and *cad*B from *E. coli* or *Salmonella enterica* serovar Typhimurium, suggesting the possibility of multiple nucleotide sequence rearrangements or lateral gene relocation. It is interesting to note that PAI II of CFT073 also has ORFs with significant similarity to these genes (Rasko et al. 2001).

9.3.4 Molecular Characteristics of PAI IV_{536}

The pathogenic island, PAI $\mathrm{IV}_{536,}$ is connected to gene coding for asparagine tRNA, *asn*T. The central part of the supposed *Yersinia* sp.'s high-pathogenicity island is present on this island. For numerous *Yersinia* strains, it has been fully sequenced (Dobrindt et al. 2002). The 5′ end of PAI IV_{536} consists of a 1-kb sequence region that has on its flanks the genes, *fyu*A and Δ*b*1978. PAI IV_{536} has a G + C composition of 57% and a size of roughly 30.2 kb. The gene cluster necessary for the desired product of the siderophore system yersiniabactin is found in PAI IV_{536}, which lacks flanking repeat structures.

9.3.5 Comparative Account of PAIs in *E. Coli* Including Uropathogenic *E. Coli*

A comparative account of pathogenic islands derived from *E. coli* strain 536 is presented in Table 9.1, while Table 9.2 provides an account of virulence factors used by UPEC strains.

The uropathogenic isolates J96 and CFT073 each have two PAIs identified (Guyer et al. 1998; Rasko et al. 2001; Swenson et al. 1996). Even though some of these islands resemble one another because they carry the same genes, they are noticeably diverse in terms of length in Kb, composition of genes and their location, point of integration in the host chromosome, and permanence. The J96 PAIs' chromosomal positions and connections differ from those of the *E. coli* 536 PAIs. When nucleotide sequence databases were searched for similar sequences against query sequences pertaining to J96 PAI, it yielded significant similarity percentages with several virulence genes, including genes encoding toxins and adhesins. Nucleotide sequences of self-transmissible plasmid, R1, Enterobacteria phage P4, and insertion DNA sequences corresponding to IS*100*, IS*630*, and IS*911* were also identified.

Table 9.1 Characteristic features of PAI I_{536} to PAI IV_{536} of *E. coli* strain 536

Pathogenic Island	Pathogenicity island characteristic tRNA coding gene	Molecular Size (kb)	GC composition (%)	Recognized or assumed virulence factors	Gene responsible for integration	Length of the direct Repetitive DNA sequence (mers)	Transposable DNA elements and its constituent genes
I_{536}	Selenocysteine-inserting tRNA gene (*selC*)	˜77 Kb	46	α-Hemolysin, fimbrial proteins, and transporters	Integrase similar to that present in Enterobacteria phage P4	16	Insertion DNA elements: IS*1–4*, IS*10*, IS*50R*, IS*100*, IS*629–630*, IS*679*, IS*911*, IS*1328*, and IS*1353*; some ORFs of Stx-encoding phage CP-933 integrated in the genome of *E. coli* O157: H7strain EDL933; and an intron associated reverse transcriptase/ maturase.
II_{536}	Leucine-inserting tRNA gene (*leuX*)	˜102 Kb	46	α-Hemolysin, adhesin similar to hemagglutinin and Hek, and fimbrial proteins	Integrase identical to that coded by P4 bacteriophage	18	Insertion DNA elements: IS*1–4*, IS*10*, IS*50R*, IS*100*, IS*629–630*, IS*679*, IS*911*, IS*1328*, and IS*1353*; some ORFs of Stx-encoding phage CP-933 integrated in the genome of *E. coli* O157: H7strain EDL933; and multidrug resistance protein, QacE.
III_{536}		˜68 Kb	47	Adhesin of S fimbriae, *sfa* gene product; adhesin of	Integrase identical to that	47	Insertion DNA elements: IS*1–4*, IS*10*, IS*50R*,

	Threonine-inserting tRNA gene (*thrW*)			autotransporter, *sap* gene product; heme receptor, HmuR; protease active on hemoglobin, similar to Tsh; proteins related to siderophore production, *iro* gene product; proteins homologous to CadA and CadB	coded by *Shigella* phage SfX		IS*100*, IS*629–630*, IS*679*, IS*911*, IS*1328*, and IS*1353*; and some ORFs of Stx-encoding phage CP-933 integrated in the genome of *E. coli* O157:H7strain EDL933.
IV_{536}	One of the four asparagine-inserting tRNA genes, *asnT*	˜30 Kb	57	Proteins related to siderophore production, Yersiniabactin	Integrase similar to that present in Enterobacteria phage P4	Absent	Absent.

Table 9.2 Virulence factors used by UPEC strains

Adhesin		
Types	**Designation**	**Function of the gene product and associated role in infections**
Type 1 fimbriae	*fim*H	To induce colony formation and subsequently produce biofilm.
Curli fimbriae	*csg*AB	To cause bacterial aggregation, create biofilms, interact with host proteins, immune system activation, and induce host cell invasion and attachment.
P fimbriae	*pap*GII, *pap*GIII	To upregulate inflammatory reactions.
S fimbriae	*sta*S	To enable adherence to cells lining the walls of intestine and urinary tract promoting access within the tissues.
F1C fimbriae	*foc*G	To promote adhering to epithelial and endothelial cell populations of kidney and urinary bladder.
Auf fimbriae	*auf*	To code for Auf fimbriae.
RTX protein TosA	*tos*A	To cause host cell adhesion, and enhance survival in disseminated infections.
Dr fimbriae	*dra*	To enable receptor binding of the epithelial cells, and also supports invasion of the host cells.
Afimbrial adhesin	*afa*	To enable binding of receptor to epithelium and enhance ability to hemagglutinate.
IrgA homolog adhesin	*iha*	To code for homologous adhesin which is genetically regulated by iron.
Invasin		
Types	**Designation**	**Function and associated role in infections**
Hemagglutinin	*fim* operon	To express type 1 pili for causing mannose-sensitive hemagglutination.
Endothelial brain invasion	*fim*	To bind and invade microvascular endothelium of the human brain.
Toxins		
Types	**Designation**	**Function and associated role in infections**
α-Hemolysin	*hly*	To express RTX toxin in order to create pores.
Cytotoxic necrotizing factor 1	Cnf1	To participate in cell necrotic process, phagocytosis resistance, and inflammatory process.
Colibactin	Clb	To induce DNA damage that will cause immune cells to apoptoses and epithelial cells to undergo senescence of cells before maturity.
Autotransporters		
Types	**Designation**	**Function and associated infections**
Antigen43	*agn*43	To get entailed in self-aggregation, adhesion, and progression of biofilm.
Secreted autotransporter toxin	Sat	To promote protein-splitting activity of serine protease stirring cytotoxicity.
Vacuolating autotransporter toxin	Vat	To facilitate epithelium colonization and induce vacuolization of host cells.

(continued)

Table 9.2 (continued)

Serine protease autotransporter	*pic*	To affect mucins, facilitate colonization of epithelial cells, and induce formation of vacuoles in the host cell.
Iron uptake		
Types	**Designation**	**Function and associated infections**
Enterobactin	*ent, fep*A	To produce and possess siderophore, enterobactin-Fe^{3+}.
Aerobactin	*aer, iuc, iut*	To produce and possess siderophore, aerobactin-Fe ions.
Salmochelin	*iro*N	To produce and possess siderophore, salmochelin-Fe ions.
ChuA, Hma	*chu, hma*	To acquire iron from hemoglobin present in blood circulation system of the host.
Sit	*sit*DCBA	To transport iron and manganese.
IreA	*ire*A	To produce homologous siderophore receptor that is regulated by concentration of iron.
Protectin		
Types	**Designation**	**Function and associated infections**
TcpC	*tcp*C	To weaken innate immune reactions of the host.
D-serine deaminase	*dsd*CXA	To consume D-serine as C, N, and energy source, thus averting D-serine-mediated bacterial growth inhibition.
TrapT	*tra*T	To suppress complement action and overcome killer potential of serum.
	Iss	To promote resistance to complement and survival of serum.
Outer membrane protein T	OmpT	To exert protamin resistance.

9.4 Genes Governing Hardiness and Virulence in Uropathogenic *E. coli*

UPEC strains are responsible for approximately 90% of urinary tract infections acquired in the community (Foxman 2014; Flores-Mireles et al. 2015). When compared to the *E. coli* K12 genome (reference strain MG1655), the genomes of CFT073, 536, and UT189 are 6–13% larger and 8–22% more abundant with open reading frames (Brzuszkiewicz et al. 2006; Chen et al. 2006; Welch et al. 2002). The main four phylogroups of UPEC isolates—A, B1, B2, and D were recognized depending on the presence of PAIs, the nature of virulence gene expression, systems controlling secretion of toxins, receptors for iron import (TonB), siderophore receptors, complex biomolecule containing both lipid and polysaccharide (LPS), capsule made of carbohydrate-containing many monosaccharide units, motility organelle, and proteins embedded in the outer membrane (OMPs) (Bien et al. 2012). A variety of virulence genes are expressed and utilized by the majority of UPEC strains to support their extraintestinal survival, although there are significant variations in the virulence factor repertoire and expression levels across other UPEC strains.

9.4.1 Lipopolysaccharide (LPS)

The extended polysaccharide string, commonly referred to as O antigen, is joined to an oligosaccharide center that is coated with fatty acids to form the LPS amphipathic molecules (Simpson et al. 2015). Lipopolysaccharides arbitrate several facets of growth phase of uropathogenic *E. coli*, such as the capacity to intensely colonize urinary bladders, shape pools, and elicit immune reactions, both general and specialized (Aguiniga et al. 2016). The reduction of LPS on the cell surface and the development of hypersensitivity to harmful compounds like polarized steroids produced in the liver and certain antimicrobials is directly proportional. LPS also provides resistance to hydrophobic antibiotics (Zhang et al. 2013).

9.4.2 Pili

The operons, *fim*, *foc*, *pap*, and *sfa*, in particular, encode Type I, F1 C pili, P, and S which are typical sticky organelles produced by UPEC. More than 10 fimbrial gene clusters can be found in a single UPEC genome (Snyder et al. 2005; Snyder et al. 2006). Several UPEC isolates encode type 1 and P pili, both are the most researched sticky organelles. The *pap* operon expresses fimbriae assembled by a chaperone-usher system, whereas the *fim* operon codes for type 1 pili in UPEC. The *pap* operon is a component of a pathogenicity island that in addition controls other supposed virulence genes, whereas the *fim* operon is constitutive in all UPEC clinical isolates. The *pap* operon, which codes for P pili, is present in both the genomes of 536 and UT189. Two copies of the CFT073 genome each have unique PAIs. Bacterial adherence to host cells is mediated by PapG, a particular adhesin protein that is found at the distal terminus of the P pilus. There are three different forms of PapG adhesin (PapG I, II, and III). Two copies of PapGII are encoded by CFT073. Both 536 and UT189 encode PapGIII (Wiles et al. 2008). In both forms of pili, which are typically a polymer, derived from two or more different types of monomers, and comprise a primary pilus protein component that makes up the stalk of the pilus and many smaller subunit peptides at the distal end, PapG and FimH serve as the true adhesins. The two domains that makeup PapG and FimH are the pilin domain, which permits copolymerization, and the carbohydrate-binding lectin domain (Kline et al. 2009).

9.4.3 Curli

The bacterial surface appendages known as curli exude protein subunits from the cell in the form of soluble monomeric proteins and have the same physical properties as amyloid fibrils. They have been linked to some degenerative disorders in humans. The development of biofilm may be aided by bacterial amyloids (Goyal et al. 2014).

The operon *csg*DEFG in UPEC encodes proteins that control curli formation. Although curly fibers are created by subunits of CsgA nucleated by CsgB, secretion of CsgA requires CsgE, CsgF, and CsgG-like auxiliary proteins (Chapman et al. 2002; Barnhart and Chapman 2006).

9.4.4 Non-pilus Adhesins

Other afimbrial adhesins are elaborated by UPEC. In fact, TosA, an adhesion molecule, is expressed in more than 30% of infectious strains (Vigil et al. 2011). The adhesion protein, Iha (regulated by the abundance of iron), intercedes adhesion to BECs in a mouse infection model, whereas FdeC is crucial in bladder and kidney colonization (Nesta et al. 2012; Johnson et al. 2005).

9.4.5 Flagella

Some EPEC strains have sticky and invasive qualities rendered by flagella, which are vital to the dynamics of biofilms (Giron et al. 2002; Pratt and Kolter 1998). According to a recent study, gene expression and regulation throughout the growth phase revealed that flagella serve various roles during biofilm formation, including adhesion, maturation, and dispersal (Nakamura et al. 2016).

9.4.6 Secreted Toxins

Different pathogenetic roles are played by UPEC toxins during infection. UPEC bacteria employ systems that are known as type I and type V secretion systems (Henderson et al. 2004). About 50% of UPEC isolates encode the type I secreted toxin—hemolysin (HlyA) (Marrs et al. 2005). The hemolysin operon, *hly*CABD, is present in only one copy in each of the UPEC isolates CFT073 and UT189, but encodes two copies in the 536 strain. When a high concentration level of the calcium-dependent 110 kDa toxin—hemolysin is attained, cell lysis occurs. Sub-lytic dosages of HlyA have been found to potently stimulate the serine/threonine kinase Akt, which is essential for the cell cycle sequence of the host, metabolism, intracellular trafficking of vesicles, survival, and signaling pathways that mediate the inflammatory responses (Wiles et al. 2008). The term "autotransporters" refers to all type V released poisons (Henderson et al. 2004). UPEC isolates frequently expressed Vat (vacuolating autotransporter toxin) and Sat (secreted autotransporter toxin) (Ewers et al. 2007; Restieri et al. 2007). While Vat and Sat are coded by the CFT073 strain, UT189 and 536 solely express VAT. CNF1 is encoded by about one-third of UPEC isolates. This UPEC spread and persistence throughout the urinary system may be aided by CNF1-mediated actions (Wiles et al. 2008).

9.4.7 Outer Membrane Vesicles (OMVs)

The OMVs that Gram-negative bacteria produce at all phases of growth may be linked to the toxins produced by UPEC rather than being secreted as bare proteins (Ellis and Kuehn 2010). Membrane vesicles are thought to be a "clever" approach for bacteria to protect their poisons and an effective means to transport them to the host cell (Wiles et al. 2008). Numerous bacterial species use OMVs to communicate with one another and with other species, exchange genetic material, cling to and invade host cells, and release toxins (Mashburn-Warren and Whiteley 2006). OMVs are hypothesized to be released by pathogenic bacteria as a way to both deliver intense ruptures of effector chemicals that can modify host actions and safeguard dangerous cargoes while they are being carried to the target host cells. Two UPEC-related toxins, HlyA and CNF1 (cytotoxic necrotizing factor 1), help OMVs in targeting cells of the host (Balsalobre et al. 2006; Davis et al. 2006; Kouokam et al. 2006).

9.4.8 Iron Acquisition

For UPEC to live in an iron-deficient environment, iron acquisition is a crucial prerequisite (Skaar 2010). Pathogens typically cause iron deficiency in their hosts. One of the intrinsic barriers against bacterial survival inside hosts is this restriction (Cassat and Skarr 2013). Host glycoproteins such as transferrin and lactoferrin bind elemental iron within the host, or it is integrated into the substituted porphyrin ring containing a ligand iron atom of hemoglobin and myoglobin. Iron uptake genes are upregulated in UPEC in response to iron deficits in order to provide appropriate amounts of intracellular iron (Garcia et al. 2011). Iron can be transported by iron transporters, ferric iron can be taken up by siderophores, and heme can be taken up by outer membrane heme receptors. All of these routes for iron absorption involve the TonB-ExbB-ExbD energy transduction system, which plays a significant role. TonB receives energy from inner membrane proteins, ExbB and ExbD, which it subsequently transfers to large receptor molecules present in the outer membrane that participate in the active iron transport, causing periplasmic migration. ATP synthesized in bacterial cells is used as a primary source of energy by ABC transporters to move iron-containing complexes across the inner membrane once they have been coupled to binding proteins in the periplasm.

9.4.9 Siderophores

Siderophores, the low-molecular chelating organic compounds specific to iron ions, utilized to forage ferric iron (Fe^{3+}), are produced by UPEC strains. These molecules bind ferric iron, and cognate outer membrane receptors detect iron–siderophore complexes. Siderophore receptors must bind to and chelate iron exterior to the cell through the TonB-mediated Fe^{3+} uptake pathway to facilitate iron import (O'Brien et al. 2016). Aerobactin, enterobactin, salmochelin, and yersiniabactin are only a few of the siderophores that are biosynthesized and taken up by UPEC strains. Aerobactin has higher amounts of iron binding than enterobactin, is stable at low pH levels, and is highly expressed (Watts et al. 2012; Valdebenito et al. 2006). Yersiniabactin, which offers resistance against copper toxicity by sequestering copper (II) within the host cell, protects against intracellular death brought on by copper stress and is essential for the formation of biofilms in urine (Chaturvedi et al. 2012). Another prototype UPEC strain, CFT073, does not have the ability to biosynthesize yersiniabactin, although some strains, such as 536 and UTI89, synthesize and use the substance. Of four siderophores, enterobactin and salmochelin are catecholates, aerobactin is hydroxamate, and yersiniabactin is a mixed-type siderophore. Additionally, Yersiniabactin binds to copper and reduces copper toxicity (Chaturvedi et al. 2012). Specific proteins made by the mammalian hosts bind siderophores and stop bacteria from reabsorbing iron-siderophore composites. As an ingredient of the general immune system, neutrophils create lipocalin-2, an enterobactin-binding molecule, and this prevents iron import to bacterial cells navigated by enterobactin (Goetz et al. 2002; Flo et al. 2004). Bacteria like UPEC and *Salmonella enterica* subspecies enterica use salmochelin, a C-glucosylated siderophore that is not identified by lipocalin-2, to get past this obstacle (Hantke et al. 2003). A quantitative metabolomics method has been used to investigate siderophore production and its relationship to the occurrence of genetic sequence encoding enzymes for biosynthesis of siderophore in archetypal strains of UPEC (Henderson et al. 2009).

9.4.10 Two-Component Signaling System

Two-component signaling systems (TCSs) have been implicated in the control of metabolic pathways by UPEC components related to colonization. The primary signal transduction routes used by bacteria to sense and react to a range of environmental variables, such as quorum sensing molecules, growth substrates, and antimicrobials, are known as two-component signaling systems (TCSs). Histidine kinase (HK), sensor of a two-component signal transduction system for accepting external input signals and a response regulator that expresses a proper change in the bacterial physiology, controls gene expression, and makes up TCSs (Stock et al. 2000). According to description, the BarA/UvrY system regulates the transition of

UPEC-associated TCSs involved in the genesis of UTIs between glycolysis and gluconeogenesis (Tomenius et al. 2006). According to Eguchi et al. (2011), the PhoQ/PhoP and EvgS/EvgA systems have been linked to acid tolerance, whereas KguS/KguR regulates the body's use of α-ketoglutarate. They do this to make UPEC's urinary tract adaption easier (Cai et al. 2013).

9.5 Omics Studies in Revealing Virulence of Uropathogenic *E. Coli*

With the help of next-generation technologies, there has been a radical change in our knowledge of biological systems. The progress in the understanding of the virulence of uropathogenic *E. coli* has been presented in Table 9.3.

9.6 Conclusion

UTI, one of the most widespread human illnesses, is caused by *E. coli* in nearly all cases. Despite this, there is a dearth of an omics perspective on the phylogenetic distribution of isolates linked to various clinical disorders. By repeatedly acquiring PAIs, distinct invasive UPEC lineages developed within the *E. coli* population. *E. coli* has many pathogenic islands, which have a significant function in bacterial genome plasticity. This bacterial species' genome has been altered by the PAIs, which additionally clarifies the creation of many pathotypes that cause various extraintestinal and intestinal disorders. The ownership of all active genes is enabled by the migration of these genomic islands in a single step, resulting in a paradigm shift in our understanding of how bacteria evolve and equip themselves to cope with new ecological niches. PAIs encode functional elements such as pathogenicity effectors, but they also include regulatory components engaged in interactions among PAIs and the surrounding genomic sequence. In addition to PAIs, this interaction may play a significant role in genome evolution and is likely to produce a variety of expression profiles. More research is needed to completely understand *E. coli* virulence and the noticeable duplication of pathogenicity modules in uropathogens.

Table 9.3 Recent Studies (2011–2022) of Uropathogenic *E. coli* (UPEC) utilizing Genomic, Transcriptomic, Proteomic, and Metabolomic tools

Method	Condition	Finding/result	Reference
Genomics			
	Comparative genomics of *Escherichia coli* ST131	The global distribution of *E. coli* ST131 clone was confirmed through comparative genome analysis of the said strain isolated from six different environmental sites in a period of 11 years. The mobile DNA sequences and recombination took a vital part in evolving this multidrug-resistant pathogen.	Petty et al. (2014)
	Comparative genome analysis of draft genome sequence of *E. coli* ST131 strain EC958	The results of this study revealed a common association between the virulence genes encoded by EC958 with that of uropathogenic *E. coli* (UPEC). A familiar transpositional mutation was identified in *fim*B locus which caused on-off genetic switch to slow down in case of type 1 fimbriae.	Totsika et al. (2011)
	Analyses of methylome in globally disseminated *E. coli* ST131 strain EC958	The full methylome of *E. coli* EC958 was analyzed; three novel recognition sites along with their cognate methyltransferase enzymes specific to only ST131 were identified. The mobile genetic elements, carrier of methyltransferase genes, could be the major driver for spread of such genes even among clonally related strains.	Forde et al. (2015)
	Analyses of 1700 draft and finished *E. coli* whole genome sequences	The overlapping relationships between strains from different pathotypes of *E. coli* were established. Proteomic analyses identified foremost genes that code for surface exposed or secreted proteins that stand for impending wide-coverage vaccine antigens, YncE.	Moriel et al. (2016)

(continued)

Table 9.3 (continued)

Method	Condition	Finding/result	Reference
	Comparative genomic analyses of *E. coli* ST131, ST38, ST405, and ST648: Belonging to ExPEC lineages	The results of this study would help to understand the evolution of ST131.	Shaik et al. (2017)
Transcriptomics			
	Co-transcriptomic analyses of UPEC strains UTI89 and 83,972 within bone marrow-derived macrophages	Identified new understandings related to host–pathogen interaction.	Mavromatis et al. (2015)
	mRNA profiling of 21 UPEC strains to create a complete transcriptional profiles	A high correlation was observed among transcriptomes and genealogies. Genes expressing both universally and exclusively were characterized from within strains belonging to every phylogenetic group.	Bielecki et al. (2014)
	Identification of novel fitness genes that were precisely expressed during human infection was carried out using comparative transcriptional analysis of cultures of uropathogenic *E. coli* strains grown in routine heterotrophic medium or human urine	Multiple new fitness-related factors exploited by UPEC in infection were discovered, along with a typical transcript sequence in urinary tract infections of women in a community caused by UPEC strains.	Subashchandrabose et al. (2014)
	Comparative transcriptomics of three distinct uropathogenic *E. coli strains,* grown in different experimental environments or media which includes Luria broth, crude urine medium (prepared by passing through bacterial filter), and live mouse infected with the uropathogen	A strong correlation in transcriptome profiles was observed for all the three strains including transcripts derived from in vivo conditions.	Frick-Chen et al. (2020)
	Thorough analyses of transcriptome profiles of uropathogenic *E. coli* strains following invasion of bladder epithelial cells in order to identify probable factors associated with virulence	New virulence-associated genes and their regulatory mechanisms were identified for the first time using transcriptomic analyses of UPEC. The small protein, MgtS, was established as a positive contributor in	Li et al. (2021)

(continued)

Table 9.3 (continued)

Method	Condition	Finding/result	Reference
		invading murine bladder epithelial cells to form colonies of uropathogenic *E. coli* strains.	
Proteomics			
	Cellular proteomes of *E. coli* collected from UTI patients' urine and fecal samples	Experimental results revealed that each of the UTI patients bears a discrete and relatively non-heterogenous load of infecting *E. coli* strain. Various structural and metabolic functions were performed by the proteins identified.	Smith et al. (2011)
	UPEC biofilm analysis	Differential protein expression profiles explained that diverse subpopulations of bacteria remain confined in the biofilms produced by UPECs.	Smith et al. (2011)
	Proteomic analyses of UPEC *E. coli* were conducted to complement genomic analyses	The molecular basis of UPEC pathogenesis was revealed.	Cash 2014
	Comparative proteomic analysis of OMPs in 54 uropathogenic *E. coli* strains	Resulted in the identification of novel proteins that were associated with virulence and also offered an outline for revealing the OMP constituent of uropathogenic *E. coli* strains.	Wurpel et al. (2015)
	Comparative proteomic analyses were used for defining the UPEC pan and core surface proteome	During growth in media composed of filter-sterilized urine, surface, and core proteome constituent of uropathogenic *E. coli* strains were revealed and a new type of fimbriae was identified.	Wurpel et al. (2016)
	Proteome analyses of UPEC *E. coli* CFT073 for characterizing hypothetical proteins	Fifty-three proteins were detected as a constituent of metabolic interactome, 8 proteins as virulent, 35 proteins as nonhomologous to humans of which three-dimensional structures of 6 proteins were modelled. These discoveries will provide novel targets for	Kaur et al. (2021)

(continued)

Table 9.3 (continued)

Method	Condition	Finding/result	Reference
		therapeutic development against the uropathogenic *E. coli* CFT073 and will aid in the molecular understanding of disease mechanisms.	
	Proteomic profiling of urine medium and J82 bladder cell culture grown UPEC *E. coli* UT189	During growth in urine and in infected J82 cell culture, significant expression of proteins linked with Fe-import and arginine metabolism was observed. There was no direct association between protein overexpression and proteins essential for infection.	Andersen et al. (2022)
Metabolomics			
	Investigating comparative metabolome during siderophore production by *E. coli* strains K12 (wild type) and UTI89 (uropathogenic)	Different metabolomic effects related to the production of various siderophore structure families were discovered by using a bacterial genetic method along with metabolome analyses.	Lv et al. (2014)
	Comparative metabolome of *E. coli* strains; HPI-bearing uropathogenic UTI89 versus wild-type K12	For fitness and favorable growth of UPEC UTI89, the presence of HPI is essential.	Yan et al. (2015)
	Comparative interactome of *E. coli* strains; uropathogenic versus and non-uropathogenic	Profound interactive metabolome perturbations were induced by urine grown uropathogenic strain in comparison to non-uropathogenic strain. Siderophore production altered the constituents of the metabolome.	Su et al. (2016)
	Identification of a new virulence-linked siderophore coded by genes of HPI in UPEC strain following metabolome assay.	Following a global metabolome study of *E. coli* strains, wild type, UTI89, UTI89ΔybtS, and UTI89ΔybtS, new siderophores were identified in the HPI-reliant biochemical path as well as a large number of novel metabolite characteristics coded by the HPI genes with	Xu et al. (2019)

(continued)

Table 9.3 (continued)

Method	Condition	Finding/result	Reference
		an apparent substrate dependency on salicylic acid.	
	Search of molecular signature for uropathogenic against urocolonizing *E. coli*	Asymptomatic bacteriuria (ASB) strains and cystitis isolates had different metabolic signatures, which led researchers to hypothesize that UPEC's definition at the molecular level may reside at the level of overall bacterial metabolism.	Eberly et al. (2020)

References

Aguiniga LM, Yaggie RE, Schaeffer AJ, Klumpp DJ (2016) Lipopolysaccharide domains modulate urovirulence. Infect Immun 84:3131–3140

Andersen S, Nawrocki A, Johansen AE et al (2022) Proteomes of uropathogenic *Escherichia coli* growing in human urine and in J82 urinary bladder cells. Proteomes 10(2):15

Balsalobre C, Silvan JM, Berglund S et al (2006) Release of the type I secreted alpha-haemolysin via outer membrane vesicles from *Escherichia coli*. Mol Microbiol 59:99–112

Barnhart MM, Chapman MR (2006) Curli biogenesis and function. Ann Rev Microbiol 60:131–147

Bielecki P, Muthukumarasamy U, Eckweiler D et al (2014) In vivo mRNA profiling of uropathogenic*Escherichia coli* from diverse phylogroups reveals common and group-specific gene expression profiles. MBio 5:e01075-14

Bien J, Sokolova O, Bozko P (2012) Role of uropathogenic *Escherichia coli* virulence factors in development of urinary tract infection and kidney damage. Int J Nephrol 2012:681473

Brzuszkiewicz E, Bruggemann H, Liesegang H et al (2006) How to become a uropathogen: comparative genomic analysis of extraintestinal pathogenic *Escherichia coli* strains. Proc Natl Acad Sci U S A 103:12879–12884

Cai W, Wannemuehler Y, Dell'anna G, Nicholson B, Barbieri NL, Kariyawasam S et al (2013) A novel two-component signaling system facilitates uropathogenic *Escherichia coli*'s ability to exploit abundant host metabolites. PLoS Pathog 9:e1003428

Cash P (2014) Proteomic analysis of uropathogenic *Escherichia coli*. Expert Rev Proteomics 11(1): 43–58

Cassat JE, Skarr EP (2013) Iron in infection and immunity. Cell Host Microbe 13(5):509–519

Chapman MR, Robinson LS, Pinkner JS, Roth R, Heuser J, Hammar M et al (2002) Role of *Escherichia coli* curli operons in directing amyloid fiber formation. Science 295:851–855

Chaturvedi KS, Hung CS, Crowley JR, Stapleton AE, Henderson JP (2012) The siderophore yersinia bactin binds copper to protect pathogens during infection. Nat Chem Biol 8:731–736

Chen SL, Hung CS, Xu J et al (2006) Identification of genes subject to positive selection in uropathogenic strains of *Escherichia coli*: a comparative genomics approach. Proc Natl Acad Sci U S A 103:5977–5982

Clermont O, Dixit OVA, Vangchhia B et al (2019) Characterization and rapid identification of phylogroup G in *Escherichia coli*, a lineage with high virulence and antibiotic resistance potential. Environ Microbiol 21:3107–3117

Dale AP, Woodford N (2015) Extra-intestinal pathogenic Escherichia coli (ExPEC): disease, carriage and clones. J Infect 71(6):615–626

Davis JM, Carvalho HM, Rasmussen SB, O'Brien AD et al (2006) Cytotoxic necrotizing factor type 1 delivered by outer membrane vesicles of uropathogenic *Escherichia coli* attenuates polymorphonuclear leukocyte antimicrobial activity and chemotaxis. Infect Immun 74:4401–4408

Dobrindt U, Blum-Oehler G, Nagy G et al (2002) Genetic structure and distribution of four pathogenicity islands (PAI I_{536} to PAI IV_{536}) of uropathogenic *Escherichia coli* strain 536. Infect Immun 70(11):6365–6372

Dozois CM, Dho-Moulin M, Bree A, Fairbrother JM, Desautels C, Curtiss R (2000) Relationship between the Tsh autotransporter and pathogenicity of avian *Escherichia coli* and localization and analysis of the Tsh genetic region. Infect Immun 68:4145–4154

Eberly AR, Beebout CJ, Tong CMC, Van Horn GT et al (2020) Defining a molecular signature for uropathogenic versus urocolonizing *Escherichia coli*: the status of the field and new clinical opportunities. J Mol Biol 432(4):786–804

Eguchi Y, Ishii E, Hata K, Utsumi R (2011) Regulation of acid resistance by connectors of two-component signal transduction systems in *Escherichia coli*. J Bacteriol 193:1222–1228

Ellis TN, Kuehn MJ (2010) Virulence and immunomodulatory roles of bacterial outer membrane vesicles. Microbiol Mol Biol Rev 74:81–94

Escobar-Páramo P, Clermont O, Blanc-Potard A-B et al (2004) A specific genetic background is required for acquisition and expression of virulence factors in *Escherichia coli*. Mol Biol Evol 21:1085–1094

Ewers C, Li G, Wilking H et al (2007) Avian pathogenic, uropathogenic, and newborn meningitis-causing *Escherichia coli*: how closely related are they? Int J Med Microbiol 297:163–176

Flo TH, Smith KD, Sato S, Rodriguez DJ, Holmes MA, Strong RK, Akira S, Aderem A (2004) Lipocalin 2 mediates an innate immune response to bacterial infection by sequestrating iron. Nature 432:917–921

Flores-Mireles AL, Walker JN, Caparon M, Hultgren SJ (2015) Urinary tract infections: epidemiology, mechanisms of infection and treatment options. Nat Rev Microbiol 13:269–284

Forde BM, Phan M-D, Gawthorne JA et al (2015) Lineage-specific methyltransferases define the methylome of the globally disseminated *Escherichia coli* ST131 clone. MBio 6:e01602-15

Foxman B (2014) Urinary tract infection syndromes: occurrence, recurrence, bacteriology, risk factors, and disease burden. Infect Dis Clin N Am 28(1):1–13

Frick-Chen AE, Sintsova A, Smith SN et al (2020) The gene expression profile of uropathogenic *Escherichia coli* in women with uncomplicated urinary tract infections is recapitulated in the mouse model. MBio 11(4):e01412-20

Garcia EC, Brumbaugh AR, Mobley HL (2011) Redundancy and specificity of *Escherichia coli* iron acquisition systems during urinary tract infection. Infect Immun 79:1225–1235

Giron JA, Torres AG, Freer E, Kaper JB (2002) The flagella of enteropathogenic *Escherichia coli* mediate adherence to epithelial cells. Mol Microbiol 44:361–379

Goetz DH, Holmes MA, Borregaard N, Bluhm ME, Raymond KN, Strong RK (2002) The neutrophil lipocalin NGAL is a bacteriostatic agent that interferes with siderophore-mediated iron acquisition. Mol Cell 10:1033–1043

Goyal P, Krasteva PV, Van Genven N, Gubellini F, Van Den Broeck I, Troupiotis-Tsailaki A et al (2014) Structural and mechanistic insights into the bacterial amyloid secretion channel CsgG. Nature 516:250–253

Guyer DM, Kao JS, Mobley HLT (1998) Genomic analysis of a pathogenicity island in uropathogenic *Escherichia coli* strain CFT073: distribution of homologous sequences among isolates from patients with pyelonephritis, cystitis, and catheter-associated bacteriuria and from fecal samples. Infect Immun 66:4411–4417

Hacker J, Kaper JB (2000) Pathogenicity islands and the evolution of microbes. Annu Rev Microbiol 54:641–679

Hantke K, Nicholson G, Rabsch W, Winkelmann G (2003) Salmochelins, siderophores of *salmonella enterica* and uropathogenic *Escherichia coli* strains, are recognized by the outer membrane receptor IroN. Proc Natl Acad Sci U S A 100:3677–3682

Henderson IR, Navarro-Garcia F, Desvaux M et al (2004) Type V protein secretion pathway: the autotransporter story. Microbiol Mol Biol Rev 68:692–744

Henderson JP, Crowley JR, Pinkner JS, Walker JN, Tsukayama P, Stamm WE, Hooton TM, Hultgren SJ (2009) Quantitative metabolomics reveals an epigenetic blueprint for iron acquisition in uropathogenic *Escherichia coli*. PLoS Pathog 5:e1000305

Hsiao WWL, Ung K, Aeschliman D et al (2005) Evidence of a large novel gene pool associated with prokaryotic genomic islands. PLoS Genet 1(5):e62

Johnson JR, Jelacic S, Schoenin LM, Clabots C, Shaikh N, Mobley HLT et al (2005) The IrgA homologue adhesin Iha is an *Escherichia coli* virulence factor in murine urinary tract infection. Infect Immun 73:965–971

Jorgensen SL, Stegger M, Kudirkiene E et al (2019) Diversity and population overlap between avian and human *Escherichia coli* belonging to sequence type 95. mSphere 4:e00333 – 18

Kaper JB, Nataro JP, Mobley HL (2004) Pathogenic *Escherichia coli*. Nat Rev Microbiol 2:123–140

Kaur H, Singh V, Kalia M, Mohan B, Taneja N (2021) Identification and functional annotation of hypothetical proteins of uropathogenic *Escherichia coli* strain CFT073 towards designing antimicrobial drug targets. J Biomol Struct Dyn 1-12. https://doi.org/10.1080/07391102.2021.2000499

Kline KA, Falker S, Dahlberg S, Normark S, Henriques-Normark B (2009) Bacterial adhesins in host-microbe interactions. Cell Host Microbe 5:580–592

Kohler C-D, Dobrindt U (2011) What defines extraintestinal pathogenic Escherichia coli? Int J Med Microbiol 301(8):642–647

Kouokam JC, Wai SN, Fallman M et al (2006) Active cytotoxic necrotizing factor 1 associated with outer membrane vesicles from uropathogenic *Escherichia coli*. Infect Immun 74:2022–2030

Le Gall T, Clermont O, Gouriou S et al (2007) Extraintestinal virulence is a coincidental by-product of commensalism in B2 phylogenetic group *Escherichia coli* strains. Mol Biol Evol 24:2373–2384

Li S, Pang Y, Zhang S, Li Q, Wang Q, Wang L, Feng L (2021) Transcriptomic analysis reveals that the small protein MgtS contributes to the virulence of uropathogenic *Escherichia coli*. Microb Pathog 152:104765

Lv H, Hung CS, Henderson JP (2014) Metabolomic analysis of siderophore cheater mutants reveals metabolic costs of expression in uropathogenic *Escherichia coli*. J Proteome Res 13:1397–1404

Marrs CF, Zhang L, Foxman B (2005) *Escherichia coli* mediated urinary tract infections: are there distinct uropathogenic *E. coli* (UPEC) pathotypes? FEMS Microbiol Lett 252:183–190

Mashburn-Warren LM, Whiteley M (2006) Special delivery: vesicle trafficking in prokaryotes. Mol Microbiol 61:839–846

Mavromatis CH, Bokil NJ, Totsika M et al (2015) The co-transcriptome of uropathogenic*Escherichia coli*-infected mouse macrophages reveals new insights into host–pathogen interactions. Cell Microbiol 17:730–746

Milkman R (1973) Electrophoretic variation in *Escherichia coli* from natural sources. Science 182: 1024–1026

Moriel DG, Tan L, Goh KGK et al (2016) A novel protective vaccine antigen from the core *Escherichia coli* genome. mSphere 1:e00326-16

Nakamura Y, Yamamoto N, Kino Y, Yamamoto N, Kamei S, Mori H et al (2016) Establishment of a multi-species biofilm model and metatranscriptomic analysis of biofilm and planktonic cell communities. Appl Microbiol Biotechnol 100:7263–7279

Nesta B, Spraggon G, Alteri C, Moriel DG, Rosini R, Veggi D et al (2012) FdeC, a novel broadly conserved *Escherichia coli* adhesin eliciting protection against urinary tract infections. MBio 3: 9

O'Brien VP, Hannan TJ, Nielsen HV, Hultgren SJ (2016) Drug and vaccine development for the treatment and prevention of urinary tract infections. Microbiol Spectr 4:42

Petty NK, Zakour NLB, Stanton-Cook M et al (2014) Global dissemination of a multidrug resistant *Escherichia coli* clone. Proc Natl Acad Sci U S A 111:5694–5699

Pratt LA, Kolter R (1998) Genetic analysis of *Escherichia coli* biofilm formation: roles of flagella, motility, chemotaxis and type I pili. Mol Microbiol 30:285–293

Rasko DA, Phillips JA, Li X, Mobley HLT (2001) Identification of DNA sequences from a second pathogenicity island of uropathogenic *Escherichia coli* CFT073: probes specific for uropathogenic populations. 184:1041–1049

Restieri C, Garriss G, Locas M-C, Dozois CM (2007) Autotransporter-encoding sequences are phylogenetically distributed among *Escherichia coli* clinical isolates and reference strains. Appl Environ Microbiol 73:1553–1562

Sarowska J, Futoma-Koloch B, Jama-Kmiecik A et al (2019) Virulence factors, prevalence and potential transmission of extraintestinal pathogenic Escherichia coli isolated from different sources: recent reports. Gut Pathog 11:10

Selander RK, Musser JM, Caugant DA et al (1987) Population genetics of pathogenic bacteria. Microb Pathog 3:1–7

Shaik S, Ranjan A, Tiwari SK et al (2017) Comparative genomic analysis of globally dominant ST131 clone with other epidemiologically successful extraintestinal pathogenic *Escherichia coli* (ExPEC) lineages. MBio 8(5):e01596-17

Simpson BW, May JM, Sherman DJ et al (2015) Lipopolysaccharide transport to the cell surface: biosynthesis and extraction from the inner membrane. Philos Trans R Soc B370(1679): 20150029

Skaar EP (2010) The battle for iron between bacterial pathogens and their vertebrate hosts. PLoS Pathog 6:e1000949

Smith A, Rooyen J-P van, Argo E, Cash P (2011) Proteomic analysis of *Escherichia coli* associated with urinary tract infections. Proteomics 11: 2283–2293

Snyder JA, Haugen BJ, Virginia Lockatell C et al (2005) Coordinate expression of fimbriae in uropathogenic *Escherichia coli*. Infect Immun 73:7588–7596

Snyder JA, Lloyd AL, Virginia Lockatell C et al (2006) Role of phase variation of type 1 fimbriae in a uropathogenic *Escherichia coli* cystitis isolate during urinary tract infection. Infect Immun 74: 1387–1393

Stock AM, Robinson VL, Goudreau PN (2000) Two-component signal transduction. Annu Rev Biochem 69:183–215

Su Q, Guan T, Lv H (2016) Siderophore biosynthesis coordinately modulated the virulence-associated interactive metabolome of uropathogenic *Escherichia coli* and human urine. Sci Rep 6:24099

Subashchandrabose S, Hazen TH, Brumbaugh AR et al (2014) Host-specific induction of *Escherichia coli* fitness genes during human urinary tract infection. Proc Natl Acad Sci U S A 111:18327–18332

Swenson DL, Bukanov NO, Berg DE, Welch RA (1996) Two pathogenicity islands in uropathogenic *Escherichia coli* J96: cosmid cloning and sample sequencing. Infect Immun 64:3736–3743

Tenaillon O, Skurnik D, Picard B, Denamur E (2010) The population genetics of commensal *Escherichia coli*. Nat Rev Microbiol 8:207–217

Tomenius H, Pernestig AK, Jonas K, Georgellis D, Mollby R, Normark S et al (2006) The *Escherichia coli* BarA-UvrY two-component system is a virulence determinant in the urinary tract. BMC Microbiol 6:27

Totsika M, Beatson SA, Sarkar S et al (2011) Insights into a multidrug resistant Escherichia coli pathogen of the globally disseminated ST131 lineage: genome analysis and virulence mechanisms. PLoS One 6:e26578

Valdebenito M, Crumbliss AL, Winkelmann G, Hantke K (2006) Environmental factors influence the production of enterobactin, salmochelin, aerobactin, and yersiniabactin in *Escherichia coli* strain Nissle 1917. Int J Med Microbiol 296:513–520

Vigil PD, Stapleton AE, Johnson JR, Hooton TM, Hodges AP, He YQ et al (2011) Presence of putative repeat-in-toxin gene tosA in *Escherichia coli* predicts successful colonization of the urinary tract. MBio 2:10

Watts RE, Totsika M, Challinor VL et al (2012) Contribution of siderophore systems to growth and urinary tract colonization of asymptomatic bacteriuria *Escherichia coli*. Infect Immun 80:333–344

Welch RA, Burland V, Plunket G, Redford P et al (2002) Extensive mosaic structure revealed by the complete genome sequence of uropathogenic *Escherichia coli*. Proc Natl Acad Sci U S A 99: 17020–17024

Wiles TJ, Kulesus RR, Mulvey MA (2008) Origins and virulence mechanisms of uropathogenic Escherichia coli. Exp Mol Pathol 85(1):11–19

Wurpel DJ, Moriel DG, Totsika M et al (2015) Comparative analysis of the uropathogenic *Escherichia coli* surface proteome by tandem mass-spectrometry of artificially induced outer membrane vesicles. J Proteome 115:93–106

Wurpel DJ, Totsika M, Allsopp LP et al (2016) Comparative proteomics of uropathogenic *Escherichia coli* during growth in human urine identify UCA-like (UCL) fimbriae as an adherence factor involved in biofilm formation and binding to uroepithelial cells. J Proteome 131:177–189

Xu G, Guo H, Haitao LV (2019) Metabolomics assay identified a novel virulence-associated siderophore encoded by the high-pathogenicity island in uropathogenic *Escherichia coli*. J Proteome Res 18(5):2331–2336

Yan L, Nie W, Lv H (2015) Metabolic phenotyping of the *yersinia* high pathogenicity island that regulates central carbon metabolism. Analyst 140:3356–3361

Zhang G, Meredith TC, Kahne D (2013) On the essentiality of lipopolysaccharide to Gram-negative bacteria. Curr Opin Microbiol 16:779–785

Genomic Islands in *Helicobacter* Species

10

Meenambiga Setti Sudharsan, Varsha K, Sowmya Hari, and Sakthiselvan Punniavan

Abstract

Horizontal gene transfer mechanisms help in the transfer of distant genes generating pathogens with different types of virulence. Genomic Islands (GI) are the evidence representing microbial genome evolution. These genes are acquired by HGT process. It determines their adaptation to the environment, compatibility, and other gene expression mechanisms. The pathogenicity islands which are a subclass of Genomic Islands contribute to the virulent nature of the pathogen. *Helicobacter pylori* known for its colonization in the stomach and the intestinal regions poses the most genetic diversity among the pathogenic species of the bacterial community. *H. pylori* has the ability to adapt to the host system by changing its genetic features. Chronic infections show the maximum genetic differential ability of *H. pylori.* The strains are categorized into type 1 and type 2 in which each secretes certain antigens and cytotoxin enhancing its virulence. Numerous computational tools have been advanced to recognize and categorize these genomic islands. One such tool is Island viewer which enables the user to access published and even unpublished GIs. It is also linked with external databases like NCBI. GIs can be predicted based on single or multiple genomes. GIs do not frequently replicate like plasmids and they serve as markers for identifying evolutionary pathways. However, real-time tracking is still under research and needs to be developed.

M. S. Sudharsan (✉) · V. K · S. Hari · S. Punniavan
Department of Bio-Engineering, School of Engineering, Vels Institute of Science, Technology and Advanced Studies, Chennai, Tamilnadu, India
e-mail: meenambiga.se@velsuniv.ac.in

I. Mani et al. (eds.), *Microbial Genomic Islands in Adaptation and Pathogenicity*,
https://doi.org/10.1007/978-981-19-9342-8_10

Keywords

Horizontal gene transfer · Genomic Islands · Pathogenicity islands · *Helicobacter pylori* · Antigens · Cytotoxins · Genomes · NCBI · Plasmids

10.1 Introduction

Genomic Islands (GIs) are a group of bacterial or archeal genes, which are acquired via horizontal gene transfer (HGT) and contributes to microbial genome evolution. These are a pool of gene sequences ranging from 10–100 kb which encode various disease causing factors, antimicrobial resistant factors, metabolic pathway regulating factors and microbial adaptations. Ambulatory genetic elements that fall under GIs include integrons, transposons, integrative and conjugative elements (ICEs), and non-segregative phage elements. These gene sequences can be detected based on the methods of acquisition of genomic island genes. Some of the known methods are via conjugation, transformation or transduction, and adaptable elements such as enzymes that catalyzes insertion of viral DNA, replicative transposition mechanism, and insertional sequences that promote relocation of GIs. Once these elements get integrated into the genome they start their evolution process through mutations, rearrangements of genomic sequences, or by insertion and deletion of ambulant genetic elements. These GI sequences differ from the genomic DNA with the percentage of GC content and the usage of codons. These genes are acquired by the process of horizontal gene transfer mechanism. HGT involves the migration of genetic materials between genetically distant organisms which is also determined as lateral gene transfer. The microbes undergo transformation process or acquire the genetic material via transduction or conjugation process. The transfer mechanisms can be vertical or horizontal in which the former one is time consuming from parent to daughter cells where as the latter is between different species or genera which further contributes to evolutionary processes. Instead of representing evolution as tree, HGT has helped to represent natural selection as a existence of interdependent organisms (Khan et al. 2000). Genes of medicinal interest favor environmental selection and they share relationships with GIs (Hacker 2002). The extent of homogeneity amid the exchanged DNA and the bacterial host, the metabolic affinity, alterations to the environment, the shift of genes and their expression, the mismatch repair, and restriction modification systems affect the movement of genes.

10.2 Genomic Islands and its Contribution to Pathogenesis

Pathogenicity islands (PAI) are a subset of genomic islands which contributes to pathogenesis and other virulent characteristics. Certain adherence factors such as fimbrial extensions in *E. coli* are encoded by the GIs which do not cause any infections in gut flora but they adhere in the urinary tract and thus become true pathogenic islands.

Virulent genes are mostly present in local movable, genetic elements which include the PAIs (Boyd and Brüssow 2002; Shankar et al. 2002). The presence of PAIs is a recent discovery after studying the mechanisms of phages and plasmids. The PAIs have unfolded from lysogenic bacteriophages and plasmids, and they form part of substantial genomic regions that are present in different types of pathogenic bacteria but this does not apply to nonpathogenic bacteria. Pathogenic bacteria consist of more than one virulent gene, which has Guanine and Cytosine contents different from other chromosomal regions.

Pathogenicity islands are mobile genes which contain sequences of transposases or integrases and are also quite unstable (Dobrindt et al. 2002). The genes which are virulent in nature and make the PAIs and they are classified into groups based on the functions they perform, they are:

- The factors responsible for adherence, enables the bacteria to adhere to host surface.
- The uptake of Fe^{3+} ions and solubilization is controlled by siderophores.
- Capsules help in preventing phagocytosis of the bacterial cells.
- The endotoxins released by Gram-negative bacteria are capable to induce the host complement mechanism and dynamic signs of inflammation are also seen.
- The exotoxins released can cause permanent impairment of eukaryotic cells by the modulation of signal transduction pathway.
- Invasins are types III and IV restriction complexes that promote the entry of bacteria into eukaryotic cells which intervene bacterial entry into eukaryotic cells and mostly interfere with the apoptotic pathways of the host and also gain entry into the non-phagocytic cells (Dobrindt et al. 2002).

10.2.1 Evolution of PAIs

After the events of HGT, chromosomal incorporation by position-specified recombination takes place which further promotes the integration of the chromosomes. These integration processes cause genetic rearrangements inducing mobile genetic elements to become GIs. Genomic islands can also be formed due to gene loss or acquisition. Immobilization of GIs can be due to inactivation or excision of genes accountable for mobility along with replication of plasmids. The genetic factors from non-virulent bacteria can be identified on extrachromosomal replicons, i.e., phages or plasmids. The autonomously transmittable sizeable plasmid pHG1 of *Ralstoniaeutropha* H16 is composed of a group of functional genes which are compassed next to movable genetic elements. These group of genes consists of sequences that are necessary for utilization of inorganic substrates, reduction of nitrate and nitrite, degradation of biomasses which consists of aromatic compounds and uptake of iron, as well as for type IV and RP4-like sex pili (Schwartz et al. 2003). The substantial number of pHG1 genes that specify integrases and transposases specify the peak of recombinational duty of plasmids, which has further promoted the collection of diverse genotypic characteristics, in that way expanding the metabolic role of the recipients (Schwartz et al. 2003).

10.2.2 Antibiotic Resistance Islands

Antibiotic resistance factors are frequently correlated with ambulatory genetic elements which are responsible for producing virulence (Paulsen et al. 2003).

10.2.3 Secondary Metabolism

Operons that encrypt enzymes which are necessary for the production of secondary metabolites are key elements for HGT because they provide an advantage of different morphological features and are not necessary to the bacterial cell. Authentication of genes that take part in the synthesis of chief products of natural medicines is becoming an integral part of microbial studies. For example, the POLYKETIDE genes are exchanged horizontally among *Streptomycetes* (Egan et al. 1998).

10.3 Identification of GIs in *Helicobacter* Species

Helicobacter pylori, which is harmful to the Human gastric tract, is studied as a distinct genetically diverse species. In an experiment with duodenal ulcer, they contrasted the genome sequence of the duodenal ulcer strain and other *H. pylori* genomes to throw light on the structural arrangement of genes and genome selection mechanisms of these *H. pylori* species (Fig. 10.1). With these experimental evidence as described in Fig. 10.2, it is subjected that *H. pylori* possess a stretched out pan-genome. Different zones of plasticity are specific for different strains which suggests different pathways of evolution (Cover and Blaser 2009). *H. pylori* forms the majority of genetically variable pathogenic bacteria since it has high mutation rates (Fernandez-Gonzalez and Backert 2014). Most of the *H. pylori* strains consist of ambiguous plasmids, and the shuffled gene concatenation of plasmids which normally take part during evolution (Ilver et al. 1998).

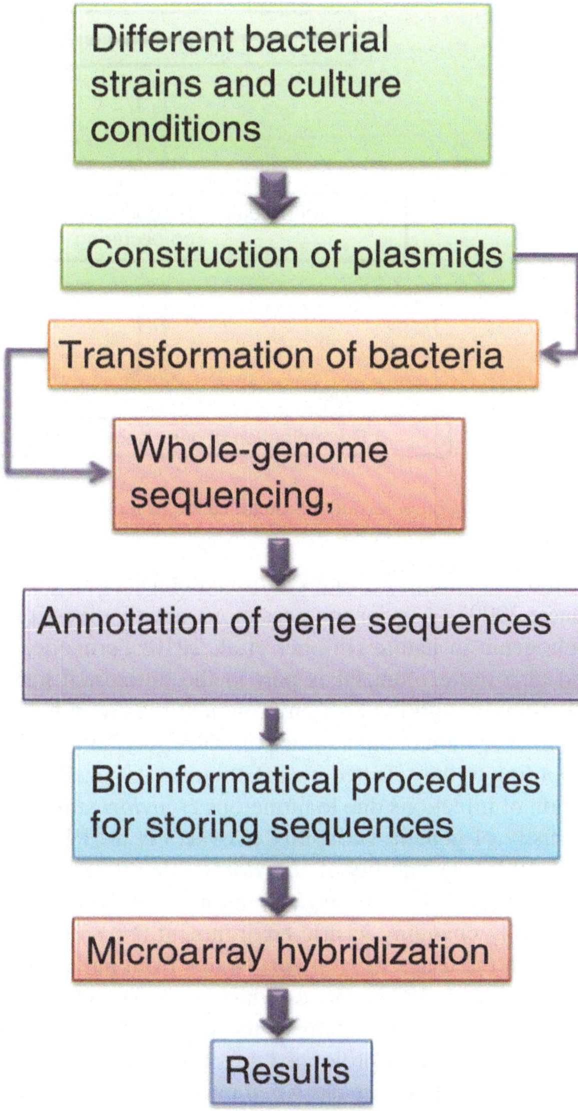

Fig. 10.1 Flow chart of identification of GIs

10.4 Adaptation and Pathogenicity of GI in *Helicobacter* Species

Helicobacter pylori is a group of bacterial species which are specific for humans and are also known for colonizing the stomach (Hooi et al. 2017). Infectious symptoms of existence of *H. pylori* are diagnosed with gastric along with duodenal pathology which includes the chronic gastritis, peptic ulcers, and even gastric cancer in the population which also depends on the variation of virulence of bacteria, genetic

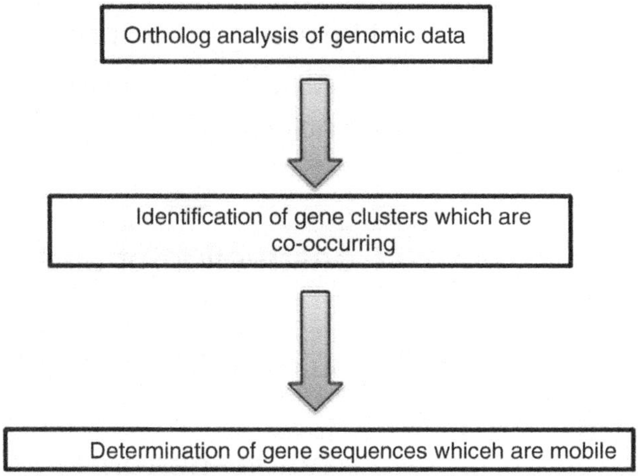

Fig. 10.2 General procedure of identification of mobile genes

attributes of the host, and environmental factors (Nr and Muller 2013; Cover and Blaser 2009). *H. pylori* possess more genetic diversity in the class of bacteria which pathogenic in nature (Fischer et al. 2010; Fernandez-Gonzalez and Backert 2014) and also more often takes part in the horizontal gene transfer (HGT) and adapts accordingly with the host environment via recombination processes (Covacci et al. 1993). The mutational abilities of DNA polymerase I help in recombination of genes according to the selected host (Fernandez-Gonzalez and Backert 2014). The combination of infections due to numerous *H. pylori* strains within one stomach shows the intensity of genetic variability (Covacci et al. 1993; Telford et al. 1994; Marchetti et al. 1995). The ability of *H. pylori* to alter its gene sequence is considered to be the key characteristic for its adaptation to different host systems, and also to the frequently changing gastric environment (Nr and Muller 2013). Current researches focus on the usage of *H. pylori* strains from patients who are infected in long terms (Covacci et al. 1993). Genes which code for outer membrane proteins change their genomic features and help in the production of diverse range of proteins. These proteins helps in the prolonged infection to the host organism (Chiapello et al. 2005).

H. pylori strains are classified into two different wide-ranging families temporarily addressed as types I and II, which are known as the bases of expression of vacuolating cytotoxin (VacA) and the CagA antigen (cytotoxin-associated gene A) (Fischer et al. 2010). Type I strain infections are seen in patients who are diagnosed with duodenal ulcers, tumors and duodenitis which with CagA and the cytotoxin which expresses together with other genes play a role in its virulence (Fischer et al. 2010; Fernandez-Gonzalez and Backert 2014; Covacci et al. 1993). The epidemiological studies are assisted by experiments in the mouse models. Type II strains are observed in patients who have gastric vandalization along with histological lesions and are similar to the biopsies from patients infected by *H. pylori* (Telford et al.

1994). The two isolates type I and type II of *H. pylori* can establish in mice and the type I strains are known to elevate gastric damages similar to the ones observed in humans (Marchetti et al. 1995).

10.5 Computational Tools Involved in GI Prediction

Distorted sequence configuration and occasional phylogenetic distribution are the two prominent attributes of horizontally transferred GIs. Nowadays, genomic islands prediction is done via computational methods which use either sequence of gene composition differences or comparative genetic approaches (Chiapello et al. 2005). Since more GIs are being predicted several databases related to GI have been developed. Islander (Mantri and Williams 2004), ICEBerg (Bi et al. 2012), and PAIDB (Yoon et al. 2007) are specific databases for GIs that have originated from tRNAs and pathogenicity islands (PAIs). Island viewer is one such database which precomputedly predicts the published GIs and unpublished genomic sequences can be submitted for analysis. If any GIs are predicted using two or more methods, then the annotations of the gene can be viewed just by clicking an image. Island viewer also links to external GI databases connected to NCBI and Joint Genome Institute (JGI). It also connects with the IslandPath which allows to probe of the features related to genomic islands for the users' choice of the genome (Mantri and Williams 2004). Based on number of inputs of genomes prediction techniques can be broadly grouped into two different types rooted in one genome and multiple genomes (Lu and Leong 2016).

10.5.1 Prediction Methods Based on One Genome

Each species develop a specific gene composition due to mutational and selection events so the horizontally transferred genes show a different composition from other species. This assumption is used to discriminate species characteristics. The codons, amino acid usage, GC content, and *k-mer* frequencies (Lawrence and Ochman 1997) are used as criteria for score comparison. Computed threshold limits are fixed and any gene frequencies crossing the limit are classified as atypical genes. They further rely on gene sequence composition which is identified on the basis of Hidden Markov rule, DNA sequence composition which uses k-mer frequencies which window based or windowless methods and GI structures.

10.5.2 Prediction Methods Focussed on Multiple Genomes

GIs are predicted based on irregular phylogenetic distribution. The comparing process involves the use of sequence alignment tools like BLAST for local alignment and MAUVE for global alignment (Darling et al. 2004). To promote genome selection, IslandPick builds a distance matrix of the whole genome and makes use of

Table 10.1 Computational tools used to predict GIs

Prediction tool	Operating format
Prediction of one genome based on gene frequencies	
SIGI-HMM (Waack et al. 2006)	Visual interface
PAI-IDA (Tu and Ding 2003)	Command line
Prediction of one genome based on DNA content	
Window based **AlienHunter** (Vernikos and Parkhill 2006)	Command line
Windowless **MJSD** (Zhang et al. 2014)	Command line
Prediction of one genome based on GI structure	
IslandPath (Arvey et al. 2009)	Web based
GIDetector (Hsiao et al. 2003)	Command line
Prediction of multiple genomes	
IslandPick (Langille et al. 2008)	Command line
tRNAcc (Che et al. 2014)	Web based

respective cut-offs to sort out adequate genomes to distinguish with the query genome. This method is completely automatic. The whole-genome pairwise positioning is done by MAUVE to get large distinct zones in the query genome. Finally, genome duplication is eliminated using BLAST and these zones are taken as assumed GIs. Different tools to predict genomic islands are given in Table 10.1.

10.6 Future Perspectives of GIs

GIs have various advantages over plasmid since they easily incorporate into the host's chromosome. They do not need to be replicated often like plasmids and a single facsimile of GI can be conferred per genome (Gaillard et al. 2008). GIs render a key role in the development of pathogenic species among bacteria. Studies on the GIs of *H. pylori* has disclosed that the HopH gene helps in the colonization of mucosal surfaces in gastrointestinal linings (Yamaoka et al. 2002). Proinflammatory signaling events can be identified via transcriptional profiling in *H. pylori*-stimulated epithelia which further helps in studies on the regulation of gene expression in different strains (Yamaoka et al. 2004). The CT dinucleotide repeats differ in different strains from different countries. These characteristics can be used for the calculation of mutation rates in different species. However the non-bioinformatical part it is difficult to visualize the GIs. Alien hunter and GI hunter predictions can give circular representations of the GIs. MTGI can give dynamic simulations of GIs in circular manner but cannot give unique interpretations. Hence, there is a need to develop these technologies further.

10.7 Conclusion

The prediction and analysis of GIs are now becoming a key part of microbial examinations. *Helicobacter* species is one the most virulent strains and is also capable of rapid mutation. GI identification is critical for the study of genomic islands in *Helicobacter* species. Recently acquired gene sequences can be probed for studies of disease outbreaks, strain mutations among patients, resistance, etc. Different computational methods can identify different features of bacterial GIs. There are no accurate GI prediction tools to present, especially for the horizontally transferred genes. The development of more comprehensive methods would further help researchers in real-time tracking of GI studies.

References

Arvey AJ, Azad RK, Raval A, Lawrence JG (2009) Detection of genomic islands via segmental genome heterogeneity. Nucleic Acids Res 37(16):5255–5266

Bi D, Xu Z, Harrison EM, Tai C, Wei Y, He X, Jia S, Deng Z, Rajakumar K, Ou HY (2012) ICEberg: a web-based resource for integrative and conjugative elements found in bacteria. Nucleic Acids Res 40(D1):D621–D626

Boyd EF, Brüssow H (2002) Common themes among bacteriophage-encoded virulence factors and diversity among the bacteriophages involved. Trends Microbiol 10(11):521–529

Che D, Wang H, Fazekas J, Chen B (2014) An accurate genomic island prediction method for sequenced bacterial and archaeal genomes. J Proteom Bioinform 7(8):214

Chiapello H, Bourgait I, Sourivong F, Heuclin G, Gendrault-Jacquemard A, Petit MA, El Karoui M (2005) Systematic determination of the mosaic structure of bacterial genomes: species backbone versus strain-specific loops. BMC Bioinformatics 6(1):1–10

Covacci A, Censini S, Bugnoli M, Petracca R, Burroni D, Macchia G, Massone A, Papini E, Xiang Z, Figura N (1993) Molecular characterization of the 128-kDa immunodominant antigen of Helicobacter pylori associated with cytotoxicity and duodenal ulcer. Proc Natl Acad Sci 90(12):5791–5795

Cover TL, Blaser MJ (2009) Helicobacter pylori in health and disease. Gastroenterology 136(6): 1863–1873

Darling AC, Mau B, Blattner FR, Perna NT (2004) Mauve: multiple alignment of conserved genomic sequence with rearrangements. Genome Res 14(7):1394–1403

Dobrindt U, Hentschel U, Kaper JB, Hacker J (2002) Genome plasticity in pathogenic and nonpathogenic enterobacteria. Pathogenicity Islands and the Evolution of Pathogenic Microbes:157–175

Egan S, Wiener P, Kallifidas D, Wellington EM (1998) Transfer of streptomycin biosynthesis gene clusters within streptomycetes isolated from soil. Appl Environ Microbiol 64(12):5061–5063

Fernandez-Gonzalez E, Backert S (2014) DNA transfer in the gastric pathogen helicobacter pylori. J Gastroenterol 49(4):594–604

Fischer W, Windhager L, Rohrer S, Zeiller M, Karnholz A, Hoffmann R, Zimmer R, Haas R (2010) Strain-specific genes of Helicobacter pylori: genome evolution driven by a novel type IV secretion system and genomic island transfer. Nucleic Acids Res 38(18):6089–6101

Gaillard M, Pernet N, Vogne C, Hagenbüchle O, van der Meer JR (2008) Host and invader impact of transfer of the clc genomic island into Pseudomonas aeruginosa PAO1. Proc Natl Acad Sci 105(19):7058–7063

Hacker J (2002) Urinary tract infection: From basic science to clinical application. In *Genes and Proteins Underlying Microbial Urinary Tract Virulence* (pp. 1–8). Springer, Boston, MA

Hooi JK, Lai WY, Ng WK, Suen MM, Underwood FE, Tanyingoh D et al (2017) Global prevalence of Helicobacter pylori infection: systematic review and meta-analysis. Gastroenterology 153(2): 420–429

Hsiao W, Wan I, Jones SJ, Brinkman FS (2003) IslandPath: aiding detection of genomic islands in prokaryotes. Bioinformatics 19(3):418–420

Ilver D, Arnqvist A, Ogren J, Frick IM, Kersulyte D, Incecik ET, Berg DE, Covacci A, Engstrand L, Borén T (1998) Helicobacter pylori adhesin binding fucosylatedhisto-blood group antigens revealed by retagging. Science 279(5349):373–377

Khan AS, Kniep B, Oelschlaeger TA, Van Die I, Korhonen T, Hacker J (2000) Receptor structure for F1C fimbriae of uropathogenic Escherichia coli. Infect Immun 68(6):3541–3547

Langille MG, Hsiao WW, Brinkman FS (2008) Evaluation of genomic island predictors using a comparative genomics approach. BMC Bioinformatics 9(1):1–10

Lawrence JG, Ochman H (1997) Amelioration of bacterial genomes: rates of change and exchange. J Mol Evol 44(4):383–397

Lu B, Leong HW (2016) Computational methods for predicting genomic islands in microbial genomes. Comput Struct Biotechnol J 14:200–206

Mantri Y, Williams KP (2004) Islander: a database of integrative islands in prokaryotic genomes, the associated integrases and their DNA site specificities. Nucleic Acids Res 32(suppl_1):D55–D58

Marchetti M, Arico B, Burroni D, Figura N, Rappuoli R, Ghiara P (1995) Development of a mouse model of helicobacter pylori infection that mimics human disease. Science 267(5204): 1655–1658

Nr S, Muller A (2013) Life in the human stomach: persistence strategies of the bacterial pathogen Helicobacter pylori. Natl Rev Microbiol 11(6):385–399

Paulsen IT, Banerjei L, Myers GSA, Nelson KE, Seshadri R, Read TD, Fouts DE, Eisen JA, Gill SR, Heidelberg JF, Tettelin H (2003) Role of mobile DNA in the evolution of vancomycin-resistant Enterococcus faecalis. Science 299(5615):2071–2074

Schwartz E, Henne A, Cramm R, Eitinger T, Friedrich B, Gottschalk G (2003) Complete nucleotide sequence of pHG1: a Ralstoniaeutropha H16 megaplasmid encoding key enzymes of H2-based lithoautotrophy and anaerobiosis. J Mol Biol 332(2):369–383

Shankar N, Baghdayan AS, Gilmore MS (2002) Modulation of virulence within a pathogenicity island in vancomycin-resistant Enterococcus faecalis. Nature 417(6890):746–750

Telford JL, Ghiara P, Dell'Orco M, Comanducci M, Burroni D, Bugnoli M, Tecce MF, Censini S, Covacci A, Xiang Z (1994) Gene structure of the Helicobacter pylori cytotoxin and evidence of its key role in gastric disease. J Exp Med 179(5):1653–1658

Tu Q, Ding D (2003) Detecting pathogenicity islands and anomalous gene clusters by iterative discriminant analysis. FEMS Microbiol Lett 221(2):269–275

Vernikos GS, Parkhill J (2006) Interpolated variable order motifs for identification of horizontally acquired DNA: revisiting the Salmonella pathogenicity islands. Bioinformatics 22(18): 2196–2203

Waack S, Keller O, Asper R, Brodag T, Damm C, Fricke WF, Surovcik K, Meinicke P, Merkl R (2006) Score-based prediction of genomic islands in prokaryotic genomes using hidden Markov models. BMC Bioinformatics 7(1):1–12

Yamaoka Y, Kikuchi S, El-Zimaity HM, Gutierrez O, Osato MS, Graham DY (2002) Importance of helicobacter pylori oipA in clinical presentation, gastric inflammation, and mucosal interleukin 8 production. Gastroenterology 123(2):414–424

Yamaoka Y, Kudo T, Lu H, Casola A, Brasier AR, Graham DY (2004) Role of interferon-stimulated responsive element-like element in interleukin-8 promoter in Helicobacter pylori infection. Gastroenterology 126(4):1030–1043

Yoon SH, Park YK, Lee S, Choi D, Oh TK, Hur CG, Kim JF (2007) Towards pathogenomics: a web-based resource for pathogenicity islands. Nucleic Acids Res 35(suppl_1):D395–D400

Zhang R, Ou HY, Gao F, Luo H (2014) Identification of horizontally-transferred genomic islands and genome segmentation points by using the GC profile method. Curr Genomics 15(2): 113–121

Genomic Islands in *Staphylococcus* 11

Zaaima AL-Jabri and Nada AL-Mebairik

Abstract

Staphylococci are Gram-positive bacteria that have successfully evolved from a normal flora with limited threats to potentially life-threatening pathogens, particularly, *Staphylococcus aureus*. Species of staphylococci have adapted to survive under selective pressure mainly due to their ability to acquire mobile genetic elements (MGEs). Methicillin-resistant *S. aureus* is a common example of this successful evolution not only in hospital setting but also in the community. Recent literature supports that Coagulase-negative staphylococci including *S. epidermidis* are the reservoir for resistance as well as virulence-associated determinants for *S. aureus*. A wide range of MGEs are present in Staphylococci including genomic islands (GI), with staphylococcal chromosome cassette (SCC*mec)* as an example of the most common GI of medical importance, found in 15–20% of the *S. aureus*. The SCC*mec* are mobile entities that have been classified, so far into 14 types. Other GIs with similar characteristics to the SCC element is the Arginine Catabolic Mobile Element (ACME) and Copper and Mercury Resistance (COMER) that form a composite island with SCC*mec* IV, which have been first described in *S. aureus* USA-300 and in *S. epidermidis* as well. Other MGEs, include Insertion sequences and Transposons, plasmids, Integrative and conjugative elements (ICEs), and bacteriophages. MGEs have a

Z. AL-Jabri (✉)
Department of Microbiology & Immunology, College of Medicine and Health Sciences, Sultan Qaboos University, Muscat, Sultanate of Oman
e-mail: zaeema@squ.edu.om

N. AL-Mebairik
Department of Clinical Laboratory Sciences, College of Applied Medical Sciences, King Saud University, Riyadh, Kingdom of Saudi Arabia
e-mail: Nalme.bairik@ksu.edu.sa

I. Mani et al. (eds.), *Microbial Genomic Islands in Adaptation and Pathogenicity*,
https://doi.org/10.1007/978-981-19-9342-8_11

significant survival advantage over their host species as these carry a wide variety of genes that confer resistance to antibiotics, heavy metals, and biocides.

Keywords

Mobile genetic elements · Staphylococci · Staphylococcal chromosome cassette (SCC*mec*)

11.1 *Staphylococcus* Species

Bacteria in the genus *Staphylococcus* are Gram-positive, cocci-shaped bacteria that are arranged in grape-like clusters. Traditionally, *Staphylococcus* species were divided into two major subtypes on the basis of their capability to produce the enzyme coagulase, which is responsible for blood plasma clotting (Foster 1996; Otto 2004). The main and the most pathogenic species, *Staphylococcus aureus*, belongs to the coagulase-positive staphylococci (CoPS) while the coagulase-negative staphylococci (CoNS) comprise most other *Staphylococcus* species. From the CoNS, *Staphylococcus epidermidis* is considered as the most important member that accounts for most of the CoNS infections (Foster 1996; Otto 2004).

11.2 *Staphylococcus aureus*

S. aureus is found as a commensal usually in the nasal carriage, on the skin, and mucous membranes. However, these bacteria are also successful as pathogens and can cause a wide range of diseases from mild skin infections to pneumonia, septicemia, and endocarditis (Malachowa and Deleo 2010; Tong et al. 2015). *S. aureus* pathogenicity and its ability to adapt under selective pressures are mostly attributed to the products of MGEs that confer virulence factors and antibiotics resistance, including the gene conferring methicillin resistance in methicillin-resistant *S. aureus* (MRSA) (Ito et al. 1999; Malachowa and Deleo 2010).

MRSA was reported shortly after the introduction of methicillin, a drug now replaced by flucloxacillin in clinical practice. Over the years, MRSA became one of the most significant causes of nosocomial infections with increasing morbidity and mortality (Ayliffe 1997; Chongtrakool et al. 2006). The healthcare-associated MRSA (HA-MRSA) shows resistance to methicillin by the acquisition of a MGE called the staphylococcal cassette chromosome *mec* (SCC*mec*) (Katayama et al. 2000; Chongtrakool et al. 2006). Additionally, MRSA was isolated from patients with no recent contact with healthcare facilities, thus labeled as community-associated MRSA (CA-MRSA) and differs from HA-MRSA as it contains various types of SCC*mec* and several virulence factors that are rarely identified in HA-MRSA, such as pore-forming toxin and the Panton-Valentine leukocidin (PVL) (Davidson et al. 2008; Herold 1998; Naas et al. 2005; Naimi 2003).

11.3 *S. epidermidis* and Other CoNS

The CoNS comprise species that normally colonize humans and they can cause infections in certain situations. These species include *S. epidermidis, S. haemolyticus, S. hominis, S. capitis, S. saccharolyticus, S. saprophyticus, S. cohnii, S. warneri*, and *S. lugdunensis* (Otto 2004). CoNS also contain species that colonize and infect animals and species that are less or non-pathogenic. However, the main and best-described species is *S. epidermidis* (Otto 2004). This species is considered a commensal bacterium on healthy skin and mucosal surfaces. Although *S. epidermidis* is less virulent than *S. aureus*, severe complications associated with indwelling medical devices can arise from this bacterium. *S. epidermidis* ability for biofilm formation in addition to the medical devices' insertion makes this bacterium a significant nosocomial pathogen, and the leading cause of surgical site infections and bloodstream infections (Cherifi et al. 2013; Lee et al. 2018; Otto 2013).

Interestingly, several previous studies discussed the role of CoNS including *S. epidermidis* as a potential reservoir for resistance-conferring genes and virulence determinants that transfer to *S. aureus* and contribute toward its diversity and pathogenicity (Hung et al. 2015; Otto 2013). For example, the *mecA* gene and the SCC*mec* elements were found and reported earlier to be more frequent in *S. epidermidis* strains in comparison to *S. aureus* strains (McManus et al. 2015; Otto 2013). Additionally, *S. epidermidis* SCC*mec* elements have DNA sequences that are homologous to these elements in *S. aureus*, however, the polymorphous structure of SCC*mec* with novel cassette chromosome recombinase (*ccr*) and *mec* gene complexes that have not been described in *S. aureus* are present in CoNS. This evidence indicates that CoNS including *S. epidermidis* may act as a pool for the SCC*mec* entities (Barbier et al. 2010; Otto 2013). Another example is the ACME mobile element which is found in *S. aureus* USA300-NAE. Some reports show that 52% of global *S. epidermidis* strains harbor the ACME mobile element. On the other hand, some investigations noted that the different types of ACMEs in *S. epidermidis* are similar to those discovered in *S. aureus* USA300. This evidence suggests that *S. epidermidis* is the origin of most ACME-associated genes (Barbier et al. 2011; Miragaia et al. 2009; Otto 2013; Onishi et al. 2013; O'Connor et al. 2018b).

11.4 Mobile Genetic Elements of *Staphylococcus* Species

Among *Staphylococcus* species, the MGEs are best described in *S. aureus* as it has been known as the most virulent species. In fact, the diversity of the MGEs in *S. aureus* contributed to *S. aureus* adaptation and evolution into successful lineages. These MGEs, including plasmids, transposons, ICEs, bacteriophages, and staphylococcal chromosome cassettes (SCCs) found to compose around 15-20% of the *S. aureus* genome (Alibayov et al. 2014; Haaber et al. 2017; Lindsay 2010). This chapter sheds light on some of these MGEs.

11.5 Staphylococcal Cassette Chromosome *mec* (SCC*mec*)

S. aureus and other CoNS show an ability to resist methicillin by the acquisition of the SCC*mec* genomic island. This MGE carries *mecA* which encodes a penicillin-binding protein named PBP2a or PBP2', which is different from the core PBP2. This PBP2a exhibits low affinity to methicillin and most semisynthetic ß-lactam antibiotics (Chongtrakool et al. 2006; Hartman and Tomasz 1984; Pinho et al. 2001). SCC*mec* is a critical mobile element as MRSA has spread worldwide and become the leading cause of both community-acquired infections and healthcare-associated infections (Davidson et al. 2008; Monecke et al. 2016; Naimi 2003; Rolo et al. 2017).

There are essential components that are usually found in the SCC*mec* element. The first one is the *mec* gene complex which contains the *mecA* gene, as well as the regulatory genes; *mecR1* and *mecI* located upstream of *mecA* and IS*431* downstream of *mecA* (Chongtrakool et al. 2006; Ito et al. 2001; IWG-SCC 2009). In addition, the other component is the *ccr* gene complex which contains the *ccrAB* or *ccrC* genes. These site-specific recombinase genes catalyze SCC*mec* element integration into a site-specific attachment sequence in the staphylococcal chromosome called the *attB* and also catalyze the excision of SCC*mec* from the same place (Chongtrakool et al. 2006; IWG-SCC 2009; Noto et al. 2008). Additionally, different accessory genes that encode virulence or resistance determinants can be found in SCC*mec* elements in areas called joining regions (J-regions) (IWG-SCC 2009; Monecke et al. 2016). Interestingly, SCC*mec* elements that lack *ccr* genes have also been reported which are known as pseudo-SCC*mec* elements, and SCC elements without the *mecA* gene but with other characteristic genes have also been identified in staphylococcal genomes (IWG-SCC 2009; Wilson et al. 2016).

There are 14 types of SCC*mec* elements (types I–XIV) which are classified based to the different combinations of *mec* gene and *ccr* gene complexes (Table 11.1).

Table 11.1 SCC*mec* types identified in *S. aureus*

SCC*mec* type	*ccr* gene complex	*mec* gene complex
I	1 (A1B1)	B
II	2 (A2B2)	A
III	3 (A3B3)	A
IV	2 (A2B2)	B
V	5 (C1)	C2
VI	4 (A4B4)	B
VII	5 (C1)	C1
VIII	4 (A4B4)	A
IX	1 (A1B1)	C2
X	7 (A1B6)	C1
XI	8 (A1B3)	E
XII	9(C2)	C2
XIII	9(C2)	A
XIV	5(C1)	A

There are four classes of the *mec* gene complex identified thus far: class A, B, C, and E; while three different *ccr* genes had been discovered: *ccr*A, B, and C. Additionally, the differences in J-regions are used for determining the SCC*mec* subtypes (Baig et al. 2018; IWG-SCC 2009; Urushibara et al. 2019; Wu et al. 2015). Currently, MRSA elements are identified by the chromosome sequence type (ST) and the SCC*mec* type. SCC*mec* types I, II, and III, comprise most of the HA-MRSA whereas CA-MRSA belongs mostly to types IV and V (Kang et al. 2015; Naimi 2003).

11.6 Arginine Catabolic Mobile Element (ACME)

The ACME is a genomic island that is found in many staphylococcal species and shows characteristics that are similar to the SCC element. It was first described in *S. aureus* USA300-NAE as well as *S. epidermidis* strain ATCC12228 (Diep et al. 2006). In *S. aureus* USA300-NAE, ACME forms a composite island with SCC*mec* IV, while in *S. epidermidis* ATCC12228, it exists as a composite island with SCC*pbp4* (Diep et al. 2006; Shore et al. 2011). This mobile element which ranges in size from 31 to 34 kb integrates into the staphylococcal chromosome at the attachment site; *attB* with direct repeat sequences at the flanks which is similar to SCC. The *ccrAB* genes encoded by SCC elements mediate the movement of ACME (Shore et al. 2011; Thurlow et al. 2013). In addition, the presence of several internal direct repeats in ACME has resulted in a stepwise pattern of assembly of this element (O'Connor et al. 2018b).

There are two gene clusters characterized in the ACME element, the *arc* operon and the *opp3* operon (Diep et al. 2008; Granslo et al. 2010). The *arc* operon comprises the regulatory gene (*argR*) and *arcABCD* genes that encode the main bacterial arginine catabolic pathway, an arginine deaminase pathway. The result of this pathway is converting arginine into ornithine, ammonia, carbon dioxide, and ATP, and consequently serves bacterial growth with arginine as the sole source of energy (Diep et al. 2008; Makhlin et al. 2007; O'Connor et al. 2018b). The *opp3* operon consists of *opp-3ABCDE* genes that encode ABC transporter systems (Diep et al. 2006; Granslo et al. 2010; Shore et al. 2011). Moreover, the ACME element has two additional associated genes; the *speG* gene encoding polyamine resistance and *copBL* genes encoding a copper export P1-type ATPase and a putative lipoprotein, respectively. The *copBL* genes are suggested to be a novel copper resistance locus (O'Connor et al. 2018a; Planet et al. 2015; Purves et al. 2018; Rosario-Cruz et al. 2019). On the whole, it is found that the presence of ACME in staphylococcal species increases their fitness and improves their capacity for skin and mucus membrane colonization (Lindgren et al. 2014; Miragaia et al. 2009; Purves et al. 2018).

ACME mobile elements are classified into three distinct types: ACME type I which contain the *arc* and *opp-3* operon, ACME type II contains only the *arc* operon, and ACME type III which contains the *opp-3* operon only (McManus et al. 2017; Shore et al. 2011; Rolo et al. 2012). Recently, two more types of *S. epidermidis* were identified. ACME type IV which carry the *arc* operon, a *kdp* operon which encodes the ABC transporter, and ACME type V which harbors both the *arc* and the *opp-3* operons, as well as the *kdp* operon (O'Connor et al. 2018a).

11.7 Copper and Mercury Resistance (COMER)

Copper and Mercury Resistance (COMER) is a novel MGE that was first described in the *S. aureus* USA300-SAE strain. Similar to ACME, this mobile element is found adjacent to SCC*mec* IV in the *S. aureus* USA300 chromosome (Planet et al. 2015). COMER element is thought to contribute to copper and mercury resistance. The presence of the copper and mercury resistance coding sequences in the COMER element support this idea. Additionally, the COMER element harbors genes encoding an abortive phage (Abi) infection system. This system is a resistance mechanism which leads to bacterial death after viral infection, thus preventing further dissemination of phages (Almebairik et al. 2020; Dy et al. 2014; Planet et al. 2015; Purves et al. 2018). It could be argued that the COMER element enhances the fitness of USA300-SAE as recently, this strain has been identified in North and South America, Europe and Gulf region (Oman) (Almebairik et al. 2020; Planet et al. 2016; Purves et al. 2018; Al-Jabri et al. 2021).

As in the ACME element, novel copper resistance genes (*copXL*) were detected in COMER, however, those two genes were associated with the *mco* gene (encodes multi-copper oxidase) in the COMER element (Almebairik et al. 2020; Planet et al. 2015; Purves et al. 2018; AL-Jabri et al. 2021). The other characteristic operon in COMER is the *mer* operon that confers mercury resistance. This operon consists of genes involved in the enzyme-mediated reduction of divalent mercury (Hg^{2+}) into the elemental form (Hg^0) that is less toxic and then volatilizes from the cell (Osborn et al. 1997). Two types of the *mer* operon have been identified: (i) a narrow-spectrum *mer* operon which confers resistance to inorganic mercurial compounds and (ii) a broad-spectrum *mer* operon which confers resistance to inorganic as well as organo-mercurial compounds (Bruce 1997). This operon encodes proteins for regulation (*merR* gene), transport, and mercuric reductase (*merA* gene). In the broad-spectrum *mer* operon, an additional protein called organomercurial lyase encoded by the *merB* gene is found (Osborn et al. 1997).

COMER elements were also detected in *S. epidermidis* isolates belonging to the ST2 clonal lineage which is associated with multidrug-resistance in hospital settings worldwide. In *S. epidermidis*, this MGE is named COMER-like element because it harbors the *mer/cop* operon as well as the *abi* gene located in COMER element of *S. aureus* USA300. However, there were other genes identified in the *S. epidermidis* COMER-like element which are lacking in COMER USA300, namely the *ars* operon and a type I restriction-modification system. Additionally, the COMER-like element is located immediately adjacent to SCC*mec* III in *S. epidermidis* chromosome, instead of SCC*mec* IV in USA300 COMER element (Almebairik et al. 2020).

11.8 The Mechanism of SCC*mec* Transfer

The excision and integration of SCC*mec* are catalyzed by the Ccr proteins. These proteins mediate the site-specific recombination events between the *attB*-specific site on the chromosome, and one in the circularized SCC*mec* named *attS* (Ito et al. 2004; Misiura et al. 2013; Wang and Archer 2010). This *attB* attachment site is terminally located in a conserved ribosomal methyltransferase gene of orfX, also known as *rlmH* (Boundy et al. 2013). When the SCC*mec* is inserted, it is flanked by direct repeat (DR) sequences and inverted repeat sequences (IRs), at both ends, referred to as *attL* and *attR*. These new pairing sites contain the *attB* sequence which is duplicated during SCC*mec* insertion in the chromosome. When the SCC*mec* excises, *attL* and *attR* sites are reconstituted and reproduce the *attB* in the chromosome and the *attS* in the circular SCC*mec* (Liu et al. 2017; Misiura et al. 2013; Wang and Archer 2010).

Chromosomal SCC*mec* excision is a significant step in its lateral horizontal transfer among *Staphylococcus* species. The excision process can occur spontaneously with a low-frequency rate, less than 10^{-4} in *S. aureus* (Ito et al. 1999; Stojanov et al. 2015). The mechanisms that trigger the excision of the SCC*mec* are still not well understood, however, some studies found that many antibiotics, including ß-lactam antibiotics, could increase the frequency of SCC*mec* excision from the chromosome and consequently increase its transfer (Higgins et al. 2009; Liu et al. 2017).

The mechanism of SCC*mec* elements transfer between staphylococci is still unknown. Some early studies suggest that the movement of SCC*mec* is via a transduction mechanism. However, these studies report conflicting conditions for successful SCC*mec* transduction (Cohen and Sweeney 1970; Scharn et al. 2013; Shafer and Iandolo 1979; Stewart and Rosenblum 1980). Cohen and Sweeney proposed that successful methicillin resistance transfer is mediated by a prophage as well as a penicillinase plasmid in the recipient cell (Cohen and Sweeney 1970). Stewart and Rosenblum suggested that recipient cells require a penicillinase plasmid only (Stewart and Rosenblum 1980). Shafer and Iandolo demonstrated the co-transduction of methicillin resistance with tetracycline resistance via a small plasmid (Shafer and Iandolo 1979). A more recent study with *S. aureus* USA300 showed the successful transduction of SCC*mec* types IV and I via bacteriophages 80a and 29. This study reported that the recipient cell and the homologs of donors require a penicillinase plasmid, in addition to recipients respecting the presence/absence of the ACME element. This study also noted the possibility of truncation, substantial deletions, or rearrangement of the SCC*mec* and ACME in the recipient during the transduction process (Scharn et al. 2013).

11.9 Plasmids

More than 90% of clinical isolates of staphylococci harbor plasmids ranging in size; however, only 5% of staphylococcal plasmids are large multiresistant conjugative plasmids. Small staphylococcal plasmids range from 1 to 10 kb in size (Malachowa and Deleo 2010; Shearer et al. 2011). On the other hand, large multiresistant plasmids of more than 15 kb in size carry antibiotic resistance, heavy metal, and biocide-resistance-conferring genes (Novick et al. 1989; Firth and Skurray 2006; Jensen and Lyon 2009; Shearer et al. 2011).

11.10 Multi-Resistant (Conjugative) Plasmids and their Mobilization System

Larger plasmids carrying multiple resistance genes (20–65 kb) are found in most staphylococci, however, lack mobilization genes (Shearer et al. 2011). In fact, there is a paucity of conjugative genes in most staphylococci. In staphylococci, the conjugative plasmids are classified based on their distinct conjugation-gene clusters. These include examples such as pSK41, pWBG749, and pWBG4 families (Kwong et al. 2017), which were identified in many countries worldwide as associated with many infections including community-acquired MRSA (Archer and Johnston 1983; Diep et al. 2008; Goering and Ruff 1983; Jaffe et al. 1982; Pérez-Roth et al. 2006). These plasmids are capable of transferring from the donor to the recipients at a relatively low frequency (Climo et al. 1996; Helinski 2022; Macrina and Archer 1993). Some conjugative plasmids like pSK41 were found to be integrated into the chromosome (Mcelgunn et al. 2002). The resistance genes are usually carried in small-sized plasmids which are cointegrated between two copies of IS to promote their conduction (Caryl et al. 2004; Climo et al. 1996; Gennaro et al. 1987). An example is IS257/IS431 found integrated within the pSK41/pGO1 plasmids (Kwong et al. 2004), harboring linezolid and high-level resistance to vancomycin (Bender et al. 2014; Clark et al. 2005). Members of pSK41-like family of plasmids carry various resistance-conferring genes including resistance to biocides and antiseptic agents (*qacC*) (Littlejohn et al. 1991), mupirocin (*mupA/ileS2*) (Morton et al. 1995; Pérez-Roth et al. 2010), MLS antibiotics [*erm(C)*] (Diep et al. 2006), trimethoprim (*dfrA*) (Evans and Dyke 1988), tetracycline [*tet(K)]* (Shearer et al. 2011), and linezolid (*cfr*) (Bender et al. 2014). The conjugative plasmids in the pWBG749 family carry penicillin, aminoglycoside as well as vancomycin resistance genes (Panesso et al. 2015; O'Brien et al. 2015; Rossi et al. 2014) and mobilized by SmpP, a putative relaxase and a distinct *oriT*. On the other hand, the conjugative plasmid pWBG637, does not harbor any resistance-conferring genes (E. E. Udo and Grubb 1990). However, pWBG637 has the ability to conjugate with other staphylococcus species including *S. aureus and S. epidermidis* as well as other Gram positives such as *Enterococcus faecalis* strains. The latter plasmid is capable of mobilizing several coresident antimicrobial resistance plasmids through conjugative transfer. The pWBG4 family of conjugative plasmids was first identified in 1985 which harbors a cointegrated Tn*554* containing *erm(A)* resistance gene with *det* conjugation-associated gene (Townsend et al. 1985, 1986; E. Udo et al. 1987). pWBG14, is another conjugative multiresistant-conferring aminoglycoside, macrolide, lincosamide, and spectinomycin resistance. The pWBG4-family plasmid (pSA737) (Shore et al. 2016) is identical to pSK73 but very different from pSK41 and pWBG749 (Néron et al. 2009; E. E. Udo et al. 1992).

11.11 Mobilization System of RC-Replicating Plasmids

Small conjugative plasmids of less than 5 kb in size usually replicate by rolling circle (RC) mechanism. These plasmids mostly harbor a single resistance-determinant and exist as multiple copies within each cell (10–60 copies) (Mojumdart and Khan 1988). Initially, there were four identified groups of plasmids in this category based on the resistance genes as follows: plasmid pT181 with *tet(k)* games encoding tetracycline resistance (Mojumdart and Khan 1988), pC194 harboring *cat* gene conferring chloramphenicol resistance (Horinouchit and Weisblum 1982a), pE194 carrying *erm(C)* conferring erythromycin resistance (Horinouchit and Weisblum 1982b), and the cryptic pSN2 plasmids (Novick et al. 1989; Walters and Dyke 2006). Each of these plasmids has a distinct replication protein namely (Rep_trans for pT181, Rep_1 for pC194, Rep_2, and RepL for pE194). Additional RC-replicating plasmids were also described which carry a mosaic of resistance determinants due to the continuous mobilization of various DNA segments in these functional modules (Novick et al. 1989; Projan and Archer 1989). Examples include RC plasmids conferring resistance to streptomycin (*str*) (Projan and Archer 1989) lincomycin [*lnu(A)*] (Brisson-Noel et al. 1988), fosfomycin (*fosB*) (Dionisio et al. 2019), quaternary ammonium compounds (*qacC* and *smr*) (Littlejohn et al. 1991), aminoglycosides (*aadD*), or bleomycin (*ble*) (McKenzie et al. 1986). Non-conjugative plasmids like pC221 are transferred via a *mobCAB* operon and origin of transfer (*oriT*) (Caryl et al. 2004; Projan and Archer 1989).

11.12 Bacteriophages

Bacteriophages are viruses that are capable of infecting bacteria. These elements play a significant role in disseminating MGEs through transduction mainly (Lindsay 2014; Xia and Wolz 2014). Bacteriophages have been demonstrated to be effective tools in biotechnology with diverse applications in therapeutics and research including alternatives to antibiotics in killing bacteria (Ul Haq et al. 2012). Phages have been shown to either act as gene transfer vehicles or carry accessory virulence-conferring genes in bacteria (Quiles-Puchalt et al. 2014b). A classic example is how bacteriophages mediate the transfer of plasmid-encoded virulence-conferring genes in *Staphylococcus aureus* (Dowell and Rosenblum 1962; Novick 1963). The range of virulence genes carried by *Staphylococcus* phages is diverse including enterotoxin A, Exfoliative toxin A, Pantheon-Valentine leucocidins (PVL), and staphylokinase (Brüssow et al. 2004). Moreover, the *Staphylococcus aureus* pathogenicity island (SaPIs) encoding superantigens utilize the help of bacteriophages to the horizontal gene transfer (Lindsay et al. 1998; Novick et al. 2010). Experimental models have attempted to demonstrate the mobility of SaPI via bacteriophages. It was shown that the SOS induction of SaPI has resulted in the recruitment of replicating phage packaging proteins to be used for their transfer in helper phage φ11(Quiles-Puchalt et al. 2014b). Bacteriophages involved in transferring genes horizontally in staphylococci are members of the order *Caudovirales* which have three families on the basis of the structure of their tails (Hatfull and Hendrix 2011; Tolstoy et al. 2018). Transduction in bacteriophages occurs mainly during the lytic cycle during which a foreign DNA or host plasmid is packaged at low frequency (Chiang et al. 2019). *Caudovirales* are mainly temperate phages that undergo lysogeny during which their genome is integrated into the host genome as prophages. Prophages become established in the bacterial lineages if they harbor advantageous survival machinery to the host, i.e., virulence or resistance-conferring genes.

Helper bacteriophages, however, are much more related to MGE packaging which happens at a very high frequency compared to generalized transduction. This is known as molecular piracy in which the prophage propagation is significantly suppressed after the integration into the host DNA (Christie and Dokland 2012). In addition, phage proteins are almost completely exploited by the MGEs for their own excision and replication, and redirection of capsid size to their own advantage (Christie and Dokland 2012). The pathway for capsid assembly and Packaging of virion DNA for the *Caudovirales* are relatively similar. The basic capsid protein (CP) is alternatively called "phage capsid fold" or "HK97 fold" (Wikoff et al. 2000). The P2/P4 paradigm serves as a classical example by which MGE are molecular piracies. P2 is a myovirus with a very small genome (33 kb) first described in the 1950s (G. Bertani 1951; L. E. Bertani 1980), and has been mainly associated with *Escherichia coli* (Nilsson et al. 2004). P4 is a Satellite bacteriophage that was initially thought to be P2-dependant MGE (Six and Klug 1973), however, later it was found to be an integrative plasmid, also known as a phasmid that is able to replicate autonomously as a plasmid and/or integrate within the genome of the host (Briani et al. 2001; Dehò and Ghisotti 2006). P4 phage does not have the ability to form infectious particles as it lacks the genes encoding structural proteins. Therefore, once a host cell with P4 becomes infected with P2, it will recruit P2 helper phage-encoded genes to package into phage particles (Six 1975).

11.13 *Staphylococcus aureus* Pathogenicity Islands (SaPIs)

Staphylococcus aureus Pathogenicity Islands (SaPIs) are chromosomally located genomic islands which are usually large in size (up to 14 kb). The first SaPIs described were reported to harbor toxic shock syndrome toxin (TSST-1) known as *tst* gene. SaPIs are usually composed of an integrase gene located at one end, a repressor gene, and a replication module each expressed by different promoters in opposing directions (Novick and Ram 2017; Penadés and Christie 2015; Viana et al. 2010). A "helper exploitation" module is located at the terminal of the genome, there is dedicated to phage interactions. However, in type 2 SaPIs, the helper module is lacking and instead, these are packaged by 80α, such as SaPIbov5, which do not change the capsid size due to the lack of *cpmA* and *cpmB* (Viana et al. 2010; Quiles-Puchalt et al. 2014a).

Moreover, similar Phage-Inducible Chromosomal Islands (PICIs) have been reported in a number of Gram-positive bacteria including *Enterococcus, Streptococcus, and Lactococcus* (Martínez-Rubio et al. 2017). It was demonstrated in *Enterococcus faecalis* strain (V583), that mitomycin C induction resulted in the formation of small capsids (Martínez-Rubio et al. 2017). PICIs were also described in Gram-negative bacteria which were similar in function, but different in the genetic composition in *Escherichia coli* and *Pasteurella multocida* (Fillol-Salom et al. 2018, 2019).

The assembly pathways of phages have evolved along the way with the evolution of MGEs. Although the phages and PICIs share a similar proto-phage ancestor, the PICIS still depend for their mobilization on the helper phages, as these lack the structural genes modules, which were either lost early in evolution or were never acquired (Dokland 2019). The mechanisms by which the capsid redirection occurs are diverse, suggesting that these structural genes encoding capsid and scaffolding proteins have been acquired horizontally at different time points. Some MGEs like P4-like elements have distinct evolutionary branches as these are more closely related to plasmids rather than phages or PICIs, however, retained their ability to redirect helper capsid assembly by a different mechanism (Briani et al. 2001).

11.14 Insertion Sequences (IS) and Composite Transposons (Tn)

Insertion sequences are a vital entity of MGEs that have long been involved in revolution of the bacterial genomes by their unique ability to transpose or alter the expression of surrounding genes (Siguier et al. 2014, 2015). These IS facilitated the recombination of transposons in plasmids as well as chromosomes (Mahillon and Chandler 1998). IS are about 2.5 kb long transposable elements (TE) composed mainly of the enzyme transposase (*tnp*) catalyzing DNA excision and transfer from the donor site to another recipient or target site. IS are diverse TEs containing short imperfect terminal inverted repeat sequences (IR) and upon insertion, short flanking directly repeated target DNA sequences (DR) are generated. Traditionally, IS can only mobilize resistance genes through composite transposons. To date, there are at least 27 different families of IS (Siguier et al. 2006, 2015) assigned in groups based on the following criteria: similarities in the sequence of the transposition enzyme (*tnp*) using Markov cluster (MCL) algorithm (Enright et al. 2002; Siguier et al. 2009), their transposition mechanism and similarities in the sequences of the ends. A complete list of families can be found in the ISfinder database (ISfinder, https://www-is.biotoul.fr/). The full description of the IS can be found in the TnCentral database (https://tncentral.proteininformationresource.org/) under Tn encyclopedia. The significance of IS in transposition of resistance-conferring genes has soon been recognized after the discovery of these elements in the 1970s (Barth et al. 1976; Hedges and Jacob 1974).

Important examples of IS are IS256 and IS257 families which play a key role in spreading resistance genes in staphylococci through various transposons (Partridge et al. 2018; Varani et al. 2021). Unit transposons are a sub-class of transposons that are flanked by IR instead of IS, a *tnp* gene (s) (which includes a transposase regulator) in addition to internal passenger genes that encode for antibiotic resistance. The latter can be exemplified by Tn*552* in Staphyloccoci which has a different transposition pathway targeting particular site(s). It is believed that Tn*552*-like elements are responsible for the dissemination of β-lactamases in staphylococci (Gregory et al. 1997). Tn*552* transposons can be found in chromosomes, however, these are mostly associated with multiresistant plasmids inserted within the *res* site of the plasmid resolution system (Berg et al. 1998; Ito et al. 2003; Paulsen et al. 1994; Rowland et al. 2002).

Most of the published literature addressing the role of MGEs in antimicrobial resistance focused on the antibiotic's efflux or inactivation, and target site modification. However, recent work has shed light on the overlooked role of heterodiploidy of metabolic genes in reducing the fitness cost in Staphylococci (Andersson 2006; Ciusa et al. 2012). For example, the dihydrofolate reductase (*dhfr*) conferring resistance to trimethoprim, is present in plasmids and conjugative elements that are located in the Tn*4003* transposon which enable the *dfrA* gene to be mobilized by IS*257* in *S. aureus* (Needham et al. 1995). Other examples include *dfrA* gene transposed by Tn*7* in *E. coli* (Barth et al. 1976), and the TE Tn*5801* harboring *dfrG* and IS*256* in various Gram-positive species (León-Sampedro et al. 2016).

Furthermore, mupirocin resistance in staphylococci has recently been attributed to an additional copy of plasmid-encoded *mupA* gene (also known as *ileS2*), which is also mobilized by IS256 (Gilbart et al. 1993; Woodford et al. 1998). Mupirocin is an antibiotic and a disinfectant that acts as a potent inhibitor of the isoleucyl tRNA synthetase and has long been used for decolonization of MRSA. However, global use of mupirocin has resulted in increased resistance by MRSA, which led to changes in the decolonization protocols (Deeny et al. 2015; Hetem and Bonten 2013). The wide use of triclosan as a disinfectant has been concerning, as it targets FabI, the NADH-dependent trans-2-enoyl-acyl carrier protein (ACP) reductase, which is involved in the bacterial fatty acid biosynthesis (Hijazi et al. 2016; Schweizer 2001). Due to absence of the eukaryotic orthologue of FabI, the selective toxicity of the drugs against the prokaryotic protein is ideal, however, there might be a reciprocal resistance to antimicrobials as well (Coelho et al. 2013; Maillard et al. 2013; Morrissey et al. 2014; Oggioni et al. 2013, 2015). In the latter case of triclosan resistance, it was demonstrated that it was due to mutations in the promoter region or the chromosomal sequence of *fabI* gene (Ciusa et al. 2012; Grandgirard et al. 2015; Heath et al. 1999; McBain et al. 2012; Oggioni et al. 2013; Slater-Radosti et al. 2001). Moreover, more than half of *S. aureus*-resistant isolates carryan additional copy of *fabI* that originated from *Staphylococcus haemolyticus*, and therefore named as *(sh-fabI*) (Ciusa et al. 2012). It was found that *sh-fabI* gene was part of a TE and mobilized by IS*1272* that belongs to the IS*1182* family and present as a truncated IS in the *mec* element in *Staphylococcus haemolyticus,* however, it is absent in *S. aureus* (Archer et al. 1994; Archer et al. 1996; Siguier et al. 2015; Tonouchi et al. 1994). Furi et al. reported the presence of two composite transposons (Tn*Sha*1 and Tn*Sha*2) facilitating the dissemination of *sh-fabI* gene, with Tn*Sha*1 mostly found in *S. aureus* and Tn*Sha*2 carried in plasmids of *S. epidermidis* and *S. haemolyticus* (Furi et al. 2016). As in the case of *iles2* and *dfrA* and *dfrG* genes, *sh-fabI* is similarly mobilized by insertion sequence and duplication of drug-target metabolic genes consequently (Furi et al. 2016). The integration mechanism of *sh-fabI* involves targeting of DNA secondary structures and generation of blunt-end deletions in these hairpin structures.

11.15 Integrative and Conjugative Elements

Integrative and Conjugative Elements (ICEs) are a group of MGEs found in a diverse range of bacteria. These elements were originally called conjugative transposons that are capable of self-transposition via conjugation. Moreover, ICEs have the ability to integrate into the host chromosome and replicate either as part of the host or self-replicating after excision (Carraro and Burrus 2015). This group can be exemplified by Tn*916*-like elements encoding for tetracycline/minocycline resistance via *tet(M)*; as well as MLS [*erm(B)*] and kanamycin/neomycin (*aphA-3*) in Tn*1545* (Cochetti et al. 2008). ICEs can mobilize resistance genes by recombination mechanisms as in transposons and phages, and conjugation mechanisms similar to plasmids. Most ICEs in literature have tyrosine recombinases to facilitate excision and integration, with much fewer examples of serine recombinases and DDE transposes (Cury et al. 2017). ICEs elements are arranged in the form of modules, with similar genetic composition shared among a number of important Tn*916*-like elements for example, Tn*5397,* Tn*6000*, and Tn*5801*, conferring tetracycline resistance via *tet(M)/tet(S)* and each have different genes for excision and integration and structure due to the various recombination events (Brouwer et al. 2010; Kuroda et al. 2001; Roberts and Mullany 2011; Tsvetkova et al. 2010). In addition, Tn*1549* harbors *vanB* that resulted in the global dissemination of resistance to vancomycin in staphylococci and enterococci (Launay et al. 2006). The mechanism of integration in these elements utilizes a tyrosine integrase that targets AT-rich regions (Sansevere and Robinson 2017).

ICE6013 is another type of ICE, however, not related to Tn*916* which was first described in ST239 strains of *S. aureus* carrying Tn*552* insertions (Smyth and Robinson 2009). A number of sub-families were subsequently identified in other staphylococcal species (Sansevere et al. 2017). ICE6013 utilizes a transposase-like enzyme for its transposition (Smyth and Robinson 2009).

11.16 Others

The occurrence of class I integrons in staphylococci has only been identified by a few studies using conventional PCR detection methods of *intI1*, with no significant evidence that integrons are associated with larger segments of mobile elements or plasmids. Apart from fragments, GenBank search of Whole genome sequences and shotgun sequences failed to identify any integrons as entities in staphylococci (accessed May 2022).

11.17 Conclusion Remarks

As staphylococci continue to evolve, our knowledge of mobile genetic elements in Gram-positive bacteria is rapidly expanding. The accessory genome of Staphylococci carries most of the antimicrobial, as well as virulence-conferring determinants. Despite the intraspecies and interspecies exchange of the mobile elements among staphylococci, there are significant variations among species and strains. For example, the successful lineages of *S. aureus* vary in their composition of MGEs of insertion sequences, genomic and pathogenicity islands, transposons, and bacteriophages. The most common example is the various evolutionary trends observed in SCC*mec* in *S. aurous* that is continuously changing practically in related ACME and COMER elements. This observation strongly supports the ability of the MGEs to evolve independently from their microbial hosts as seen in phylogenies constructed for these elements in many studies to facilitate our understanding of their emergence.

References

Alibayov B, Baba-Moussa L, Sina H, Zdeňková K, Demnerová K (2014) Staphylococcus aureus mobile genetic elements. Mol Biol Rep 41:5005–5018. https://doi.org/10.1007/s11033-014-3367-3

Al-Jabri Z, Al-Shabibi Z, Al-Bimani A, Al-Hinai A, Al-Shabibi A, Rizvi M (2021) Whole genome sequencing of methicillin-resistant Staphylococcus epidermidis clinical isolates reveals variable composite SCCmec ACME among different STs in a tertiary Care Hospital in Oman. Microorganisms 9(9):1824

Almebairik N, Zamudio R, Ironside C, Joshi C, Ralph JD, Roberts AP, Gould IM, Morrissey JA, Hijazi K, Oggioni MR (2020) Genomic stability of composite SCCmec ACME and COMER-like genetic elements in Staphylococcus epidermidis correlates with rate of excision. Front Microbiol 11:166. https://doi.org/10.3389/fmicb.2020.00166

Andersson DI (2006) The biological cost of mutational antibiotic resistance: any practical conclusions? Curr Opin Microbiol 9(5):461–465. https://doi.org/10.1016/j.mib.2006.07.002

Archer GL, Johnston JL (1983) Self-transmissible plasmids in staphylococci that encode resistance to aminoglycosides. Antimicrob Agents Chemother. https://journals.asm.org/journal/aac

Archer GL, Niemeyer DM, Thanassi JA, Pucci MJ (1994) Dissemination among Staphylococci of DNA sequences associated with methicillin resistance. Antimicrob Agents Chemother 38:3

Archer GL, Thanassi JA, Niemeyer DM, Pucci MJ (1996) Characterization of IS1272, an insertion sequence-like element from staphylococcus haemolyticus. Antimicrob Agents Chemother 40:4

Ayliffe GAJ (1997) The progressive intercontinental spread of methicillin-resistant *Staphylococcus aureus*. Clin Infect Dis 24(Supplement_1):S74–S79. https://doi.org/10.1093/clinids/24.Supplement_1.S74

Baig S, Johannesen TB, Overballe-Petersen S, Larsen J, Larsen AR, Stegger M (2018) Novel SCC mec type XIII (9A) identified in an ST152 methicillin-resistant Staphylococcus aureus. Infect Genet Evol 61:74–76. https://doi.org/10.1016/j.meegid.2018.03.013

Barbier F, Lebeaux D, Hernandez D, Delannoy A-S, Caro V, François P, Schrenzel J, Ruppé E, Gaillard K, Wolff M, Brisse S, Andremont A, Ruimy R (2011) High prevalence of the arginine catabolic mobile element in carriage isolates of methicillin-resistant Staphylococcus epidermidis. J Antimicrob Chemother 66:29–36. https://doi.org/10.1093/jac/dkq410

Barbier F, Ruppé E, Hernandez D, Lebeaux D, Francois P, Felix B, Desprez A, Maiga A, Woerther P, Gaillard K, Jeanrot C, Wolff M, Schrenzel J, Andremont A, Ruimy R (2010) Methicillin-resistant coagulase-negative staphylococci in the community: high homology of SCCmec IVa between *Staphylococcus epidermidis* and major clones of methicillin-resistant *Staphylococcus aureus*. J Infect Dis 202:270–281. https://doi.org/10.1086/653483

Barth PT, Datta N, Hedges RW, Grinter NJ (1976) Transposition of a deoxyribonucleic acid sequence encoding trimethoprim and streptomycin resistances from R483 to other replicons. *J Bacteriol* 125:3

Bender J, Strommenger B, Steglich M, Zimmermann O, Fenner I, Lensing C, Dagwadordsch U, Kekulé AS, Werner G, Layer F (2014) Linezolid resistance in clinical isolates of Staphylococcus epidermidis from German hospitals and characterization of two cfr-carrying plasmids. J Antimicrob Chemother 70(6):1630–1638. https://doi.org/10.1093/jac/dkv025

Berg T, Firth N, Apisiridej S, Hettiaratchi A, Leelaporn A, Skurray RA (1998) Complete nucleotide sequence of pSK41: evolution of staphylococcal conjugative multiresistance plasmids. J Bacteriol 180:17

Bertani G (1951) *Studies on Lysogenesis I. The mode of phage liberation by lysogenic Escherichia Coli1*. 293–300

Bertani LE (1980) Genetic interaction between the nip1 mutation and genes affecting integration and excision in phage P2. Mol Gen Genet MGG 178(1):91–99. https://doi.org/10.1007/BF00267217

Boundy S, Safo MK, Wang L, Musayev FN, O'Farrell HC, Rife JP, Archer GL (2013) Characterization of the Staphylococcus aureus rRNA methyltransferase encoded by orfX, the gene containing the staphylococcal chromosome cassette mec (SCCmec) insertion site. J Biol Chem 288:132–140. https://doi.org/10.1074/jbc.M112.385138

Briani F, Dehò G, Forti F, Ghisotti D (2001) The plasmid status of satellite bacteriophage P4. Plasmid 45(1):1–17). Academic Press Inc. https://doi.org/10.1006/plas.2000.1497

Brisson-Noel A, Delrieull P, Samainli D, & Courvalins P (1988) The journal of biological chemistry inactivation of lincosaminide antibiotics in staphylococcus identification of lincosaminide 0-nucleotidyltransferases and comparison of the corresponding resistance genes* (Vol. 263, Issue 31)

Brouwer MSM, Mullany P, Roberts AP (2010) Characterization of the conjugative transposon Tn6000 from Enterococcus casseliflavus 664.1H1 (formerly Enterococcus faecium 664.1H1). FEMS Microbiol Lett 309(1):71–76. https://doi.org/10.1111/j.1574-6968.2010.02018.x

Bruce KD (1997) Analysis of mer gene subclasses within bacterial communities in soils and sediments resolved by fluorescent-PCR-restriction fragment length polymorphism profiling. Appl Environ Microbiol 63:4914–4919. https://doi.org/10.1128/aem.63.12.4914-4919.1997

Brüssow H, Canchaya C, Hardt W-D (2004) Phages and the evolution of bacterial pathogens: from genomic rearrangements to lysogenic conversion. Microbiol Mol Biol Rev 68(3):560–602. https://doi.org/10.1128/mmbr.68.3.560-602.2004

Carraro N, Burrus V (2015) The dualistic nature of integrative and conjugative elements. Mob Genet Elem 5(6):98–102. https://doi.org/10.1080/2159256x.2015.1102796

Caryl JA, Smith MCA, Thomas CD (2004) Reconstitution of a staphylococcal plasmid-protein relaxation complex in vitro. J Bacteriol 186(11):3374–3383. https://doi.org/10.1128/JB.186.11.3374-3383.2004

Cherifi S, Byl B, Deplano A, Nonhoff C, Denis O, Hallin M (2013) Comparative epidemiology of Staphylococcus epidermidis isolates from patients with catheter-related Bacteremia and from healthy volunteers. J Clin Microbiol 51:1541–1547. https://doi.org/10.1128/JCM.03378-12

Chiang YN, Penadés JR, Chen J (2019) Genetic transduction by phages and chromosomal islands: the new and noncanonical. PLoS Pathog 15:8. https://doi.org/10.1371/journal.ppat.1007878

Chongtrakool P, Ito T, Ma XX, Kondo Y, Trakulsomboon S, Tiensasitorn C, Jamklang M, Chavalit T, Song J-H, Hiramatsu K (2006) Staphylococcal cassette chromosome *mec* (SCC *mec*) typing of methicillin-resistant *Staphylococcus aureus* strains isolated in 11 Asian countries: a proposal for a new nomenclature for SCC *mec* elements. Antimicrob Agents Chemother 50:1001–1012. https://doi.org/10.1128/AAC.50.3.1001-1012.2006

Christie GE, Dokland T (2012) Pirates of the Caudovirales. Virology 434(2):210–221. https://doi.org/10.1016/j.virol.2012.10.028

Ciusa ML, Furi L, Knight D, Decorosi F, Fondi M, Raggi C, Coelho JR, Aragones L, Moce L, Visa P, Freitas AT, Baldassarri L, Fani R, Viti C, Orefici G, Martinez JL, Morrissey I, Oggioni MR (2012) A novel resistance mechanism to triclosan that suggests horizontal gene transfer and demonstrates a potential selective pressure for reduced biocide susceptibility in clinical strains of Staphylococcus aureus. Int J Antimicrob Agents 40(3):210–220. https://doi.org/10.1016/j.ijantimicag.2012.04.021

Clark NC, Weigel LM, Patel JB, Tenover FC (2005) Comparison of Tn1546-like elements in vancomycin-resistant Staphylococcus aureus isolates from Michigan and Pennsylvania. Antimicrob Agents Chemother 49(1):470–472. https://doi.org/10.1128/AAC.49.1.470-472.2005

Climo MW, Sharma VK, Archer GL (1996) Identification and characterization of the origin of conjugative transfer (oriT) and a gene (nes) encoding a single-stranded endonuclease on the staphylococcal plasmid pGO1. J Bacteriol 178:16. https://journals.asm.org/journal/jb

Cochetti I, Tili E, Mingoia M, Varaldo PE, Montanari MP (2008) erm(B)-carrying elements in tetracycline-resistant pneumococci and correspondence between Tn1545 and Tn6003. Antimicrob Agents Chemother 52(4):1285–1290. https://doi.org/10.1128/AAC.01457-07

Coelho JR, Carriço JA, Knight D, Martínez JL, Morrissey I, Oggioni MR, Freitas AT (2013) The use of machine learning methodologies to analyse antibiotic and biocide susceptibility in Staphylococcus aureus. PLoS One 8:2. https://doi.org/10.1371/journal.pone.0055582

Cohen S, Sweeney HM (1970) Transduction of methicillin resistance in *Staphylococcus aureus* dependent on an unusual specificity of the recipient strain. J Bacteriol 104:1158–1167. https://doi.org/10.1128/jb.104.3.1158-1167.1970

Cury J, Touchon M, Rocha EPC (2017) Integrative and conjugative elements and their hosts: composition, distribution and organization. Nucleic Acids Res 45(15):8943–8956. https://doi.org/10.1093/nar/gkx607

Davidson AL, Dassa E, Orelle C, Chen J (2008) Structure, function, and evolution of bacterial ATP-binding cassette systems. Microbiol Mol Biol Rev 72:317–364. https://doi.org/10.1128/MMBR.00031-07

Deeny SR, Worby CJ, Tosas Auguet O, Cooper BS, Edgeworth J, Cookson B, Robotham JV (2015) Impact of mupirocin resistance on the transmission and control of healthcare-associated MRSA. J Antimicrob Chemother 70(12):3366–3378. https://doi.org/10.1093/jac/dkv249

Dehò G, Ghisotti D (2006) The satellite phage P4. The Bacteriophages:391–408

Diep BA, Gill SR, Chang RF, Phan TH, Chen JH, Davidson MG, Lin F, Lin J, Carleton HA, Mongodin EF, Sensabaugh GF, Perdreau-Remington F (2006) Complete genome sequence of USA300, an epidemic clone of community-acquired meticillin-resistant Staphylococcus aureus. Lancet 367(9512):731–739. https://doi.org/10.1016/S0140-6736(06)68231-7

Diep BA, Stone GG, Basuino L, Graber CJ, Miller A, des Etages SA, Jones A, Palazzolo-Ballance AM, Perdreau-Remington F, Sensabaugh GF, DeLeo FR, Chambers HF (2008) The arginine catabolic mobile element and staphylococcal chromosomal cassette mec linkage: convergence of virulence and resistance in the USA300 clone of methicillin-resistant Staphylococcus aureus. J Infect Dis 197(11):1523–1530. https://doi.org/10.1086/587907

Dionisio F, Zilhão R, Gama JA (2019) Interactions between plasmids and other mobile genetic elements affect their transmission and persistence. Plasmid 102:29–36). Academic Press Inc. https://doi.org/10.1016/j.plasmid.2019.01.003

Dokland T (2019) Molecular piracy: redirection of bacteriophage capsid assembly by mobile genetic elements. Viruses 11(11):1003. https://doi.org/10.3390/v11111003

Dowell CE, Rosenblum ED (1962) Serology and transduction in staphylococcal phage. J Bacteriol 84(5):1071–1075. https://doi.org/10.1128/jb.84.5.1071-1075.1962

Dy RL, Przybilski R, Semeijn K, Salmond GPC, Fineran PC (2014) A widespread bacteriophage abortive infection system functions through a type IV toxin–antitoxin mechanism. Nucleic Acids Res 42:4590–4605. https://doi.org/10.1093/nar/gkt1419

Enright AJ, van Dongen S, Ouzounis CA (2002) An efficient algorithm for large-scale detection of protein families. Nucleic Acids Res 30:7. www.ensembl.org

Evans J, Dyke KGH (1988) Characterization of the conjugation system associated with the Staphylococcus aureus plasmid pJE1. Microbiology 134(1):1–8

Fillol-Salom A, Bacarizo J, Alqasmi M, Ciges-Tomas JR, Martínez-Rubio R, Roszak AW, Cogdell RJ, Chen J, Marina A, Penadés JR (2019) Hijacking the hijackers: Escherichia coli Pathogenicity Islands redirect helper phage packaging for their own benefit. Mol Cell 75(5):1020–1030.e4. https://doi.org/10.1016/j.molcel.2019.06.017

Fillol-Salom A, Martínez-Rubio R, Abdulrahman RF, Chen J, Davies R, Penadés JR (2018) Phage-inducible chromosomal islands are ubiquitous within the bacterial universe. ISME J 12(9): 2114–2128. https://doi.org/10.1038/s41396-018-0156-3

Firth N, Skurray RA (2006) Genetics: accessory elements and genetic exchange. In: Fischetti VA, Novick RP, Ferretti JJ, Portnoy DA, Rood JI (eds) Gram-positive pathogens, 2nd edn. Wiley. https://doi.org/10.1128/9781555816513.ch33

Foster T (1996) Medical microbiology, 4th edn. University of Texas Mediacl Branch at Galveston

Furi L, Haigh R, Al Jabri ZJH, Morrissey I, Ou HY, León-Sampedro R, Martinez JL, Coque TM, Oggioni MR (2016) Dissemination of novel antimicrobial resistance mechanisms through the insertion sequence mediated spread of metabolic genes. Front Microbiol 7(JUN). https://doi.org/10.3389/fmicb.2016.01008

Gennaro ML, Kornblum J, Novick RP (1987) A site-specific recombination function in Staphylococcus aureus plasmids. J Bacteriol 169:6. https://journals.asm.org/journal/jb

Gilbart J, Perry CR, Slocombe B (1993) High-level mupirocin resistance in Staphylococcus aureus: evidence for two distinct Isoleucyl-tRNA synthetases. Antimicrob Agents Chemother 37:1

Goering RV, Ruff EA (1983) Comparative analysis of conjugative plasmids mediating gentamicin resistance in Staphylococcus aureus. Antimicrob Agents Chemother 24:3. https://journals.asm.org/journal/aac

Grandgirard D, Furi L, Ciusa ML, Baldassarri L, Knight DR, Morrissey I, Largiadèr CR, Leib SL, Oggioni MR (2015) Mutations upstream of fabI in triclosan resistant Staphylococcus aureus strains are associated with elevated fabI gene expression. BMC Genomics 16:1. https://doi.org/10.1186/s12864-015-1544-y

Granslo HN, Klingenberg C, Fredheim EGA, Rønnestad A, Mollnes TE, Flægstad T (2010) Arginine catabolic mobile element is associated with low antibiotic resistance and low pathogenicity in Staphylococcus epidermidis from neonates. Pediatr Res 68:237–241. https://doi.org/10.1203/PDR.0b013e3181eb01e0

Gregory PD, Lewis RA, Curnock SP, Dyke KGH (1997) Studies of the repressor (BlaI) of β-lactamase synthesis in *Staphylococcus aureus*. Mol Microbiol 24(5):1025–1037. https://doi.org/10.1046/j.1365-2958.1997.4051770.x

Haaber J, Penadés JR, Ingmer H (2017) Transfer of antibiotic resistance in Staphylococcus aureus. Trends Microbiol 25:893–905. https://doi.org/10.1016/j.tim.2017.05.011

Hartman BJ, Tomasz A (1984) Low-affinity penicillin-binding protein associated with r-lactam resistance in Staphylococcus aureus. J Bacteriol 158:513–516. 0021-9193/84/050513-04 $02.00/0

Hatfull GF, Hendrix RW (2011) Bacteriophages and their genomes. Curr Opin Virol 1(4): 298–303). Elsevier B.V. https://doi.org/10.1016/j.coviro.2011.06.009

Heath RJ, Rubin JR, Holland DR, Zhang E, Snow ME, Rock CO (1999) Mechanism of triclosan inhibition of bacterial fatty acid synthesis. J Biol Chem 274(16):11110–11114. https://doi.org/10.1074/jbc.274.16.11110

Hedges RW, Jacob AE (1974) Transposition of ampicillin resistance from RP4 to other replicons. Mol Gen Genet MGG 132(1):31–40

Helinski DR (2022) A brief history of plasmids. EcoSal Plus:eESP-0028

Herold BC (1998) Community-acquired methicillin-resistant Staphylococcus aureus in children with no identified predisposing risk. JAMA 279:593. https://doi.org/10.1001/jama.279.8.593

Hetem DJ, Bonten MJM (2013) Clinical relevance of mupirocin resistance in Staphylococcus aureus. In. J Hosp Infect 85(4):249–256. https://doi.org/10.1016/j.jhin.2013.09.006

Higgins PG, Rosato AE, Seifert H, Archer GL, Wisplinghoff H (2009) Differential expression of *ccrA* in methicillin-resistant *Staphylococcus aureus* strains carrying staphylococcal cassette chromosome *mec* type II and IVa elements. Antimicrob Agents Chemother 53:4556–4558. https://doi.org/10.1128/AAC.00395-09

Hijazi K, Mukhopadhya I, Abbott F, Milne K, Al-Jabri ZJ, Oggioni MR, Gould IM (2016) Susceptibility to chlorhexidine amongst multidrug-resistant clinical isolates of Staphylococcus epidermidis from bloodstream infections. Int J Antimicrob Agents 48(1):86–90. https://doi.org/10.1016/j.ijantimicag.2016.04.015

Horinouchit S, Weisblum B (1982a) Nucleotide sequence and functional map of pC194, a plasmid that specifies inducible chloramphenicol resistance. J Bacteriol

Horinouchit S, Weisblum B (1982b) Nucleotide sequence and functional map of pE194, a plasmid that specifies inducible resistance to macrolide, lincosamide, and streptogramin type B antibiotics. J Bacteriol 150:2

Hung W-C, Chen H-J, Lin Y-T, Tsai J-C, Chen C-W, Lu H-H, Tseng S-P, Jheng Y-Y, Leong KH, Teng L-J (2015) Skin commensal staphylococci may act as reservoir for Fusidic acid resistance genes. PLoS One 10:e0143106. https://doi.org/10.1371/journal.pone.0143106

Ito T, Katayama Y, Asada K, Mori N, Tsutsumimoto K, Tiensasitorn C, Hiramatsu K (2001) Structural comparison of three types of staphylococcal cassette chromosome mec integrated in the chromosome in methicillin-resistant Staphylococcus aureus. Antimicrob Agents Chemother 45:14

Ito T, Katayama Y, Hiramatsu K (1999) Cloning and nucleotide sequence determination of the entire *mec* DNA of pre-methicillin-resistant *Staphylococcus aureus* N315. Antimicrob Agents Chemother 43:1449–1458. https://doi.org/10.1128/AAC.43.6.1449

Ito T, Ma XX, Takeuchi F, Okuma K, Yuzawa H, Hiramatsu K (2004) Novel type V staphylococcal cassette chromosome *mec* driven by a novel cassette chromosome recombinase, *ccrC*. Antimicrob Agents Chemother 48:2637–2651. https://doi.org/10.1128/AAC.48.7.2637-2651.2004

Ito T, Okuma K, Ma XX, Yuzawa H, Hiramatsu K (2003) Insights on antibiotic resistance of Staphylococcus aureus from its whole genome: genomic island SCC. Drug Resist Updat 6(1): 41–52). Churchill Livingstone. https://doi.org/10.1016/S1368-7646(03)00003-7

IWG-SCC, I.W.G. on the C. of S.C.C.E. (2009) Classification of Staphylococcal Cassette Chromosome *mec* (SCC *mec*): Guidelines for Reporting Novel SCC *mec* Elements. Antimicrob. Agents Chemother. 53, 4961–4967. https://doi.org/10.1128/AAC.00579-09

Jaffe HW, Sweeney HM, Weinstein RA, Kabins SA, Nathan C, Cohen S (1982) Structural and phenotypic varieties of gentamicin resistance plasmids in hospital strains of Staphylococcus aureus and coagulase-negative staphylococci. Antimicrob Agents Chemother 21:5. https://journals.asm.org/journal/aac

Jensen SO, Lyon BR (2009) Genetics of antimicrobial resistance in Staphylococcus aureus. Future Microbiol 4(5):565–582. https://doi.org/10.2217/fmb.09.30

Kang CK, Cho JE, Choi YJ, Jung Y, Kim N-H, Kim C-J, Kim TS, Song K-H, Choe PG, Park WB, Bang J-H, Kim ES, Park KU, Park SW, Kim N-J, Oh M, Kim HB (2015) agr dysfunction affects staphylococcal cassette chromosome *mec* type-dependent clinical outcomes in methicillin-resistant Staphylococcus aureus Bacteremia. Antimicrob Agents Chemother 59:3125–3132. https://doi.org/10.1128/AAC.04962-14

Katayama Y, Ito T, Hiramatsu K (2000) A new class of genetic element, Staphylococcus Cassette Chromosome *mec* , Encodes Methicillin Resistance in *Staphylococcus aureus*. Antimicrob Agents Chemother 44:1549–1555. https://doi.org/10.1128/AAC.44.6.1549-1555.2000

Kuroda M, Ohta T, Uchiyama I, Baba T, Yuzawa H, Kobayashi I, Cui L, Oguchi A, Aoki K, Nagai Y, Lian J, Ito T, Kanamori M, Matsumaru H, Maruyama A, Murakami H, Hosoyama A, Mizutani-Ui Y, Takahashi NK et al (2001) Whole genome sequencing of meticillin-resistant Staphylococcus aureus. Lancet 357(9264):1225–1240. https://doi.org/10.1016/S0140-6736(00)04403-2

Kwong SM, Ramsay JP, Jensen SO, Firth N (2017) Replication of staphylococcal resistance plasmids. Front Microbiol 8(NOV) Frontiers Media S.A. https://doi.org/10.3389/fmicb.2017.02279

Kwong SM, Skurray RA, Firth N (2004) Staphylococcus aureus multiresistance plasmid pSK41: analysis of the replication region, initiator protein binding and antisense RNA regulation. Mol Microbiol 51(2):497–509. https://doi.org/10.1046/j.1365-2958.2003.03843.x

Launay A, Ballard SA, Johnson PDR, Grayson ML, Lambert T (2006) Transfer of vancomycin resistance transposon Tn1549 from Clostridium symbiosum to Enterococcus spp. in the gut of gnotobiotic mice. Antimicrob Agents Chemother 50(3):1054–1062. https://doi.org/10.1128/AAC.50.3.1054-1062.2006

Lee JYH, Monk IR, Gonçalves da Silva A, Seemann T, Chua KYL, Kearns A, Hill R, Woodford N, Bartels MD, Strommenger B, Laurent F, Dodémont M, Deplano A, Patel R, Larsen AR, Korman TM, Stinear TP, Howden BP (2018) Global spread of three multidrug-resistant lineages of Staphylococcus epidermidis. Nat Microbiol 3:1175–1185. https://doi.org/10.1038/s41564-018-0230-7

León-Sampedro R, Novais C, Peixe L, Baquero F, Coque TM (2016) Diversity and evolution of the Tn5801-tet(M)-like integrative and conjugative elements among Enterococcus, streptococcus, and Staphylococcus. Antimicrob Agents Chemother 60(3):1736–1746. https://doi.org/10.1128/AAC.01864-15

Lindgren JK, Thomas VC, Olson ME, Chaudhari SS, Nuxoll AS, Schaeffer CR, Lindgren KE, Jones J, Zimmerman MC, Dunman PM, Bayles KW, Fey PD (2014) Arginine deiminase in Staphylococcus epidermidis functions to augment biofilm maturation through pH homeostasis. J Bacteriol 196:2277–2289. https://doi.org/10.1128/JB.00051-14

Lindsay JA (2010) Genomic variation and evolution of Staphylococcus aureus. Int J Med Microbiol 300:98–103. https://doi.org/10.1016/j.ijmm.2009.08.013

Lindsay JA (2014) Staphylococcus aureus genomics and the impact of horizontal gene transfer. Int J Med Microbiol 304(2):103–109. https://doi.org/10.1016/j.ijmm.2013.11.010

Lindsay JA, Ruzin A, Ross HF, Kurepina N, Novick RP (1998) The gene for toxic shock toxin is carried by a family of mobile pathogenicity islands in *Staphylococcus aureus*. Mol Microbiol 29(2):527–543. https://doi.org/10.1046/j.1365-2958.1998.00947.x

Littlejohn TG, DiBerardino D, Messerotti LJ, Spiers SJ, Skurray RA (1991) Structure and evolution of a family of genes encoding antiseptic and disinfectant resistance in Staphylococcus aureus. Gene 101(1):59–66. https://doi.org/10.1016/0378-1119(91)90224-Y

Liu P, Wu Z, Xue H, Zhao X (2017) Antibiotics trigger initiation of SCCmec transfer by inducing SOS responses. Nucleic Acids Res 45:3944–3952. https://doi.org/10.1093/nar/gkx153

Macrina FL, & Archer GL (1993) Conjugation and broad host range plasmids in streptococci and staphylococci. In *Bacterial conjugation* (pp. 313–329). Springer

Mahillon J, Chandler M (1998) Insertion Sequences. Microbiol Mol Biol Rev 62:3

Maillard JY, Bloomfield S, Coelho JR, Collier P, Cookson B, Fanning S, Hill A, Hartemann P, McBain AJ, Oggioni M, Sattar S, Schweizer HP, Threlfall J (2013) Does microbicide use in consumer products promote antimicrobial resistance? A critical review and recommendations for a cohesive approach to risk assessment. Microb Drug Resist 19(5):344–354. https://doi.org/10.1089/mdr.2013.0039

Makhlin J, Kofman T, Borovok I, Kohler C, Engelmann S, Cohen G, Aharonowitz Y (2007) *Staphylococcus aureus* ArcR controls expression of the arginine deiminase operon. J Bacteriol 189:5976–5986. https://doi.org/10.1128/JB.00592-07

Malachowa N, Deleo FR (2010) Mobile genetic elements of Staphylococcus aureus. Cell Mol Life Sci 67(18):3057–3071. https://doi.org/10.1007/s00018-010-0389-4

Martínez-Rubio R, Quiles-Puchalt N, Martí M, Humphrey S, Ram G, Smyth D, Chen J, Novick RP, Penadés JR (2017) Phage-inducible islands in the Gram-positive cocci. ISME J 11(4):1029–1042. https://doi.org/10.1038/ismej.2016.163

McBain AJ, Forbes S, Latimer J (2012) Reply to "lack of evidence for reduced fitness of clinical staphylococcus aureus isolates with reduced susceptibility to triclosan.". Antimicrob Agents Chemother 56(11):6072. https://doi.org/10.1128/AAC.01515-12

Mcelgunn CJ, Zahurul M, Bhuyian A, Sugiyama M (2002) Integration analysis of pSK41 in the chromosome of a methicillin-resistant Staphylococcus aureus K-1. J Basic Microbiol 42

McKenzie T, Hoshino T, Tanaka T, Sueoka N (1986) The nucleotide sequence of pUB110: some salient features in relation to replication and its regulation. Plasmid 15(2):93–103. https://doi.org/10.1016/0147-619X(86)90046-6

McManus BA, Coleman DC, Deasy EC, Brennan GI, O'Connell B, Monecke S, Ehricht R, Leggett B, Leonard N, Shore AC (2015) Comparative genotypes, staphylococcal cassette chromosome mec (SCCmec) genes and antimicrobial resistance amongst Staphylococcus epidermidis and Staphylococcus haemolyticus isolates from infections in humans and companion animals. PLoS One 10(9):e0138079. https://doi.org/10.1371/journal.pone.0138079

McManus BA, O'Connor AM, Kinnevey PM, O'Sullivan M, Polyzois I, Coleman DC (2017) First detailed genetic characterization of the structural organization of type III arginine catabolic mobile elements harbored by Staphylococcus epidermidis by using whole-genome sequencing. Antimicrob Agents Chemother 61(10):e01216–e01217. https://doi.org/10.1128/AAC.01216-17

Miragaia M, de Lencastre H, Perdreau-Remington F, Chambers HF, Higashi J, Sullam PM, Lin J, Wong KI, King KA, Otto M, Sensabaugh GF, Diep BA (2009) Genetic diversity of arginine catabolic mobile element in Staphylococcus epidermidis. PLoS One 4:e7722. https://doi.org/10.1371/journal.pone.0007722

Misiura A, Pigli YZ, Boyle-Vavra S, Daum RS, Boocock MR, Rice PA (2013) Roles of two large serine recombinases in mobilizing the methicillin-resistance cassette SCC *mec*: roles of the SCCmec recombinases. Mol Microbiol 88:1218–1229. https://doi.org/10.1111/mmi.12253

Mojumdart M, Khan SA (1988) Characterization of the tetracycline resistance gene of plasmid pT181 of Staphylococcus aureus. J Bacteriol 170:12. https://journals.asm.org/journal/jb

Monecke S, Jatzwauk L, Müller E, Nitschke H, Pfohl K, Slickers P, Reissig A, Ruppelt-Lorz A, Ehricht R (2016) Diversity of SCCmec elements in Staphylococcus aureus as observed in South-Eastern Germany. PLoS One 11:e0162654. https://doi.org/10.1371/journal.pone.0162654

Morrissey I, Oggioni MR, Knight D, Curiao T, Coque T, Kalkanci A, Martinez JL, Baldassarri L, Orefici G, Yetiş Ü, Rödger HJ, Visa P, Mora D, Leib S, Viti C (2014) Evaluation of epidemiological cut-off values indicates that biocide resistant subpopulations are uncommon in natural isolates of clinically-relevant microorganisms. PLoS One 9:1. https://doi.org/10.1371/journal.pone.0086669

Morton TM, Linda JJ, Patterson J, Archer GL (1995) Characterization of a conjugative staphylococcal mupirocin resistance plasmid. Antimicrob Agents Chemother 39:6

Naas T, Fortineau N, Spicq C, Robert J, Jarlier V, Nordmann P (2005) Three-year survey of community-acquired methicillin-resistant Staphylococcus aureus producing Panton-valentine leukocidin in a French university hospital. J Hosp Infect 61:321–329. https://doi.org/10.1016/j.jhin.2005.01.027

Naimi TS (2003) Comparison of community- and health care–associated methicillin-resistant Staphylococcus aureus infection. JAMA 290:2976. https://doi.org/10.1001/jama.290.22.2976

Needham C, Noble WC, Dyke KGH (1995) The staphylococcal insertion sequence IS257 is active. Plasmid 34

Néron B, Ménager H, Maufrais C, Joly N, Maupetit J, Letort S, Carrere S, Tuffery P, Letondal C (2009) Mobyle: a new full web bioinformatics framework. Bioinformatics 25(22):3005–3011. https://doi.org/10.1093/bioinformatics/btp493

Nilsson AS, Karlsson JL, Haggård-Ljungquist E (2004) Site-specific recombination links the evolution of P2-like coliphages and pathogenic enterobacteria. Mol Biol Evol 21(1):1–13. https://doi.org/10.1093/molbev/msg223

Noto MJ, Kreiswirth BN, Monk AB, Archer GL (2008) Gene acquisition at the insertion site for SCC *mec* , the Genomic Island conferring methicillin resistance in *Staphylococcus aureus*. J Bacteriol 190:1276–1283. https://doi.org/10.1128/JB.01128-07

Novick RP (1963) Analysis by transduction of mutations affecting penicillinase formation. J Gen Microbiol 33:121–136. https://doi.org/10.1099/00221287-33-1-121

Novick RP, Christie GE, Penadés JR (2010) The phage-related chromosomal islands of gram-positive bacteria. Nat Rev Microbiol 8(8):541–551. https://doi.org/10.1038/nrmicro2393

Novick RP, Iordanescu S, Projan SJ, Kornblum J, Edelman I (1989) pT181 plasmid replication is regulated by a countertranscript-driven transcriptional attenuator. Cell 59(2):395–404. https://doi.org/10.1016/0092-8674(89)90300-0

Novick RP, Ram G (2017) Staphylococcal pathogenicity islands—movers and shakers in the genomic firmament. Curr Opin Microbiol 38:197–204). Elsevier Ltd. https://doi.org/10.1016/j.mib.2017.08.001

O'Brien FG, Ramsay JP, Monecke S, Coombs GW, Robinson OJ, Htet Z, Alshaikh FAM, Grubb WB (2015) Staphylococcus aureus plasmids without mobilization genes are mobilized by a novel conjugative plasmid from community isolates. J Antimicrob Chemother 70(3):649–652. https://doi.org/10.1093/jac/dku454

O'Connor AM, McManus BA, Coleman DC (2018a) First description of novel arginine catabolic mobile elements (ACMEs) types IV and V harboring a kdp operon in Staphylococcus epidermidis characterized by whole genome sequencing. Infect Genet Evol 61:60–66. https://doi.org/10.1016/j.meegid.2018.03.012

O'Connor AM, McManus BA, Kinnevey PM, Brennan GI, Fleming TE, Cashin PJ, O'Sullivan M, Polyzois I, Coleman DC (2018b) Significant enrichment and diversity of the staphylococcal arginine catabolic mobile element ACME in Staphylococcus epidermidis isolates from subgingival peri-implantitis sites and periodontal pockets. Front Microbiol 9:1558. https://doi.org/10.3389/fmicb.2018.01558

Oggioni MR, Coelho JR, Furi L, Knight DR, Viti C, Orefici G, Martinez J-L, Freitas AT, Coque TM, Morrissey I (2015) Send orders for reprints to reprints@benthamscience.ae significant differences characterise the correlation coefficients between biocide and antibiotic susceptibility profiles in Staphylococcus aureus. Curr Pharm Des 21

Oggioni MR, Furi L, Coelho JR, Maillard JY, Martínez JL (2013) Recent advances in the potential interconnection between antimicrobial resistance to biocides and antibiotics. Expert Rev Anti-Infect Ther 11(4):363–366. https://doi.org/10.1586/eri.13.16

Onishi M, Urushibara N, Kawaguchiya M, Ghosh S, Shinagawa M, Watanabe N, Kobayashi N (2013) Prevalence and genetic diversity of arginine catabolic mobile element (ACME) in clinical isolates of coagulase-negative staphylococci: identification of ACME type I variants in Staphylococcus epidermidis. Infect Genet Evol 20:381–388. https://doi.org/10.1016/j.meegid.2013.09.018

Osborn AM, Bruce KD, Strike P, Ritchie DA (1997) Distribution, diversity and evolution of the bacterial mercury resistance (*mer*) operon. FEMS Microbiol Rev 19:239–262. https://doi.org/10.1111/j.1574-6976.1997.tb00300.x

Otto M (2004) Virulence factors of the coagulase-negative staphylococci. Front Biosci 9:841. https://doi.org/10.2741/1295

Otto M (2013) Coagulase-negative staphylococci as reservoirs of genes facilitating MRSA infection: Staphylococcal commensal species such as Staphylococcus epidermidis are being recognized as important sources of genes promoting MRSA colonization and. BioEssays 35: 4–11. https://doi.org/10.1002/bies.201200112

Panesso D, Planet PJ, Diaz L, Hugonnet JE, Tran TT, Narechania A, Munita JM, Rincon S, Carvajal LP, Reyes J, Londoño A, Smith H, Sebra R, Deikus G, Weinstock GM, Murray BE, Rossi F, Arthur M, Arias CA (2015) Methicillin-susceptible, vancomycin-resistant staphylococcus aureus, Brazil. Emerg Infect Dis 21(10):1844–1848. https://doi.org/10.3201/eid2110.141914

Partridge SR, Kwong SM, Firth N, Jensen SO (2018) Mobile genetic elements associated with antimicrobial resistance. Clin Microbiol Rev 31(4):e00088–17. https://doi.org/10.1128/CMR.00088-17

Paulsen IT, Gillespie MT, Littlejohn TG, Hanvivatvong O, Rowland S-J, Dyke KGH, Skurray RA (1994) Characterisation of sin, a potential recombinase-encoding gene from Staphylococcus aureus. Gene 141(1):109–114. https://doi.org/10.1016/0378-1119(94)90136-8

Penadés JR, Christie GE (2015) The phage-inducible Chromosomal Islands: a family of highly evolved molecular parasites. Ann Rev Virol 2:181–201). Annual Reviews Inc. https://doi.org/10.1146/annurev-virology-031413-085446

Pérez-Roth E, Kwong SM, Alcoba-Florez J, Firth N, Méndez-Álvarez S (2010) Complete nucleotide sequence and comparative analysis of pPR9, a 41.7-kilobase conjugative staphylococcal multiresistance plasmid conferring high-level mupirocin resistance. Antimicrob Agents Chemother 54(5):2252–2257. https://doi.org/10.1128/AAC.01074-09

Pérez-Roth E, López-Aguilar C, Alcoba-Florez J, Méndez-Álvarez S (2006) High-level mupirocin resistance within methicillin-resistant Staphylococcus aureus pandemic lineages. Antimicrob Agents Chemother 50(9):3207–3211. https://doi.org/10.1128/AAC.00059-06

Pinho MG, Filipe SR, de Lencastre H, Tomasz A (2001) Complementation of the essential peptidoglycan transpeptidase function of penicillin-binding protein 2 (PBP2) by the drug resistance protein PBP2A in *Staphylococcus aureus*. J Bacteriol 183:6525–6531. https://doi.org/10.1128/JB.183.22.6525-6531.2001

Planet PJ, Diaz L, Kolokotronis S-O, Narechania A, Reyes J, Xing G, Rincon S, Smith H, Panesso D, Ryan C, Smith DP, Guzman M, Zurita J, Sebra R, Deikus G, Nolan RL, Tenover FC, Weinstock GM, Robinson DA, Arias CA (2015) Parallel epidemics of community-associated methicillin-resistant *Staphylococcus aureus* USA300 infection in North and South America. J Infect Dis 212:1874–1882. https://doi.org/10.1093/infdis/jiv320

Planet PJ, Diaz L, Rios R, Arias CA (2016) Global spread of the community-associated methicillin-resistant *Staphylococcus aureus* USA300 Latin American variant. J Infect Dis 214:1609–1610. https://doi.org/10.1093/infdis/jiw418

Projan SJ, Archer GL (1989) Mobilization of the relaxable Staphylococcus aureus plasmid pC221 by the conjugative plasmid pGOl involves three pC221 loci. J Bacteriol 171:4. https://journals.asm.org/journal/jb

Purves J, Thomas J, Riboldi GP, Zapotoczna M, Tarrant E, Andrew PW, Londoño A, Planet PJ, Geoghegan JA, Waldron KJ, Morrissey JA (2018) A horizontally gene transferred copper resistance locus confers hyper-resistance to antibacterial copper toxicity and enables survival of community acquired methicillin resistant *Staphylococcus aureus* USA300 in macrophages: *Staphylococcus aureus* copper resistance and innate immunity. Environ Microbiol 20:1576–1589. https://doi.org/10.1111/1462-2920.14088

Quiles-Puchalt N, Carpena N, Alonso JC, Novick RP, Marina A, Penadés JR (2014a) Staphylococcal pathogenicity island DNA packaging system involving cos-site packaging and phage-encoded HNH endonucleases. Proc Natl Acad Sci U S A 111(16):6016–6021. https://doi.org/10.1073/pnas.1320538111

Quiles-Puchalt N, Martínez-Rubio R, Ram G, Lasa I, Penadés JR (2014b) Unravelling bacteriophage φ11 requirements for packaging and transfer of mobile genetic elements in Staphylococcus aureus. Mol Microbiol 91(3):423–437. https://doi.org/10.1111/mmi.12445

Roberts AP, Mullany P (2011) Tn916-like genetic elements: a diverse group of modular mobile elements conferring antibiotic resistance. FEMS Microbiol Rev 35(5):856–871. https://doi.org/10.1111/j.1574-6976.2011.00283.x

Rolo J, Miragaia M, Turlej-Rogacka A, Empel J, Bouchami O, Faria NA, Tavares A, Hryniewicz W, Fluit AC, de Lencastre H, the CONCORD Working Group (2012) High genetic diversity among community-associated Staphylococcus aureus in Europe: results from a Multicenter study. PLoS One 7:e34768. https://doi.org/10.1371/journal.pone.0034768

Rolo J, Worning P, Boye Nielsen J, Sobral R, Bowden R, Bouchami O, Damborg P, Guardabassi L, Perreten V, Westh H, Tomasz A, de Lencastre H, Miragaia M (2017) Evidence for the evolutionary steps leading to mecA-mediated β-lactam resistance in staphylococci. PLoS Genet 13:e1006674. https://doi.org/10.1371/journal.pgen.1006674

Rosario-Cruz Z, Eletsky A, Daigham NS, Al-Tameemi H, Swapna GVT, Kahn PC, Szyperski T, Montelione GT, Boyd JM (2019) The copBL operon protects Staphylococcus aureus from copper toxicity: CopL is an extracellular membrane–associated copper-binding protein. J Biol Chem 294:4027–4044. https://doi.org/10.1074/jbc.RA118.004723

Rossi F, Diaz L, Wollam A, Panesso D, Zhou Y, Rincon S, Narechania A, Xing G, di Gioia TSR, Doi A, Tran TT, Reyes J, Munita JM, Carvajal LP, Hernandez-Roldan A, Brandão D, van der Heijden IM, Murray BE, Planet PJ et al (2014) Transferable vancomycin resistance in a community-associated MRSA lineage. N Engl J Med 370(16):1524–1531. https://doi.org/10.1056/nejmoa1303359

Rowland S-J, Stark Marshall W, Boocock Martin R (2002) Sin recombinase from Staphylococcus aureus:synaptic complex architecture and transposon targeting. Mol Microbiol 44(3):607–619. https://doi.org/10.1046/j.1365-2958.2002.02897.x

Sansevere EA, Luo X, Park JY, Yoon S, Seo KS, Robinson DA (2017) Transposase-mediated excision, conjugative transfer, and diversity of ICE6013 elements in Staphylococcus aureus. J Bacteriol 199:8. https://doi.org/10.1128/JB.00629-16

Sansevere EA, Robinson DA (2017) Staphylococci on ICE: overlooked agents of horizontal gene transfer. Mob Genet Elem 7(4):1–10. https://doi.org/10.1080/2159256x.2017.1368433

Scharn CR, Tenover FC, Goering RV (2013) Transduction of staphylococcal cassette chromosome *mec* elements between strains of Staphylococcus aureus. Antimicrob Agents Chemother 57: 5233–5238. https://doi.org/10.1128/AAC.01058-13

Schweizer HP (2001) Triclosan: a widely used biocide and its link to antibiotics. FEMS Microbiol Lett 202(1):1–7. https://doi.org/10.1111/j.1574-6968.2001.tb10772.x

Shafer WM, Iandolo JJ (1979) Genetics of staphylococcal enterotoxin B in methicillin- resistant isolates of Staphylococcus aureust. Infect Immun 25:902–911. https://doi.org/0019-9567/79/09-0902/10$02.00/0

Shearer JES, Wireman J, Hostetler J, Forberger H, Borman J, Gill J, Sanchez S, Mankin A, LaMarre J, Lindsay JA, Bayles K, Nicholson A, O'Brien F, Jensen SO, Firth N, Skurray RA, Summers AO (2011) Major families of multiresistant plasmids from geographically and epidemiologically diverse staphylococci. G3: Genes, Genomes Genetics 1(7):581–591. https://doi.org/10.1534/g3.111.000760

Shore AC, Lazaris A, Kinnevey PM, Brennan OM, Brennan GI, O'Connell B, Feßler AT, Schwarz S, Coleman DC (2016) First report of cfr-carrying plasmids in the pandemic sequence type 22 methicillin-resistant Staphylococcus aureus staphylococcal cassette chromosome mec type IV clone. Antimicrob Agents Chemother 60(5):3007–3015. https://doi.org/10.1128/AAC.02949-15

Shore AC, Rossney AS, Brennan OM, Kinnevey PM, Humphreys H, Sullivan DJ, Goering RV, Ehricht R, Monecke S, Coleman DC (2011) Characterization of a novel arginine catabolic Mobile element (ACME) and staphylococcal chromosomal cassette *mec* composite Island with significant homology to Staphylococcus epidermidis ACME type II in methicillin-resistant

Staphylococcus aureus genotype ST22-MRSA-IV. Antimicrob Agents Chemother 55:1896–1905. https://doi.org/10.1128/AAC.01756-10
Siguier P, Gagnevin L, Chandler M (2009) The new IS1595 family, its relation to IS1 and the frontier between insertion sequences and transposons. Res Microbiol 160(3):232–241. https://doi.org/10.1016/j.resmic.2009.02.003
Siguier P, Gourbeyre E, Chandler M (2014) Bacterial insertion sequences: their genomic impact and diversity. FEMS Microbiol Rev 38(5):865–891. https://doi.org/10.1111/1574-6976.12067
Siguier P, Gourbeyre E, Varani A, Ton-Hoang B, Chandler M (2015) Everyman's guide to bacterial insertion sequences. Microbiol Spectrum 3:2. https://doi.org/10.1128/microbiolspec.mdna3-0030-2014
Siguier P, Perochon J, Lestrade L, Mahillon J, Chandler M (2006) ISfinder: the reference centre for bacterial insertion sequences. Nucleic Acids Res 34(Database issue). https://doi.org/10.1093/nar/gkj014
Six EW (1975) The helper dependence of satellite bacteriophage P4: which gene functions of bacteriophage P2 are needed by P4? Virology 67(1):249–263. https://doi.org/10.1016/0042-6822(75)90422-5
Six EW, Klug CAC (1973) Bacteriophage P4: a satellite virus depending on a helper such as prophage P2. Virology 51(2):327–344. https://doi.org/10.1016/0042-6822(73)90432-7
Slater-Radosti C, van Aller G, Greenwood R, Nicholas R, Keller PM, Dewolf WE, Fan F, Payne DJ, Jaworski DD (2001) Original articles biochemical and genetic characterization of the action of triclosan on Staphylococcus aureus. J Antimicrob Chemother 48
Smyth DS, Robinson DA (2009) Integrative and sequence characteristics of a novel genetic element, ICE6013, in Staphylococcus aureus. J Bacteriol 191(19):5964–5975. https://doi.org/10.1128/JB.00352-09
Stewart GC, Rosenblum ED (1980) Transduction of methicillin resistance in Staphylococcus aureus: recipient effectiveness and beta-lactamase production. Antimicrob Agents Chemother 18:424–432. https://doi.org/10.1128/AAC.18.3.424
Stojanov M, Moreillon P, Sakwinska O (2015) Excision of staphylococcal cassette chromosome mec in methicillin-resistant Staphylococcus aureus assessed by quantitative PCR. BMC Res Notes 8:828. https://doi.org/10.1186/s13104-015-1815-3
Thurlow LR, Joshi GS, Clark JR, Spontak JS, Neely CJ, Maile R, Richardson AR (2013) Functional modularity of the arginine catabolic mobile element contributes to the success of USA300 methicillin-resistant Staphylococcus aureus. Cell Host Microbe 13:100–107. https://doi.org/10.1016/j.chom.2012.11.012
Tolstoy I, Kropinski AM, Brister JR (2018) Bacteriophage taxonomy: an evolving discipline. In: Azeredo J, Sillankorva S (eds) Bacteriophage therapy: from lab to clinical practice. Springer, New York, pp 57–71. https://doi.org/10.1007/978-1-4939-7395-8_6
Tong SYC, Davis JS, Eichenberger E, Holland TL, Fowler VG (2015) Staphylococcus aureus infections: epidemiology, pathophysiology, clinical manifestations, and management. Clin Microbiol Rev 28:603–661. https://doi.org/10.1128/CMR.00134-14
Tonouchi N, Tsuchida T, Yoshinaga F, Horinouchi S, Beppu T (1994) A host–vector system for a cellulose-producing acetobacter strain. Biosci Biotechnol Biochem 58(10):1899–1901. https://doi.org/10.1271/bbb.58.1899
Townsend DE, Ashdown N, Grubb WB, Pearman JW, Annear DI (1985) Genetics and epidemiology of methicillin-resistant Staphylococcus aureus isolated in a Western Australian hospital. Med J Aust 142(2):108–111
Townsend DE, Bolton S, Ashdown N, Annear DI, Grubb WB (1986) Conjugative, staphylococcal plasmids carrying hitch-hiking transposons similar to Tn554: intra-and interspecies dissemination of erythromycin resistance. Aust J Exp Biol Med Sci 64(4):367–379
Tsvetkova K, Marvaud JC, Lambert T (2010) Analysis of the mobilization functions of the vancomycin resistance transposon Tn1549, a member of a new family of conjugative elements. J Bacteriol 192(3):702–713. https://doi.org/10.1128/JB.00680-09

Udo E, Townsend DE, Grubb WB (1987) A conjugative staphylococcal plasmid with no resistance phenotype. FEMS Microbiol Lett 40(2–3):279–283. https://doi.org/10.1111/j.1574-6968.1987.tb02039.x

Udo EE, Grubb WB (1990) Conjugal transfer of plasmid pWBG637 from Staphylococcus aureus to Staphylococcus epidermidis and Streptococcus faecalis. FEMS Microbiol Lett 72(1–2): 183–187. https://doi.org/10.1111/j.1574-6968.1990.tb03886.x

Udo EE, Wei M-Q, Grubb WB (1992) Conjugative trimethoprim resistance in Staphylococcus aureus. FEMS Microbiol Lett 97(3):243–248. https://doi.org/10.1111/j.1574-6968.1992.tb05470.x

Ul Haq I, Chaudhry WN, Akhtar MN, Andleeb S, Qadri I (2012) Bacteriophages and their implications on future biotechnology: a review. Virol J 9. https://doi.org/10.1186/1743-422X-9-9

Urushibara N, Aung MS, Kawaguchiya M, Kobayashi N (2019) Novel staphylococcal cassette chromosome mec (SCCmec) type XIV (5A) and a truncated SCCmec element in SCC composite islands carrying speG in ST5 MRSA in Japan. J Antimicrob Chemother dkz406. https://doi.org/10.1093/jac/dkz406

Varani A, He S, Siguier P, Ross K, Chandler M (2021) The IS6 family, a clinically important group of insertion sequences including IS26. Mob DNA 12:1). BioMed Central Ltd. https://doi.org/10.1186/s13100-021-00239-x

Viana D, Blanco J, Tormo-Más MÁ, Selva L, Guinane CM, Baselga R, Corpa JM, Lasa Í, Novick RP, Fitzgerald JR, Penadés JR (2010) Adaptation of Staphylococcus aureus to ruminant and equine hosts involves SaPI-carried variants of von Willebrand factor-binding protein. Mol Microbiol 77(6):1583–1594. https://doi.org/10.1111/j.1365-2958.2010.07312.x

Walters JA, Dyke KGH (2006) Characterization of a small cryptic plasmid isolated from a methicillin-resistant strain of Staphylococcus aureus. FEMS Microbiol Lett 71(1–2):55–63. https://doi.org/10.1111/j.1574-6968.1990.tb03798.x

Wang L, Archer GL (2010) Roles of CcrA and CcrB in excision and integration of staphylococcal cassette chromosome *mec* , a *Staphylococcus aureus* Genomic Island. J Bacteriol 192:3204–3212. https://doi.org/10.1128/JB.01520-09

Wikoff WR, Liljas L, Duda RL, Tsuruta H, Hendrix RW, & Johnson JE (2000) Topologically linked protein rings in the bacteriophage HK97 Capsid. https://www.science.org

Wilson LK, Coombs GW, Christiansen K, Grubb WB, O'Brien FG (2016) Characterization of a novel staphylococcal cassette chromosome composite island from community-associated MRSA isolated in aged care facilities in Western Australia. J Antimicrob Chemother 71: 3372–3375. https://doi.org/10.1093/jac/dkw317

Woodford N, Watson AP, Patel S, Jevon M, Waghorn DJ, Cookson BD (1998) Heterogeneous location of the mupA high4evel mupirocin resistance gene in staphylococcus aureus. *J Med Microbiol-Vbl* 47

Wu Z, Li F, Liu D, Xue H, Zhao X (2015) Novel type XII staphylococcal cassette chromosome *mec* harboring a new cassette chromosome recombinase, CcrC2. Antimicrob Agents Chemother 59: 7597–7601. https://doi.org/10.1128/AAC.01692-15

Xia G, Wolz C (2014) Phages of Staphylococcus aureus and their impact on host evolution. Infect Genet Evol 21:593–601. https://doi.org/10.1016/j.meegid.2013.04.022

Genomic Islands in *Pseudomonas* Species 12

Maytiya Konkit and Van Thai Than

Abstract

Pseudomonas is a Gammaproteobacteria, Gram negative which normally found in natural habitat such as soil, water and air. Especially, in marine habitat and environmental such as ocean, deep sea, coral reef and Hydrothermal vents, *Pseudomonas* also found in several species and play the important role in decomposition the macro-organic and inorganic substances and then utilized to soluble substances. Marine *Pseudomonas* is well-known as a halophilic-heterotroph bacteria owing to the ability to produce potential enzymes, for example, proteolytic enzyme and chitinase to degrade the biopolymer in marine environment and release the biomolecule which is used as the food nutrient for other marine organisms in food chain cycle. Some species of *Pseudomonas* were isolated from marine environmental origins; sea water, seashore, Antarctica and rainbow trout (*Oncorhynchus mykiss*) and they found to have the closely relationship which show in the phylogenetic tree analysis. Not only the intra genetic relationship within the marine *Pseudomonas* species but also the closely genetic interaction and relationship with other marine macroorganism such as sponges, coral reefs, fishes and algae in the term of symbiosis for the survival and growth supporting to each other. Many marine bioactive compounds; pyrrole (2,3,4-tribromo-5(1'hydroxy,2′,4′-dibromophenyl)), moiramides, and zafrin were

M. Konkit
Division of Microbiology, Faculty of Science and Technology, Nakhon-Pathom Rajabhat University, Nakhon-Pathom, Thailand

V. T. Than (✉)
Faculty of Applied Sciences, International School, Vietnam National University, Hanoi, Viet Nam

Center for Biomedicine and Community Health, International School, Vietnam National University, Hanoi, Viet Nam
e-mail: thaitv@vnuis.edu.vn

I. Mani et al. (eds.), *Microbial Genomic Islands in Adaptation and Pathogenicity*,
https://doi.org/10.1007/978-981-19-9342-8_12

found to produce from marine *pseudomonas* and can inhibit the growth of pathogenic bacteria; *Staphylococcus aureus*, *Streptococcus pyogenes*, and *Escherichia coli*.

Keywords

Pseudomonas · Marine environment · Genomic islands

12.1 Genus *Pseudomonas*

Pseudomonas is a bacteria that belongs to Phylum Proteobacteria, Class Gammaproteobacteria, Order Pseudomonadales, Family Pseudomonadaceae within Genus *Pseudomonas (*Euzéby 1997*)*. These bacteria are Gram negative, rod, or slightly curved rod shape, not helical, motile with one and several polar flagella. The size of the cell is around 0.5–1.0 × 1.5–5.0 μm (Palleroni 2015). They are non-spore forming bacteria, non-fermentative, catalase and oxidase positive, other biochemical characterization was showed in Table 12.1 (Bennasar et al. 1996; Stanier et al. 1966; Fiedler et al. 2022; Zhao et al. 2022; Kong et al. 2022; Hauth et al. 2022). Most species do not collect the reserve material in the form of polyhydroxybutyrate granules but produce polyhydroxyalkanoate with the higher monomer lengths of carbon in four atoms. They are aerobic and usually use oxygen rather than nitrate as a last electron acceptor in cellular respiration metabolism. The fatty acid components of this bacterial genus were the hydroxylated fatty acids, $C_{10:03OH}$, $C_{12:0}$, and $C_{12:0\ 2OH}$, and the major respiratory quinone was ubiquinone Q-9. The percentage of GC content is 58–69 (Palleroni 2015). Pigmentation is one typical characteristic of this genus because they frequently express the color as a normal cellular composition. Several species in this genus have the dark brown color

Table 12.1 Phenotypic characteristics of *Pseudomonas* species strains 1, *Pseudomonas aeruginosa* (Stanier et al., 1966); 2, *Pseudomonas canavaninivorans* (Hauth et al., 2022); 3, *Pseudomonas tumuqiensis* (Kong et al., 2022); 4, *Pseudomonas allivorans* (Zhao et al., 2022); 5, *Pseudomonas rustica* (Fiedler et al., 2022)

Characteristics	1	2	3	4	5
Hydrolysis of					
Gelatin	+	–	–	+	+
Starch	–	ND	–	ND	–
Utilization of					
Glucose	+	+	+	+	+
Maltose	–	+	+	–	–
Mannitol	+	+	+	+	+
Citrate	+	+	+	+	+
Denitrification	+	ND	+	–	–
% Mol G + C	67	61	63.5	59	58.9

ND not detected/not reported, +, positive; –, negative

colonies owing to the increase concentration of cytochrome c in their cells (Palleroni 2015). Six color pigments produced by the type species of this genus include with four phenazines; pyocyanine which produced by *Pseudomonas aeruginosa* a blue color act as a redox secondary metabolite (Lau et al. 2004), pyorubin a red color pigment which reported in patient of scalp infected with *Pseudomonas aeruginosa* (Rajyaguru et al. 2014), chlororaphin a crystal pigment of green and yellow found in *Pseudomonas chlororaphis* (Haynes and Rhodes 1962) and oxiphenazin (Jessen 1965). *Pseudomonas* blue protein; a blue copper protein from *Pseudomonas aeruginosa* (Tang et al. 1968) and pyoverdine a siderophore which can be complex with free ferric iron produced by *Pseudomonas syringae* (Jülich et al. 2001). The pigment profiles are important in morphological characterization and identification of *Pseudomonas* genus (Shelly et al. 1980). *Pseudomonas* normally grow in the minimal media with the addition of nitrate or ammonium ion as a nitrogen source and can use organic as a carbon source (Palleroni 2015). Sodium chloride (NaCl) has been found to be a requirement chemical element in the growth of *Pseudomonas elongata* (Anzai et al. 2000) and *Pseudomonas halophila* (Fendrich 1988). The NaCl requirement is the one technical usage to characterize the marine eukaryotic cell (Palleroni 2015). The poteintial of growth in the simple media at the various of organic compounds has bring to the broad nutritional representation of a huge number of strain, together with a phenotyping data is appropriate for taxonomic studies by statistical method (Palleroni 2015). The suitable growth temperature for almost species in this genus is 28 °C. Rare species can grow at 4 °C which can call as psychrophilic bacteria such as *Pseudomonas psychrotolerans* (Hauser et al. 2004) whereas some species grow at 55 °C; thermophilic bacteria such as *Pseudomonas thermotolerans (*Manaia and Moore 2002*)*. Many species are aerobe and are found to resist various numbers of antimicrobial agents. This topic is the specific medical importance since almost members of this group are well-known as an opportunistic human pathogen and frequently found to isolate from human clinical samples. The most clinical bacterial infection case reported found that *Pseudomonas aeruginosa* (Hugh and Leifson 1964) has been known as a human bacteria pathogen which related with cystic fibrosis (Finnan et al. 2004) and causing a various types of infections in reconciled hosts (Head and Yu 2004; Poirel et al. 2004). It can cause the infection disease in plants (Buysens et al. 1996; Vivanco 2004) and animals (Prevatt et al. 2004; Yeruham et al. 2004). The notable antibiotic using against the growth of *Pseudomonas* is the β-lactam (Smith et al. 1994). The antimicrobial resistance against *Pseudomonas* occurs due to the β-lactamase which is present in penicillinase (Furth, 1975) and inducible cephalosporinase (Sabath et al. 1965; Sykes and Matthew 1976). The antimicrobial enzyme activity production bring *Pseudomonas* can resist in many types of antibiotics by lowing in permeability in cell wall and the systematic efflux pumps (Strateva and Yordanov 2009).

12.2 Ecological and Marine Habitat of *Pseudomonas*

Ecological systems and habitats are the objects that are widely used for the study of the dissemination of microorganisms in the world. The various types of natural source habitats are recommended for screen and isolation of *Pseudomonas* species. The suitable condition is aerobe, adequate oxygen supply with a neutral organic pH, and mesophilic in temperature (Palleroni 2015). The species members of this genus are resisted in almost the main natural environments such as earth land, fresh water and marine as well as in the familiar connection with animals and plants (Spiers et al. 2000). *Pseudomonas aeruginosa* has been reported and found to isolate from the contaminated hospital distilled water (Favero et al. 1971). Some species of *Pseudomonas* are present in low amount on their ecology. A total of 57 strains of fluorescent-producing *Pseudomonas* species were isolated by (Den Dooren de Jong 1926) and found that 23 isolates have the main origin from soil and later identified as *Pseudomonas putida* whereas the other strains were isolated from water. In recent years, there have more available data with the respect to the habitats and suggested that the influent of fluorescent producing *Pseudomonas* species are growth rapidly and frequently found in plant rhizosphere, where they help to against other plant pathogenic bacteria and induce the effect on plant growth (Kloepper et al. 1980; Sarniguet et al. 1995).

In the center among three major biosphere habitats, 70% is the marine which is the most cover of earth's surfaces (Dinerstein et al. 2019). There is the largest habitat area for microorganism. Marine microorganisms are flourish not only in the surface water of the sea but also in the deep depths from coastal to the limit of the shore and from the normal oceanic to the special related places such as blue waters of coral reefs and gray to black vapor of hot thermal vents at the sea ground (Qasim 1998). Freshwater and marine origin were different in several character including saltiness, median temperature, dept. and nutrient component (Waisel 2012). Huge numbers of microorganisms in the mass of water basically point of the high nutrient level in the water (Redfield 1934). Freshwater sources are the are the plenty variable in the resources and suitable in growth supplements condition for microbial lived. The respiration of oxygen supply and demand of microorganism appear in the marine environment and the equanimity between photosynthesis and respiration reaction which control the natural circle of oxygen, carbon and other compounds (Aryal et al. 2015). The representative lake and pound are the examples to present the various zones and types of microbial diversity found in the freshwater. The littoral zone is concluded with the shore where consider a place for marine plant vegetation and the sunlight also infiltrate throughout this zone. The second, limnetic zone which position is next to the littoral zone compoes with the top of open water area far from the shore. This zone cover with the fullfill of oxygen and can found *Psedomonas*, *Cytophaga*, *Caulobacter* and *Hyphomicrobium* (Másmela-Mendoza and Lizarazo-Forero 2021). The profundal zone is the next zone deeper from limnetic zone. The last, the benthic zone is composed of sediment and located at the bottom.

The ocean is well known as a high-pressure cold storage with a deep down volume of 1100 meters at 3 °C with a vicinity of 1000 atm (Prescott et al. 2002). However, ocean estuaries are one the place of marine environments that have higher nutrient levels and composed of huge numbers of microorganisms than other shoreline environments (Tortoraand and Funke 2008). On the other hand, low nutrient concentration of water brings the microorganism grow on static surfaces and in the form of particle. In this reason, microorganism can enter to the nutrient than freely suspended (Aryal et al. 2015). *Pseudomonas aeruginosa* is an example of *Pseudomonas* species which was isolated from open ocean environments (Khan et al. 2007). This species can survived at higher concentrations of NaCl and was at least resistant to 7 antibiotics; cefazolin, cefuroxime sodium, cefpodoxime-proxetil, ampicillin, amoxicillin-clavulanate, ampicillin-sulbactam, and tetracycline. Interestingly, there have other *Pseudomonas* species names; *Pseudomonas marincola* YsY11 and *Pseudomonas oleovorans* T9AD which are found to isolate from marine sediment of Pacific Ocean (Wang et al. 2021a).

Deep sea level is amount approximately 92% of the sea surface and covered with 60% of water and more than 2000 meters deep. The deep-sea microorganisms, especially heterotrophs take the advantage of plenty of organic carbon or soluble organic matter at maximum concentration (Jannasch 1967). *Pseudomonas profundi* is the one species of the genus *Pseudomonas* that was isolated from deep-sea water in the western Pacific Ocean. This strain can utilize several organic carbon sources such as glutamic acid, lactic acid, propionic acid, and acetic acid (Sun et al. 2018). *Pseudomonas bauzanensis* DN13–1 is a heterotrophic nitrifying-aerobic denitrifying bacterium isolated from deep-sea sediment. This strain is capable to removing nitrogen in the form of NH_4^+, NO_3^-, and NO_2^- through nitrification and denitrification cycles (Zhang et al. 2020). *Pseudomonas stutzeri* 273 is a strong mercuric-resistant marine bacteria isolated from the deep-sea sediment in the East China Sea (Zheng et al. 2021). *Pseudomonas nanhaiensis* is a lipase-producing bacteria isolated from deep-sea sediment of the South China Sea (Pang et al. 2021). Lipase production by genus *Pseudomonas* can apply in food technology and chemical industry (Pandey et al. 1999).

Coral reefs area are the reatricted to depthless-water ecosystem in the sea and mean as a splendid and multiplex communities of marine microorganism. These organisms can build, modify, or retain the shore environment by the development of $CaCO_3$ structure (Das et al. 2006). Reef sediments are the main sources of nutrients such as phosphorus and nitrogen utilized by bacteria. Coral creates mucus that acts an important role in reef metabolism as a source of nutrient and promotes bacterial activity (Richman et al. 1975). The coral mucus comprises polysaccharides and protein (Meikle et al. 1988) which provides beneficial growth substances for microorganism. *Pseudomonas* were found to be the one bacterial community related to the Caribbean coral "*Montastraea franksi*" in the 39% proportion from all culture isolates characterized by 16S rDNA sequencing (Rohwer et al. 2001).

Hydrothermal vents are individual habitats that gives limited or complete energy and nutrient flows essential to supporting the various microbial communities that are allocated together with the temperature range and decreased compound gradients corresponding with the transformation from anaerobic and aerobic conditions (Canganella 2001). The main microorganism that supports the animals living around volcanic warm vents is chemoautotrophic microorganisms (Cambon-Bonavita et al. 2002). Twenty one whole genome sequencing of *Pseudomonas* strains isolated from an active Kolumbo submarine volcanic hydrothermal vent at Greece show the genes which couple provide the adaptation in stress environmental condition and antibiotic multi-drug resistant (Bravakos et al. 2021).

Cold marine environment is the habitat that depended on the temperature in ranges from 4 to 5 °C. The temperature is reduced and changed with the dept. (Poli et al. 2017). Psychrophile is the cold-liking lived microorganism that grows at temperatures between 0 and 20 °C and has adaptation to low nutrients (Siddiqui et al. 2013). *Pseudomonas* is one notable genus of Psychrophilic microorganisms such as *Pseudomonas antarctica*, *Pseudomonas meridiana*, and *Pseudomonas proteolytica* (Reddy et al. 2004). The adaptation appliance of Psychrophile in cold environments is due to the single-stranded RNA which is well known as "cold shock protein" (Csp) (Jones and Inouye 1994; Kawahara et al. 2001; Cavicchioli et al. 2002).

12.3 Marine *Pseudomonas* Diversity and Important Role

The new finding of marine bacteria including *Pseudomonas* has noticeably increasing in aspiration throughout the decade with the breakthrough of high output sequencing analysis (Sunagawa et al. 2015). The diversity spreading among bacterial species are unequal with a small amount of dominant strains reported by using conventional sequencing and culturable methods (Pedrós-Alió 2012). Due to the innovator (Sogin et al. 2006), a dramatic number of reports show the infrequent portion of marine microorganism diversity over huge location and temporal scale (Szabó et al. 2007; Elshahed et al. 2008; Youssef and Elshahed 2009; Vergin et al. 2013). The study of marine microorganism diversity is necessary to aim to understand the relationship between community and figure of distribution. In marine diversity, the dominant microorganism is Gram-negative bacteria which has a proportion of 90% and unique character (ZoBell 1946). The Gram negative can flourish in drastic ocean and better to survive because of the composition of cell wall. The component of Gram negative cell wall; Lipopolysaccharide (LPS) is composed of three parts: O antigen, lipid A, and a core region which act as a trigger for the immune system and are related to causing the disease in marine organisms (Anwar and Choi 2014).

The salinity marines are the natural home for a huge sort of halophilic and halotolerant microorganisms, especially bacteria that adapt to growth over the various concentration of salt (Poli et al. 2017). The diversity of halophilic bacteria was reported to isolate from sea sand, sea algae, and sea sediment (Menezes et al. 2010). The culture-dependent study in Santa Pola saltern as a model of hypersaline environments (the sea salt concentration of 15% and 30%) found that tolerable halophilic bacteria and some haloarchaea are leading as heterotrophs. The first reported genusesus were *Pseudomonas-Alteromonas-Alcaligenes* group and they grew in the appearance of 10–30% salt concentration (Ventosa et al. 2014). In the hypersaline condition, *Pseudomonas* the succed special protein that are steady and active in the existence of salts (Edbeib et al. 2017).

Pseudomonas was mentioned as a halophilic bacteria and leading as a heterotroph in marine environments (Fendrich 1988; Ventosa et al. 2014). The heterotrophic marine bacteria have been modified themselves to live in the extreme marine environmental such as high rise of pressure, saltiness and cold temperature (Das et al. 2006). This heterotroph plays an important role to stimulate the organic decomposition and mineralization step in sediments and deliver the solute organic and inorganic substances (Arndt et al. 2013). *Pseudomonas* can produce the proteolytic enzyme to decompose the protein cycle in the water and obey as a necessary food for various marine organisms (Rheinheimer 1992). Chitin, the second plentiful polysaccharide in the environmental is found to manufacture by various marine organism such as cell wall material of algae and chlorophyte (Mulisch 1993), exoskeleton structure of copepods and marine invertebrates (Gooday 1990). Although, this biopolymer is hard to degrade (Kirchman and White 1999) as the study on fossils sustained with chitin (Stankiewicz et al. 1997). *Pseudomonas* has been reported to produce chitinase enzyme which helps to degrade chitin in marine environments (Das et al. 2006). Belonging to Gram negative bacteria and act as a heterotroph and halophilic, *Pseudomonas* genus was found in the various marine environment and has the interesting in the diversity.

In the current time of writing found that *Pseudomonas* has the 496 species and 23 subspecies (Parte 2018). We found that 40 out of 496 species were isolated from marine environments such as sea surface, seawater, seashore, sediment, fish, green algae, and sponge. In this study, the 16S rRNA gene sequence of 40 marine-originated strains were obtained from the EzTazon-e server (https://www.ezbiocloud.net/taxonomy) (Kim et al. 2012). The evolutionary distances between the *Pseudomonas* strains were created by using the neighbour-joining method (Jukes and Cantor 1969). The phylogenetic tree was reconstructed using the neighbour-joining method (Saitou and Nei 1987). Clade and branch were supported in the neighbour-joining tree using the resampling (1000 replicated) bootstrap method (Felsenstein 1985). The phylogenetic tree shows the relationship between *Pseudomonas* species that were isolated from marine environment (Fig. 12.1). Interestingly, found that the *Pseudomonas* strain which was isolated from a similar marine environment has a close relationship and belongs to the same clade of the phylogenetic tree. *Pseudomonas litoralis* 2SM5^T (Pascual et al. 2012) and *Pseudomonas neustonica* SSM26^T (Jang et al. 2020) were both found to be isolated from seawater. *Pseudomonas pohangensis* DSM 17875^T (Weon et al. 2006) and *Pseudomonas taeanensis* MS-3^T (Lee et al. 2010) were also isolated from seashore. *Pseudomonas arcuscaelestis* V1, *Pseudomonas tructae* SNU WT1^T, *Pseudomonas piscium* P50^T and *Pseudomonas mucoides* P154a^T were found to isolate from rainbow trout (*Oncorhynchus mykiss*) as well as *Pseudomonas piscium* P50^T and *Pseudomonas mucoides* P154a^T which found to isolated from rainbow trout in same time same condition by (Duman et al. 2021a). Moreover, in the clade that contained *Pseudomonas gregormendelii* CCM 8506 (Kosina et al. 2016), *Pseudomonas Prosekii*

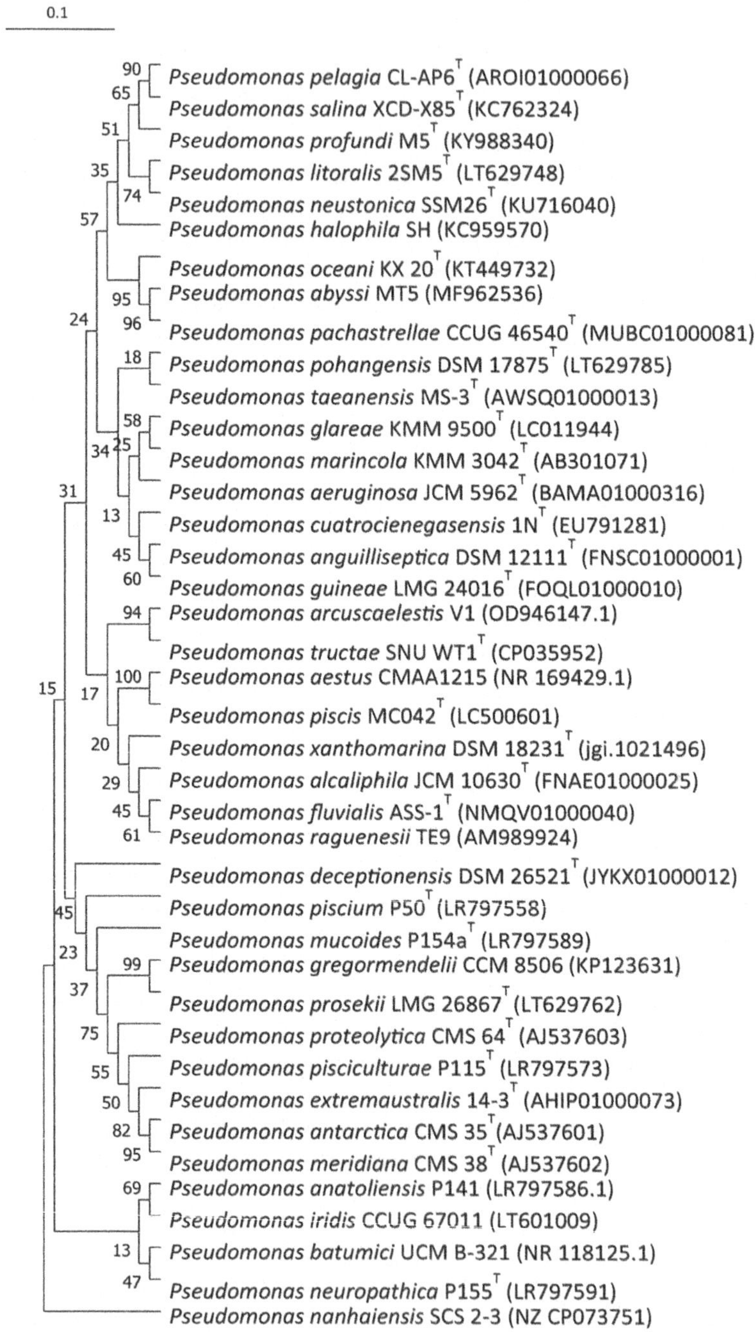

Fig. 12.1 Neighbour-joining phylogenetic tree based on nearly complete 16S rRNA gene sequences from *Pseudomonas* species which were isolated in marine environment. Bootstrap values are shown based on a neighbour-joining analysis of 1000 resampled datasets. Bar represents 0.1 substitutions per nucleotide position

LMG 26867^T (Kosina et al. 2013), *Pseudomonas proteolytica* CMS 64^T (Reddy et al. 2004), *Pseudomonas extremaustralis* 14-3^T (López et al. 2009), *Pseudomonas antarctica* CMS35^T, and *Pseudomonas meridiana* CMS 38^T (Reddy et al. 2004) have the close relationship and all strains were isolated from Antarctica marine environment.

12.4 *Pseudomonas* and Marine Macroorganisms

Normally, bountifulness and the appearance scheme surveyed of microorganisms is different from the traditional relationship that prevails for macroorganism (Gaston et al. 2000). The transition of microorganisms from uncommonness to dominant was driven by the variable environment, habitat differentiation, or random events (Shade and Gilbert 2015; Shade et al. 2014; Lynch and Neufeld 2015). However the important role of rare microorganism sitll not known yet, the documentation repoted can support that the long to short term ecological habitat condition could help them grow and along with communities (Shade and Gilbert 2015; Alonso-Sáez et al. 2014; Aanderud et al. 2015). Interestingly, the impact of macroorganism–microorganism cooperation (McFall-Ngai et al. 2013; McFall-Ngai 2015) has an effect on the ecological and energetic of the rare microbial taxa, and some reports suggest that macroorganisms could perform as a specific habitat for rare microorganisms in marine environments (Sunagawa et al. 2015; Frias-Lopez et al. 2002; Taylor et al. 2004; Hao et al. 2015; Weiland-Bräuer et al. 2015). Most of the studied reported stated that the microorganism and macroorganism interaction were pointed on theirrelationship in diets, development and survival rate (McFall-Ngai 2015; Nayak 2010a; Nayak 2010b; Clements et al. 2014). Marine macroorganism (host associated microbiota) plays the important role in marine bacterial biodiversity (microbiome) such as be able to advantageous the environmental supporting the growth of infrequent marine microorganism and perform as a habitat for them (Troussellier et al. 2017). Marine host-associated microbiota and microbiome reaction in the marine environment may refer to a symbiosis situation which means marine host and microbiome are associated in the metabolism and immunity interaction (Apprill 2017).

Fish is the one dominant microorganism host found in the marine environment. The interactions between fish and microbiome are almost studied in the intestinal, skin, and mucus which was concentrated on the advantageous point for fish (Nayak 2010a, b; Clements et al. 2014). For example, microbiome in the skin and mucus of the killifish help the host resistant to the change in marine environment (Apprill 2017; Larsen et al. 2015). The highest abundance of the microbiome in the gastrointestinal of fish is around 800 OTUs (Givens et al. 2015). There are several reports that various *Pseudomonas* species were isolated from fish, especially in rainbow trout (*Oncorhynchus mykiss*) such as *Pseudomonas arcuscaelestis (*Mulet et al. 2021*)*, *Pseudomonas tructae* (Oh et al. 2019), *Pseudomonas piscium* (Duman et al. 2021a), *Pseudomonas mucoides* (Duman et al. 2021a), *Pseudomonas pisciculturae* (Duman et al. 2021a), and *Pseudomonas neuropathica* (Duman et al.

2021a). For the other fishes, *Pseudomonas piscis* was found to be isolated from Murray cod (*Maccullochella peelii peelii*) (Liu et al. 2020). The studied report by (Duman et al. 2021b) found that *Pseudomonas anatoliensis* and *Pseudomonas iridis* were isolated from fish farms in Turkey during 2012–2018. *Pseudomonas anguilliseptica* is the one of *Pseudomonas* species that was isolated from eels (*Anguilla japonica*) (Wakabayashi and Egusa 1972; Wiklund and Bylund 1990).

Sponge is a marine organism which has been reported to the living in symbiosis type with the bacteria, especially in the family of Gammaproteobacteria (Taylor et al. 2007; Hentschel et al. 2012; Schmitt et al. 2007). The study by Vacelet and Donadey (1977) showed that the total amount of sponge weight comprised 38% weight of bacteria cells. The sponge gives the appropriate condition for the bacteria's growth and ability (Webster et al. 2010). The bacteria (microbiome) involve in the nitrogen and phosphorus cycles (Bayer et al. 2008; Radax et al. 2012) and facilitate sponges to survive in phosphorus decreasing marine habitats (Colman 2015). The microbiome is closely relate with the sponge in the term of secondary metabolite manufacture and develop the defense oppose the predator of sponge host (Taylor et al. 2007). The highest abundance of the microbiome related to sponge is around 17,800 OTUs which was higher than in fish (Reveillaud et al. 2014). *Pseudomonas pachastrellae* is the only one of the *Pseudomonas* species that was reported to isolate from marine sponge (Romanenko et al. 2005).

Algae is well-known as a primary producer in the food chain in marine environments due to its photosynthesis reaction (Field et al. 1998). Algae and bacteria can live together and have a symbiosis relationship, which can call algae as phytoplankton and bacteria as bacterioplankton (Sarmento and Gasol 2012). The plently of the studies reported that bacterioplankton play an important role in the growth promotion to phytoplankton (Amin et al. 2015; Gonzalez et al. 2000; Kim et al. 2014; Seyedsayamdost et al. 2011). Bacterioplankton decays the marine organic substances and supports plant growth promotion through the nutrient mechanism (Philippot et al. 2013). The study by (Croft et al. 2005) showed that bacterioplankton provides vitamin B12 to phytoplankton and phytoplankton gives carbon fixation back to bacterioplankton. Phytoplankton does not have nitrogen fixation on its own but can supply organic carbon to bacterioplankton although, bacterioplankton gives back the dissolved organic nitrogen and carbon (Cho et al. 2015). The previous studied confirmed that bacterioplankton can transfer the important gene that help the phytoplankton to resist with the extreme environmental (Schönknecht et al. 2013). The bacterioplankton associated with phytoplankton showed a pretty high levels with 17,779 OTUs (Brodie et al. 2016). *Pseudomonas pelagia* is one of genus *Pseudomonas* that was isolated from Antarctic green algae (*Pyramimonas gelidicola*) (Hwang et al. 2009).

Coral reefs are recognized as a role model of macroorganism–microorganism interaction in the marine environment because of the hosting photosynthesis reaction which provides the organic nutrient (Muscatine et al. 1981). The microbiome can release reactive oxygen species (ROS) which helps to protect the coral reef from the stress of high temperature and light (Diaz et al. 2016; Zhang et al. 2015). Mucus Coral microbiome showed a high abundance in OTUs but a lower sponge

(Sunagawa et al. 2015). The reported studied from (Carlos et al. 2013) found that bacteria in genus *Sphingomonas*, *Pseudomonas* with some family Gammaproteobacteria have the interaction with coral reefs especially in mucus organ.

12.5 Bioactive Compounds and Enzyme Kinetics Production Isolated from Marine *Pseudomonas*

Currently, there are various studies and reports of bioactive compounds and enzyme kinetics production which were isolated from marine *Pseudomonas*. Several species of marine *Pseudomonas* species are famous and play a significant role in bioactive compound production (Isnansetyo and Kamei 2009). However in the current time of writing, there has few of studies report of bioactive compounds isolated from marine *Pseudomonas*. Marine environment is the one of large natural habitats for *Pseudomonas* diversity thus, some unknown bioactive compounds from marine *Pseudomonas* have not been reported yet.

Pseudomonas bromoutilis is the first marine *Pseudomonas* that was reported to produce the bioactive compound and this strain was isolated from *Thalassia* from tropical water in the region of Puerto Rico (Burkholder et al. 1966). The antibiotic, pyrrole (2,3,4-tribromo-5(1'hydroxy,2′,4′-dibromophenyl)) was produced from this strain and has the ability against the growth of pathogenic bacteria and fungi such as *Staphylococcus aureus*, *Diplococcus pneumoniae*, *Streptococcus pyogenes*, and *Candida albicans* (Burkholder et al. 1966). *Pseudomonas* sp. 102–3 which was isolated from seawater from La Jolla, California tide poll found to have three antimicrobial activity compounds: 4-hydroxy-benzaldehyde, 2-n-heptyl-4-quinolinol, and 2-n-pentyl-4-quinolinol against the growth of *Vibrio anguillarum*, *Vibrio harveyi*, *Staphylococcus aureus*, and *Candida albicans* (Wratten et al. 1977). *Pseudomonas fluorescens* is the one of marine *Pseudomonas* which was isolated from the surface of tunicates. This strain can produce three distinctive compounds: moiramides A,B,C and andrimid which are the new member of antimicrobial activity called pseudopeptide pyrrolidinedione (Needham et al. 1994). This pyrrolidinedione has antimicrobial activity against the growth of pathogenic bacteria: *Staphylococcus aureus*, *Streptococcus pneumoniae*, and *Escherichia coli* (Freiberg et al. 2006). *Pseudomonas stuzeri* CMG1030 found in the marine habitat can produce the new antimicrobial compound name zafrin which has the ability to inhibit the growth of *Staphylococcus aureus* and *Salmonella typhi* (Uzair et al. 2008).

Pseudomonas sp. 1531-E7 can produce quinolones which was found to inhibit the growth of herpes simplex virus-1 and Gram-positive bacteria; *Staphylococcus aureus* and this strain was isolated from *Homophymia* sp., a marine sponge (Bultel-Poncé et al. 1999) as well as *Pseudomonas aeruginosa* which also found to isolate from *Isodictya setifera*, an Antarctic sponge have ability to produce the antimicrobial substances: diketopiperazines and phenazine alkaloids which have the ability against the growth of *Bacillus cereus*, *Micrococcus luteus*, and *Staphylococcus aureus* (Jayatilake et al. 1996). Not only *Pseudomonas* which was isolated from the sponge but also *Pseudomonas* sp. PB2 associated with *Suberites domuncula*, a sponge has an extraction compound which acts as hemolytic, cytotoxic, and antimicrobial activity to inhibit the growth of *Staphylococcus aureus*, *Staphylococcus epidermidis*, *Escherichia coli*, and *Candida albicans* (Tamegai et al. 1997). *Pseudomonas* sp. WAK-1 is the marine *pseudomonas* that was isolated from *Undaria pinnatifida*, brown seaweed. This strain has a sulfated polysaccharide compound against the growth of herpes simplex virus-1 and is anticancer (Matsuda et al. 1999). *Pseudomonas* sp. A6–5 which was isolated from the Antarctic near the Chinese Great wall station has the ability to produce the novel bioactive metabolite (Liu et al. 2022).

Not only the bioactive compound is against the growth of pathogenic bacteria but also the plant growth-promoting activity and whitening agent produced from marine *Pseudomonas*. Plant growth-promoting bacteria (PGPB) is the group of bacteria that can support the growth of the plant in straight and incidental ways. In a straight way, PGPB help to release the plant hormone in the form of indole acetic acid (IAA), gibberellins, auxins, and cytokines (Patten and Glick 2002). Although, PGPB also helps in nitrogen fixation (Kennedy and Tchan 1992) and phosphate solubilization in plant (Richardson 2001; Banerjee and Yasmin 2002). On the other hand in an incidental way, PGPB can help to produce iron chelators, siderophores, and cyanides (Glick 2012; Ahmad et al. 2008). *Pseudomonas fluorescens* and *Pseudomonas aeruginosa* are the two species of *Pseudomonas* that have been reported to be the PGPB (Ganeshan and Manoj 2005). *Pseudomonas* spp. Olive green (OG) strain was isolated from the seawater of the Gulf of Khambhat and can produce ammonia, IAA, phosphate solubilization, siderophore and hydrocyanic acid to stimulate plant growth (Goswami et al. 2013). The study suggested the advantage of marine *Pseudomonas* as a potential source of new skin-whitening agents. *Pseudomonas* sp. that was isolated from sandbar near Gangwha Island in Korea has the potential to decrease the pigmentation of Melan-a cells, human melanocytes, skin culturable, and in vivo zebrafish (Kang et al. 2011).

Pseudomonas species have been observed to produced enzyme to reduce the chemical reaction in cell in the term of surviving in the extreme marine environmental. Psychrophilic pseudomonas; *Pseudomonas putida*, which was isolated from the Arctic continent, can produce the antifreeze protein (AFPs) (Kawahara et al. 2001) which has important use in industrial production (Cavicchioli et al. 2002; Feller and Gerday 2003). *Pseudomonas* sp. ID1 is the cold modified microorganism from a marine sediment in Antarctica which can produce exopolysaccharide; glucose, galactose, and fructose (Matsuyama et al. 2015; Carrión et al. 2015). This EPS created a high amount of emulsion and cryoprotection further used as an important role of cold protectant in the food and pharmaceutical industries (Carrión et al. 2015).

12.6 Pathogenicity of *Pseudomonas* sp. in Marine Habitats

Despite the fact that *Pseudomonas* infection has been studied since the nineteenth century, its pathogenic mechanisms still remain limited. The *Pseudomonas aeruginosa (P. aeruginosa)*, a major pathological species of the genus *Pseudomonas*, is known as an important opportunistic pathogen cause of human healthcare related to acute and chronic localized infections (Doring et al. 1971; Jurado-Martin et al. 2021). Among the major nosocomial pathogens, hospitalized patients infected with *P. aeruginosa* are associated with high morbidity and mortality, especially those who are neutropenia and immunocompromised (Kang et al. 2003; CDC 2019). There are major numbers of virulence factors, including the cell-associated and extracellular virulence factors, that have been determined to better understand the pathogen and host interactions, as well as to improve the treatment outcomes of these infections (Doring et al. 1971; Jurado-Martin et al. 2021; Naik et al. 2021; Prasad et al. 2020; Rigane et al. 2020; van't Wout et al. 2015).

The cell-associated virulence factors are included the epithelial cell and mucin adherence factors, and the LPS-associated pathogenesis endotoxin factors (Ramphal et al. 1996; Azghani et al. 2002). The extracellular virulence factors that are produced via the type I (alkaline protease), type II (ETA, lipases, LasA and LasB elastases, protease IV, phospholipase C), and type III (exoenzyme S, –T, and -Y; and exotoxin U) secretion systems of the bacterium (Jurado-Martin et al. 2021; Van Delden 2004). Normally, the main pathogenesis symptoms of the disease such as tissue necrosis, vascular system dissemination and inflammatory responses et al., have been known available by the activations between these different virulent factors (Naik et al. 2021).

Until recently, there are very few *Pseudomonas* species that were isolated from marine environments. The *P. aeruginosa* were isolated from open ocean environments (Khan et al. 2007; Kimata et al. 2004), in which the *P. profundi, P. marincola, P. bauzanensis, P. stutzeri, P. nanhaiensis*, and *P. oleovorans* were isolated from deep-sea water or marine sediment of Pacific Ocean (Sun et al. 2018; Zheng et al. 2021; Pang et al. 2021; Wang et al. 2021b; Zhang et al. 2019). The *P. aeruginosa* that were isolated from open ocean were known for better survival and suggesting that marine environments could be a potential reservoir of *P. aeruginosa (*Khan et al. 2007*)*. Therefore, more extensive studies about the characterization of the origin, physiology, genetic exchanges, and the ability of human infection of these isolates are needed.

12.7 Future Perspectives

Genus *Pseudomonas* is the one of rare marine microorganism found to help in the decomposition of organic substances for other organisms to use as a food and growth activity. The symbiosis interaction system between *Pseudomonas* and other organisms such as coral reef, algae and fish help us to understand the potential and beneficial of *Pseudomonas* in marine environmental and indicated that those marine environmental are plenty of the organic substances and abundant of marine organisms. Moreover, marine *Pseudomonas* found to play an important role to produce novel bioactive compounds which can inhibit the growth of pathogenic bacteria. Not only the production of novel bioactive compounds from marine *Pseudomonas* but also the enzyme production which is important use in industrial production. The discovery of novel marine *Pseudomonas* in the future will bring new sources of bioactive compounds and enzymes which can be used in medical and industrial applications.

References

Aanderud ZT et al (2015) Resuscitation of the rare biosphere contributes to pulses of ecosystem activity. 6:24

Ahmad F, Ahmad I, Khan MJMR (2008) Screening of free-living rhizospheric bacteria for their multiple plant growth promoting activities. 163(2):173–181

Alonso-Sáez L et al (2014) Winter bloom of a rare betaproteobacterium in the Arctic Ocean. 5:425

Amin S et al (2015) Interaction and signalling between a cosmopolitan phytoplankton and associated bacteria. 522(7554):98–101

Anwar MA, Choi SJMD (2014) Gram-negative marine bacteria: structural features of lipopolysaccharides and their relevance for economically important diseases. 12(5):2485–2514

Anzai Y et al (2000) Phylogenetic affiliation of the pseudomonads based on 16S rRNA sequence. Int J Syst Evol Microbiol 50(4):1563–1589

Apprill AJFIMS (2017) Marine animal microbiomes: toward understanding host–microbiome interactions in a changing ocean. 4:222

Arndt S et al (2013) Quantifying the degradation of organic matter in marine sediments: a review and synthesis. 123:53–86

Aryal S, Karki G, Pandey SJNJOB (2015) Microbial diversity in freshwater and marine environment. 3(1):68–70

Azghani AO et al (2002) Pseudomonas aeruginosa outer membrane protein F is an adhesin in bacterial binding to lung epithelial cells in culture. Microb Pathog 33(3):109–114

Banerjee MR, Yasmin L (2002) Sulfur oxidizing rhizobacteria: an innovative environment friendly soil biotechnological tool for better canola production. Proceeding of AGROENVIRON. Cairo, Egypt, pp 1–7

Bayer K, Schmitt S, Hentschel UJEM (2008) Physiology, phylogeny and in situ evidence for bacterial and archaeal nitrifiers in the marine sponge Aplysina aerophoba. 10(11):2942–2955

Bennasar A et al (1996) 16S rRNA gene sequence analysis relative to genomovars of pseudomonas stutzeri and proposal of Pseudomonas balearica sp. nov. 46(1):200–205

Bravakos P et al (2021) Genomic adaptation of Pseudomonas strains to acidity and antibiotics in hydrothermal vents at Kolumbo submarine volcano, Greece. 11(1):1–12

Brodie J et al (2016) Characterising the microbiome of Corallina officinalis, a dominant calcified intertidal red alga. 92(8):fiw110

Bultel-Poncé V et al (1999) Metabolites from the sponge-associated Bacterium pseudomonas species. 1:4

Burkholder PR, Pfister RM, Leitz FHJAM (1966) Production of a pyrrole antibiotic by a marine bacterium. 14(4):649–653

Buysens S et al (1996) Involvement of pyochelin and pyoverdin in suppression of Pythium-induced damping-off of tomato by Pseudomonas aeruginosa 7NSK2. Appl Environ Microbiol 62(3): 865–871

Cambon-Bonavita MA et al (2002) A novel polymer produced by a bacterium isolated from a deep-sea hydrothermal vent polychaete annelid. 93(2):310–315

Canganella FJAOM (2001) Hydrothermal vent ecosystems and representative hyperthermophilic microorganisms. 51(1):11–28

Carlos C, Torres TT, Ottoboni LMJSR (2013) Bacterial communities and species-specific associations with the mucus of Brazilian coral species. 3(1):1–7

Carrión O, Delgado L, Mercade EJCP (2015) New emulsifying and cryoprotective exopolysaccharide from Antarctic Pseudomonas sp. ID1. 117:1028–1034

Cavicchioli R et al (2002) Low-temperature extremophiles and their applications. 13(3):253–261

CDC (2019) Antibiotic resistance threats in the United States. Centers for Disease Control and Prevention

Cho D-H et al (2015) Enhancing microalgal biomass productivity by engineering a microalgal–bacterial community. 175:578–585

Clements KD et al (2014) Intestinal microbiota in fishes: what's known and what's not. Wiley Online Library

Colman AS (2015) Sponge symbionts and the marine P cycle. Proc Natl Acad Sci U S A 112(14): 4191–4192

Croft MT et al (2005) Algae acquire vitamin B12 through a symbiotic relationship with bacteria. 438(7064):90–93

Das S, Lyla P, and Khan SAJCS (2006) Marine microbial diversity and ecology: importance and future perspectives. 1325–1335

Den Dooren de Jong LE (1926) Bijdrage tot de kennis van het mineralisatieproces. Rotterdam Nijgh & Van Ditmr's Uitgevers-Mij MAMXXVI

Diaz JM et al (2016) Species-specific control of external superoxide levels by the coral holobiont during a natural bleaching event. 7(1):1–10

Dinerstein E et al (2019) A global deal for nature: guiding principles, milestones, and targets. 5(4): eaaw2869

Doring G et al (1971) Virulence factors of Pseudomonas aeruginosa. Antibiot Chemother 1987(39): 136–148

Duman M et al (2021a) *Pseudomonas piscium* sp. nov., *Pseudomonas pisciculturae* sp. nov., *Pseudomonas mucoides* sp. nov. and *Pseudomonas neuropathica* sp. nov. isolated from rainbow trout. 71(3):004714

Duman M et al (2021b) Pseudomonas anatoliensis sp. nov and Pseudomonas iridis sp. nov. isolated from fish. 44(3):126198

Edbeib MF et al (2017) In silico characterization of a novel dehalogenase (DehHX) from the halophile Pseudomonas halophila HX isolated from Tuz Gölü Lake, Turkey: insights into a hypersaline-adapted dehalogenase. 67(5):371–382

Elshahed MS et al (2008) Novelty and uniqueness patterns of rare members of the soil biosphere. 74(17):5422–5428

Euzéby JP (1997) List of bacterial names with standing in nomenclature: a folder available on the internet. Int J Syst Evol Microbiol 47(2):590–592

Favero M et al (1971) Pseudomonas aeruginosa: growth in distilled water from hospitals. 173(3999):836–838

Feller G, Gerday CJNRM (2003) Psychrophilic enzymes: hot topics in cold adaptation. 1(3): 200–208

Felsenstein JJE (1985) Confidence limits on phylogenies: an approach using the bootstrap. 39(4): 783–791

Fendrich C (1988) *Halovibrio variabilis* gen. nov. sp. nov., *Pseudomonas halophila* sp. nov. and a new halophilic aerobic coccoid Eubacterium from Great Salt Lake, Utah, USA. Systematic and Applied Microbiology 11(1):36–43

Fiedler G et al (2022) Pseudomonas rustica sp. nov., isolated from bulk tank raw milk at a German dairy farm. 72(4):005288

Field CB et al (1998) Primary production of the biosphere: integrating terrestrial and oceanic components. 281(5374):237–240

Finnan S et al (2004) Genome diversity of Pseudomonas aeruginosa isolates from cystic fibrosis patients and the hospital environment. J Clin Microbiol 42(12):5783–5792

Freiberg C et al (2006) Novel bacterial acetyl coenzyme A carboxylase inhibitors with antibiotic efficacy in vivo. 50(8):2707–2712

Frias-Lopez J et al (2002) Partitioning of bacterial communities between seawater and healthy, black band diseased, and dead coral surfaces. 68(5):2214–2228

Furth AJ (1975) Purification and properties of a constitutive beta-lactamase from *Pseudomonas aeruginosa* strain Dalgleish. Biochim Biophys Acta 377(2):431–443

Ganeshan G, Manoj Kumar A (2005) Pseudomonas fluorescens, a potential bacterial antagonist to control plant diseases. J Plant Interact 1(3):123–134

Gaston KJ et al (2000) Abundance–occupancy relationships. 37:39–59

Givens CE et al (2015) Comparison of the gut microbiomes of 12 bony fish and 3 shark species. 518:209–223

Glick BRJS (2012) Plant growth-promoting bacteria: mechanisms and applications. 2012

Gonzalez LE, Bashan YJA, Microbiology E (2000) Increased growth of the microalga Chlorella vulgaris when coimmobilized and cocultured in alginate beads with the plant-growth-promoting bacterium Azospirillum brasilense. 66(4):1527–1531

Gooday GW (1990) *The ecology of chitin degradation*, in *Advances in microbial ecology*, Springer. p. 387–430

Goswami D et al (2013) Plant growth promoting potentials of Pseudomonas spp. strain OG isolated from marine water. 8(4):281–290

Hao W et al (2015) Bacterial communities associated with four ctenophore genera from the German Bight (North Sea). 91(1):1–11

Hauser E, Kämpfer P, Busse H-J (2004) Pseudomonas psychrotolerans sp. nov. Int J Syst Evol Microbiol 54(5):1633–1637

Hauth F et al (2022) Pseudomonas canavaninivorans sp. nov., isolated from bean rhizosphere. 72(1):005203

Haynes WC, Rhodes LJ (1962) Comparative taxonomy of crystallogenic strains of Pseudomonas aeruginosa and Pseudomonas chlororaphis. J Bacteriol 84(5):1080–1084

Head NE, Yu H (2004) Cross-sectional analysis of clinical and environmental isolates of Pseudomonas aeruginosa: biofilm formation, virulence, and genome diversity. Infect Immun 72(1): 133–144

Hentschel U et al (2012) Genomic insights into the marine sponge microbiome. 10(9):641–654

Hugh R, Leifson E (1964) The proposed neotype strains of Pseudomonas aeruginosa (Schroeter 1872) Migula 1900. Int J Syst Evol Microbiol 14(2):69–84

Hwang CY et al (2009) Pseudomonas pelagia sp. nov., isolated from a culture of the Antarctic green alga Pyramimonas gelidicola. 59(12):3019–3024

Isnansetyo A, Kamei Y (2009) Bioactive substances produced by marine isolates of Pseudomonas. J Ind Microbiol Biotechnol 36(10):1239–1248

Jang GI et al (2020) Pseudomonas neustonica sp. nov., isolated from the sea surface microlayer of the Ross Sea (Antarctica). 70(6):3832–3838

Jannasch HW (1967) Growth of marine bacteria at limiting concentrations of organic carbon in seawater 1. Limnol Oceanogr 12(2):264–271

Jayatilake GS et al (1996) Metabolites from an Antarctic sponge-associated bacterium, Pseudomonas aeruginosa. 59(3):293–296

Jessen O (1965) Pseudomonas aeruginosa and other green fluorescent pseudomonads: a taxonomic study. Munksgaard

Jones PG, Inouye MJMM (1994) The cold-shock response—a hot topic. 11(5):811–818

Jukes T and Cantor C (1969). Evolution of protein molecules. In "Mammalian Protein Metabolism" Ed by HN Munro

Jülich M et al (2001) The structure of the pyoverdin isolated from various Pseudomonas syringae pathovars. Zeitschrift für Naturforschung C 56(9–10):687–694

Jurado-Martin I, Sainz-Mejias M, McClean S (2021) Pseudomonas aeruginosa: an audacious pathogen with an adaptable arsenal of virulence factors. Int J Mol Sci 22:6

Kang CI et al (2003) Pseudomonas aeruginosa bacteremia: risk factors for mortality and influence of delayed receipt of effective antimicrobial therapy on clinical outcome. Clin Infect Dis 37(6): 745–751

Kang HY, Yoon T-J, Lee GJ (2011) Whitening effects of marine pseudomonas extract. Ann Dermatol 23(2):144–149

Kawahara H et al (2001) Relationship between antifreeze protein and freezing resistance in Pseudomonas putida GR12-2. 43(5):365–370

Kennedy IR, Tchan Y-T (1992) Biological nitrogen fixation in non-leguminous field crops: recent advances. Plant Soil 141(1):93–118

Khan NH et al (2007) Isolation of Pseudomonas aeruginosa from open ocean and comparison with freshwater, clinical, and animal isolates. Microb Ecol 53(2):173–186

Kim B-H et al (2014) Role of rhizobium, a plant growth promoting bacterium, in enhancing algal biomass through mutualistic interaction. 69:95–105

Kim O-S et al (2012) Introducing EzTaxon-e: a prokaryotic 16S rRNA gene sequence database with phylotypes that represent uncultured species. 62(Pt_3):716–721

Kimata N et al (2004) Pseudomonas aeruginosa isolated from marine environments in Tokyo Bay. Microb Ecol 47(1):41–47

Kirchman DL, White JJAME (1999) Hydrolysis and mineralization of chitin in the Delaware estuary. 18(2):187–196

Kloepper JW et al (1980) Enhanced plant growth by siderophores produced by plant growth-promoting rhizobacteria. 286(5776):885–886

Kong D et al. (2022) *Pseudomonas Tumuqiensis Sp. Nov., Isolated From Greenhouse Soil*

Kosina M et al (2013) Pseudomonas prosekii sp. nov., a novel psychrotrophic bacterium from Antarctica. 67(6):637–646

Kosina M et al (2016) Description of pseudomonas gregormendelii sp. nov., a novel psychrotrophic bacterium from James Ross Island, Antarctica. 73(1):84–90

Larsen AM et al (2015) Community structure of skin microbiome of gulf killifish, Fundulus grandis, is driven by seasonality and not exposure to oiled sediments in a Louisiana salt marsh. 70(2):534–544

Lau GW et al (2004) The role of pyocyanin in Pseudomonas aeruginosa infection. Trends Mol Med 10(12):599–606

Lee D-H et al (2010) Pseudomonas taeanensis sp. nov., isolated from a crude oil-contaminated seashore. 60(12):2719–2723

Liu J-M et al. (2022) Bioactive compounds from the Antarctic bacterium pseudomonas SP. A6-5. 1–3

Liu Y et al (2020) Pseudomonas piscis sp. nov., isolated from the profound head ulcers of farmed Murray cod (Maccullochella peelii peelii). 70(4):2732–2739

López NI et al (2009) Pseudomonas extremaustralis sp. nov., a poly (3-hydroxybutyrate) producer isolated from an Antarctic environment. 59(5):514–519

Lynch MD, Neufeld JDJNRM (2015) Ecology and exploration of the rare biosphere. 13(4): 217–229

Manaia CM, Moore ER (2002) Pseudomonas thermotolerans sp. nov., a thermotolerant species of the genus Pseudomonas sensu stricto. Int J Syst Evol Microbiol 52(6):2203–2209

Másmela-Mendoza JE, Lizarazo-Forero LMJUS (2021) Denitrifying bacteria in the limnetic zone of Lake Tota, Colombia. 26(1)

Matsuda M, Shigeta S, Okutani KJMB (1999) Antiviral activities of marine Pseudomonas polysaccharides and their oversulfated derivatives. 1(1):68–73

Matsuyama H et al (2015) Psychrobacter oceani sp. nov., isolated from marine sediment. 65(Pt_5): 1450–1455

McFall-Ngai M et al (2013) Animals in a bacterial world, a new imperative for the life sciences. 110(9):3229–3236

McFall-Ngai MJ (2015) Giving microbes their due–animal life in a microbially dominant world. J Exp Biol 218(12):1968–1973
Meikle P, Richards G, Yellowlees DJMB (1988) Structural investigations on the mucus from six species of coral. 99(2):187–193
Menezes CB et al (2010) Microbial diversity associated with algae, ascidians and sponges from the north coast of São Paulo state, Brazil. 165(6):466–482
Mulet M et al (2021) Pseudomonas arcuscaelestis sp. nov., isolated from rainbow trout and water. 71(7):004860
Mulisch M (1993) Chitin in protistan organisms: distribution, synthesis and deposition. Eur J Protistol 29(1):1–18
Muscatine L et al (1981) Estimating the daily contribution of carbon from zooxanthellae to coral animal respiration 1. 26(4):601–611
Naik P et al (2021) Virulence factors in multidrug (MDR) and Pan-drug resistant (XDR) Pseudomonas aeruginosa: a cross-sectional study of isolates recovered from ocular infections in a high-incidence setting in southern India. J Ophthalmic Inflamm Infect 11(1):36
Nayak SK (2010a) Probiotics and immunity: a fish perspective. Fish Shellfish Immunol 29(1):2–14
Nayak SKJAR (2010b) Role of gastrointestinal microbiota in fish. 41(11):1553–1573
Needham J et al (1994) Andrimid and moiramides AC, metabolites produced in culture by a marine isolate of the bacterium Pseudomonas fluorescens: structure elucidation and biosynthesis. 59(8): 2058–2063
Oh WT et al (2019) Pseudomonas tructae sp. nov., novel species isolated from rainbow trout kidney. 69(12):3851–3856
Palleroni NJ (2015) *Pseudomonas.* Bergey's manual of systematics of archaea and bacteria 1–1
Pandey A et al (1999) The realm of microbial lipases in biotechnology. 29(2):119–131
Pang Y et al (2021) Pseudomonas nanhaiensis sp. nov., a lipase-producing bacterium isolated from deep-sea sediment of the South China Sea. 114(11):1791–1804
Parte AC (2018) LPSN–list of prokaryotic names with standing in nomenclature (bacterio. Net), 20 years on. Int J Syst Evol Microbiol 68(6):1825–1829
Pascual J et al (2012) Pseudomonas litoralis sp. nov., isolated from Mediterranean seawater. 62(2): 438–444
Patten CL, Glick BR (2002) Role of Pseudomonas putida indoleacetic acid in development of the host plant root system. Appl Environ Microbiol 68(8):3795–3801
Pedrós-Alió CJA (2012) The rare bacterial biosphere. Annu Rev Mar Sci 4:449–466
Philippot L et al (2013) Going back to the roots: the microbial ecology of the rhizosphere. 11(11): 789–799
Poirel L et al (2004) Nosocomial outbreak of extended-spectrum β-lactamase SHV-5-producing isolates of Pseudomonas aeruginosa in Athens, Greece, Antimicrobial agents and chemotherapy. 48(6):2277–2279
Poli A et al (2017) Microbial diversity in extreme marine habitats and their biomolecules. 5(2):25
Prasad ASB et al (2020) Pseudomonas aeruginosa virulence proteins pseudolysin and protease IV impede cutaneous wound healing. Lab Investig 100(12):1532–1550
Prescott L, Harley J, Klein DJM (2002) Food and industrial microbiology. 5:125–964
Prevatt AR et al (2004) Hearing loss with semicircular canal transection and Pseudomonas aeruginosa otitis media. Otolaryngol Head Neck Surg 131(3):248–252
Qasim SZ (1998) Glimpses of the Indian Ocean. Universities Press
Radax R et al (2012) Ammonia-oxidizing archaea as main drivers of nitrification in cold-water sponges. 14(4):909–923
Rajyaguru N et al (2014) Pyorubin producing pseudomonas scalp infection. Infect Dis Clin Pract 22(4):e91–e92
Ramphal R, Arora SK, Ritchings BW (1996) Recognition of mucin by the adhesin-flagellar system of Pseudomonas aeruginosa. Am J Respir Crit Care Med 154(4 Pt 2):S170–S174

Reddy GS et al (2004) Psychrophilic pseudomonads from Antarctica: Pseudomonas Antarctica sp. nov., Pseudomonas meridiana sp. nov. and Pseudomonas proteolytica sp. nov. 54(3): 713–719

Redfield AC (1934) On the proportions of organic derivatives in sea water and their relation to the composition of plankton, vol 1. University Press of Liverpool Liverpool

Reveillaud J et al (2014) Host-specificity among abundant and rare taxa in the sponge microbiome. 8(6):1198–1209

Rheinheimer G (1992) The influence of environmental factors on the development of microorganisms. Aquat Microbiol 4:111–147

Richardson AEJFPB (2001) Prospects for using soil microorganisms to improve the acquisition of phosphorus by plants. 28(9):897–906

Richman S et al (1975) The rate of mucus production by corals and its assimilation by the coral reef copepod Acartia negligens 1. 20(6):918–923

Rigane E et al (2020) Characterization of putative virulence factors of Pseudomonas aeruginosa strain RBS isolated from a saltern, Tunisia: effect of metal ion cofactors on the structure and the activity of LasB. Biomed Res Int 2020:6047528

Rohwer F et al (2001) Diversity of bacteria associated with the Caribbean coral Montastraea franksi. 20(1):85–91

Romanenko LA et al (2005) Pseudomonas pachastrellae sp. nov., isolated from a marine sponge. 55(2):919–924

Sabath L, Jago M, Abraham EJBJ (1965) Cephalosporinase and penicillinase activities of a β-lactamase from Pseudomonas pyocyanea. 96(3):739–752

Saitou N, Nei MJMB (1987) The neighbor-joining method: a new method for reconstructing phylogenetic trees. 4(4):406–425

Sarmento H, Gasol JMJEM (2012) Use of phytoplankton-derived dissolved organic carbon by different types of bacterioplankton. 14(9):2348–2360

Sarniguet A et al (1995) The sigma factor sigma s affects antibiotic production and biological control activity of Pseudomonas fluorescens Pf-5. 92(26):12255–12259

Schmitt S et al (2007) Vertical transmission of a phylogenetically complex microbial consortium in the viviparous sponge Ircinia felix. 73(7):2067–2078

Schönknecht G et al (2013) Gene transfer from bacteria and archaea facilitated evolution of an extremophilic eukaryote. 339(6124):1207–1210

Seyedsayamdost MR et al (2011) The Jekyll-and-Hyde chemistry of Phaeobacter gallaeciensis. 3(4):331–335

Shade A, Gilbert JAJ (2015) Temporal patterns of rarity provide a more complete view of microbial diversity. Trends Microbiol 23(6):335–340

Shade A et al (2014) Conditionally rare taxa disproportionately contribute to temporal changes in microbial diversity. 5(4):e01371–e01314

Shelly DC, Quarles J, Warner IM (1980) Identification of fluorescent Pseudomonas species. Clin Chem 26(8):1127–1132

Siddiqui KS et al (2013) Psychrophiles. 41:87–115

Smith M et al (1994) Susceptibility of Pseudomonas pseudomallei to some newer b-lactam antibiotics and antibiotic combinations using time-kill studies. J Antimicrob Chemother 33(1): 145–150

Sogin ML et al (2006) Microbial diversity in the deep sea and the underexplored "rare biosphere". 103(32):12115–12120

Spiers AJ, Buckling A, Rainey PB (2000) The causes of Pseudomonas diversity. Microbiology 146(10):2345–2350

Stanier RY, Palleroni NJ, Doudoroff MJM (1966) The aerobic pseudomonads a taxonomic study. 43(2):159–271

Stankiewicz BA et al (1997) Preservation of chitin in 25-million-year-old fossils. 276(5318): 1541–1543

Strateva T, Yordanov D (2009) Pseudomonas aeruginosa–a phenomenon of bacterial resistance. J Med Microbiol 58(9):1133–1148
Sun J et al (2018) Pseudomonas profundi sp. nov., isolated from deep-sea water. Int J Syst Evol Microbiol 68(5):1776–1780
Sunagawa S et al (2015) Structure and function of the global ocean microbiome. 348(6237): 1261359
Sykes RB, Matthew M (1976) The β-lactamases of gram-negative bacteria and their role in resistance to β-lactam antibiotics. J Antimicrob Chemother 2(2):115–157
Szabó KÉ et al (2007) Importance of rare and abundant populations for the structure and functional potential of freshwater bacterial communities. 47(1):1–10
Tamegai H et al (1997) A denitrifying bacterium from the deep sea at 11 000-m depth. 1(4): 207–211
Tang S-PW, Coleman JE, Myer YP (1968) Conformational studies of copper proteins: pseudomonas blue protein and polyporus laccase. J Biol Chem 243(16):4286–4297
Taylor MW et al (2004) Host specificity in marine sponge-associated bacteria, and potential implications for marine microbial diversity. 6(2):121–130
Taylor MW et al (2007) Sponge-associated microorganisms: evolution, ecology, and biotechnological potential. 71(2):295–347
Tortoraand G, Funke B (2008) Microbiology: an introduction. Pearson Benjamin Cummings, San Francisco
Troussellier M et al (2017) Sustaining rare marine microorganisms: macroorganisms as repositories and dispersal agents of microbial diversity. 8:947
Uzair B et al (2008) The isolation, purification and biological activity of a novel antibacterial compound produced by Pseudomonas stutzeri. 279(2):243–250
Vacelet J, Donadey C (1977) Electron microscope study of the association between some sponges and bacteria. J Exp Mar Biol Ecol 30(3):301–314
Van Delden C (2004) Virulence factors in pseudomonas aeruginosa. In: Ramos J-L (ed) Virulence and gene regulation. Springer US, Boston, MA, pp 3–45
Ventosa A et al (2014) The Santa Pola saltern as a model for studying the microbiota of hypersaline environments. 18(5):811–824
Vergin KL et al (2013) Spatiotemporal distributions of rare bacterioplankton populations indicate adaptive strategies in the oligotrophic ocean. 71(1):1–13
Vivanco J (2004) *Pseudomonas aeruginosa-Plant Root Interactions*. Pathogenicity
Waisel Y (2012) Biology of halophytes. Elsevier
Wakabayashi H, Egusa SJBJSSF (1972) Characteristics of a Pseudomonas sp. from an epizootic of pond-cultured eels (Anguilla japonica). 38(6):577–587
Wang SZ et al (2021b) Complete genome sequences of two pseudomonas species isolated from marine environments of the Pacific Ocean. Microbiol Resour Announc 10:16
Wang S-Z et al (2021a) Complete genome sequences of two pseudomonas species isolated from marine environments of the Pacific Ocean. 10(16):e01062-19
Webster NS et al (2010) Deep sequencing reveals exceptional diversity and modes of transmission for bacterial sponge symbionts. 12(8):2070–2082
Weiland-Bräuer N et al (2015) Composition of bacterial communities associated with Aurelia aurita changes with compartment, life stage, and population. 81(17):6038–6052
Weon H-Y et al (2006) Pseudomonas pohangensis sp. nov., isolated from seashore sand in Korea. 56(9):2153–2156
Wiklund T, Bylund G (1990) Pseudomonas anguilliseptica as a pathogen of salmonid fish in Finland. Dis Aquat Org 8(1):13–19
van't Wout EF et al (2015) Virulence factors of Pseudomonas aeruginosa induce both the unfolded protein and integrated stress responses in airway epithelial cells. PLoS Pathog 11(6):e1004946
Wratten SJ et al (1977) Antibiotic metabolites from a marine pseudomonad. 11(3):411–414
Yeruham I et al (2004) Four-year survey of urinary tract infections in calves in Israel. Vet Rec 154(7):204–206

Youssef NH, Elshahed MSJTIJ (2009) Diversity rankings among bacterial lineages in soil. 3(3): 305–313

Zhang F et al (2015) Phosphorus sequestration in the form of polyphosphate by microbial symbionts in marine sponges. 112(14):4381–4386

Zhang M et al (2019) Nitrogen removal characteristics of a versatile heterotrophic nitrifying-aerobic denitrifying bacterium, Pseudomonas bauzanensis DN13-1, isolated from deep-sea sediment. Bioresour Technol 305:122626

Zhang M et al (2020) Nitrogen removal characteristics of a versatile heterotrophic nitrifying-aerobic denitrifying bacterium, Pseudomonas bauzanensis DN13-1, isolated from deep-sea sediment. 305:122626

Zhao M et al (2022) Pseudomonas alliivorans sp. nov., a plant-pathogenic bacterium isolated from onion foliage in Georgia, USA. 45(1):126278

Zheng R et al (2021) MerF is a novel regulator of deep-sea Pseudomonas stutzeri flagellum biogenesis and motility. Environ Microbiol 23(1):110–125

ZoBell, C.E.J.M (1946) m.A.m.o.h., *Marine microbiology. A monograph on hydrobacteriology*

Genomic Islands in *Klebsiella pneumoniae* 13

Suraj Shukla, Purvi Joshi, Pinal Trivedi, Oluwatosin Akinwotu, and Devarshi Gajjar

Abstract

Genomic Islands (GI) of *Klebsiella pneumoniae* include integrative and conjugative elements (ICEs), prophages, integrons, and transposons belonging to a group of genetic elements transferred horizontally and have integrated into the genome of *K. pneumoniae.* Integrative and conjugative elements of *K. pneumoniae* (ICE*Kp*) are flanked by direct repeats, encode the yersiniabactin (*ybt*) locus, a mobilization locus-type 4 secretion system (T4SS), and other variable regions based on which they are classified into 14 types (ICE*Kp*1–14). Their sizes range from 75–200 kb and their chromosomal insertion site is mostly one of the four tRNA-Asn sites. Each *K. pneumoniae* genome can harbor one to six prophages; accounting for 0.1–8% of the genome. The site of phage integration could be either the tRNA or ABC transporter permease SapC. Class I integrons are the most commonly found integrons in *K. pneumoniae.* They contain three essential components for the capture of external genes: an integrase, attI site, and an outwardly oriented promoter (Pc) that controls transcription of the captured genes. Conjugative transposons (CTn) in *K. pneumoniae* are associated with resistance (Tn916 and Tn6009) and hypervirulence (Tn6497).

S. Shukla · P. Joshi · P. Trivedi · D. Gajjar (✉)
Department of Microbiology and Biotechnology Centre, Faculty of Science, The Maharaja Sayajirao University of Baroda, Vadodara, India
e-mail: devarshi.gajjar-microbio@msubaroda.ac.in

O. Akinwotu
Department of Microbiology and Biotechnology Centre, Faculty of Science, The Maharaja Sayajirao University of Baroda, Vadodara, India

Environmental and Biotechnology Unit, Department of Microbiology, University of Ibadan, Ibadan, Nigeria

I. Mani et al. (eds.), *Microbial Genomic Islands in Adaptation and Pathogenicity*,
https://doi.org/10.1007/978-981-19-9342-8_13

Keywords

Integrative and conjugative elements · Prophages · Integrons · Transposons · *K. pneumoniae*

13.1 Introduction

Klebsiella spp. are non-motile, Gram-negative, encapsulated, bacteria found as commensals (on human mucosal surfaces) as well as in the environment. In the last two decades, a particular species (*Klebsiella pneumoniae*) has caused havocs by causing life-threatening diseases. Further, the situation has become uncontrollable as it is a frequent source of hospital-acquired pneumonia and the second most important cause of other nosocomial infections including urinary tract infections (Russo and Marr 2019). The virulence and antibiotic resistance of *K. pneumoniae* are the main factors leading to fatal outcomes. One of the major concerns for *K. pneumoniae* is that it is the reservoir of antimicrobial resistance (AMR) genes, and it efficiently spreads AMR in many other *Enterobacteriaceae* (Navon-Venezia et al. 2017). Continuous surveillance studies have indicated that resistance in *K. pneumoniae* has increased in the last few years and hence it contributes majorly to the burden of antibiotic resistance. It has been grouped as one of the ESKAPE pathogens and happens to be one of the critical priority pathogens listed by WHO (Mogasale et al. 2021). Though *K. pneumoniae's* capability to acquire genes (resistance and virulence) is marvelous, *Klebsiella* strains have so far shown a distinct demarcation of resistance (i.e., Carbapenem resistance *K. pneumoniae* [CRKP] strains) and virulence (hypervirulence *K. pneumoniae* [hvKP] strains). However, recent years have noticed a convergence (CR-hvKP strains) of these two kinds of traits and the situation seems threatening (Rodrigues et al. 2022; Lam et al. 2019; Yang et al. 2021). The worldwide occurrence of multidrug-resistant clinical strains is a result of the acquisition of AMR genes on mobile genetic elements (mostly plasmids) followed by the spread of these lineages. Horizontal gene transfer is the most important phenomenon that aids in the acquisition of AMR genes, and the emergence of multiple phenotypes is owed to the accumulation of gene arrays on plasmids, transposons, integrons, integrative and conjugative elements (ICEs), and prophages. Most of this mobilizable DNA when integrated into the bacterial genome is referred to as a genomic island (GI). A stretch of DNA on the bacterial genome having the following common features are GIs (Langille et al. 2010): (1) their size is between 10 and 200 kb; (2) their GC content and codon usage differ from the rest of the genome; (3) they are commonly incorporated at the tRNA genes (tDNAs); (4) the direct repeats that flank them, correspond to the 3′ portion of the tDNA; (5) they, by and large, have integrases that help in the island integration or excision; (6) few carry other mobility genes such as transposases or factors that contribute to conjugation; and (7) they normally carry genes conferring new metabolic proficiencies to the respective host.

K. pneumoniae GIs coding virulence and antibiotic resistance-related determinants are grouped under (1) Integrative conjugative elements (abbreviated as ICE*Kp*), (2) Prophages, (3) Conjugative transposons (CTs), and (4) Integrons. Though ICEs and prophages qualify to be GIs (according to their size range) here we are attempting to compile all information regarding all elements integrated into the genome of *K. pneumoniae*.

13.2 Integrative and Conjugative Elements—*Kp* (ICE*Kp*)

ICE*Kp* is a self-transmissible GI, and its excision occurs due to gene *xis*. An extrachromosomal circular intermediate is a prerequisite for mobilization to the recipient cells. The process requires integrase (*int*) and direct repeats (17 bp) at both ends. The virB1, mobB, and oriT are needed for mobilization. Integration occurs at attO sites present in four tRNA-Asn copies in the chromosome (Lin et al. 2008; Lery et al. 2014). *K. pneumoniae* chromosome region containing the tRNA-Asn sites with incorporated yersiniabactin ICE*Kp* elements is shown in Fig. 13.1. The hotspots for ICE*Kp* insertion are highlighted in the figure and occur inside four tRNA-Asn sites, which are denoted by green colored blocks. Coding sequences are represented by arrows, which are labeled with the gene symbol or the product.

In *K. pneumoniae*, ICE*Kp* mobilizes the yersiniabactin (*ybt*) locus, and its extensive genomic characterization using a large number of strains (n = 2499) identified 17 diverse *ybt* lineages and 14 ICE*Kp* structural variants (Lam et al. 2018). Each ICE*Kp* comprises (1) an integrase (P4-like); (2) the *ybt* locus (29 kb); (3) the oriT transfer origin (14 kb), virB-type4 secretion system (T4SS), and mobBC proteins (mobilization); and (4) genes at the right end (variable region) which were utilized to classify the ICE into 14 separate structures.

ICE*Kp* integration was identified at all four tRNA-Asn sites with varying frequencies. Sites 1, 3, and 4 showed 35.7, 44.7, and 19.5% integration, respectively, while site 2 had only one integration. Most *ybt* lineages had several ICE*Kp* integration sites, indicating that ICE*Kp* variations do not target particular tRNA-Asn copies. Yersiniabactin, along with other siderophores are important for bacterial virulence as

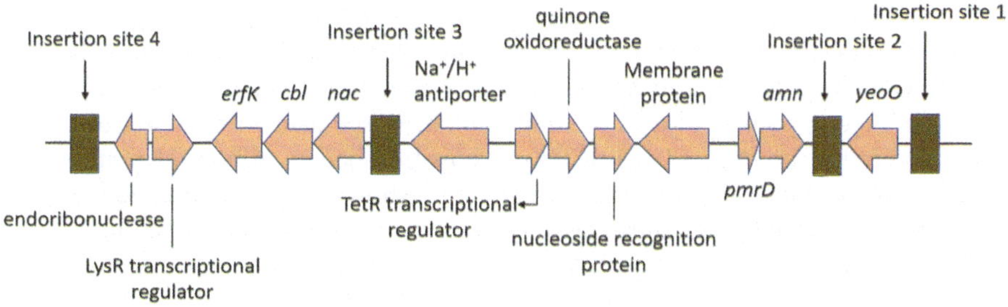

Fig. 13.1 *K. pneumoniae* chromosome region with tRNA-Asn insertion sites for yersiniabactin ICE*Kp*

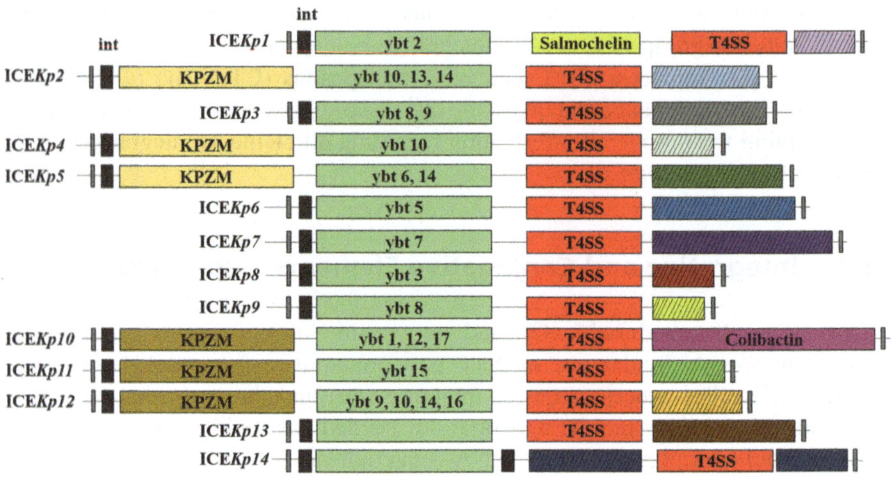

Fig. 13.2 Integrative and conjugative elements of *K. pneumoniae* (ICE*Kp*). Inverted repeats (gray boxes at ends), Integrase gene (black), Yersiniabactin synthesis locus *ybt* (light green, labeled with the most prevalent associated *ybt* lineage), immobilization module (red), Zn2+/Mn2+ module (brown: generally present; light brown: seldom present), diverse gene contents specific to each ICE*Kp* structure (presented in a unique colors with cross-line)

they scavenge iron from host proteins, thereby increasing the chance of survival within the host (Ramirez et al. 2014; Gorrie et al. 2017; Runcharoen et al. 2017). Yersiniabactin is present in approximately one-third of clinical strains, particularly with strains isolated from bacteremia and systemic infections (Lin et al. 2008; Holt et al. 2015). The siderophore enterobactin is produced by many clinical isolates of *K. pneumoniae*, but human lipocalin-2 (Lcn2) inhibits its scavenging mechanisms. Lcn2 binds to ferric and aferric enterobactin with high affinity (Goetz et al. 2002) following which an inflammatory response is induced (Bachman et al. 2009). Yersiniabactin functions importantly in invasive infections as it avoids Lcn2 binding and also avoids the inflammatory response. Thus, it enhances bacterial persistence in the host (Bachman et al. 2009, Bachman et al. 2011; Holden et al. 2016; Lawlor et al. 2007). The *ybt* locus was initially discovered in the Yersinia high pathogenicity island (HPI), and variations in additional *Enterobacteriaceae* species (Wami et al. 2021) are reported, along with *K. pneumoniae*, where *ybt* is found within ICE*Kp*. The very first reported ICE in *K. pneumoniae* was ICE*Kp*1 in 2008 (Lin et al. 2008) and with the comparison of a large number of sequence data 14 other variants have been reported (Lam et al. 2018). ICE*Kp* acquisition occurs in both cKp and hvKp strains of *K. pneumoniae* population. Figure 13.2 shows the diagrammatic representation of the 14 ICE*Kp* variants, classified as distinct structures (Lam et al. 2018).

The common elements of all ICE*Kp* (1–14) are the inverted repeats, the integrase, the bt locus, and the T4SS—mobilization module. In addition to these, ICE*Kp*s (2, 4, 5, 10, 11, 12) have the Zn^{2+}/Mn^{2+} metabolism module (KpZM); while the module is absent in the rest. The salmochelin (iro) locus is only present in ICE*Kp*1 and the colibactin locus is only present in ICE*Kp*10. Colibactin is genotoxic and hybrid non-ribosomal peptide polyketide that not only crosslinks with DNA but also causes double-strand DNA breaks in host cells (Vizcaino and Crawford 2015). It was firstly discovered in *E. coli* (Nougayrède et al. 2006), but it is now found in 3.5–4% *K. pneumoniae* isolates (Putze et al. 2009; Lam et al. 2018) where it was shown to cause DNA breaks in HeLa cells (Putze et al. 2009). The absence of colibactin is related to the reduction in dissemination to the blood and organs, e.g., liver, spleen, and brain (Lu et al. 2017). Colibactin-positive *K. pneumoniae* is very widespread in Taiwan, where it is present in 17–25% of cases of non-abscess infections and is strongly linked to K1 strains (mainly ST23) (Huang et al. 2012; Dalmasso et al. 2015) Further, all ICE*Kp*'s have a variable region and some of these are hypothetical proteins (not mentioned in Fig. 13.2) whose functions are yet to be known. In the following section, a summary of predominantly found ICE*Kp* (1, 2, 3, 4, 5, 10, 12, and 14) is given.

13.2.1 ICE*Kp*1

The first ICE in *K. pneumoniae* (named ICE*Kp*1) was described by Lin et al. (2008). It is a 76-kb region in a hvKP strain NTUH-K2044 and harbors genes for the biosynthesis of siderophores; yersiniabactin and salmochelin. The unique genes in the variable region include a transporter protease, mucoid phenotype regulator, methyltransferase (Sam-dependent), three transposases, and two hypothetical proteins. The role of ICEKp1 in hvKp pathogenesis was shown in this study as it was found to be more prevalent in hvKp strains (38/42) than cKp strains (5/32). Along with the yersinia pathogenicity island, another region similar to the virulence plasmid pK2044 and genes homologous to salmochelin (iro) and the capsular polysaccharide regulator *rmpA* biosynthesis were also present. Later ICE*Kp*nRJF293, a highly syntenic ICE to ICE*Kp*1 was reported from a hvKp strain RJF293 belonging to ST374 and K2 serotypes (Wang et al. 2018). ICE*Kp*nRJF293 is a 56-kb region incorporated into a tRNA-Asn locus and also contained yersiniabactin gene cluster, a type IV secretion system but lacked salmochelin (iroBCDN) gene cluster. Remarkably, ICE*Kp*nRJF293 contains a unique 10 Kb region at the tRNA-distal end, which encodes a restriction modification system, an ABC transporter, two transposases, and one hypothetical protein (Shen et al. 2019). documented a sequence type 35 (ST35) hypervirulent *Klebsiella pneumoniae* strain (RJY9645) that produced NDM-5 and was isolated from the blood of a patient who underwent a liver transplant. Apart from ICE*Kp*1 (75.4 kb region), additional four chromosomally borne ICE variants were identified, including two type VI secretion system (T6SS) loci (23.1 and 27.1 kb) and two prophages (21.4 and 67 kb). The chromosomal integration of ICE*Kp*1 and the acquisition of the blaNDM-5-carrying plasmid may have contributed to the formation of CR-hvKp strain RJY9645. Though, subsequent reports documented that ICE*Kp*1 was not representative of ICE*Kp* homologs present in the majority of other hvKp strains, recent isolated reports on the convergence of strains having both virulence and resistance are troublesome.

13.2.2 ICE*Kp*2

ICE*Kp*2, a member of the PAPI family, was reported in 2019 in a *K. pneumoniae* strain (HS11286) from China (Farzand et al. 2019). It was present along with ICE*Kp*1 in the same isolate. A 34-Kb Zn^{2+} and Mn^{2+} metabolism module abbreviated as KpZM was a part of the conserved region along with the *ybt* locus and T4SS locus. The variable region consisted of thymidylate synthase, adenylate kinase, TIR domain protein, and nine hypothetical proteins. In the same study, authors examined 1000 *Klebsiella* genomes and found that ICE*Kp*1 and ICE*Kp*2 are present individually and co-occurred (150 out of 1000 isolates). The occurrence was ICE*Kp*1 (500 out of 1000) and ICE*Kp*2 (300 out of 1000). The element was present in sequence types ST11, ST258, and ST512 of *Klebsiella pneumoniae* from the USA, the UK, and Asia. This was the first evidence of two integrative and conjugative elements interacting with one another. The study showed, that in an isolate with two elements (i.e., ICE*Kp*1 and ICE*Kp*2), ICE*Kp*2 clearly affected the mobility of plasmid positively driven by ICEKp1. It was proposed that Mob2ATPase of ICE*Kp*2 may be a factor for the conjugation of ICE*Kp*1.

13.2.3 ICE*Kp*3

The conserved region of ICE*Kp*3 contains the *ybt* locus and T4SS locus, while the variable region has genes for restriction endonuclease, phosphatase, reverse transcriptase, DDE endonuclease, and five hypothetical proteins. Shankar et al. (2020) reported the *ybt*9 locus located in ICE*Kp*3 in two MDR hypervirulent isolates of sequence type (ST23). In the global collection, isolates of lineage CG23-I are accompanied by *ybt*1 located on ICE*Kp*10 while other sub-lineages either lack ICE*Kp* or carry *ybt*8/9 on ICE*Kp*3. Moreover, the CG23-II isolates produced aerobactin and salmochelin but not colibactin. In a recent review, an elaborate summary of global incidence of hypervirulent and carbapenem-resistant *Klebsiella pneumoniae* showed that ICE*Kp*3 was predominantly found in strains from China, Singapore, and India while only two reports from UK and Canada were noted. According to a stool metagenomic analysis done by Molton et al. (2021), ICE*Kp*3 was found in 2 isolates (out of 24) with *ybt*9 lineage and one of these two isolates also had the clb3 gene.

13.2.4 ICE*Kp*4 and ICE*Kp*12

Apart from the common conserved regions (KpZM, *ybt*, T4SS), the variable region in ICE*Kp* 4 has the enzyme (transposase), a transporter (ABC), a restriction endonuclease (Type I), a DNA methyltransferase and a hypothetical protein. In one of our studies, a pan drug-resistant strain (DJ) had a *ybt*10 placed on ICE*Kp*4 (Rodrigues et al. 2022). The phylogenetic origins of this strain were investigated within the global diversity of CG147 using publicly available genome sequences of isolates from 2002 to 2018 ($n = 217$). The three main branches of CG147 were ST147, ST273, and ST392. First, a group of 29 genomes emerged in the year 2007, that showed the presence of *ybt*16/ICE*Kp*12. Second, a group of 22 genomes appeared in the year 2009 having *ybt*10/ICE*Kp*4. Further, the *ybt*; ICE*Kp* was rarely detected among ST392 and ST273 genomes. Despite a high diversity of ICE observed among ST147 isolates, *ybt*16; ICE*Kp*12 and *ybt*10; ICE*Kp*4 were two predominant variants found in ST147 genomes and overall it was found in 53% of ST147 genomes. Recently, a CTX-M-15-producing *K. pneumoniae* (TIES-4900 strain) was isolated from an urban Brazilian river. TIES-4900 strain was of sequence type ST15, had a yersiniabactin locus on ICE*Kp*4, the K locus was KL24 (wzi-24), and had O1v1 locus (Cardoso et al. 2022). The authors validated the virulent behavior of TIES-4900 strain in the insect (*Galleria mellonella*) infection model and concluded that the convergence of resistome and virulome in the high-risk clone ST15 is a critical issue, which could be contributing to severe infections in humans, and persistence and adaptation to aquatic environments impacted by anthropogenic activities like hospital and urban discharges. In a recent study on 17 *K. pneumoniae* isolates from wild animals found that six isolates harbored 4 distinct *ybt* lineages (*ybt*1, *ybt*5, *ybt*9, and *ybt*16) harbored on different integrative conjugative elements (ICE*Kp* 1, 3, 6, and 12, respectively) (Chiaverini et al. 2022). ICE*Kp*1/3 was present in approximately 50% of clinical isolates studied in the UK and a global study (Farzand et al. 2019).

13.2.5 ICE*Kp*5

The variable region of ICE*Kp*5 has helicases, thiamine biosynthesis, 2 patatin-like phospholipases, and 6 hypothetical proteins. ICE*Kp*5 appears to be prevalent in Asia and Southeast Asia. To understand the genomic features of Kp ST231 lineage and compare our isolates M2 and M6 (collected from patients with Urine Infection in Gujarat, India) with the ST231 genomes worldwide, we performed comparative genomic analysis using $n = 95$ publicly available genomes of ST231 lineage, collected between 2010 and 2018. The *ybt*14; ICE*Kp*5 was the most prevalent (79.4%; 77/97) in ST231 lineage (Desai 2021). ICE*Kp*5 was recently found in nine XDR isolates collected from bloodstream infections belonging to ST2095–K64 serotype from South India (Shankar et al. 2020). All nine isolates had the ICE*Kp*5 integrated into the chromosome that carried yersiniabactin (*ybt14*).

13.2.6 ICE*Kp*10

ICE*Kp*10 possesses the bacterial genotoxin—colibactin (clb) cluster in addition to the rest of the elements. It was first described by Lai et al. (2014) as a 208-kb chromosomal region with ideal characteristics of a genomic island in *K. pneumoniae* 1084 strain. This 208 kb genomic island was named KPHPI208 (*Klebsiella pneumoniae* high pathogenicity island 208) which also composed 7 other genomic modules (GMs). GM1 contained genes ~100% identical to the pks colibactin gene cluster reported in *E. coli* IHE3034. The other modules were predicted to be having functions like integration, conjugation, yersiniabactin production, microcin production, and some unknown functions. Later, Struve et al. (2015) mapped the evolutionary profile of hypervirulent K1 isolates belonging to clonal complex 23 (CC23), and found ICE*Kp*s similar to 208 kb genomic island mentioned above. Homologs of ICE*Kp*1 were detected in 24 of CG23 isolates as well as in the ST260 CG23 hybrid strain. Though the yersiniabactin cluster was constant, the center region (containing salmochelin and *rmpA* genes) was missing in the ICE region of CC23-related isolates except for NTUH-K2044. Furthermore, in all CC23-related isolates the six ORFs in the third region of ICE*Kp*1 encoding hypothetical proteins were swapped by a 50-kb segment encoding the polyketide genotoxin colibactin. Hence, the ICE region of all CC23 isolates studied by Struve et al. (2015) resembled the ICE described in the Taiwanese ST23 liver abscess strain 1084 (Lai et al. 2014). It is also observed that in the 3 non-CG23 hvKp strains studied, ICE*Kp*10 was poorly conserved, with 2 of the 3 strains possessing only genes that encoded yersiniabactin. Such ICE's having the presence of colibactin along with yersiniabactin are now designated as ICE*Kp*10 (Lam et al. 2018). Their comparative analysis of CG23 genomes ($n = 97$) elucidated that the 81 members of sublineage CG23-I had acquired ICE*Kp*10, which contained genes that encode yersiniabactin and colibactin. This event was estimated to occur in the year 1928, which was followed by the global population expansion of CG23-I. In a recent study, nearly 375,000 bacterial genome sequences were screened to correlate the diversity and evolution of yersiniabactin and colibactin carrying ICEs (i.e., ICE*Kp*10 in case of *Kp*) (Wami et al. 2021). Interestingly, the colibactin-*ybt* carrying ICE was detected only in *E. coli, Klebsiella* species, and *Citrobacter koseri*. To find if the frequency of the colibactin gene cluster is constrained to particular lineages of *E. coli* and *Klebsiella* species, the sequence types of the corresponding *E. coli* and *Klebsiella* species isolates were also analyzed. The clb gene cluster was enriched in a relatively meager group of *E. coli* STs (12/11,537 STs), *K. aerogenes* STs (2/214 STs), and *K. pneumoniae* STs (6/5237 STs), respectively. In *K. pneumoniae*, all ST3 isolates were clb-positive, and over 75% of the examined ST23 and ST234 isolates had the colibactin gene cluster. However, a lower number of the *K. pneumoniae* isolates of sequence types (ST11, ST258, and ST48) had the clb gene cluster. Though the percentage of clinical isolates processing ICE*Kp*10 is minor, a recent study showed the presence of ICE*Kp*10 in a *K. pneumoniae* ST66/K2 strain isolated from a community-acquired infection (Rodrigues et al. 2020). The four *K. pneumoniae* isolates (from the same patient) exhibited a positive string test, i.e., a hypermucoviscous phenotype and a susceptible antimicrobial profile. Phylogenetic analysis pointed out that the SB5881 strain was close to AJ210 strain (ST66/K2 serotype reported from Australia) which did not harbor ICE*Kp*. The authors describe a worrisome clinical presentation of a typical community-acquired invasive infection caused by *K. pneumoniae* strain that had spread to multiple organs. The dissemination was attributed to the high pathogenic potential due to the acquisition of virulence plasmids and the genomic island (ICE*Kp*10).

13.3 Prophages

Prophages are bacteriophages that have integrated into the bacterial chromosome, can enable horizontal gene transfer and contain important genetic information for the bacteria (Saltzman 2003). Prophages integrate into the bacterial genome and use host machinery for their replication. Genome analysis studies have emphasized on mosaicism in phage genomes suggesting presence of different regions corresponding to different evolutionary histories due to horizontal transfer of genes (Dion et al. 2020). The integration of prophages in the bacterial genome causes degradation of the phage genome or transposition of genes into the host which might lead to toxin production and antibiotic resistance hence making the bacteria more virulent and resistant. The presence of prophage also contributes to fitness and evolution of bacteria (Marques et al. 2021).

Several groups studying prophages have reported that all prophages isolated from chromosomal DNA of *K. pneumoniae* belong to four families of the order *Caudovirales* whose members are characterized by non-enveloped phages that are tailed and have icosahedral heads containing double-stranded DNA. The majority of phages found in *K. pneumoniae* belong to the family *Myoviridae* possessing straight, long contractile tail with a large variation in genome size ranging from 33 to 244 kb. Phages belonging to *Siphoviridae* and *Podoviridae* families have also been observed in *K. pneumoniae*. *Siphoviridae* phages are characterized by long, flexible, non-contractile tail with genome of about 50 kb while *Podoviridae* phages have short, non-contractile tails with genome varying between 40 and 42 kb (Marques et al. 2021). Most studies on prophages classified them as intact or complete phage and defective/incomplete/questionable using the tool PHASTER. Intact phages have a complete sequence of the reference phage and it indicates that integration has been recent (Marques et al. 2021). On the other hand, defective or questionable phage often lack essential phage function (Maxwell 2017) and are indicative of the integration of phage into the bacterial genome. A study has shown that the inductive frequency of AMR carrying phages decreased in presence of antibiotics and hence frequently phages become defective and are inherited in the bacterial genome (Bobay et al. 2014; Wendling et al. 2021).

Kondo et al. (2021) performed a comparative study between prophages from pathogens, *Enterococcus faecium, Staphylococcus aureus, Klebsiella pneumoniae, Acinetobacter baumannii, Pseudomonas aeruginosa, and Enterobacter* spp. which are grouped under ESKAPE pathogens. The study involved analysis of

408 *K. pneumoniae* strains and other ESKAPE pathogens. The results reveal that 20.9% of total strains of *K. pneumoniae* encoded AMR genes with is the highest proportion in the ESKAPE pathogen group. On the contrary, only 1.2% and 0.3% proportion of prophages harbored virulence factor (VF) genes and both AMR and VF genes, respectively, which is the lowest among the ESKAPE population under study (Kondo et al. 2021). While *Klebsiella* prophages carried the AMR genes, these do not belong to the high-priority AMR genes (e.g., carbapenemases).

13.3.1 Integration of Phages in Genome

In lysogenic cycle, integration of phage in bacterial genome is an extremely crucial step. Previous records have shown that prophages integrate site specifically in the genome. It is observed that prophages encoding tyrosine integrase integrate adjacent to host tRNA, and one probable reason for this integration is the affinity of temperate phage toward palindromic sequences present near that region (Bobay et al. 2013).

In the study by Marques et al. (2021), they analyzed upstream and downstream regions of bacteriophage insertion site and found that maximum prophages integrated between genes clusters involved in metabolic pathways, transcriptional regulators transporters, tRNA genes, protein synthesis, transferases, recombinant proteins, membrane proteins, and ribosome biogenesis. Bleriot et al. (2020) and Baliga et al. (2021) have also obtained similar results. *K. pneumoniae* prophages and their site of integration with additional remarks are listed in Table 13.1. And Antimicrobial resistance, Virulence, and genes regarding phage defense associated with *K. pneumoniae* are listed in Table 13.2.

13.4 Integrons

Integrons can be defined as genetic systems of bacteria that detain and express gene cassettes. They usually have an *intI* gene that encodes an enzyme known as integrase and via site-specific recombination that catalyzes the excision or incorporation of gene cassettes, a site for recombination (attI), along with a promoter that controls inserted gene cassettes's expression (Mazel 2006) (Fig. 13.3).

IntI integrase amino acid sequences have been used to divide integrons into different "classes," with those harboring intI1 being classified as "class 1," intI2 as "class 2," intI3 as "class 3," and so on. IntI1, intI2, and intI3 are most often accompanied on mobile genetic elements, while intI4 and rests were discovered in association with chromosomal integrons (Deng et al. 2015). Integrons are assembly platforms that use site-specific recombination to include exogenous open reading frames (ORFs) and by assuring their correct expression alter them to functional genes. Three components have so far been discovered to be crucial for the capture of foreign genes in all integrons: an *intI* gene that encodes a tyrosine-recombinase integrase, a main recombination site (attI), and an outwardly oriented promoter (Pc) that controls transcription of the acquired genes (Hall and Collis 1995). Gene

Table 13.1 Site of integration of bacteriophage in the *K. pneumoniae* genome

Sr. no.	Phages	Site of integration	Note
1	ST405-OXA48phi1.2, ST15-VIM1phi2, ST437-OXA245phi4.1, ST101-KPC2phi6.1, ST147-VIM1phi7.2, ST405-OXA48phi1.3, ST11-VIM1phi8.1, ST101-KPC2phi6.2, ST13-OXA48phi12.1, ST512-KPC3phi13.1, ST13-OXA48phi12.2, ST512-KPC3phi13.6 ST258-KPC3phi16.1, ST13-OXA48phi12.5	Before or after intact Host tRNA	Commonly tRNA-arg is found before prophage
2	ST11-OXA245phi3.1, ST340-VIM1phi10.2, ST437-OXA245phi4.2, ST11-VIM1phi8.4, ST512-KPC3phi13.2, ST11-OXA48phi15.3, ST258-KPC3phi16.2	Between intact genes of TerT transcriptional regulator and transporter intact genes	Genes remained intact after phage integration
3	ST405-OXA48phi1.1, ST16-OXA48phi5.2, ST11-OXA245phi3.2, ST846-OXA48phi9.1	Adjacent to bacterial transcription regulator	ST16-OXA48phi5.2, ST846-OXA48phi9.1 Disruption of adjacent genes due to phage integration

(continued)

Table 13.1 (continued)

Sr. no.	Phages	Site of integration	Note
4	ST16-OXA48phi5.1, ST846-OXA48phi9.2, ST974-OXA48phi18, ST11-VIM1phi8.2,	Integration between *sapB* and *sapC* intact gene of *sapABCDEF* operon coding for ATP binding cassette (ABC transporter)	
5	ST16-OXA48phi5.3, ST340-VIM1phi10.1, ST11-VIM1phi8.3, ST11-OXA48phi15.1, ST512-KPC3phi13.5	Immediately after an intact Protease	
6	ST101-KPC2phi6.3, ST13-OXA48phi12.3, ST147-VIM1phi7.1, ST15-OXA48phi14, ST13-OXA48phi12.4	Next to gene coding for an unknown protein	Integration of ST15-OXA48phi14 phage caused truncation of gene
7	ST16-OXA48phi5.4	After a sensor domain-containing diguanylate cyclase	Disruption of adjacent genes due to phage integration

cassettes (Gc) typically contain a promoter less open reading frame (orf) and a recombination site attC (Also known as the element of 59-base) for integration. They can occur in the form of free circular molecules or as integrons (Hall et al. 1999). Integrons are highly mobile as they are placed on transposons, plasmids, and pathogenicity islands, allowing them to be transferred across bacteria. The nucleotide sequence of the integrase gene has classified integrons into five types (Guérin et al. 2011). The most common integrons are class 1, and are found in *K. pneumoniae* and other gram-negative clinical isolates (Lima et al. 2014).

13.4.1 Integrons Associated with Antibiotic Resistance

Two conserved segments, the 3′ conserved segment (3′ CS) and the 5′ conserved segment (5′ CS), together with internal gene cassettes (antimicrobial resistance

Table 13.2 Antimicrobial resistance, Virulence, and Phage defense genes associated with prophages in *K. pneumoniae*

Strain's accession number/name of prophage	Most common phage/closely related phage	Gene present in prophage/protein coded by prophage
AMR genes (Kondo et al. 2021)		
Kp-AP018748	Escher_RCS47	*$bla_{CTX-M-15-1}$, aac(6′)-Ib_1, blaTEM-1A_1, tet(D)_1, dfrA14_5, ant(3″)-Ia_1, qnrB1_1, aac(6′)-Ib-cr_1, bla_{OXA-1}_1, catB3_1*
Kp-CP008797	Entero_P1	*$bla_{TEM-105\_1}$, $bla_{TEM-105\_1}$, sul1_5, aadA2_1, aac(3)-Ib_1*
Kp-CP009876, Kp-CP015382	Entero_186	*bla_{KPC-2_1}*
Kp-CP011578	Entero_186	*$bla_{CTX-M-15\_1}$*
Kp-CP018140	Entero_mEp237	*aac(6′)-Ib-cr_1, bla_{OXA-1_1}, catB3_1, aac(3)-IIa_1*
Kp-CP018447, Kp-CP018450	Entero_P2	*oqxB_1, oqxA_1*
Kp-CP018816	Escher_HK639	*sul1_5, aadA2_1, aac(3)-Ib_1*
Kp-CP018883, Kp-CP018885, Kp-CP020071, Kp-CP020837, Kp-CP021539, Kp-CP043047	Entero_P1	*sul1_5, aadA2_1, aac(3)-Ib_1*
Kp-CP022023	Salmon_SJ46	*sul2_2, aadA2_1, dfrA12_8, ant (3″)-Ia_1*
Kp-CP022882, Kp-CP022997, Kp-CP023722, Kp-CP023933, Kp-CP023941, Kp-CP024191, Kp-CP024521, Kp-CP024528, Kp-CP024535, Kp-CP024570, Kp-CP024563, Kp-CP024556, Kp-CP024549, Kp-CP025951, Kp-CP026130, Kp-CP026132, Kp-CP026149, Kp-CP026145, Kp-CP026140, Kp-CP026136, Kp-CP027068, Kp-CP028548, Kp-CP028542, Kp-CP029384, Kp-CP031721, Kp-CP032163, Kp-CP032207, Kp-CP033954, Kp-CP034123, Kp-CP034415, Kp-CP036300, Kp-CP036365, Kp-CP036371, Kp-CP041373	Entero_phi80	*sul1_5, aadA2_1, aac(3)-Ib_1*
Kp-CP023949	Salmon_RE_2010	*mdf(A)_1*
Kp-CP025456	Entero_phi80 Salmon_Fels_2	*sul1_5, aadA2_1, aac(3)-Ib_1* *oqxA_1*
Kp-CP025461, Kp-CP027146, Kp-CP028180	Escher_HK639	*sul1_5, aadA2_1, aac(3)-Ib_1*
Kp-CP026159, Kp-CP028787, Kp-CP037963, Kp-CP041099, Kp-CP043932, Kp-CP011624, Kp-CP013322	Entero_P4	*oqxB_1, oqxA_1*
Kp-CP026177	Entero_mEp235	*oqxB_1, oqxA_1*

(continued)

Table 13.2 (continued)

Strain's accession number/name of prophage	Most common phage/closely related phage	Gene present in prophage/protein coded by prophage
Kp-CP028583, Kp-CP033396	Entero_phi80 Salmon_Fels_2	*sul1_5, aadA2_1, aac(3)-Ib_1* *mdf(A)_1*
Kp-CP028797	Salmon_Fels_2 Entero_Tyrion	*oqxB_1, oqxA_1* *mdf(A)_1*
Kp-CP029738	Escher_RCS47	$bla_{\text{SHV-12\_1}}$
Kp-CP031800	Salmon_RE_2010	*mdf(A)_1*
Kp-CP033625	Salmon_Fels_2	*oqxB_1, oqxA_1*
Kp-CP034249	Escher_HK639	*sul1_5, aadA2_1, aac(3)-Ib_1*
Kp-CP034327	Salmon_Fels_2	*oqxA_1*
Kp-CP036305	Entero_phi80 Salmon_Fels_2	*sul1_5, aadA2_1, aac(3)-Ib_1* *oqxB_1, oqxA_1*
Kp-CP036320, Kp-CP036327	Salmon_RE_2010	*mdf(A)_1*
Kp-CP040533, Kp-CP040539, Kp-CP040545, Kp-CP033960	Entero_phi80 Salmon_Fels_2	*sul1_5, aadA2_1, aac(3)-Ib_1* *oqxB_1*
Kp-CP042481	Entero_P4	*oqxA_1*
Virulence genes (Bleriot et al. 2020)		
ST512-KPC3phi13.1		Invasion-associated protein B T4SS
ST258-KPC3phi16.1, ST512-KPC3phi13.6, ST437-OXA245phi4.1		Transferase-kinase
ST13-OXA48phi12.5, ST16-OXA48phi5.2, ST13-OXA48phi12.3, ST405-OXA48phi1.3, ST101-KPC2phi6.3, ST15-VIM1phi2.1, ST11-VIM1phi8.2		MarR family of transcriptional regulators
Genes regarding phage defense (Bleriot et al. 2020)		
ST405-OXA48phi1.2 ST16-OXA48phi5.3		RelBE-like TA proteins
ST11-VIM1phi8.3 ST846-OXA48phi9.2		HigBA-like TA modules
ST512-KPC3phi13.6 ST437-OXA245phi4.1		CRISPR-associated Endoribonuclease Cas2
ST846-OXA48phi9.2		Putative anti-CRISPR/ Cas9 protein, AcrIIC3-like
ST13-OXA48phi12.3		TerB protein from The operon *terZABCDEF*

genes), make up the class 1 integrons's structure (Lima et al. 2014). In *K. pneumoniae* Class 2 integrons are occasionally discovered (Odumosu et al. 2013). According to Firoozeh et al. (2019), the most common cassettes were 1000–1500 bp long *aadA1* and *dfrA1-sat1* cassette arrays. Meanwhile, class

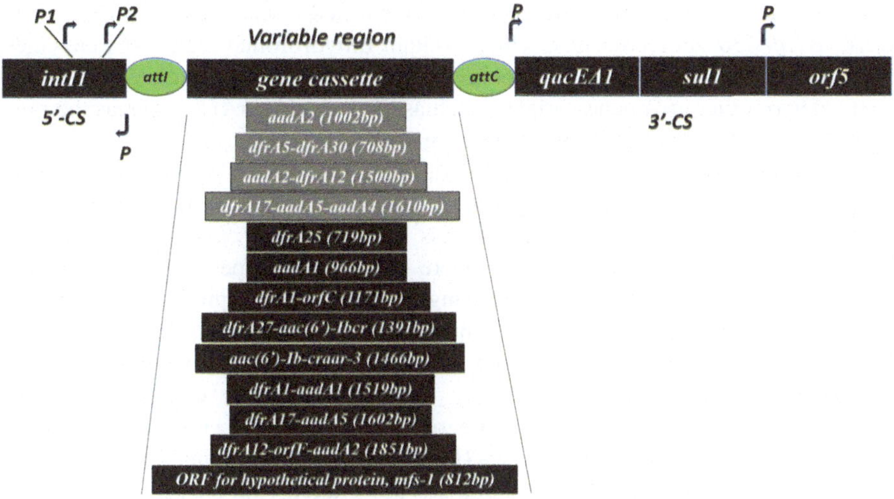

Fig. 13.3 A class 1 integron is represented in this diagram (Deng et al. 2015). P1 promoter for gene cassette transcription, P2 another promoter that is often inactive, an int gene for integrase, an attI1 integration site, moderately deleted gene *qacE* that encodes resistance against quaternary ammonium compounds (QACs), sulfonamide resistance gene *sulI*, orf5 uncharacterized function, P promoters for the *sulI and qacEΔ1* genes. An integrase recognizes the attC sequence on the gene cassette. Gene cassette which is a variable region of the class 1 integron. Some gene cassettes are mentioned below in gray (Firoozeh et al. 2019) and black (Li et al. 2013) color boxes

3 integrons have only been found in a few strains of *K. pneumoniae*. Correia et al. (2003) described a natural *K. pneumoniae* plasmid p22K9 that had a 2863-bp long class 3 integron that included an intI3 integrase gene, two (Pint and Pc) promoter areas, an attI3 recombination site, a cassette of $bla_{\text{GES-1}}$ gene, and a fused cassette of $bla_{\text{OXA-10}}$-type/*aac(6)-Ib* gene (Correia et al. 2003). Many different resistance gene cassettes are carried by class 1 integrons, the majority of which hold the *aadA* gene, which confers streptomycin/spectinomycin resistance. It has been shown that the distribution of class 1 integrons carrying different aadA alleles is widespread (Deng et al. 2015). In addition, the *dfrA* cassette arrays, which encode trimethoprim resistance, are typically seen in both class 1 and 2 integrons (Kiiru et al. 2013).

Firoozeh et al. (2019) studied clinical isolates of MDR *K. pneumoniae* (MDR*Kp*) (n = 150) from specimens such as urine, wounds, blood, respiratory tract samples, CSF, and catheters were used to isolate *K. pneumoniae* in Iran and identified class 1, 2, and 3 integrons. All of the MDRKp strains n = 150 (100%) had class 1 integrons and *K. pneumoniae* n = 55 (36.66%) had class 2 integrons. IntI-positive strains were used for sequencing indicated that the cassette arrays of class 1 integron included ten different array groups ranging from A to J, consisting of (1610 bp, 1500 bp, 1002 bp and 708 bp integrons) and gene cassettes were identified and shown in Fig. 13.3. Whereas, four separate groups of cassette array (1000 bp and 1500 bp integrons) were discovered, ranging from a to d in class 2 integron which harbored gene cassettes were as follows: (no cassette; *aadA1; dfrA1-sat1; aadA1, dfrA1-sat1*).

708 bp arrays were the most prevalent type identified in class 1 integrons, and the *dfrA5* & *dfrA30* gene cassettes, which contain dihydrofolate reductases enzymes, were identified. Class 1 integron-positive *K. pneumoniae* strains also have a high frequency of other *dfrA* gene variants, such as *dfrA12* and *dfrA17*. whereas the most common cassettes in class 2 integrons were 1000–1500 bp.

In another study by Li et al. (2013), they studied *K. pneumoniae* isolates ($n = 176$) of patients from tertiary care hospitals. The isolates found positive for class 1 integron contained ten different class 1 integron gene cassette arrays ranging between 700 bp and 1860 bp, which were classified as types I–X and shown in Fig. 13.3. There were no ESBL-expressing gene cassettes or proteins connected to carbapenem resistance detected. The majority of *K. pneumoniae* isolates contained a 1171-bp integron with the *dfrA1* and orfC genes (type I), which was the most prevalent integron gene cassette array seen. Additionally, compared to class 1 integron-negative isolates, class 1 integron-positive isolates showed resistance to a significantly greater number of drugs (Li et al. 2013). Class 1 integrons are highly prevalent in Gram-negative bacteria, and this association with the presence of MDR is significant (Wu et al. 2012; Li et al. 2013). Other investigations have found a high prevalence of integron-positive MDR*Kp* (Gruteke et al. 2003; Wu et al. 2012). Integrons may provide a selective advantage to strains residing in environments where selected pressures are induced by antibiotic abuse, such as hospitals, explaining the high occurrence of integrons in MDR strains.

In *Klebsiella* species, Salimizand et al. (2013) reported a *dfrA17* variation. The genes *dfrA17, dfrA12, dfrA1, dfrA25,* and *dfrA27* were found in class 1 integron cassette arrays in *K. pneumoniae* strains of intI1-positive in China (Li et al. 2013; Cao et al. 2014). The *dfrA17* and *dfrA12* variants have been detected in Gram-negative bacteria carrying class 1 integrons in the United States (Adams-Sapper et al. 2012), indicating that these variants are prevalent across class 1 integron cassettes around the world. Some *K. pneumoniae* strains can produce $bla_{\text{NDM-1}}$carbapenemase and have a class 1 integron with the following configuration in their genome and plasmid (Cortés-Ortíz et al. 2021).

ISMRK (imipenem-susceptible but meropenem-resistant *Klebsiella*) is a term used by Kayama et al. (2015) to describe isolates that were extended-spectrum beta-lactamase (ESBL) positive and displayed a contradictory pattern of resistance, being extremely resistant to nearly all antibiotics of beta-lactam except imipenem. The class 1 integron, In722, has a cassette containing the MBL gene $bla_{\text{IMP-6}}$, and pKPI-6, a 47-kb self-transmissible plasmid, which also had the ESBL gene $bla_{\text{CTX-M-2}}$ in ISMRK bacteria. Isolates of ISMRK have a phenotype called "stealth" that is undetectable with imipenem when IMP-6 (Shigemoto et al. 2012) and CTXM-2 are combined. In pKPI-6, there are three acquired extra DNA insertions that carry resistance genes: an integron region containing $bla_{\text{IMP-6}}$, a Tn1721 segment containing *tetA* and *tetR*, and a stability operon region harboring $bla_{\text{CTX-M-2}}$ (Fig. 13.4).

Kondo et al. (2021) discovered that the integron cassette array is similar to other AMR gene combinations seen in prophage area, and used the INTEGRALL database to investigate the integrons in these prophage regions. They discovered that certain *K. pneumoniae* prophage sites had integrase belonging to Class1 and cassette

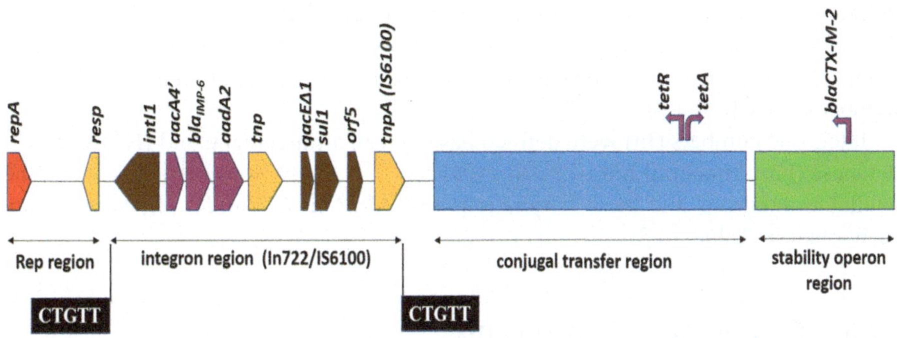

Fig. 13.4 Representative image of pKPI-6 plasmid (Kayama et al. 2015). The ORFs of the rep region and integron regions are symbolized by Pentagons, the genes that have been annotated are colored based on the expected gene function as follows: antimicrobial resistance genes, pink; conjugation genes, sky blue (in conjugation transfer regions); transposons, yellow; integrons, brown and plasmid maintenance genes, red

Table 13.3 List of different Integrons harboring AMR genes in cassette array in *Klebsiella pneumoniae*

Integron class	Integron number (In)	AMR genes in cassette array	Year	Source	Accession number
1	**In722**[a]	*aacA4'-3*, bla_{IMP-6}, *aadA2*, *sul1*	2012	Japan	AB616660
1	**In719**[a]	*sul1*, *aadA2*, *dfrA12D6*	2011	n.m.	CP003225
1	–	*dfrA17*, *aadA5*	2009	Russia	GQ896493
1	–	*aacA4*	2009	Russia	GQ924771
1	**In560**[a]	*dfrA30b*	2011	Libya	HE613853
1	**In578**[a]	*sul1*, *cmlA1l*, *aadA1e*, *ereC*, *arr-2*	2011	Kenya	JN157804
1	**In27**[a]	*dfrA12*, *aadA2*, *sul1*	2011	n.m.	JN233704
1	**In191**[a]	*dfrA14b*	2012	Czech Republic	JX424423
1	**In27**[b], **In191**[b]	$bla_{CTX-M-15\_1}$, *aac(6')-Ib_1*, bla_{TEM-1A_1}, *tet(D)_1*, *dfrA14_5*, *ant(3")-Ia_1*, *qnrB1_1*, *aac(6')-Ib-cr_1*, bla_{OXA-1_1}, *catB3_1*	2016	Thailand	AP018748
1	**In127**[b]	$bla_{TEM-105\_1}$, $bla_{TEM-105\_1}$, *sul1_5*, *aadA2_1*, *aac(3)-Ib_1*	2012	USA	CP008797
1	**In127**[b], **In610**[b]	*sul1_5*, *aadA2_1*, *aac(3)-Ib_1*	2014	China	CP026130
1	**In1680**[b], **In610**[b]	*sul1_5*, *aadA2_1*, *aac(3)-Ib_1*	2014	China	CP026145

Note—n.m – not mentioned, In[a] – From Moura et al. (2009), In[b] – From Kondo et al. (2021)

arrays for antimicrobial resistance (AMR) (Table 13.3). These distinctive areas including AMR genes cassette arrays were referred to as integron cassette arrays, i.e., integron-associated prophages. Additionally, they found that all phage regions with an integron included three or greater than three AMR genes, but those lacking an integron contained less number of AMR genes. These results showed that compared to other groups, prophages carrying integrons had a significantly more number of AMR genes.

Integrons numbers (In) were defined based on an arrangement, and INTEGRALL database (http://integrall.bio.ua.pt/?list) (Moura et al. 2009) were retrieved for all the available integrons and their association with AMR in *Klebsiella pneumoniae* are mentioned in Table 13.3.

13.5 Conjugative Transposons

Vertical transmission of conjugative transposons (CTns) occurs through chromosomal replication and partitioning (Wright and Grossman 2016). It is challenging to determine the original host for any conjugative transposon since a species of bacterium that has been initially identified as containing a novel CTn may not be the species from which the CTn developed (Scott 2002). Conjugative transposons can move into a new host by transposition. They are capable of conjugative transfer into new hosts without being mediated by plasmids (Tomich et al. 1979). CTns are known for their heterogeneity in form and function, thereby conferring the adaptive features and evolution in *Klebsiella pneumoniae*. Conjugative transposons as well as other genomic Islands are integrated within the chromosome and are regarded as important as conjugative plasmids involved in the transfer of chromosomal-borne genes among diverse bacterial species (Scott 2002) using the self-encoded transmission machinery or the type IV secretion system (T4SS) that is conjugation machinery (Wozniak and Waldor 2010; Johnson and Grossman 2015).

Usually, the CTns identified in the environment often code for resistance to heavy metals, and aromatic compounds and also encode functions such as Nitrogen fixation; mobile catabolic genes encoding degradation of xenobiotic compounds.

13.5.1 Antibiotic Resistance

CTns, reportedly hosting cascades of genes encoding Antibiotic resistance have been detected in quite a lot of pathogenic strains of *K. pneumonia* (Soge et al. 2008; Roberts and Mullany 2011). CTns are known to encode essential functions that enhance the survival of bacteria under specific environmental conditions as seen in Antibiotic resistance. Many bacteria including *Klebsiella pneumoniae* can adapt to any environment either by introducing a compensatory mutation in genes or by conditioning the expression of the resistance genes. Here, we discuss the most commonly found CTns associated with resistance (Tn916 and Tn6009) and hypervirulence (Tn6497).

13.5.1.1 Tn916

Tn916 is a 16.4-kb broad-host-range conjugative transposon originally discovered in *Enterococcus faecalis* (Rice 1998). It confers resistance to tetracycline via *tet(M)*. This transposon has been detected in various bacteria including *K. pneumoniae* (Soge et al. 2008). It is a self-transmissible genomic island usually associated with the chromosome and also found on certain plasmids (Rice 1998). Two transposon-encoded proteins; Xis-Tn and Int-Tn are required for the excessive recombination. Although the latter alone is enough for integration (Storrs et al. 1991). In some cases, the active integrase of both the donor and the receiver is necessary for the conjugative transposition of Tn916 (Storrs et al. 1991).

13.5.1.2 Tn6009

This is a novel, 17.8 kb size, non-composite conjugative transposon which belongs to the Tn916 family. It contains a Tn916 element which is incorporated with a functional inorganic mercury resistance (*merA*) that sits upstream of the conjugation module (Roberts and Mullany 2011). The *mer* genes and the *tet(M)* genes are directly related, and 24 orfs of the Tn916 are linked to a distinct 37-bp sequence that comes before the *merA, merB,* and *merT*, among other *mer* genes. These features make it unique (Soge et al. 2008). The successful demonstration of the conjugative transfer of Tn6009 from *Klebsiella pneumoniae* to *Enterococcus faecalis* (Soge et al. 2008) subsequently conferred its resistance to mercury and tetracycline due to the actions of the *merA* and *tetM* genes, respectively.

13.5.1.3 Tn6497

A transposon called Tn6497 was discovered in the hypervirulent strain of *Klebsiella pneumoniae* 11492's high pathogenicity island (HPI). IS903D, the colibactin gene cluster (clbABHIJKLMNOPQ), and the yersiniabactin gene cluster are all present (fyuA, *ybt*ETU, irp1, irp2, *ybt*APQXS) (Shen et al. 2019).

13.6 Concluding Remarks

GIs contribute to the genomic plasticity of *K. pneumoniae*. ICE*Kp* acts as a reservoir for virulence genes and is more stably integrated compared to others. The polylysogenic property of *K. pneumoniae* helps many prophages to reside on one genome and is intimately associated with virulence, resistance, evolution, and fitness. Antibiotic resistance genes primarily accumulate due to integrons and transposons. There is fast information generated about the GIs from the whole genome sequencing data and much of the data is lying without experimental proof of concept. There is a need to deepen our understanding through functional analysis. Certain pressing questions to be addressed are (1) under which conditions do the ICEs express? (2) can antibiotics induce prophages? And (3) association between prophages and integrons residing in them. It is also necessary to understand the situations in which the horizontal transfer of GIs occurs. Future functional translational studies should be designed to heighten our understanding of GIs in *K. pneumoniae*.

References

Adams-Sapper S, Sergeevna-Selezneva J, Tartof S et al (2012) Globally dispersed mobile drug-resistance genes in Gram-negative bacterial isolates from patients with bloodstream infections in a US urban general hospital. J Med Microbiol 61:968–974. https://doi.org/10.1099/jmm.0.041970-0

Bachman MA, Miller VL, Weiser JN (2009) Mucosal lipocalin 2 has pro-inflammatory and iron-sequestering effects in response to bacterial enterobactin. PLoS Pathog 5. https://doi.org/10.1371/journal.ppat.1000622

Bachman MA, Oyler JE, Burns SH et al (2011) Klebsiella pneumoniae yersiniabactin promotes respiratory tract infection through evasion of lipocalin 2. Infect Immun 79:3309–3316. https://doi.org/10.1128/IAI.05114-11

Baliga P, Shekar M, Kallappa GS (2021) Genome-wide identification and analysis of chromosomally integrated putative prophages associated with clinical Klebsiella pneumoniae strains. Curr Microbiol 78:2015–2024. https://doi.org/10.1007/s00284-021-02472-2

Bleriot I, Trastoy R, Blasco L et al (2020) Genomic analysis of 40 prophages located in the genomes of 16 carbapenemase-producing clinical strains of Klebsiella pneumoniae. Microbial Genomics 6:1–18. https://doi.org/10.1099/mgen.0.000369

Bobay LM, Rocha EPC, Touchon M (2013) The adaptation of temperate bacteriophages to their host genomes. Mol Biol Evol 30:737–751. https://doi.org/10.1093/molbev/mss279

Bobay LM, Touchon M, Rocha EPC (2014) Pervasive domestication of defective prophages by bacteria. Proc Natl Acad Sci U S A 111:12127–12132. https://doi.org/10.1073/pnas.1405336111

Cao X, Xu X, Zhang Z, et al (2014) Molecular characterization of clinical multidrug-resistant Klebsiella pneumoniae isolates

Cardoso B, Esposito F, Fontana H et al (2022) Genomic analysis of a Kpi (pilus system)-positive and CTX-M-15-producing Klebsiella pneumoniae belonging to the high-risk clone ST15 isolated from an impacted river in Brazil. Genomics 114:378–383. https://doi.org/10.1016/j.ygeno.2021.12.007

Chiaverini A, Cornacchia A, Centorotola G et al (2022) Phenotypic and genetic characterization of Klebsiella pneumoniae isolates from wild animals in Central Italy. Animals 12:1347. https://doi.org/10.3390/ani12111347

Correia M, Boavida F, Grosso F et al (2003) Molecular characterization of a new class 3 integron in Klebsiella pneumoniae. Antimicrob Agents Chemother 47:2838–2843. https://doi.org/10.1128/AAC.47.9.2838-2843.2003

Cortés-Ortíz IA, Juárez-Gómez JC, Cu-Quijano C et al (2021) Klebsiella pneumoniae blaNDM-1 carrying a class 1 integron causing a hospital outbreak in a Mexican attention center. J Infect Dev Ctries 15:657–664. https://doi.org/10.3855/JIDC.12996

Dalmasso G, Cougnoux A, Delmas J et al (2015) The bacterial genotoxin colibactin promotes colon tumor growth by modifying the tumor microenvironment. Gut Microbes 5:675–680. https://doi.org/10.4161/19490976.2014.969989

Deng Y, Bao X, Ji L et al (2015) Resistance integrons: class 1, 2 and 3 integrons. Ann Clin Microbiol Antimicrob 14

Desai S (2021) Antibiotic resistance and virulence factors of pathogenic klebsiella species (Order No. 29064963). Available from Dissertations & Theses @ Maharaja Sayajirao University of Baroda. (2645889215)

Dion MB, Oechslin F, Moineau S (2020) Phage diversity, genomics and phylogeny. Nat Rev Microbiol 18:125–138

Farzand R, Rajakumar K, Zamudio R et al (2019) ICE*Kp*2: description of an integrative and conjugative element in Klebsiella pneumoniae, co-occurring and interacting with ICE*Kp*1. Sci Rep 9. https://doi.org/10.1038/s41598-019-50456-x

Firoozeh F, Mahluji Z, Khorshidi A, Zibaei M (2019) Molecular characterization of class 1, 2 and 3 integrons in clinical multi-drug resistant Klebsiella pneumoniae isolates. Antimicrob Resist Infect Control 8. https://doi.org/10.1186/s13756-019-0509-3

Goetz DH, Holmes MA, Borregaard N, Bluhm ME, Raymond KN, Strong RK (2002) The neutrophil lipocalin NGAL is a bacteriostatic agent that interferes with siderophore-mediated iron acquisition. Mol Cell 10(5):1033–1043

Gorrie CL, Mirceta M, Wick RR, et al (2017) Gastrointestinal carriage is a major reservoir of K. pneumoniae infection in intensive care patients

Gruteke P, Goessens W, van Gils J et al (2003) Patterns of resistance associated with integrons, the extended-spectrum β-lactamase SHV-5 gene, and a multidrug efflux pump of Klebsiella pneumoniae causing a nosocomial outbreak. J Clin Microbiol 41:1161–1166. https://doi.org/10.1128/JCM.41.3.1161-1166.2003

Guérin E, Jové T, Tabesse A et al (2011) High-level gene cassette transcription prevents integrase expression in class 1 integrons. J Bacteriol 193:5675–5682. https://doi.org/10.1128/JB.05246-11

Hall RM, Collis CM (1995) MicroReview Mobile gene cassettes and integrons: capture and spread of genes by site-specific recombination

Hall RM, Collis CM, KIM MJ, Partridge SR, Recchia GD, Stokes HW (1999) Mobile gene cassettes and integrons in evolution. Annals New York Acad Sci 870(1):68–80

Holden VI, Breen P, Houle S et al (2016) Klebsiella pneumoniae siderophores induce inflammation, bacterial dissemination, and HIF-1α stabilization during pneumonia. MBio 7. https://doi.org/10.1128/mBio.01397-16

Holt KE, Wertheim H, Zadoks RN et al (2015) Genomic analysis of diversity, population structure, virulence, and antimicrobial resistance in Klebsiella pneumoniae, an urgent threat to public health. Proc Natl Acad Sci U S A 112:E3574–E3581. https://doi.org/10.1073/pnas.1501049112

Huang WK, Chang JWC, See LC et al (2012) Higher rate of colorectal cancer among patients with pyogenic liver abscess with Klebsiella pneumoniae than those without: An 11-year follow-up study. Color Dis 14. https://doi.org/10.1111/j.1463-1318.2012.03174.x

Johnson CM, Grossman AD (2015) Integrative and conjugative elements (ICEs): what they do and how they work. Annu Rev Genet 49:577–601. https://doi.org/10.1146/annurev-genet-112414-055018

Kayama S, Shigemoto N, Kuwahara R et al (2015) Complete nucleotide sequence of the IncN plasmid encoding imp-6 and CTX-M-2 from emerging carbapenem-resistant Enterobacteriaceae in Japan. Antimicrob Agents Chemother 59:1356–1359. https://doi.org/10.1128/AAC.04759-14

Kiiru J, Butaye P, Goddeeris BM, Kariuki S (2013) Analysis for prevalence and physical linkages amongst integrons, ISEcp1, ISCR1, Tn21 and Tn7 encountered in Escherichia coli strains from hospitalized and non-hospitalized patients in Kenya during a 19-year period (1992–2011)

Kondo K, Kawano M, Sugai M (2021) Distribution of antimicrobial resistance and virulence genes within the prophage-associated regions in nosocomial pathogens. mSphere 6. https://doi.org/10.1128/msphere.00452-21

Lai YC, Lin AC, Chiang MK et al (2014) Genotoxic Klebsiella pneumoniae in Taiwan. PLoS One 9. https://doi.org/10.1371/journal.pone.0096292

Lam MMC, Wick RR, Wyres KL et al (2018) Genetic diversity, mobilisation and spread of the yersiniabactin-encoding mobile element ICE*Kp* in klebsiella pneumoniae populations. Microbial Genomics 4. https://doi.org/10.1099/mgen.0.000196

Lam MMC, Wyres KL, Wick RR et al (2019) Convergence of virulence and MDR in a single plasmid vector in MDR Klebsiella pneumoniae ST15. J Antimicrob Chemother 74:1218–1222. https://doi.org/10.1093/jac/dkz028

Langille MGI, Hsiao WWL, Brinkman FSL (2010) Detecting genomic islands using bioinformatics approaches. Nat Rev Microbiol 8:373–382

Lawlor MS, O'connor C, Miller VL (2007) Yersiniabactin is a virulence factor for Klebsiella pneumoniae during pulmonary infection. Infect Immun 75(3):1463–1472

Lery LMS, Frangeul L, Tomas A et al (2014) Comparative analysis of Klebsiella pneumoniae genomes identifies a phospholipase D family protein as a novel virulence factor. BMC Biol 12. https://doi.org/10.1186/1741-7007-12-41

Li B, Hu Y, Wang Q et al (2013) Structural diversity of class 1 Integrons and their associated gene cassettes in Klebsiella pneumoniae isolates from a Hospital in China. PLoS One 8. https://doi.org/10.1371/journal.pone.0075805

Lima AMDS, de Melo MES, Alves LC et al (2014) Investigation of class 1 integrons in Klebsiella pneumoniae clinical and microbiota isolates belonging to different phylogenetic groups in Recife, State of Pernambuco. Rev Soc Bras Med Trop 47:165–169. https://doi.org/10.1590/0037-8682-0021-2014

Lin TL, Lee CZ, Hsieh PF et al (2008) Characterization of integrative and conjugative element ICE*Kp*1-associated genomic heterogeneity in a Klebsiella pneumoniae strain isolated from a primary liver abscess. J Bacteriol 190:515–526. https://doi.org/10.1128/JB.01219-07

Lu MC, Chen YT, Chiang MK et al (2017) Colibactin contributes to the hypervirulence of pks+ K1 CC23 Klebsiella pneumoniae in mouse meningitis infections. Front Cell Infect Microbiol 7. https://doi.org/10.3389/fcimb.2017.00103

Marques AT, Tanoeiro L, Duarte A et al (2021) Genomic analysis of prophages from klebsiella pneumoniae clinical isolates. Microorganisms 9. https://doi.org/10.3390/microorganisms9112252

Maxwell KL (2017) The anti-CRISPR story: a battle for survival. Mol Cell 68:8–14

Mazel D (2006) Integrons: agents of bacterial evolution. Nat Rev Microbiol 4:608–620

Mogasale VV, Saldanha P, Pai V et al (2021) A descriptive analysis of antimicrobial resistance patterns of WHO priority pathogens isolated in children from a tertiary care hospital in India. Sci Rep 11. https://doi.org/10.1038/s41598-021-84293-8

Molton JS, Lee IR, Bertrand D et al (2021) Stool metagenome analysis of patients with Klebsiella pneumoniae liver abscess and their domestic partners. Int J Infect Dis 107:1–4. https://doi.org/10.1016/j.ijid.2021.04.012

Moura A, Soares M, Pereira C et al (2009) INTEGRALL: a database and search engine for integrons, integrases and gene cassettes. Bioinformatics 25:1096–1098. https://doi.org/10.1093/bioinformatics/btp105

Navon-Venezia S, Kondratyeva K, Carattoli A (2017) Klebsiella pneumoniae: a major worldwide source and shuttle for antibiotic resistance. FEMS Microbiol Rev 41:252–275

Nougayrède JP, Homburg S, Taieb F et al (2006) Escherichia coli induces DNA double-strand breaks in eukaryotic cells. Science 313:848–851. https://doi.org/10.1126/science.1127059

Odumosu BT, Adeniyi BA, Chandra R (2013) Analysis of integrons and associated gene cassettes in clinical isolates of multidrug resistant Pseudomonas aeruginosa from Southwest Nigeria. Ann Clin Microbiol Antimicrob 12. https://doi.org/10.1186/1476-0711-12-29

Putze J, Hennequin C, Nougayrède JP et al (2009) Genetic structure and distribution of the colibactin genomic island among members of the family Enterobacteriaceae. Infect Immun 77:4696–4703. https://doi.org/10.1128/IAI.00522-09

Ramirez MS, Xie G, Johnson S et al (2014) Genome sequences of two carbapenemase-resistant Klebsiella pneumoniae ST258 isolates. Genome Announc 2. https://doi.org/10.1128/genomeA.00558-14

Rice LB (1998) MINIREVIEW Tn916 family conjugative transposons and dissemination of antimicrobial resistance determinants

Roberts AP, Mullany P (2011) Tn916-like genetic elements: a diverse group of modular mobile elements conferring antibiotic resistance. FEMS Microbiol Rev 35:856–871

Rodrigues C, D'Humières C, Papin G et al (2020) Community-acquired infection caused by the uncommon hypervirulent klebsiella pneumoniae ST66-K2 lineage. Microbial Genomics 6:1–5. https://doi.org/10.1099/mgen.0.000419

Rodrigues C, Desai S, Passet V et al (2022) Genomic evolution of the globally disseminated multidrug-resistant Klebsiella pneumoniae clonal group 147. Microb Genom 8. https://doi.org/10.1099/mgen.0.000737

Runcharoen C, Moradigaravand D, Blane B et al (2017) Whole genome sequencing reveals high-resolution epidemiological links between clinical and environmental Klebsiella pneumoniae. Genome Med 9. https://doi.org/10.1186/s13073-017-0397-1

Russo TA, Marr CM (2019) Hypervirulent Klebsiella pneumoniae

Salimizand H, Shahcheraghi F, Kalantar E, Badmasti F (2013) Molecular characterization of class 1 integrons and gene cassettes in multidrug resistant (MDR) Klebsiella spp. isolated from hospitalized and outpatients in Iran, 2009. Iran J Microbiol 5(1):48

Saltzman WM (2003) Lateral DNA transfer: mechanisms and consequences. By Frederic Bushman. Cold Spring Harbor (New York): Cold Spring Harbor Laboratory Press. ISBN: 0–87969–603–6 (hc); 0–87969–621–4 (pb). 2002. Q Rev Biol 78:89–90. https://doi.org/10.1086/377835

Scott KP (2002) The role of conjugative transposons in spreading antibiotic resistance between bacteria that inhabit the gastrointestinal tract

Shankar C, Jacob JJ, Vasudevan K et al (2020) Emergence of multidrug resistant hypervirulent ST23 Klebsiella pneumoniae: multidrug resistant plasmid acquisition drives evolution. Front Cell Infect Microbiol 10. https://doi.org/10.3389/fcimb.2020.575289

Shen Z, Gao Q, Qin J, et al (2019) Emergence of an NDM-5-producing hypervirulent Klebsiella pneumoniae sequence Type 35 Strain with chromosomal integration of an integrative and conjugative element, ICE*Kp*1. https://doi.org/10.1128/AAC

Shigemoto N, Kuwahara R, Kayama S et al (2012) Emergence in Japan of an imipenem-susceptible, meropenem-resistant Klebsiella pneumoniae carrying bla IMP-6. Diagn Microbiol Infect Dis 72:109–112. https://doi.org/10.1016/j.diagmicrobio.2011.09.019

Soge OO, Beck NK, White TM et al (2008) A novel transposon, Tn 6009, composed of a Tn 916 element linked with a Staphylococcus aureus mer operon. J Antimicrob Chemother 62:674–680. https://doi.org/10.1093/jac/dkn255

Storrs MJ, Poyart-Salmeron C, Trieu-Cuot P, Courvalin P (1991) Conjugative transposition of Tn916 requires the excisive and integrative activities of the transposon-encoded integrase. J Bacteriol 173:4347–4352. https://doi.org/10.1128/jb.173.14.4347-4352.1991

Struve C, Roe CC, Stegger M et al (2015) Mapping the evolution of hypervirulent Klebsiella pneumoniae. MBio:6. https://doi.org/10.1128/mBio.00630-15

Tomich PK, An FY, Damle SP, Clewell DB (1979) Plasmid-related transmissibility and multiple drug resistance in Streptococcus faecalis subsp. zymogenes Strain DS16

Vizcaino MI, Crawford JM (2015) The colibactin warhead crosslinks DNA. Nat Chem 7:411–417. https://doi.org/10.1038/nchem.2221

Wami H, Wallenstein A, Sauer D et al (2021) Insights into evolution and coexistence of the colibactin-and yersiniabactin secondary metabolite determinants in enterobacterial populations. Microbial Genomics 7. https://doi.org/10.1099/MGEN.0.000577

Wang X, Xie Y, Li G, Liu J, Li X, Tian L et al (2018) Whole-Genome-Sequencing characterization of bloodstream infection-causing hypervirulent Klebsiella pneumoniae of capsular serotype K2 and ST374. Virulence 9(1):510–521

Wendling CC, Refardt D, Hall AR (2021) Fitness benefits to bacteria of carrying prophages and prophage-encoded antibiotic-resistance genes peak in different environments. Evolution (N Y) 75:515–528. https://doi.org/10.1111/evo.14153

Wozniak RAF, Waldor MK (2010) Integrative and conjugative elements: mosaic mobile genetic elements enabling dynamic lateral gene flow. Nat Rev Microbiol 8:552–563

Wright LD, Grossman AD (2016) Autonomous replication of the conjugative transposon Tn916. J Bacteriol 198:3355–3366. https://doi.org/10.1128/JB.00639-16

Wu K, Wang F, Sun J et al (2012) Class 1 integron gene cassettes in multidrug-resistant Gram-negative bacteria in southern China. Int J Antimicrob Agents 40:264–267. https://doi.org/10.1016/j.ijantimicag.2012.05.017

Yang X, Dong N, Chan EWC et al (2021) Carbapenem resistance-encoding and virulence-encoding conjugative plasmids in Klebsiella pneumoniae. Trends Microbiol 29:65–83

Molecular Insights into Genomic Islands and Evolution of *Vibrio cholerae* 14

Atul Vashist, Jyoti Verma, Lekshmi Narendrakumar, and Bhabatosh Das

Abstract

Vibrio cholerae is a Gram-negative, motile, slightly curved bacteria that produces an enterotoxin, resulting in a potentially fatal acute watery diarrheal disease known as cholera. The bacterium has the potential to cause endemic and pandemic outbreaks. Over the years *V. cholerae* has established itself as a successful pathogen owing to its ability to acquire genomic islands (GIs) and other mobile genetic elements (MGEs) linked with virulence factors, metabolic functions, and antibiotic resistance genes (ARGs) through horizontal gene transfer (HGT). The combination of whole genome sequencing and traditional phenotypic and molecular fingerprinting methods helped in understanding the genome dynamics and virulence potency of newly evolved *V. cholerae* strains. This chapter summarizes the overall clinical spectrum, pathogenesis, epidemiology, and basic composition of GIs in *V. cholerae*. We also included our brief understanding of various tools commonly used to study the dynamics of GIs and their functional potency.

Keywords

Antibiotic resistance · Biofilm · CRISPR-Cas · Genomic islands · Mobile genetic elements · Vaccines

A. Vashist · J. Verma · L. Narendrakumar · B. Das (✉)
Molecular Genetics Laboratory, Department of Infection and Immunology, Translational Health Science and Technology Institute, Faridabad, Haryana, India
e-mail: bhabatosh@thsti.res.in

I. Mani et al. (eds.), *Microbial Genomic Islands in Adaptation and Pathogenicity*,
https://doi.org/10.1007/978-981-19-9342-8_14

14.1 Introduction

Vibrio the motile, vibrating, Gram-negative, slightly curved bacteria ubiquitously found in a diverse range of environmental reservoirs including freshwater rivers, estuarine, aquatic, and marine habitats (Baker-Austin et al. 2018; Baker-Austin et al. 2017). These environmental reservoirs are responsible for bacterial transmission to the human host via contaminated food or water (Kaysner and Hill 2014). Vibrionaceae exemplifies various distinct prototypes of facultative and emergent pathogens. Out of more than 100 described *Vibrio* spp., only ~12 species are known to elicit infections in humans which have been further classified into two categories: cholera and non-cholera infections (Baker-Austin et al. 2018). The acute diarrheal disease caused by toxigenic *V. cholerae*, an enterotoxin-producing bacteria is the aetiological agent of a potentially fatal acute watery diarrheal syndrome known as Cholera (Finkelstein 2011, Clemens et al. 2017). It was first identified microscopically by Pacini in 1854 followed by its first isolation by Robert Koch in 1883 from Egyptian patients (Lippi et al. 2016). Since antiquity, Cholera has plagued humans with the first reported incidence of cholera-like symptoms recorded in primeval Indian medical scriptures in the Sanskrit language around fifth century BC ago. Historically, the Ganges delta in India has been considered as the original reservoir to spread Cholera disease across the globe leading to seven pandemics so far during the last two centuries (Kanungo et al. 2022). Although the risk of Cholera infection varies globally from country to country with minor cases in Australia and America as compared to Asia, Africa, and Sub-Saharan Africa majorly remain an epicenter of various cholera epidemics (Mengel et al. 2014; Jensen et al. 2021; D'Mello-Guyett et al. 2022). Thus, cholera continues as a public health threat of concern predominantly affecting the countries with poor resources, unhygienic sanitation conditions, and unavailability of clean drinking water. In addition to this, natural disasters such as earthquakes and societal crises are also known to precipitate cholera epidemics or pandemics.

V. cholerae has emerged as a successful pathogen owing to its ability to acquire major factors regulating its pathogenesis and virulence expressed either as a part of mobile genetic elements (MGEs) or in the core genome through a step-by-step process known as horizontal gene transfer (HGT) from other bacterial strains in the environment (Bik et al. 1995). For example, cholera toxin (CT) is known to be derived from a lysogenic phage CTXϕ (Faruque et al. 1998) whereas toxin-coregulated pilus (TCP), which aids in colonization in the intestinal lumen of humans, is expressed as part of Vibrio Pathogenicity Island-1 (VPI-1) (Kumar et al. 2020). However, other factors, which include but are not limited to virulence regulator ToxR which is localized in the inner membrane (Herrington et al. 1988), N-acetylglucosamine-binding protein A (GbpA), encoded in the core genome of *V. cholerae* function as an adhesin responsible for attachment of pathogen to host epithelial cells in the intestine (Sakib et al. 2018, Wong et al. 2012). The HGT is also known as a possible mechanism for the growing antimicrobial resistance in *Vibrio* species (Sun et al., 2019). Environmental *Vibrio* particularly nonpathogenic strains of *V. parahaemolyticus* usually carrying Antimicrobial Resistance (AMR) genes are considered to possibly transmit a set of drug resistance traits often called as "resistome" through horizontal gene transfer to pathogenic strains living in the same environment (D'Costa et al. 2006; D'Costa et al. 2011). This chapter provides comprehensive insights into the structure and evolutionary dynamics of Genome Islands (GIs) in *Vibrio cholerae* to define their role in pathogenesis, antimicrobial resistance. and vaccine development.

14.1.1 Taxonomic Classification of *Vibrio cholerae*

V. cholerae, part of *Vibrionaceae* family is closely related to *Aeromonadaceae* besides *Aeromonas, Phobacterium, and Plesiomonas* species and belongs to the Gamma subdivision of the phylum *Proteobacteria* (Ruimy et al. 1994). It is a comma-shaped Gram-negative bacteria. In rich and minimal media, the bacterium is motile by monotrichous polar flagella. The flagella are confined in a sheath in continuity with the lipopolysaccharide (LPS) rich outer membrane of the bacterium. Like all other Gram-negative organisms, the LPS of *V. cholerae* abundantly present all over the cell surface, not only helps bacteria to survive in presence of hydrophobic agents and detergents but also imparts major antigenic variability (Chatterjee and Chaudhuri 2006). Basically, LPS comprises three well-defined sections, (1) lipid bilayer of the outer membrane made up of lipid A, (2) the core oligosaccharide; and, (3) the serotype-specific O-antigen (Chatterjee and Chaudhuri 2006; Bertani and Ruiz 2018). The wide heterogeneity in *Vibrio* species is observed due to differences in their chemical constitution of the immunodominant O-antigen of lipopolysaccharide (LPS) (Aydanian et al. 2011), ability to produce cholera enterotoxin (Pérez-Reytor et al. 2018), and potential to cause pandemics (Finkelstein 2011; Hu et al. 2016; Kanungo et al. 2022). So far based on heat-stable O-antigen ~200 distinguished O serogroups have been discovered out of which strains assigned to O1 and O139 serogroups from natural inhabitants of aquatic ecosystems turn out to be facultative human pathogens causing severe disease and responsible for various epidemics and pandemics cholera outbreaks. Hence, these serogroups are categorized under pandemic genome (PG) group which consists of phylogenetically related strains of *V. cholerae* (Chun et al. 2009). The serological classification divides O1 serogroup into two main serotypes viz. *V. cholerae* O1 Inaba and *V. cholerae* O1 Ogawa (Finkelstein 2011). While *V. cholerae* Ogawa remains the most prevalent serotype which expresses the A and B antigens along with low levels of C antigen, *V. cholerae* Inaba strains are known to express exclusively the A and C antigens (Mandal et al. 2011; Banerjee et al. 2014). A third rare and unstable serotype known as Hikojima is known to express all three antigens, A, B, and C (Karlsson et al. 2014). These three serotypes undergo dynamic interconversion due to mutation in the WbeT methyl transferase which modifies Ogawa O-specific polysaccharide by introducing a 2-O-methyl group in its nonreducing terminal saccharide (Karlsson et al. 2014; Alam et al. 2016; Stroeher et al. 1992). Strains that are prone to an elevated frequency of conversion are grouped under Hikojima serotype. Considering the O-antigen as possible receptor for bacteriophage, this serogroup conversion may be an evolutionary strategy to escape predation by

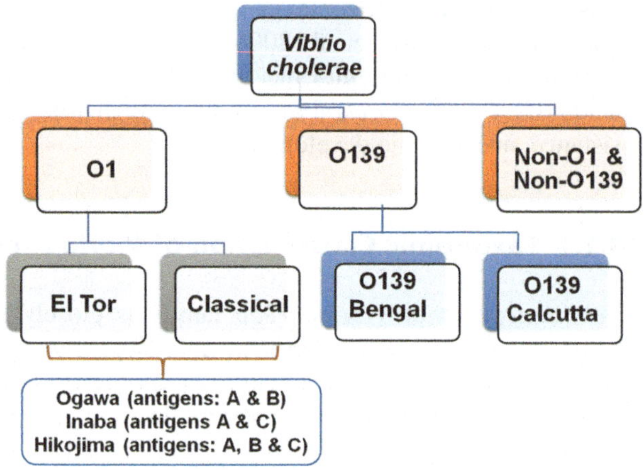

Fig. 14.1 Classification of *V. cholerae* based on serotype and biotype

environmental phage (Seed et al. 2012). Based on variations in phenotypic and genetic traits, *V. cholerae* O1 has been also categorized into two biotypes namely classical and El Tor which can be either Ogawa or Inaba serotypes (Nair et al. 2002; Kaper et al. 1995). The classification of *V. cholerae* based on serotypes and biotypes is depicted in Fig. 14.1. The global spread of the classical biotype starting from the Indian subcontinent is considered to be a major reason for the first six cholera pandemics happened between 1817 and 1923 (Siddique and Cash 2014), whereas the substitution of classical biotype by El Tor biotype in 1961 resulted in the onset of the seventh and current pandemic cholera outbreak in Sulawesi, Indonesia (Hu et al. 2016). Later in 1992, a new, non-O1 *V. cholerae* strain was assigned as *V. cholerae* O139 with origin in India rapidly spread across most parts of Asia. *V. cholerae* O139 shared a considerable homology with O1 El Tor strains except for the O1-antigen biosynthetic genes cluster and presence of a polysaccharide capsule (Faruque et al. 2003). The unique composition of O-antigen-enabled O139 strains to become epidemic through populations already immune to *V. cholerae* O1 strains (Aydanian et al. 2011).

Non-agglutinating vibrio (NAGs) strains belonging to non-O1 and non-O139 classes are devoid of genes encoding cholera toxins and are reported to sporadically cause gastroenteritis or/and sometimes septicemia but not induce cholera disease in humans. Thus, NAGs do not impact public health significantly (Morris 2014, Dutta et al. 2013, Chowdhury et al. 2016, Igere et al. 2022).

14.1.2 Clinical Spectrum and Pathogenesis of *V. cholerae*

V. cholerae is an exemplary noninvasive mucosal pathogen entering the host majorly through contaminated water or food. Depending on host susceptibility and bacterial load, cholera may persist from several hours to 3–5 days (Nelson et al. 2009). The high sensitivity of *V. cholerae* to acidic pH restricts its growth under hostile low gastric pH conditions resulting in the requirement of a higher infectious dose of $\sim 10^8$

bacilli to cause a successful infection inside the host (Kaper et al. 1995, Cash et al. 1974). *V. cholerae* infection may lead to both asymptomatic intestinal colonization and symptomatic disease. While fever is not a common symptom, stomach cramps, discomfort in abdomen and vomiting are observed as usual symptoms in the early stages of cholera (Li 2015; Nelson et al. 2009). More complicated symptoms like severe fluid volume depletion due to diarrhea, hypovolemia, metabolic acidosis, and profuse rice watery stools with output reaching as high as 1 l/h, may lead to circulatory collapse and death in severe infections of cholera also known as "cholera gravis." Rice water stool carries approximately 10^{10} and 10^{12} bacteria/liter responsible for subsequent transmission. Unlike, asymptomatic patients, who transiently shed vibrios in their stool, symptomatic patients may start shedding bacteria even before the onset of illness continues for 1–2 weeks. Hypovolemia in serious cholera patients often results in symptoms like dryness in mouth, sunken eyes, and a condition of hands also known as "washer woman's hands" in which a decrease in skin turgor results in formations of wrinkles in hand and feet with cold clammy skin. Patients often complain of muscle cramping due to decay in levels of potassium and calcium ions and eventually become apathetic and lethargic. Loss of stool bicarbonate and poor perfusion due to acidosis and lactic acidosis, respectively, may result in uneasiness in breathing reflecting compensatory hyperventilation also known as "Kussmaul breathing" which is characterized as perturbation in levels of electrolytes, water, *Oncotic Pressure*, and intracellular redox status (Harris et al. 2012). Decreased urine output may result in acute tubular necrosis leading to renal failure under severe diarrheal conditions of cholera in children (Vakrani and Nambakam 2021). Cholera in children may result in severe hypoglycemia and sometimes coma due to depletion of glycogen stores and disturbed gluconeogenesis (Harris et al. 2012, Clemens et al. 2017, Kanungo et al. 2022). Hypervolemia in children may sometimes lead to additional comorbidity like pneumonia which may prove to be fatal sometimes. In cholera endemic regions, the risk of mortality in children is ~10 times greater than in adults (Williams and Berkley 2018; Ali et al. 2012). Although a defined link between pregnancy and risk of cholera infection has not been yet established, increased risk of miscarriages, preterm deliveries, and stillbirths have been associated in pregnant women with cholera (Ciglenecki et al. 2013). Any delay in therapeutic intervention in cholera may result in 50–70% mortality, whereas the timely implementation of rehydration therapy can significantly decrease the mortality to <0.5% (Harris et al., 2012, Lippi et al. 2016, Kanungo et al. 2022, Nelson et al. 2009).

The likelihood of onset of cholera disease depends on several factors including route of infection, previous exposure to infections and host–pathogens interactions. The severity of disease is found to be independent of serotypes of *V. cholerae* and exhibits similar clinical manifestations in case of infection with either *V. cholerae* O1 or O139 strains. The function of the gut microbiota in bacterial colonization and severity of disease has been recently appreciated in defining the clinical outcomes of *V. cholerae* infection (Qin et al. 2020; Hsiao and Zhu 2020; Cho et al. 2021; Cho et al. 2022; Barrasso et al. 2022). In particular species from Prevotella and Bifidobacterium genera may confer protection in cholera by modulating the expression levels of virulence factors of *V. cholerae* (Levade et al. 2021). *V. cholerae* exploits signal molecules called autoinducers associated with quorum sensing to facilitate biofilm formation which induces virulence genes and production of secondary metabolites for pathogen survival in the presence of bile salts and hypoxic niches in small intestine (Mashruwala and Bassler 2020; Laj et al. 2022).

Cholera as a disease is established by a multifactorial process that begins with bacterial colonization inside the lumen of the intestine which is mediated by a type IVb pili subclass filamentous structure known as toxin coregulated pilus (TCP). Animal studies revealed TCP to be an essential virulence factor for pathogenesis of *V. cholerae* inside the host (Pérez-Reytor et al. 2018; Shi et al. 2022). Subsequent to bacterial colonization, the bacteria moves toward the epithelial cells by chemotaxis, where it secretes cholera toxin (CT), the predominant virulence factor, known to perturb ion transport inside the gut epithelium, resulting in the bulk intestinal efflux of water causing debilitating diarrhea also known as "rice water" stools often accompanied with vomiting. Apart from CT, two additional secreted toxins namely Zonula occludens toxin (ZoT) and the accessory cholera enterotoxin (Ace) that contribute to pathophysiology of *V. cholerae* by acting on tight junctions of epithelial cells thereby altering the host intestinal permeability (Pérez-Reytor et al. 2018). Presence of ZoT with a similar function has been noticed in *V. parahemolyticus* (Castillo et al. 2018; Pérez-Reytor et al. 2020; Prithvisagar et al. 2021).

Structurally, CT is a molecular complex (~85 kDa), made of one A-subunit (~27.2 kDa) and five B subunits (each of CT B subunit is 11.6 kDa) which are expressed as part of *ctxAB* operon (Bharati and Ganguly, 2011). CT A-subunit is an adenosine diphosphate (ADP)-ribosyl transferase with two catalytic polypeptides viz. CT-A1 and CT-A2. C-terminal A2 fragment interacts non-covalently with five B subunits via disulfide bond to form the holotoxin, CT-AB_5. Each of the five B subunits of CT-AB_5 bind to five molecules of ganglioside GM1 molecules present as receptors on lipid rafts fractions (detergent-insoluble membrane microdomains) of the target cells in the lumen of intestine. Endocytosed holotoxin is then subjected to degradasome pathway in endoplasmic reticulum which results in translocation of CT-A subunit into cytosol after it gets dissociated from the pentameric B subunit. Refolding of CT-A1 polypeptides in the cytosol facilitates its binding to the cell membrane via α-subunits of stimulatory G proteins to inject a portion of the A-subunit (A1) in the small intestine. Internalized A1 subunit is then passed through the Golgi apparatus for ADP-ribosylation of the adenylate cyclase in the cytosol resulting in cleavage of ATP to cAMP to increase levels of cAMP in the host epithelial cells of intestine. Activation of protein kinase A owing to increased cAMP levels, results in phosphorylation of CFTR, chloride channel proteins, impeding sodium chloride absorption followed by an ATP-dependent efflux of chloride ions which triggers secretion of substantial fluid along with bicarbonate, sodium, and potassium ions into the small intestine. This in turn disrupts the resorptive capacity of the large intestine, causing acute diarrhea. The overall mechanism of pathogenesis of *V. cholerae* in cholera patients is depicted in Fig. 14.2.

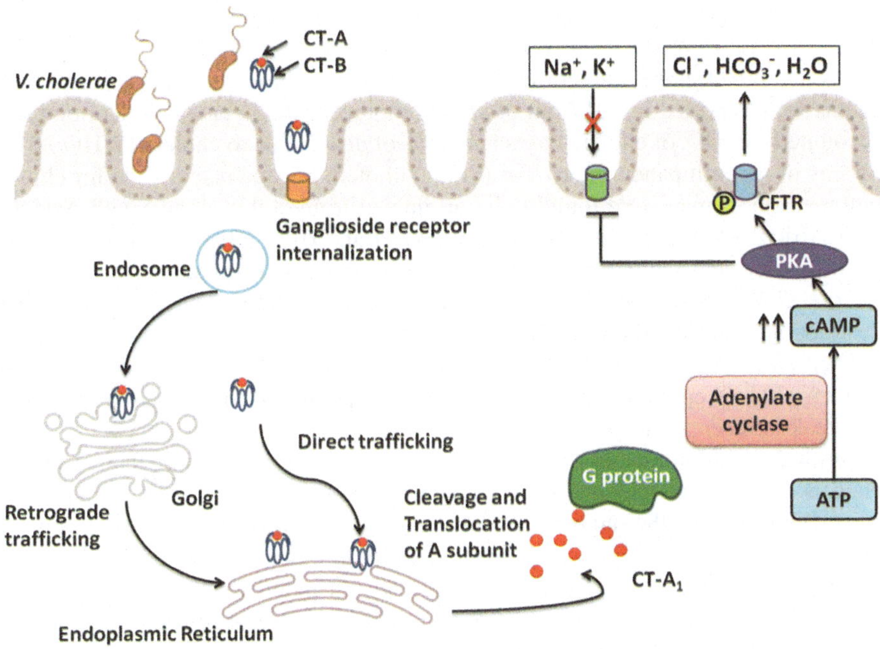

Fig. 14.2 Mechanism of pathogenesis of *V. cholerae* in cholera patients

14.1.3 Epidemiology of Pathogenic *V. cholerae*

With a predisposition to cause epidemics and potential to lead pandemics, in a large part of Asia and more recently across Africa and Haiti, cholera continues to remain endemic. As per WHO guidelines, various African regions including Rift Valley region of Central Africa, reporting confirmed cholera cases for at least three of the past 5 years have been declared as cholera endemic regions. A spatial regression model has been applied to reported cholera incidence data to estimate the global burden of disease in these endemic countries. Since cholera is considered an outcome of sociodemographic variables including poverty, poor hygiene, and infrastructure, many countries avoid the reporting of actual incidence of cholera cases as this may negatively impact on their trade and tourism. Moreover, lack of specificity in clinically defining confirmed cholera cases affecting systematic laboratory confirmation poses another limitation towards our understanding of the true global burden of cholera. Despite the limitations in surveillance systems, around 323, 369 cholera cases, and 857 deaths were notified to WHO from 24 countries across the globe in the year 2020. However, these numbers could be misleading and maybe a false representation of the global estimate of the disease because of underreporting of the cholera cases and deaths. A total of 99 countries declared a minimum of one cholera case spreading locally, with >26 million reported cases and > 10,000 deaths reported. Interestingly, the majority of deaths and cases were reported in Africa and the Middle East (WHO 2020; WHO 2017; Kanungo et al. 2022).

As per WHO global cholera report, more than a million cholera cases and 5654 deaths were reported in 2017 out of which unprecedented Yemen cholera outbreak accounted for 84% of cholera cases and 41% of deaths due to cholera (WHO 2017). Strains of seventh pandemic El Tor (7PET) cholera lineage responsible for cholera outbreaks in Yemen were found to be originated from outbreaks in South Asia and East Africa (Weil et al., 2019, Ramamurthy et al. 2019). The basic infrastructure, health, water, and sanitation in Yemen have been devastated badly due to ongoing internal civil war and also airstrikes by the Saudi-led coalition. Other than war-based destruction, natural calamities have also been linked with outbreaks of cholera in Yemen with a death toll of 639 reported so far (Ng et al. 2020). This makes the Yemen cholera epidemic as the biggest and expeditious transmission of cholera outbreak in recent history.

With the reblooming cholera outbreak in 2010 after a gap of more than a century, Haiti is also now considered to be cholera endemic (Lantagne et al. 2014). In the early 1960s, during the ongoing seventh cholera pandemic, the classical strain was replaced by El Tor biotype which spread rapidly throughout Indonesia followed by the Philippines, Malaysia, and Taiwan and later in Cambodia, Thailand, Singapore, and India. This pandemic spread its roots to Pakistan, Nepal, Brunei, Afghanistan, Iran, Hong Kong, Laos, and Myanmar (Deen et al. 2020). Many countries in Asia continued to report *V. cholerae* El Tor-mediated cholera cases for many decades. However, in late 1992, O139 Bengal, an unappreciated serogroup of *V. cholerae* led to cholera outbreak in India and Bangladesh (Deen et al. 2020; Nair et al. 2002; Faruque et al. 2003; Das et al. 2016). Although *V. cholerae* O139 continued to spread in other parts of Asia till 1993, it eventually disappeared and was very less frequently reported. Currently, *V. cholerae* O1 El Tor continues to remain the cause of virtually all cholera cases (Hu et al. 2016, Piret and Boivin 2021). Notably, the clinical manifestations in the host remain indistinguishable in case of cholera caused by either O1 or O139 strains responding similarly to given treatment (Morris Jr et al. 1995; Bhattacharya et al. 1993).

Hybrid strains of *V. cholerae* exhibiting El Tor phenotype known to cause severe dehydration possibly due to increased expression of classical cholera toxin, were identified in the early 1990s which later displaced the typical El Tor biotype in Bangladesh have other sites in 2002 (Nair et al. 2006; Nair et al. 2002). As per data between 2008 and 2012, India and Bangladesh were estimated to have the highest number of cholera cases and deaths annually (Ali et al. 2015; Deen et al. 2020; WHO 2017).

Recent data from Integrated Disease Surveillance Program at district level of 24 of 36 states in India, 13 states were designated as endemic during the period 2010–2015 revealing 27,615 cholera cases (Ali et al. 2017). From 1997 to 2006, North-eastern states and Andaman Nicobar Islands accounted for 91% of all cholera cases reported at least a year (Kanungo et al. 2010).

A continuous evolution of genetically and phenotypically pathogenic *V. cholerae* strains has been noted across Asia (Jain et al. 2013). However, many countries in

Asia, despite their risk for cholera due to significant seasonal transmission, were hesitant to officially recognize cholera and did not report any cholera cases in 2017 outbreaks. On the other hand, many countries like China and Vietnam in the Asian region that encountered terrible cholera epidemics in the past proved to be successful in efficient control and management of cholera outbreaks (Didelot et al. 2015; Ahmed et al. 2018). This is evident by the fact that since 2012, no cases of cholera have been reported from Vietnam so far.

Ecological studies in cholera endemic areas highlighted the role of environmental factors such as climate and surface temperature of oceans and sea also modulate cholera epidemics (Jutla et al. 2013, Xu et al. 2014, Sakib et al. 2018, Colwell 1996). However, human-to-human transmission due to immigration remains the major factor than the climate conditions for large-scale spread of 7 PET strains.

14.2 Mobile Genetic Elements in *V. cholerae*

MGEs constitute the accessory genome of bacteria and include phages, genomic islands, plasmids, insertion sequence (IS) element, integrons, integrative conjugative elements (ICEs), and transposon-like elements. The genome of *V. cholerae* is dynamic and has evolved over the decades. MGEs have played a pivotal role in its genome dynamics, evolution, and pathogenesis. A schematic diagram of the ***V. cholerae*** genome and various MGEs present in most of the clinical isolates have been shown in chromosome (a) and the genetic organization of different MGEs (b) is shown in Fig. 14.3.

14.2.1 Types of MGEs

Phages

CTX phage is the filamentous phage that produces the deadly cholera toxin that leads to acute diarrheal illness. It consists of a long circular single-stranded (ss) + ve strand genome of 7 kb size that comprises two modules namely Core (4.6 kb in size) and Repeat Sequence 2 (RS2) (2.4 kb in size). The three genes viz. *rstR*, *rstA*, and *rstB* that are part of RS2 locus are known to respectively control the regulation of its genes, replication, and integration of CTXϕ. The core genome of CTXϕ contains seven structural genes, out of which 5 genes viz. *psh, cep* (Core encoded pillin), *pIII*^{CTX}, *ace* (accessory cholera enterotoxin), *zot* (zonula occludens) play a central role in phage assembly and morphogenesis (Waldor and Mekalanos 1996). Psh, Cep, pIII^{CTX}, and Ace proteins act as structural proteins. According to *rstR* gene sequences and the host bacteria, CTXϕ can be divided into 4 classes (I) CTXϕ^{class} (source: *V. cholerae* O1 classical biotype strains), (II) CTXϕ^{ET} (source: *V. cholerae* O1 El Tor biotype strains), (III) CTXϕ^{calc} (source: *V. cholerae* O139 strains), and (IV) CTXϕ^{Env} (source: *V. cholerae* non-O1, non-O139 strains) (Pant et al. 2020b).

In order to permanently integrate at the dimer resolution site (*dif*) of the *V. cholerae* chromosome, CTXϕ utilizes XerC and Xer, two tyrosine recombinases

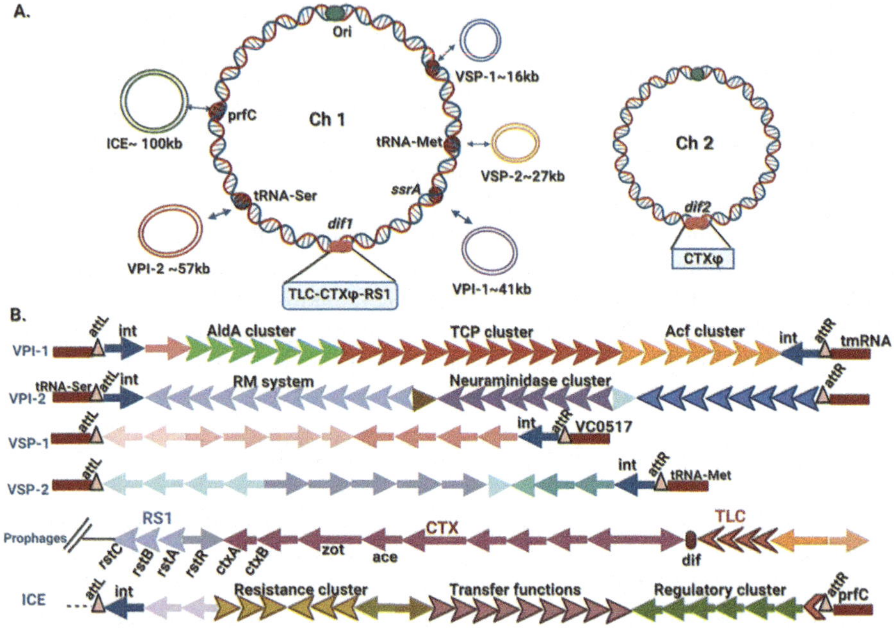

Fig. 14.3 Schematic diagram of the *V. cholerae* genome and various Mobile Genetic Elements present in most of the clinical isolates have been shown in chromosome (**a**) and the genetic organization of different MGEs (**b**)

originated from host. The cholera toxin is encoded by two CTX genes viz. *ctxA* and *ctxB*. Following the phage infection into *V. cholerae* cells, the integration of ssDNA of CTXϕ genome occurs either through site-specific recombination at 28 bp *dif* site of the chromosome or through conversion into double-stranded DNA (dsDNA) form for the rolling circle replication and the transcription of genes involved in the replication of phage and its morphogenesis (Das 2014). It has been noted that different *V. cholerae* isolates derived from environment or patients get lysogenized singularly or jointly at the dif loci (Chun et al. 2009).

A study reported the presence of a modified CTXΦ prophage array from the original El Tor strain in the five clinical *V. cholerae* isolates of Kolkata (India) between 2007 and 2011, which led to defective replication of the phage. Since the 1970s, the epidemic strains of *V. cholerae* produced CTXΦ phage particles, however, around 2010, the epidemic strains lost the ability to produce CTX, which is suggestive of the changing dissemination patterns of the cholera pathogen (Ochi et al. 2021).

An advanced genetic tool has been developed by Das et al., a novel vector pBD62 based on the CTXϕ genome fragment (Das et al. 2014). The following characteristics of this vector are distinctive: irreversible site-specific integration at the dif site of the chromosome; a wide host range because the vector's attachment site (*attP*) is compatible with a number of bacterial species using the XerC and XerD systems; the vector carries a regulatory promoter called pBAD (arabinose inducible); and can be used to tag one of the chromosomes in the host with multiple chromosomes. Several Proteobacteria species of bacteria contain distinct sequences that are compatible with the pBD62 attachment sequence, hence, this novel vector is apt for genetic engineering and functional genomics with an advantage of stable integration, absence of selection pressure, and expression of proteins detrimental to the growth of cells.

The other additional phages that mediate the HGT along with CTXϕ in *V. cholerae* include 493φ, fs1 and fs2 phages, KSF-1, RS1 and TLC satellite phages, and, VGJ, VSK, VSKK phages (Faruque and Mekalanos 2012). KSF-1φ and VGJφ are known to provide proteins for capsid as well as for the assembly and packaging of the genomes of RS1 and other satellite phage genomes, hence facilitating the lateral gene transfer of the satellite phages. A brief description of the phages and their functions has been provided in Table 14.1.

Plasmids are the extrachromosomal, circular, and self-replicating DNA molecules, which may or may not carry genes for its mobility (Smillie et al. 2010). In addition to being frequently associated with heavy metals and ARG (Antibiotic resistance gene) cassettes, they may also contain additional MGEs including transposons, integrons, and Integrative Conjugative Elements (ICE) (Ceccarelli et al. 2006) (Rivard et al. 2020). Since the 1960s, various investigations have shown that the genomes *V. cholerae* clinical and environmental strains from serogroups O1, O139, and non-O1, non-O139 contain plasmids (Amaro et al. 1988). The reported plasmids from various isolates of *V. cholerae* range in size (4–200 kb), dissemination methods (conjugative or non-conjugative), and encoded functionality (virulence, pathogenesis, antibiotic, and heavy metal resistance).

Insertion sequence (IS) elements are autonomous mobile elements with two terminal inverted repeats (IRs) at the flanking sides, two direct repeated sequences (DRs) and one or two transposase encoding genes that facilitate its transfer (Siguier et al. 2015). IS elements are divided into various families according to how closely the amino acid composition of the transposase gene matches (Mahillon and Chandler 1998). The IS element can inactivate or modulate the gene function depending on its site of insertion and proximity to the promoter region. As a result, IS elements have been crucial to *V. cholerae*'s evolutionary dynamics. The cholera outbreak in Haiti during the year 2010 lasted for 2 years and the genome sequences of the isolates had 207 insertions of five different ISs, as identified by PanISa, software for IS detection (Couchoud et al. 2020).

Transposons were initially identified as the MGEs responsible for antibiotic resistance. They can integrate into the plasmids or chromosomes at the non-homologous target site. However, some transposons integrate at the selective target sites while others show less selectivity. They are flanked by inverted repeat (IR) sequences and encode a transposase protein which mediates its integration and excision. Transposons have been linked to reports of drug resistance as well as other functions like metabolic plasticity and resistance to heavy metals. They may have an insertion sequence (IS) element along with a metal or antibiotic resistance cassette.

Table 14.1 Different phages that mediate horizontal gene transfer in *V. cholerae*

Phage	Genome size (kb)	Receptor	Function	References
CTXΦ	6.9	Toxin coregulated pilus (TCP)	Filamentous bacteriophage that produces cholera toxin responsible for the pathogenesis in the host.	Waldor and Mekalanos (1996)
RS1Φ	3	TCP/ mannose sensitive hemagglutinin (MSHA) pilus	A satellite phage that utilizes CTXφ-encoded proteins for the formation of RS1 phage particles. RS1 encodes the gene for an anti-repressor protein RstC that regulates CTXφ replication and transmission.	Faruque and Mekalanos (2003)
TLCΦ (Toxin Linked Cryptic)	5.3	MSHA	The satellite phage genome has a sequence which is similar to the dif recombination sequence that helps in chromosome dimer resolution during cell division. Lysogeny by this phage generates a functional dif site in dif defective strains and leads to stable integration of CTXφ genome.	Hassan et al. (2010)
VGJΦ	7.5	MSHA	VGJφ integrates into the same chromosomal attachment site as CTXφ and enters into a lysogenic state.	Campos et al. (2003)
KSF-1Φ	7.1	MSHA	The filamentous phage that mediates the packaging of RS1.	Faruque et al. (2005)
fs-2Φ	8.6	MSHA	Filamentous phage that forms turbid plaques on V. cholerae O139 and O1 El Tor biotype strains. It has a 715 nucleotide fragment located in its large intergenic region, which is homologous to a part of region RS2 of CTXφ. It produces satellite phage TLCφ particles, hence acts as a helper phage.	Ikema and Honma (1998)
CP-T1	43.5	O-antigen	A generalized transducing phage of *V. cholerae* that helps in lateral transfer of chromosomal segments among *V. cholerae* strains.	Seed et al. (2011)

ICEs are the self-transmissible MGEs that are mosaics of plasmids and phages, as like the plasmids they can get transferred through conjugation and similar to the phages they can integrate and replicate with the host chromosome (Wozniak et al. 2009). ICEs are known to significantly contribute to the emergence of antibiotic resistance in *V. cholerae* (Waldor et al. 1996). The ICEs contain distinct modules with different functions such as module for integration/excision, replication/DNA processing, DNA secretion, regulation, and other auxiliary functions (Das et al. 2020). The genes responsible for the alternate catabolic processes, genes for resistance to heavy metals, virulence factors, or genes encoding toxin protein and AMR genes are frequently found in the auxiliary modules. After integration, the circular intermediates may be created when the ICEs get excised from the chromosome. These circular intermediates are then conjugated to nearby bacteria and passed on to them.

In *V. cholerae*, ICEs belonging to the SXT/R391 family can be found. Since they all encode a similar integrase, they are categorized as the ICE family (Int). The tyrosine recombinase Int facilitates 5′ end integration of the conserved chromosomal gene *prfC* (peptide chain release factor 3) (*attB*) (Hochhut and Waldor 1999). During the excision process, Int along with a recombination directionality factor named Xis, is involved in the reverse recombination reaction between *attL* and *attR* which are respectively present on the left and right ends of the integrated element. This leads to the reconstitution of *attP* and *attB*. Further, the tra genes express proteins involved in DNA processing for transfer, formation of mating pair and conjugation machinery (Wozniak et al. 2009). SXT represents an ICE of ~100 Kb in size which was first reported in a non-O1 serogroup O139 *V. cholerae* responsible for cholera epidemic in the Indian subcontinent during the year 1992. The resistance genes for several antibiotics including sulfamethoxazole and trimethoprim (abbreviated as SXT), chloramphenicol, and streptomycin used for treatment of cholera, are present in SXT locus. Most *V. cholerae* O1 serogroup clinical isolates from Asia and Africa were reported to have a comparable ICE (Wozniak et al. 2009).

Integrons are the genetic system that utilizes site-specific recombination to capture the exogenous ORFs and the integrated cassettes are converted to functional genes with the help of active transcription machinery (Hall and Collis 2006). The fundamental modules include an integrase encoding gene (*int*), a functional promoter, and a specific integration site. The integrase enzyme mediates the site-specific integration between the host chromosome (attC) and the single-stranded folded exogenous cassette (attI). Based on their genetic composition, the integrons can be categorized as: (i) the mobile integrons and (ii) the superintegrons. The mobile integrons are involved in the dissemination of ARGs since they are often associated with the MGEs (transposons, insertion sequences, conjugative plasmids, and ICEs). However, the superintegrons are a “distinct type” of integron with the characteristic features: (i) 20 cassettes or more should be in the respective array, (ii) a single type (over 80% identical) of *attC* site (59-be) is predominantly present, and (iii) the integron is not linked with other MGEs (Hall et al. 2007). In *V. cholerae*, superintegron is part of Ch2 and is not associated with any MGE. The first *V. cholerae* strain N16961 with its entire genome sequenced was reported to harbor the superintegron with more than 210 ORFs, which encoded hypothetical or toxin-antitoxin functions. Mostly the genes encoded by the superintegron are involved in fitness and survival under stress conditions (Hall et al. 2007).

Genomic islands are large clusters of genes acquired by HGT and carry genes required for fitness, pathogenesis, and AMR. Their sizes ranges from 4.5 to 600 kb and have direct repeats, specific integration sites, and presence of mobility genes

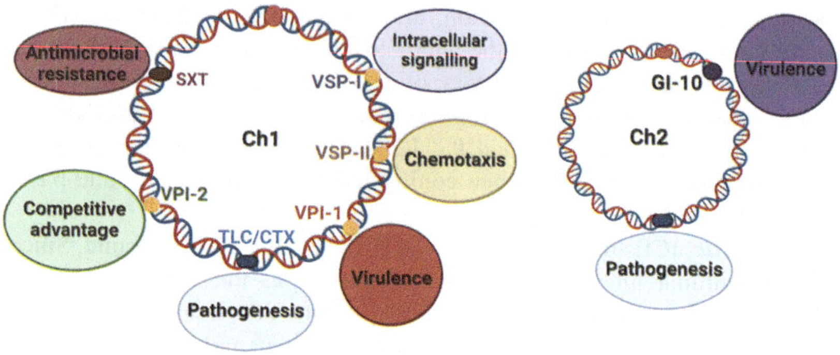

Fig. 14.4 Schematic representation of different functional traits encoded by the Genomic Islands of *V. cholerae*

encoding integrases and transposases as one of its defining characteristics (Bellanger et al. 2014). The sequence composition of the GIs differs from the core genome, with variable GC content, different codon usages, and dinucleotide frequency as indicators of their location (Juhas et al. 2009). The characteristic feature of GIs is the presence of flanking direct repeats (DR) sequences and are mostly found near the tRNA genes. They mostly harbor phage genes, genes required for conjugation, insertion element (IS), integrases, and transposons that mediate their mobilization (Juhas et al. 2009). The functional traits encoded by the GIs have been summed up in Fig. 14.4.

14.2.2 Types of Genomic Islands

The GIs are divided according to their functional role: (i) metabolic islands (MIs) involved in metabolic processes; (ii) resistance islands (RIs), encoding metal or antibiotic resistance proteins; and (iii) symbiotic islands (SIs), that can carry out various functions in different environments (Hacker and Carniel 2001). They are crucial to the bacterial genomes' adaptability and evolution. GIs are mostly associated with the transmission of pathogenic and AMR factors (Juhas et al. 2009). The seventh pandemic O1 El Tor biotype strains of *V. cholerae* has replaced the O1 classical biotype from the sixth pandemic. The transition from environment to pathogenic strains has been the result of acquisition of various genomic islands such as *Vibrio* pathogenicity islands (VPI-1 and 2) and the switch from the sixth to the seventh pandemic has been because of the *Vibrio* seventh pandemic islands (VSP-1 and 2) (Hu et al. 2016). The transfer of genes linked to virulence, such as the type III secretion system (T3SS) of non-O1/non-O139 *V. cholerae* strains is known to be mediated via GIs (Dziejman et al. 2005).

14.2.2.1 *Vibrio* Pathogenicity Island-1 (VPI-1)

VPI-1 is the most well-characterized GI, since it encodes various virulence factors, aids in the uptake of CTXϕ and production of cholera toxin as well as aids in the better colonization of *V. cholerae* epidemic isolates that facilitate the pathogenesis of the disease cholera. All the pathogenic strains of *Vibrio* harbor VPI-1 along with CTXϕ in their genome. It is a 41.3-kb essential virulence gene cluster found in the epidemic strains of the Sixth and Seventh Cholera pandemic, respectively. VPI-1 encompasses *VC0817–VC0847* of reference *V. cholerae* El Tor strain N16961 and encodes 31 genes with both known and hypothetical functions. The site of integration is tmRNA (*ssrA*) loci, which is a frequently used site for the integration PAI in *Vibrio* species (Karaolis et al. 1998). It has terminal phage-like attachment sites (*att*) and the integrase gene, Transposase like gene (*vpiT*), toxin-coregulated pilus (TCP), the accessory colonization factor (*acf*), which plays an important role in the colonization, and the virulence regulators ToxT and TcpPH (Karaolis et al. 1998; Kovach et al. 1996). The main factor impacting cholera pathogenesis is VPI-1, the transcriptional activator ToxT stimulates the transcription of CT by directly interacting with the promoters of *ctxA*, *ctxB*, *acfs*, *aldA*, and *tcpA* via the direct repeats (TTTTGAT) known as Tox boxes.

Earlier VPI-1 was proposed to be encoded by a novel filamentous phage that mediated its movement and inter-strain transfer (Karaolis et al. 1999). However, further studies defied this finding (Faruque et al. 2003). It was shown that VPI-1 has the ability to excise and form extrachromosomal circular excision product (pVPI-1) from its chromosomal insertion site (Rajanna et al. 2003). The excision is independent of *recA* and occurs through site-specific recombination. The analysis of genome sequence showed that the VPI-1 is conserved in all the epidemic and pandemic strains of *V. cholerae*, while it was either present or absent in non O1- non-O139 strains (Chun et al. 2009). However, in both biotypes, there were observed to be approximately 483 polymorphic nucleotides (Karaolis et al. 2001). There was observed to be the most variation at the nucleotide level (22.5%) and at the protein level (16.9%) in the *tcpA* gene that encodes the type IV pilus, which acts as a receptor for CTXΦ. In the El Tor strain, AcfD, an auxiliary colonization factor involved in intestinal colonization, had a long open reading frame (ORF) (Kumar et al. 2020). On the same chromosomal location as VPI-1, Labbate et al. reported a GI named as GI*Vch*S12 in a non-O1/O139 strain of *V. cholerae* that harbors an integrase gene having 100% protein and 94% nucleotide identity as to the VPI-1 integrase, however, the attachment *(att)* sites were 100% identical to the ones found in VPI-1. However, the TCP and other auxiliary genes were found to be absent. A type VI secretion system (T6SS) and a CRISPR-Cas element were both present in GI*Vch*S12, and other *V. cholerae* genomes also had similar GIs (Labbate et al. 2016).

14.2.2.2 *Vibrio* Pathogenicity Island-2 (VPI-2)

VPI-2 is a 57.3-kb gene cluster that harbors 52 ORFs and encompasses *VC1758* to *VC1809* on the reference genome of N16961(Jermyn and Fidelma Boyd 2002). It is integrated at a transfer RNA (tRNA)-serine locus (*VC1757.1*) flanked by direct repeats and is found in the pandemic isolates (Jermyn and Fidelma Boyd 2005). It encodes a P4-like integrase (*VC1758*), a Mu phage-like region, a restriction modification (RM) system, a sialic acid metabolism region, and a neuraminidase (*VC1784*) that functions as a glycosyl hydrolase and releases sialic acid from sialoglycoconjugates to uncover GM1 gangliosides, which is the receptor for cholera toxin (Galen et al. 1992). The neuraminidase could also constitute the mucinase complex, which hydrolyzes the intestinal mucus mediating the movement of bacterium in the epithelium (Reen et al. 2006). The increased expression of integrase, *intV2* (VC1758), and the recombination directionality factor (RDFs), *vefA* (VC1785) mediates the excision of VPI-2.

VPI-2 appears to act as a defense island, harboring three unique defense systems clustered together—Zorya, R/M, and DdmDE. These defense systems prevent phage infection and eliminate the plasmids, which has led to the emergence of 7PET strains over the sixth pandemic strains (Jaskólska et al. 2022).

14.2.2.3 *Vibrio* Seventh Pandemic Island-I (VSP-I)

The gene cluster VSP-I is present in the *V. cholerae* El Tor strains of seventh cholera pandemic and O139 serogroup isolates. It is a 16-kb long region from ORFs *VC0175* to *VC0185* (Dziejman et al. 2002). The G + C content of the entire cluster is 40% as compared to 47% of the whole genome.

It has also been reported to have two distinct phage defense systems, namely the DncV–CapV CBASS and DcdV systems (Jaskólska et al. 2022). It was shown by ChIP-seq and RNA-seq that a VPI-1-encoded short RNA reduced the expression of a formerly unknown transcription factor (VspR) of VSP-1. VspR is involved in modulating the expression of various VSP-1 genes including a gene that codes for a novel class of dinucleotide cyclase (DncV). DncV contributes to the production of a hybrid cyclic AMP-GMP molecule necessary for effective intestinal colonization and for reducing the chemotaxis of *V. cholerae*, a trait linked to increased infectiousness. This depicts VSP-1 as a pathogenicity island in *V. cholerae* and links its emergence to seventh cholera pandemic strains (Davies et al. 2012).

14.2.2.4 *Vibrio* Seventh Pandemic Island-II (VSP-II)

VSP-II is a 26.9-kb gene cluster spanning *VC0490* to *VC0516* of the reference N16961 genome and encodes 30 ORFs (Taviani et al. 2010). It encodes a P4-like integrase (*VC0516*) and is integrated at a tRNA-methionine locus with flanking direct repeats. VSP-II encodes type IV pilin, an AraC-like transcriptional regulator, two methyl-accepting chemotaxis proteins, and a DNA repair protein. The complete function of VSP-II is poorly understood, since most of the encoded genes are not functional in standard lab conditions which implies an unknown signal from host or the environment. Both the host and environmental factors can cause zinc starvation stress. The well-conserved Zur repressor in *V. cholerae* controls the expression of a group of genes in response to the stress of zinc deprivation.VSP-II encodes Zur-regulated congregation factors, which include the transcriptional activator VerA (Vibrio energy taxis regulator A). VerA induces the expression of the chemotaxis receptor AerB (aerotaxis B) that carries out the oxygen-dependent congregation and energy taxis. This implicates the role of VSP-II genes in chemotaxis and motility during zinc deprivation (Murphy et al. 2021).

El Tor isolates from various continents, including Asia, Africa, and Latin America, have been reported to contain a number of VSP-II variations. Understanding the genetic lineages implicated in the cholera epidemic's global expansion has also benefited from the characterization of VSP-II types. An investigation using whole genome sequencing (WGS) of *V. cholerae* O1 strains isolated from cholera patients in Kolkata between 2007 and 2014 revealed the heterogeneity of VSP-II based on single nucleotide polymorphisms (SNPs) that showed the changes of dominant strains in recent years in Kolkata (Imamura et al. 2017).

VSP-II was reported to harbor DdmABC that plays an important role in defense against phage infection and stability of small plasmids. The defense module *ddmABC*, which functions in conjunction with VPI-2, has been crucial to the development of 7PET, the most productive lineage of *V. cholerae* to date (Jaskólska et al. 2022).

V. cholerae also harbors PLEs (phage-inducible chromosomal island-like elements) as a defense tool against predatory phages (O'Hara et al. 2017). Gram-negative pathogenic bacteria harbor a type-3 secretion system (T3SS) that mediates their pathogenesis. The T3SS genes were first discovered in the *Vibrio* species *V. parahaemolyticus* with two sets, T3SS1 and T3SS2, and later, in the genome of *V. cholerae* strain AM-19226. This indicated the existence of T3SS genes related to *V. parahaemolyticus* T3SS in VPI-2 GI (Dziejman et al. 2005). Infant mouse model depicted its role in intestinal colonization and thus acting as a virulence determinant. T3SS genes are distributed among *Vibrio* species and are implied to be transferred by HGT mechanisms. These genes are localized on the VPI-2 region, which excises to form circular intermediates and leads to horizontal transfer of the T3SS-related genes (Morita et al. 2013).

It has been reported that a fraction of *V. cholerae* strains belonging to the non-O1/non-O139 serogroup transmit disease through T3SS-mediated pathways that are encoded by genes on a ~ 50 kb genomic island. T3SS GI contains genes that code for the T3SS structural apparatus, effector proteins, and the ToxR homologous VttRA and VttRB, which are transmembrane transcriptional regulators. Additionally, the transcriptional control of stress responses, type 6 secretion (T6SS), chemotaxis, and motility is regulated by VttRA and VttRB (Chaand and Dziejman 2013).

The *V. cholerae* non-O1/O139 strain S24, which was found in an estuarine river in Sydney, Australia, was identified to contain a novel GI that could insert into *recA*. The GI encodes (I) a *recA* gene that is phylogenetically divergent from the disrupted host *recA*; (II) a DNA polymerase V encoding *umuDC* operon; and (III) hypothetical proteins encoding genes and proteins with DNA processing domains (Rapa et al. 2015). The GI possessed excision and integration function. It also provided protection against the action antibiotics bleomycin and ciprofloxacin. A recent study has reported the presence of a genomic island named GI-10 in *V. cholerae* (Herrera et al. 2022). Motility-associated killing factor (MakA) which is a distinct component of GI-10 was reported earlier as a motility-associated secreted toxin from *V. cholerae* (Nadeem et al. 2021). MakA is a component of an operon that also contains the genes for MakB and MakE, two other possible alpha-pore-forming toxins (PFTs) that could function as a three-component cytotoxin against phagocytic cells. Hence, this gene cluster may provide fitness and virulence potential to *Vibrio* under different conditions.

14.3 *Vibrio cholerae* CRISPR-Cas System

Being a natural inhabitant of the environmental water bodies, *V. cholerae* is exposed to several bacteriophages that can infect the bacterium and hijack the bacterial replication machinery for their propagation and survival. The knowledge of phage–*V. cholerae* interaction dates back to the identification of bacteriophages itself when M. E. Hankin in 1896 observed reduced cholera cases along certain regions of the banks of the Ganga river. The reduction in cholera cases was attributed to the reduction in *V. cholerae* count in certain regions of Ganges due the abundance of a virus-like particle (Letchumanan et al. 2016). Further, evolutionary studies have revealed that many of the toxin genes of *V. cholerae* have been acquired from lysogenic phages that have historically infected the bacterium and then became a part of the *V. cholerae* genome (Das et al. 2010). Thus, there are phages that help the bacterium to evolve by increasing their survival fitness. On the other hand, there are also phages that lyse and kill the bacterium. All bacteria including *V. cholerae* have devised various mechanisms like restriction modification systems and phage receptor modifications to evade phage infections that can cause bacterial lysis. Another recently identified defense mechanism against phage infection is by "Clustered Regularly Interspaced Short Palindromic Repeat" (CRISPR) and CRISPR associated (Cas) system which includes microbial nucleases that can degrade the genomic elements of the phage thereby protecting the bacterium (Mojica et al. 2005; Barrangou and van der Oost 2012). These systems constitute the bacterial adaptive immunity that can recognize foreign genetic material as that of phages and neutralize them. Another major mechanism is through the MGE, phage-inducible chromosomal island-like element (PLE). When a phage infects the *V. cholerae*, the PLE gets excised from the chromosome, replicates itself and gets packaged into the viral capsid and prevents the phage elements from assembling into new virions. In turn, phages have also adopted strategies to defy the bacterial CRISPR-Cas defense systems.

14.3.1 General Structure of CRISPR-Cas

The three major components of the bacterial CRISPR-Cas system include a leader sequence, a set of *cas* genes and the CRISPER array which constitute 1–100 direct repeats of about 25–35 bp differentiated by 25–35 bp long spacers. The spacers are generally sequences from the phage genome that has previously infected the bacterium that codes memory and protects the bacterial cells from future directions (Hille and Charpentier 2016). The *cas* genes code for Cas proteins which are important for recognition of the phage genetic element, addition of new spacer sequences, and nuclease activity (Nuñez et al. 2014). Further, based on the differences in the arrangement of *cas* loci and Cas protein, the CRISPR-Cas system in different bacteria are classified into two broad groups and six different types. Further based on differences at the molecular levels, the six types (I–VI) are further subdivided into different subtypes (e.g., I-A, I-B, I-C, I-D, I-E, I-U, II-A, II-B, II-C, III-A, III-B, III-C, III-D, V-A, V-B, V-C, V-D, V-E, V-U, VI-A, VI-B, and VI-C). Most of the types recognize DNA while type III systems recognize both DNA and RNA (Samai et al. 2015). It has been identified that different bacteria can harbor different subtypes of CRISPR-Cas system and it can be used as a typing method to distinguish different biotypes of bacteria. Upon the recognition of a foreign phage nucleic acid, the CRISPR loci transcribes to form CRISPR RNA (crRNA). This transcribed small RNA will be the memory molecules that have complementary sequences to that of the phage nucleic acid sequences that have once infected the bacterium. This crRNA will guide the CRISPR protein toward the phage nucleic acid and facilitate its cleavage. Type I system, in addition to the target DNA also known as the protospacer, requires an additional motif known as the protospacer adjacent motif (PAM) (2–5 bps long) upstream of the protospacer sequence for its effective action (van der Oost et al. 2014). However, not all CRISPR types (type III) require the PAM for their action.

14.3.2 CRISPR-Cas in Clinical and Environmental *V. cholerae* Isolates

In pandemic *V. cholerae,* classical strains possessed the CRISPR-Cas type I subtype E system located on a 17-kb genomic island known as the GI 24 (Chakraborty et al. 2009). The CRISPR-Cas type I subtype E system was identified to identify *Vibrio* phage X29/phi2. Box et al. showed that the GI 24 of classical strains could be transferred to El Tor strains through simple transformation (Box et al. 2016; Labbate et al. 2016). The acquisition of GI 24 of classical *V. cholerae* in El Tor biotype enabled the bacterium to resist CP-T1 phage predation. However, in majority of the seventh pandemic El Tor isolates, CRISPR-Cas type I subtype F system was identified in a 29-kb genomic island known as the *Vibrio* pathogenicity island -6 (Carpenter et al. 2017). However, this VPI-6 was absent in the N16961 strain. In environmental *V. cholerae* isolates, the CRISPR-Cas system was located in GI*Vch*S12 which had close homology to that of the *V. cholerae* O1 VPI-1 that possessed the TCP. The GI (GI*Vch*S12) was also identified to harbor genes for type VI secretion system in *V. cholerae* O1/O139 (Labbate et al. 2016). Recent advancements in whole genome sequencing and comparative genomics revealed additional variations in the CRISPR-Cas types and subtypes. A variant CRISPR-Cas type I subtype F system known as the CRISPR-Cas IFv was identified and a mini CRISPR-Cas type I subtype F system was identified located at the Tn7 transposon of *V. cholerae* El Tor strains. This CRISPR-Cas type I subtype F system located on the Tn7 transposon was observed to be integrated at different chromosomal locations in different *V. cholerae* strains and the Tn7 transposon was also identified to possess a restriction modification system (McDonald et al. 2019). Interestingly, *Vibrio paraheamolyticus* strains were also identified to possess Tn7 transposons that harbored CRISPR-Cas type I subtype F systems. Additionally, hybrid types of CRISPR-Cas systems were identified which, unlike containing conventional type-subtype combinations, possessed different type-subtype combinations (McDonald et al. 2019). Analysis of the variations in CRISPR-Cas system using all *V. cholerae* strains in the NCBI database revealed that there were different variant CRISPR-Cas systems apart from the conventional CRISPR-Cas system identified in prototype *V. cholerae* strains. The variations were identified in both *cas* genes arrangements and spacer sequences and some non-choleragenic strains were identified to possess additional genes that codes for hypothetical proteins within the cas gene operon. An in silico study of different *V. cholerae* strains revealed a hybrid CRISPR-Cas with type III-B/I-F system incorporated in the chromosome II.

14.3.3 Resistance Against CRISPR-Cas System

About 97% of the identified CRISPR-Cas systems were identified to be present with MGEs. It is today well established that phages infecting *V. cholerae* and CRISPR-Cas system of the pathogen is co-evolved. CRISPR-Cas loci have been identified in many phages as well. The phage ICIP, which infects *V. cholerae* has been identified to carry CRISPR-Cas loci to evade the protection of the bacterial host (Seed et al. 2013). Additionally, phages JSF5 and JSF6 were also identified to encode CRISPR-Cas loci that were different from that identified in the ICP1 phage. Further, the ICP1 phages have also gained CRISPR systems that target the PLEs in *V. cholerae* apart from its CRISPR-Cas. A chimeric endonuclease coded by the ICP1 phage CRISPR was identified to have a similar sequence with that of the PLE origin protein which targets the protein and cleaves it thereby inhibiting PLE production. Interestingly, ICP1 phages that do not have the adaptive CRISPR-Cas system were also identified to inhibit the PLE production (Barth et al. 2021). A recent study conducted by Naser et al. analyzed different phages and *V. cholerae* strains from cholera patients to study their co-evolution on the basis of structural and functional characteristics of CRISPR-Cas (Naser et al. 2017). The CRISPR-Cas-positive phages were identified to be evolutionarily fit and counteracted the defense of *V. cholerae.* With the evolution of the CRISPR-Cas system in *V. cholerae,* especially at the spacer sequences region, the phages have also been identified to gain identical sequences. These identical sequences have helped evade the CRISPR-Cas recognition of the *V. cholerae*. Phages acquire these identical spacer sequences by different mechanisms such as genomic rearrangements, or by infecting bacterial hosts that closely resemble the actual host. Today, bacteriophage therapy is an emerging and strong field that is studied to control antibiotic-resistant pathogens. However, the diverse CRISPR-Cas systems and their variants are posing a threat to the field even before it could be fully utilized. *V. cholerae* is a pathogen that resists multiple phages and this property has been used to differentiate between different biotypes of the bacterium. Understanding of the CRISPR-Cas system types and subtypes in the pathogen and the phage-mediated resistance to this immunity could help scientist and researchers to engineer vibrio phages with identical spacers and CRISPR-Cas systems that could be used to control cholera.

14.4 Tools to Study Evolution of *V. cholerae* and Its Genomic Islands

The advent of diagnostic tools and techniques took place in the late 1890s toward the end of the fifth pandemic, which enabled researchers to study the evolution of *V. cholerae* strains in the following pandemics. The various methods employed to understand the evolution of *V. cholerae* are summarized below:

Phenotyping Fingerprinting

Serological categorization of the cholera-causing Vibrios, which uses the anti-sera of heat-killed microorganisms, is one of the phenotypic techniques. These are based on the variations in the sugar moiety of the somatic "O" antigen (Rahaman et al. 2015). The hemagglutinating property of *Vibrios* is also used for the biotyping of classical and El Tor strains as El Tor but not classical *V. cholerae* agglutinate sheep or chicken red blood cells. Other phenotypic tests that are used to differentiate between the biotypes of *V. cholerae* include the Voges–Proskauer test and resistance to polymyxin B. Further, phage typing is also used to differentiate the serotype and biotype of *V. cholerae*. Antimicrobial susceptibility has also been useful in characterization of *V. cholerae* as a changing AMR pattern has been observed in cholera epidemiology.

Molecular Fingerprinting

Numerous DNA-based techniques are used to molecularly type *V. cholerae*. Amplification and Sanger sequencing of toxin gene CtxB allows the differentiation of classical, El tor and Haitian strains of *V. cholerae*. Other genes such as hlyA, toxR, and rstB also differ at specific nucleotides allowing the use of these genes to differentiate between the isolates. Phylogenetic differentiation of the isolates at strain level could be achieved by techniques such as pulsed-field gel electrophoresis (PFGE) and multi-locus variable tandem repeat analysis (MLVA) (Rahaman et al. 2015). With the advancement in the next generation sequencing facilities, Whole genome sequencing of the isolates has also become a tool for differentiating the pathogen at the strain level.

14.4.1 Tools to Study GIs

Next generation sequencing has been crucial to infer the presence of GIs in various bacteria and various tools have been developed for their prediction and analysis. The in silico tools are based on prediction and can be categorized into two groups: (i) sequence composition analysis and (ii) comparative genomic analysis (Lu and

Leong 2016). Most commonly used tools for GI prediction have been discussed below:

Alien Hunter
This technique was created at the Sanger Institute in the UK and is based on Interpolated Variable Order Motifs (IVOMs) (Vernikos and Parkhill 2006). It uses sequence composition analyses, such as GC content variation, repeating dinucleotides, and aberrant codon frequency, to try to find unusual regions in the entire genome. An IVOM score defines the variability of the target region. To find potential GIs, the genomic areas with a score above or below the cutoff are evaluated (Lu and Leong 2016). Alien Hunter does not require a preexisting annotation for the predictions, hence, can be used for genome sequences that have no reference.

Predict Bias
Predict Bias tool was developed in the Bioinformatics lab of Devi Ahila Vishwavidyalya, Indore, India. It analyses the sequence composition and predicts primarily the transposases and integrases to identify the genomic locations of GIs. It further uses VFPD (A profile database of virulence factors) to understand the relationship between GI and pathogenicity. It utilizes the RPS-BLAST (Reversed Position Specific—Basic Local Alignment Search Tool) for the prediction from GBK files (Pundhir et al. 2008).

GI Hunter
This tool was developed by the Bioinformatics Laboratory of the University of Pennsylvania, East Stroudsburg. It utilizes the sequence composition, intergenic distance, tRNA genes, and other genes with high levels of expression, phages information, mobile genes with integrase, and transposases for the IVOM methodology and subsequent analysis (Che et al. 2014). It is an ideal tool to predict GI from bacteria as well as archaea.

Zisland Explorer
This tool was developed at Tianjin University, Bioinformatics Center, China that utilizes a non-supervised algorithm-dependent annotation tool apt for automated targeting. It utilizes the GC + Profile software (Zhang et al. 2014) and divides the entire genome sequence into shorter fragments for further analysis. This approach differentiates the homogeneity or heterogeneity of sequences and presents a plot of GC content, highlighting the candidate GIs (Wei et al. 2016).

Genomic Island Prediction Software GIPSy
This tool is an updated version of the Pathogenicity Island Prediction Software (PIPS) (Soares et al. 2012), for the identification of pathogenic GIs in bacterial genomes. However, in the updated version, GIPSy identifies other candidate regions and classifies them according to their biological functions. The prediction requires a reference genome and is based on the deviation of the GC content, tRNA, virulence factors, mobility genes, resistance, and other metabolic genes (Soares et al. 2016).

Islandviewer4

This tool was developed at Brinkman Lab, Simon Fraser University in Canada. It includes a database of GIs in bacterial and archaea organisms. This predictive tool uses three integrated methodologies: (i) genomic comparison using IslandPick, (ii) analysis of sequence composition using score-based prediction of genomic islands—Hidden Markov Model (SIGI-HMM) and (iii) IslandPath-DIMOB for identification of atypical sequences and mobility-related genes. This tool does not utilize the reference genome before the prediction, however, after the results, the user can select another genome for comparison (Bertelli et al. 2017).

A novel in vitro-based method has been developed to study the stability of genomic islands in *V. cholerae* (Kumar et al. 2018). A vector-based genetic tool with selectable and counter-selectable markers is employed to flag the GI in the chromosome and its stability can be measured under in vitro and in vivo conditions.

14.5 Antimicrobial Resistance in *V. cholerae*

The plasticity of the pathogen genome enabled the acquisition of ~40% of the genes from other pathogens cohabiting with them either in the gut or the environmental niche. Over the past decade, increased inappropriate use of antibiotics has led to the emergence of multidrug resistant (MDR) *V. cholerae* in many low- and middle-income countries (Hazen et al. 2010; Thapa Shrestha et al. 2015; Gupta et al. 2016). Until 1960s, a broad range of antibiotics including ampicillin, kanamycin, trimethoprim, sulfamethoxazole, chloramphenicol, erythromycin, streptomycin, gentamicin, tetracycline, and azithromycin were effective against cholera reducing the morbidity in both children and adults. However, in 1979, the first report of MDR *V. cholerae* was reported from Bangladesh which was resistant to multiple antibiotics including tetracycline, ampicillin, kanamycin, streptomycin, and Co-trimoxazole (Glass et al. 1980). Later, the emergence of *V. cholerae* O139 in 1992 also resulted in a distinct pattern of resistance unlike that previously reported in O1 pandemic clones (Garg et al. 2000). The resistance genes of *V. cholerae* have been largely accounted for by Mobile Genetic Elements (MGE) though few are associated with mutations in proto-resistant genes (genes which otherwise have a metabolic function, but can confer drug resistance upon attaining mutation). As discussed in the earlier sections of the chapter, there are different MGEs like plasmids, transposons, integrons and superintegrons such as the SXT constins (conjugable, self-transmissible, integrating element) reported in *V. cholerae* and these MGEs have been pivotal in increasing its antibiotic resistance (Hazen et al. 2010; Pant et al. 2020a). Often large superintergrons can form genome islands like that of the SXT constin which have become widespread in Asian *V. cholerae*. The SXT constin which is a Ctn (conjugative transposon)-like element carrying resistant determinants to sulfamethoxazole, trimethoprim, chloramphenicol, and streptomycin was first detected in *V. cholerae* O139, and was soon found in all clinical cases of *V. cholerae*. Further studies on the SXT revealed that very similar elements were found in the genome of *Providencia alcalifaciens* and *Providencia rettgeri* (Coetzee et al. 1972; Hochhut et al. 2001). Most of the MGEs detected in *V. cholerae* could be evolutionarily traced back to other bacteria of the same genus or other. With the advent of the next generation sequencing techniques, a bigger picture of the ARG evolution and transmissibility of MGEs were obtained. Recently, novel genomic islands containing more than one antibiotic resistance gene (ARGs) were reported from clinical non-O1/O139 *V. cholerae* isolates causing cholera-like diarrhea in Kolkata, India (Morita et al. 2020).

14.5.1 Drugs and Their Targets Used for the Treatment of Cholera

Though cholera is a self-limiting infection that requires only fluid rehydration therapy, antibiotics have been recommended for pregnant women and patients with severe diarrhea and comorbidities. In the early 1940s, streptomycin and chloramphenicol were considered to be effective drugs to treat cholera (Olejnik and Davidovitch 1951). Since the 1960s, tetracycline and the macrolide antibiotic, erythromycin have become the frontline drugs for the treatment of cholera patients (Williams and Berkley 2018). However, with the exorbitant reports of tetracycline and erythromycin resistance in both O1 and O139 *V. cholerae*, doxycycline and azithromycin have become prevalent for the treatment of severe cholera (Nelson et al. 2011). It is important that the drug used for the treatment of the infection should be determined by the local antibiotic resistance pattern of the pathogen and alternative drugs should be used if doxycycline or azithromycin resistance is reported. Apart from these antibiotics, β-lactam antibiotics such as ampicillin (penicillin antibiotic), meropenem (carbapenem), and ceftriaxone (cephalosporin) which inhibits the cell wall peptidoglycan synthesis of the pathogen are used in many cholera endemic countries. Other antibiotics which have been reported to be used in the treatment of cholera are spiramycin (macrolide), aminoglycoside antibiotic streptomycin, quinolone antibiotics nalidixic acid, norfloxacin, and ciprofloxacin (De 2021). While macrolide and aminoglycoside inhibit the bacterial protein synthesis, the quinolone antibiotics inhibit bacterial DNA replication by binding to DNA gyrase.

Though doxycycline and azithromycin remain the drug of choice to treat cholera globally, the antibiotic regimen for treating extraintestinal *V. cholerae* infections such as septicemia, wound infections, ear infections, cellulitis, meningitis, urinary tract infections (UTI) varies from region to region, and also the site of infection such as blood, ear, skin, and urinary tract (Hao et al. 2015; De 2021) (Chowdhury et al. 2016). In most cases, extraintestinal infections are caused by non-O1/O139 *V. cholerae* and have been identified to be sensitive to most antibiotics. Antibiotics like meropenem, piperacillin-tazobactam, ofloxacin, levofloxacin, and ciprofloxacin are commonly used for treating extraintestinal infections caused by non-O1/O139 strains of *V. cholerae* (Chowdhury et al. 2016). Unlike for the treatment of cholera, there are no definitive guidelines for the treatment of *V. cholerae* extraintestinal infections and the treatment primarily depends upon the antimicrobial susceptibility pattern of the bacteria causing the infection. While single antibiotic therapy is commonly used for non-O1/O139 *V. cholerae* extraintestinal infections, dual antibiotics or combinatorial therapy has been used to treat severely ill patients. However, for both toxigenic *V. cholerae* O1/O139 and non-O/O139 strains of *V. cholerae*, antibiotics are not recommended to be used for prophylaxis and are suggested to be treated during severe infection.

14.5.2 Genomics of Antimicrobial Resistance and Its Mechanisms in *Vibrio cholerae*

The different mechanisms of antibiotic resistance and major genes contributing to the resistance in *V. cholerae* are listed in Table 14.2. In Gram-negative bacteria, especially *V. cholerae*, resistance to a single antibiotic can be due to multiple mechanisms. For example, *V. cholerae*, resistance to tetracycline is contributed by two major mechanisms; active efflux of antibiotics and production of ribosomal protection proteins (encoded by *tet* genes). However, there has been also evidence that reduced susceptibility of tetracycline among clinical strains is due to target site mutation, decreased drug permeability, and enzymatic degradation of the antibiotic (Ahmadi 2021). Apart from the resistance gained by chromosomal mutation and acquired SNP, the most prevalent mechanism of AMR acquisition is through horizontal gene transfer. Jain et al. (2016) reported the emergence of tetracycline-resistant *V. cholerae* O1 isolates in India by the acquisition of a plasmid carrying Major Facilitator Superfamily (MFS) efflux pumps (Jain et al. 2016). Further, many of the genes like *tet*M that code for ribosomal protection proteins were found to be associated with conjugative transposons Tn*916*, Tn*1545* that can facilitate its transfer to other microbes. The transposons were identified to be genetically similar to those found in *Shigella flexneri* and *Enterococcus faecalis*. Previously, there has been a report of large outbreaks caused by tetracycline-resistant *V. cholerae* from Tanzania owing to a megaplasmid belonging to the C incompatibility complex that codes resistance to multiple antibiotics like tetracycline, ampicillin, sulfonamides, chloramphenicol, kanamycin, and streptomycin. Such plasmids belonging to the C incompatibility complex have been previously observed in *Pseudomonas, Klebsiella, Serretia, Providencia*, and *Proteus* (Towner et al. 1980).

Chromosome 2 of *V. cholerae* has been identified to be more rich in antimicrobial resistance genes as compared to chromosome 1 which harbors majorly genes associated with metabolism, survival, and virulence. Chromosome 2 possesses GIs that are formed by acquired AMR genes which have got integrated into the genome over the evolutionary process of the pathogen. A few major MGEs predominant in the spread of AMR in *V. cholerae* are presented below.

SXT In 1992, a 99.5-kb integrative conjugative element (ICE) from the SXT family harboring gene that codes resistance for sulfamethoxazole and trimethoprim was discovered in *V. cholerae* O139 strains isolated from India and was designated SXT^{MO10} (Waldor et al. 1996). From then, there have been various reports of SXT variants (SXT^{ET}/ ICEVchInd1, SXT^{LAOS}/ ICE*Vch*Lao1, ICEVchVie1) carrying

Table 14.2 Different mechanisms of antibiotic resistance and major genes contributing to the resistance in *V. cholerae*

Resistance mechanism	Genes involved in resistance	MGE associated	Antibiotic class/ antibiotic	References
V. cholerae efflux system	1. **MATE** (multidrug and toxic compound extrusion) efflux system: *norM, vcrM, vcmA, vcmB, vcmD, vcmH and vcmN*	Nil (chromosomally encoded)	Fluoroquinolone, aminoglycosides, ethidium bromide and Hoechst 33342	Mohanty et al. (2012)
	2. **RND** (resistance-nodulation-cell division) efflux system: *vexAB, vexCD* (*breAB*), *vexEF*, *vexGH*, *vexIJK* and *vexLM*	Nil (chromosomally encoded)	Antibiotics (benzylpenicillin, erythromycin, and polymyxin) and detergents	Bina et al. (2008)
	3. **MFS** (major facilitator superfamily): *emr*D-3, *vce*AB	Nil (chromosomally encoded)	Linezolid, tetraphenylphosphonium chloride, rifampin, minocycline, erythromycin, trimethoprim, chloramphenicol oxytetracycline, tetracycline, nalidixic acid and florfenicol	Woolley et al. (2005), Smith et al. (2009); Varela et al. (2013), Jain et al. (2016)
	4. **SMR** (small multidrug resistance): *qac*E, *qac*F, *qac*G, *qac*H, *qac*I, *qac*L *sug*E	Plasmids, class I integrons	Quaternary ammonium compounds	Slipski et al. (2021)
	5. **ABC** (ATP-binding cassette): *vca*M	Nil (Chromosomally encoded)	Tetracycline and fluoroquinolones	Huda et al. (2003), Lu et al. (2018)
Inactivation of antibiotics	1. ***Hydrolysis:*** *blaNDM-1, blaCMY-2, blaTEM-1, blaCTX-M-1, blaDHA-1, blaSAR-1,*	Plasmids, ICEs	Beta-lactam	Reid and Amyes (1986), Mandal et al. (2012), Folster et al. (2014)
	2. **ADP-ribosylation**: *aad*A1, *aad*A16, *arr3*	Plasmids, class 1 integron	Streptomycin, Spectinomycin	Folster et al. (2014)
	3. **Nucleotidylation:** *ant(3″)-1a*	Plasmid, class 1 integron	Aminoglycosides	Dalsgaard et al. (2000)

	4. ***Phosphorylation***: *mphA*	Plasmid, class 1 integron	Macrolide	Folster et al. (2014)
	5. ***Acetylation:*** *aac(3)-IIa, aac(3)-Id, aac(6′)-Ib*	Plasmid, class 1 integron	Aminoglycosides	Ahmed (2004)
Enzymatic modification of drug target	1. **LPS modification**: *alm*EFG	Chromosomal mutation	Polymyxins	Ahmed (2004), Henderson et al. (2014)
	2. **DNA gyrase, topoisomerase modification:** *gyr*A, *par*C	Chromosomal mutation	Quinolone antibiotics	Quilici et al. (2010)

different antibiotic resistance genes such as chloramphenicol (*flo*R), streptomycin (*str*A and *str*B), sulfamethoxazole (*sul2*), trimethoprim (*dfr*A18 or *dfr*A1), and tetracycline (*tet*A and *tet*R) reported. Apart from *V. cholerae*, SXT elements have been also identified in other Gram-negative pathogens such as *V. fluvialis*, *E. coli*, and *Proteus rettgeri* highlighting its role in the transfer of AMR genes within and across bacterial species (Ahmed et al. 2005).

Integrons Apart from the SXT element, other ICEs and integrons have been reported in *V. cholerae* conferring antibiotic resistance. A large number of resistance genes such as *qnr*VC, *sul*1, *sul2*, *cat*1, *aph*, *tet*G, *dfr*A1, *dfr*A15, *aad*A1, *aad*A2, *aar*-3, *aac*A4 have been associated with integrons. Usually, integrons are present in association with transposons. An array of different transposons like Tn*21*, Tn*1403*, Tn*1404*, Tn*1696*, Tn*1412*, and Tn*2000* has been found to be associated with these integrons (Partridge et al. 2001). These transposons in turn carry insertion sequences (IS elements) that code for transposases. There are over 100 classes of integrons identified in Gram-negative bacteria of which class 1 and class 2 are dominant in *V. cholerae*. *V. cholerae* possessing class 1 integron carrying *aad*A2 and *aad*A7 were isolated from environmental biotopes of Brazil while environmental biotopes of India, Bangladesh, and Ghana were identified to possess class 2 integron possessing *dfr*A1, *sat*1, and *aad*A1 genes (Ahmed et al. 2006). Also, there are class 4 integrons or superintegrons (SI) that form a part of the pathogen genome which not just harbors ARG but also toxin–antitoxin systems. In *V. cholerae*, the SI is located in chromosome 2 which has 216 ORFs. The SI carries chloramphenicol acetyltransferase and fosfomycin resistance protein. However, most of the genes located on the SI encode for hypothetical proteins. Recently, Carraro et al. discovered the presence of a mobilizable genomic island (GI) MGIVchHai6 in a non-O1/ O139 *V. cholerae* isolate retaining resistance genes for β-lactam antibiotics, SXT, tetracycline, chloramphenicol, and streptomycin/spectinomycin and mercury resistance (Carraro et al. 2016).

Plasmids Plasmids constitute another large group of MGEs that aid in the dissemination of ARGs between *V. cholerae* and other pathogens. During recent years, many novel plasmids carrying ARGs were identified in outbreak clones of *V. cholerae* and non-O1/ O139 environmental strains. Many of these plasmids have genetic composition that facilitates their transfer among environmental and clinical isolates. Two novel cryptic plasmids pSDH-1 and pSDH-2 were identified to be widely present among environmental *V. cholerae* isolates from Haiti (Ceccarelli et al. 2017). The same team had previously identified p3iANG, a large conjugative plasmid possessing Class 1 integrons encoding *dfr*A15, *bla*P1, *sul2*, *tet*G, and *qac*H-*aad*A8 gene cassette encompassing 19 kb genetic cluster from *V. cholerae* circulating in Angola (Ceccarelli et al. 2006). Further, the size and genetic composition of plasmids identified in *V. cholerae* has been found to be diverse. Small plasmids were identified to carry β-lactam-resistant genes in clinical *V. cholerae* strains isolated from Africa. The plasmid-encoded β-lactamase gene genetically similar to that of the β-lactamase gene found in *K. pneumonia* isolated from the same region (Ceccarelli et al. 2006; Ismail et al. 2011). This report served as a strong proof of ARG dissemination across bacterial species by plasmids. On the other hand, in another study from China, MDR *V. cholerae* O139 isolates were identified to have mega plasmids belonging to the IncA/C family harboring more than 10 ARGs (Wang et al. 2015). Recently, two large plasmids pVC1 and pVC2 were reported to be present in *V. cholerae* strains isolated from India possessing genes that code resistance to rifampicin, ciprofloxacin, tetracycline, neomycin, aztreonam, β-lactams, chloramphenicol, and aminoglycosides (Wang et al. 2015; Verma et al. 2019). Further, plasmid bearing *bla*NDM-1 was reported from a cholera patient from India (Mandal et al. 2012) in 2012, while dissemination of similar plasmids encoding *bla*NDM-1 among bacterial pathogens isolated from sewage and other environmental bodies were reported earlier (Walsh et al. 2011). This highlighted the role of plasmids in the dissemination of ARGs among pathogens and environment being the reservoir for such transmission. Interestingly, reports of multiple MGEs responsible for AMR coexisting in the same *V. cholerae* are also not sparse. Wang et al., had reported the presence of SXT, IncA/C plasmids and a novel integron in clinical strains of *V. cholerae* isolated from 1998 (Wang et al. 2016). Thus, MGEs serve as the backbone of *V. cholerae* evolution with respect to antibiotic resistance and understanding genetic organization and their dissemination patterns is of utmost importance to combat the looming AMR burden in the pathogen.

14.5.3 Genetic Factors for Biofilm Formation in *Vibrio cholerae*

Biofilms are film-like self-assembled bacterial structures formed by almost 90% of bacteria. These biofilms are formed in the environment and clinical settings and help the bacteria to tolerate multiple stresses like nutrient deprivation, antibiotics, and host immunity. Both uni-species and multispecies bacterial biofilms have been reported. While in the environment, both uni and multispecies biofilms are equally prevalent while, in clinical settings, uni-species bacterial biofilms are predominant. Having a dual life cycle, one in the environment and the other within the host intestine, *V. cholerae* is known to form biofilm in both conditions (Silva and Benitez 2016). Recent studies revealed that *V. cholerae* biofilm formed in vitro, in the environment and within the host intestine vary significantly in structure (Teschler et al. 2015; Silva and Benitez 2016). Unlike the monolayer biofilms formed in vitro, the matured, 3D structured biofilm formed in the environment and in host vary significantly in their global transcriptome profiles mainly owing to the signals received by the bacterial cells in these complex niches. In the environment, *V. cholerae* biofilms protect the pathogen from various stresses such as nutrient deprivation, extreme temperatures, oxidative stress, bacteriophage predation, and protozoan grazing. It also helps the pathogen in transmission by adhering to phytoplanktons and zooplanktons (de Magny et al. 2011). Whereas, in the host intestine, the biofilm formation protects the pathogen while surpassing the acidic pH of the stomach, high osmolarity, nutrient and iron deprivation, and host immunity (de Magny et al. 2011; Silva and Benitez 2016). The biofilm structure also increases

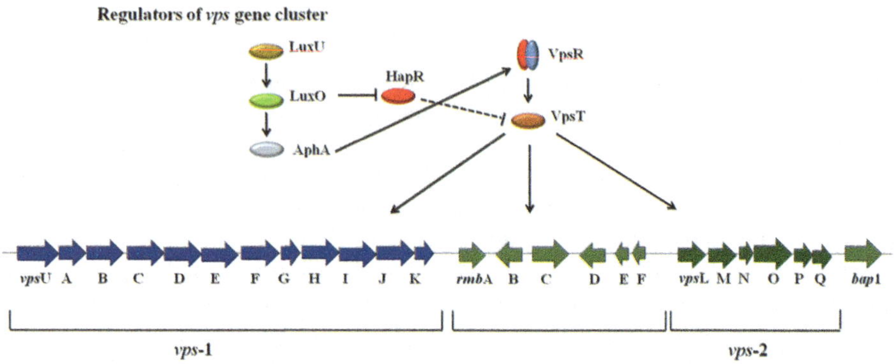

Fig. 14.5 Schematic represention of major proteins and their regulation of the *vps* gene cluster in *V. cholerae*

the transmission of the disease as the number of bacteria within the biofilm structure is high and attains the minimum number for the infectious dose (Faruque et al. 2006; Bridges and Bassler 2021). Biofilms can resist up to 150 times more antibiotics than that in the planktonic state and *V. cholerae* in the biofilm state are hypervirulent (Gallego-Hernandez et al. 2020). Thus, biofilm forms a very important state in the *V. cholerae* life cycle.

The steps involved in biofilm formation are multifactorial and involve many genes. Biofilm formation begins with initial attachment of the bacterial cells to the biotic or abiotic surface. This is facilitated by *V. cholerae* type IV pili such as the toxin-coregulated pili (TCP) and mannose-sensitive hemagglutinin (MSHA). Next, the *V. cholerae* produce the extra polysaccharide required for micro-colony formation that further matures to complex biofilms. One of the main components of *V. cholerae* biofilm is the *V. cholerae* polysaccharide (VPS). The VPS is produced by the expression of 18 genes that are organized into two gene clusters. The gene cluster-1, known as the *vps* gene cluster1 comprises 12 genes (*vps*A-K, U) and the second gene cluster, *vps*-2 comprises 6 genes (*vps*L–Q). Both vps-1 and -2 are separated by another gene cluster *rmb* comprising 6 genes, those codes for matrix proteins such as RmbA, RmbC, RbmBDEF, and Bap1. Together, all three gene clusters are known as the *V. cholerae* biofilm matrix cluster (VcBMC) and are present in chromosome 1 (VC0916, VC0917–27, and VC0934–9) of *V. cholerae* spanning a 30-kb genome region (Teschler et al. 2015; Gallego-Hernandez et al. 2020). Though these gene clusters primarily help in biofilm formation in *V. cholerae*, a multitude of other genes are also important in biofilm formation and maintenance by directly or indirectly regulating the *vps* clusters. Few major proteins and their regulation of the *vps* gene cluster are schematically represented in Fig. 14.5. Further, concentrations of secondary messenger molecules such as cyclic di-GMP have also been identified to have a role in *V. cholerae* biofilm formation (Zhu and Mekalanos 2003). The whole process of biofilm formation in *V. cholerae* is regulated by signals from the environment and also the bacteria among themselves through a process known as quorum sensing. HapR acts as the master quorum-sensing regulator in *V. cholerae* repressing the expression of *vps* genes at high cell density. Interestingly, this HapR is truncated in classical strains of *V. cholerae* while conserved in El Tor biotype except for a few like the N16961 strain (Ball et al. 2017). HapR also regulates the TCP in *V. cholerae* which is important for both virulence and biofilm formation of the pathogen in vivo (Gao et al. 2020). The tcp gene cluster is located in a large genomic island known as the *Vibrio* pathogenicity island (VPI-1) in chromosome 1 of *V. cholerae* and is important for the initial attachment of the pathogen to the host intestinal microvilli. Many frameshift mutations have also been reported in the *hap*R gene of recent outbreak strains and this could be correlated to the varied pathogenicity, environmental persistence, and biofilm formation of these strains. Hence, the interplay between the core genomes such as the *lux*O, *hap*R, and the genes in the acquired genomic island such as the *tcp*A, have been noticed during the pathogenicity and biofilm formation of *V. cholerae* (Haycocks et al. 2019; Gao et al. 2020). Recent studies have demonstrated that there is increased transcription of genes associated with biofilm formation in the El Tor strains as compared to the classical *V. cholerae* strains (Kostiuk et al. 2022). Apart from these major proteins, which act as activators and repressors of biofilm formation, various other proteins such as the flagellin proteins and proteins responsible for the synthesis and degradation of c-di-GMP (*cdgD, vieSAB*) have been known to play important roles in the biofilm formation of *V. cholerae*. However, most of these genes are located in the core genome of the pathogen.

14.6 Role of Genomic Islands in Development of Cholera Vaccines

The quest for cholera vaccine had begun in the early 1880s as there were severe pandemics of the disease that claimed thousands of lives. Sir Robert Koch and his team was the first to notice that patients who survived the disease got protected from acquiring the disease again during the same outbreak. This opened hopes for the development of cholera vaccine and in 1885, Spanish physician Dr. Jaime Ferrán who worked with Louis Pasteur came up with the first cholera vaccine which he used to mass vaccinate over 50,000 people in Spain during cholera pandemic (Pollitzer and Burrows 1955). However, the composition of Dr. Jaime Ferrán's cholera vaccine is unknown and is assumed to be an attenuated whole-cell *V. cholerae*. Later, in 1893, Sawtschenko and Sabolotny used killed *V. cholerae* culture broth to vaccinate people. The vaccine proved to be effective in eliciting protection against cholera but it required high doses of *V. cholerae* and hence large amounts of killed culture broth.

However, over the years, different variants of *V. cholera*e were identified from different outbreaks and the cholera vaccines used initially against the classical strains of *V. cholerae* proved to be less effective against these variants. Also, another finding that the protection of cholera vaccines was only for a short period lessened its popularity. Today, cholera vaccines are not mandatory and are only used as an adjunct to cholera prevention WASH programs. Vaccines are only used by travelers as a prophylactic measure before their travel to a cholera-endemic region. Unlike vaccines for other infectious diseases such as mumps and measles, oral cholera vaccines (OCVs) are more popular than intramuscular or intravenous vaccination. OCVs majorly consisted of either whole bacterial cell killed or attenuated which consisted of the bacterial lipopolysaccharide (LPS) O-antigen that elicits strong mucosal immunity with secretory IgA produced locally in the intestine or the nontoxic component of cholera toxin, the B pentamer. Svennerholm et al., in the 1980s had reported that a single oral or intramuscular immunization with purified cholera toxin B subunit induced the same amount of IgA and IgG antibodies (Svennerholm et al. 1984). The cholera toxin B subunit is coded by the *ctx*B gene which is located in the *ctx*AB operon within the *Vibrio* pathogenicity island 1 (VPI-1). Construction of overexpression systems that can produce a large amount of cholera toxin B subunit is an important step for including cholera toxin B subunit in the vaccine. Maximal synergistic immune protection was observed when a combination of bacterial LPS and cholera toxin B subunit was used (Svennerholm and Holmgren 1976; Svennerholm et al. 1984). Additionally, considering the variants, bivalent vaccines were introduced that consisted of more than one *V. cholerae* variant LPS like the *V. cholerae* O1 and O139.

Dukoral® was the first monovalent OCV approved by the World Health Organization (WHO) in 2001 (WHO Publication 2010). Dukoral® composed of a combination of heat killed *V. cholerae* O1 Inaba strain Cairo 48 and Ogawa strain Cairo 50 and formalin killed Inaba strain Phil 6973 and Ogawa strain Cairo 50 in addition to recombinant cholera toxin B subunit. Another OCV that uses a combination of *V. cholerae* whole cell and recombinant cholera toxin B subunit is the OraVacs that is only licensed in China and the Philippines (Lopez et al. 2014). Recently, a live OCV named VA 1.4 was developed by the Indian Council of Medical Research using a nontoxic *V. cholerae* O1 strain with *ctx*B gene insertion and its safety and immunogenicity were studied (Kanungo et al. 2014). A few other monovalent OCVs that were developed and under various phases of clinical trials are Vaxchora (FDA approved) and Hillchol® (Mosley 2nd et al. 2017; Chowdhury et al. 2021). Majority of the OCVs deploy only killed or attenuated whole bacterial cells. Shanchol™, another WHO-prequalified cholera vaccine, consisted of only *V. cholerae* whole cells that gave protection against both O1 and O139 serotypes but did not contain the recombinant cholera toxin B subunit. Shanchol™ consisted of heat and formalin killed *V. cholerae* O1, Ogawa, and Inaba strains and also additionally formalin killed *V. cholerae* O139 strain (Bhattacharya et al. 2013). Other bivalent OCVs that only use killed or attenuated whole cells are Euvichol®/ Euvichol-Plus® and mORC-Vax™ (Odevall et al. 2018; Shaikh et al. 2020). Recently, studies on utilizing *V. cholerae* outer membrane vesicles that contain several antigens have gained pace (Balhuizen et al. 2021). Different promising oral cholera vaccines in use and under clinical trials, their composition, dosage, and immune protection are summarized in Table 14.3.

Table 14.3 List of oral cholera vaccines used for cholera prophylaxis

Vaccine type	Vaccine	Composition	Dosage	Booster dose	Route of Administration	Age range	Protection	WHO authorization status
1. Killed whole cell vaccine								
a. Monovalent	Dukoral®	*V. cholerae* O1 serogroup and recombinant cholera toxin B subunit	Adults: 2 doses at 7–14 days interval Children >5: 3 doses	Required	Orally with bicarbonate buffer	2 years and above	2 years	Prequalified since 2001
	OraVacs	*V. cholerae* O1 serogroup and recombinant cholera toxin B subunit	3 doses	Required	Not recommended	2 years and above	3 years	Licensure in China and Philippines only
b. Bivalent	Shanchol™	*V. cholerae* O1 and O139	2 dose at 14 day interval	Not recommended	Oral	Above 1 year	3 years	Prequalified since 2011
	Euvichol®/ Euvichol-Plus®	*V. cholerae* O1 and O139	2 dose at 14 day interval	Not recommended	Oral	Above 1 year	3 years	Euvichol®- since 2015 Euvichol-Plus®- since 2017
	mORC-Vax™	*V. cholerae* O1 and O139	2 dose at 14 day interval	Not recommended	Oral	Above 2 years	3 years	Licensure of IVI-reformulated vaccine in Vietnam in 2009
	Cholvax	*V. cholerae* O1 and O139	2 dose at 14 day interval	Not recommended	Oral	Above 1 year	No data available	Clinical phase III/IV studies

(continued)

Table 14.3 (continued)

Vaccine type	Vaccine	Composition	Dosage	Booster dose	Route of Administration	Age range	Protection	WHO authorization status
2. Live attenuated								
a. Monovalent	Vaxchora	*V. cholerae* strain cvd 103-hgr, live	Single dose	Not recommended	Orally with bicarbonate buffer	Effective in adults 18–65 years	No data available	FDA approved since 2016
	Hillchol®	*Vibrio cholerae* O1 El Tor Hikojima strain (MS1568)	2 dose	Not recommended	Oral	Above 1 year	No data available	Clinical phase I/II studies
	VA 1.4	*V. cholerae* O1 serogroup and recombinant cholera toxin B subunit	Single dose	Not recommended	Orally with bicarbonate buffer	18–65 years	No data available	Clinical trial registered

14.7 Future Perspectives and Conclusions

V. cholerae as a pathogen is continuously evolving via mutations and lateral gene transfer to maintain the fitness and survival of particular clones to continue their dominance during various pandemics. Global estimates suggest ~29 million cases are recorded annually with a total death toll of 95,000 patients in cholera endemic regions. However, following the Global Roadmap to 2030 for the global elimination of cholera, sincere efforts by many countries like China and Vietnam resulted in a drastic decrease in the incidence of cholera cases in their region. Improved infrastructure, universal accessibility of safe drinking water, sanitation, and hygiene (WASH) play a crucial role in the development of national cholera plans for a significant reduction in global burden of cholera. Development of newer vaccines and integration of WHO approved Oral Cholera Vaccines (OCVs) in routine immunization schedules along with adopting WASH strategies could be a game changer for disrupting the transmission of cholera across the globe.

Despite being a well-studied representative genome, the evolutionary mechanisms remain elusive to understand the etiology of emergence of seventh pandemic strains. Moreover, the emerging drug resistance in *V. cholerae* is posing a serious threat to the global control and management of cholera outbreaks. The genetic traits responsible for imparting drug resistance can travel across the *Vibrio* species via horizontal transfer of genetic elements. Therefore, it becomes imperative to identify the bacterial strategies responsible for exhibiting antimicrobial resistance (AMR) not only in pathogenic *V. cholerae* but also in other environmental non-pathogenic strains which could serve as reservoirs of AMR genes.

The advent of next generation sequencing technologies facilitated the annotation of complete genomes and performing of comparative genetics studies at population level by making genome sequencing fast and affordable. Accurate prediction of GI boundaries is necessary to define GI regions in the genome, IslandPick program can easily do so by utilizing the data from reference genomes and applying comparative genomics approaches. Whole Genome Sequencing (WGS) platforms have proved to be instrumental in unraveling single nucleotide polymorphism (SNP) across various strains to establish evolutionary networks among closely related seventh pandemic strains.

In conclusion, this chapter summarizes the overall clinical spectrum, pathogenesis, epidemiology, and major focus on the basic structure of Genomic Islands in *V. cholerae* and various tools to study GIs. It is anticipated that combination of WGS with traditional phenotypic and molecular fingerprinting methods may prove to be revolutionary for improved mobilome discovery, epidemiology, and development of efficient theranostic tools to address challenges of AMR in *V. cholerae*. Hence, the role of GIs in antimicrobial resistance and development of oral cholera vaccines advocates GIs as an important drug target to combat pathogenic MDR *V. cholerae* strains.

Comparative analysis of pan-genome sequences may prove to be beneficial for not only understanding bacterial pathogenesis but also in selecting ideal pathogenic isolates as best vaccine candidates to target cholera endemic regions in developing countries. On other hand, the mechanism of shuffling genomic traits across variants of phages inside host cells still remains an unsolved mystery. Also, the reasons for the preferential integration of MGEs at tRNA loci needs to be further explored. Therefore, the plethora of sequencing data needs to be translated to answer fundamental questions associated with the pathogenicity and evolution of *V. cholerae.*

Acknowledgments BD is thankful to Dept. of Biotechnology (DBT), Govt. of India for research funding (No. BT/NBM0018/01/17). LN is thankful to the Dept. of Biotechnology (DBT), Govt. of India for the MK Bhan Fellowship.

References

Ahmadi MH (2021) Global status of tetracycline resistance among clinical isolates of *Vibrio cholerae*: a systematic review and meta-analysis. Antimicrob Resist Infect Control 10(1):115

Ahmed AM (2004) New aminoglycoside acetyltransferase gene, aac(3)-Id, in a class 1 integron from a multi resistant strain of Vibrio fluvialis isolated from an infant aged 6 months. J Antimicrob Chemother:947–951. https://doi.org/10.1093/jac/dkh221

Ahmed AM, Kawaguchi F, Shimamoto T (2006) Class 2 integrons in *Vibrio cholerae*. J Med Microbiol:643–644. https://doi.org/10.1099/jmm.0.46378-0

Ahmed AM, Shinoda S, Shimamoto T (2005) A variant type of *Vibrio cholerae*SXT element in a multidrug-resistant strain of Vibrio fluvialis. FEMS Microbiol Lett:241–247. https://doi.org/10.1016/j.femsle.2004.11.012

Ahmed MU, Baquilod M, Deola C, Tu ND, Anh DD, Grasso C, Gautam A, Hamzah WM, Heng S, Iamsirithaworn S, Kadim M, Kar SK, Le Thi Quynh M, Lopez AL, Lynch J, Memon I, Mengel M, Long VN, Pandey BD, Quadri F, Saadatian-Elahi M, Gupta SS, Sultan A, Sur D, Tan DQ, Ha HTT, Hein NT, Lan PT, Upreti SR, Endtz H, Ganguly NK, Legros D, Picot V, Nair GB (2018) Cholera prevention and control in Asian countries. BMC Proc 12(Suppl 13):62–62. https://doi.org/10.1186/s12919-018-0158-1

Alam MT et al (2016) Major shift of toxigenic V. cholerae O1 from Ogawa to Inaba serotype isolated from clinical and environmental samples in Haiti. PLoS Negl Trop Dis:e0005045. https://doi.org/10.1371/journal.pntd.0005045

Ali M et al (2012) The global burden of cholera. Bull World Health Organ:209–218. https://doi.org/10.2471/blt.11.093427

Ali M et al (2015) Updated global burden of cholera in endemic countries. PLoS Negl Trop Dis 9(6):e0003832

Ali M et al (2017) Identification of burden hotspots and risk factors for cholera in India: an observational study. PLoS One:e0183100. https://doi.org/10.1371/journal.pone.0183100

Amaro C et al (1988) R plasmids in environmental *Vibrio cholerae* non-O1 strains. Appl Environ Microbiol 54(11):2771–2776

Aydanian A et al (2011) Genetic diversity of O-antigen biosynthesis regions in *Vibrio cholerae*. Appl Environ Microbiol:2247–2253. https://doi.org/10.1128/aem.01663-10

Baker-Austin C et al (2017) Non-cholera vibrios: the microbial barometer of climate change. Trends Microbiol:76–84. https://doi.org/10.1016/j.tim.2016.09.008

Baker-Austin C et al (2018) Vibrio spp. infections. Nat Rev Dis Primers 4(1):8

Balhuizen MD, Veldhuizen EJA, Haagsman HP (2021) Outer membrane vesicle induction and isolation for vaccine development. Front Microbiol 12:629090

Ball AS, Chaparian RR, van Kessel JC (2017) Quorum sensing gene regulation by LuxR/HapR master regulators in Vibrios. J Bacteriol 199:19. https://doi.org/10.1128/JB.00105-17

Banerjee R et al (2014) Dynamics in genome evolution of *Vibrio cholerae*. Infect Genet Evol:32–41. https://doi.org/10.1016/j.meegid.2014.01.006

Barrangou R, van der Oost J (2012) CRISPR-Cas systems: RNA-mediated adaptive immunity in bacteria and archaea. Springer Science & Business Media

Barrasso K et al (2022) Impact of a human gut microbe on *Vibrio cholerae* host colonization through biofilm enhancement. elife. https://doi.org/10.7554/elife.73010

Barth ZK, Nguyen MH, Seed KD (2021) A chimeric nuclease substitutes a phage CRISPR-Cas system to provide sequence-specific immunity against subviral parasites. elife 10. https://doi.org/10.7554/eLife.68339

Bellanger X et al (2014) Conjugative and mobilizable genomic islands in bacteria: evolution and diversity. FEMS Microbiol Rev:720–760. https://doi.org/10.1111/1574-6976.12058

Bertani B, Ruiz N (2018) Function and biogenesis of lipopolysaccharides. EcoSal Plus. https://doi.org/10.1128/ecosalplus.esp-0001-2018

Bertelli C et al (2017) IslandViewer 4: expanded prediction of genomic islands for larger-scale datasets. Nucleic Acids Res:W30–W35. https://doi.org/10.1093/nar/gkx343

Bharati K, Ganguly NK (2011) Cholera toxin: a paradigm of a multifunctional protein. Indian J Med Res 133(2):179–187

Bhattacharya SK et al (1993) Clinical profile of acute diarrhoea cases infected with the new epidemic strain of *Vibrio cholerae* O139: designation of the disease as cholera. J Infect:11–15. https://doi.org/10.1016/0163-4453(93)93488-p

Bhattacharya SK et al (2013) 5 year efficacy of a bivalent killed whole-cell oral cholera vaccine in Kolkata, India: a cluster-randomised, double-blind, placebo-controlled trial. Lancet Infect Dis 13(12):1050–1056

Bik EM et al (1995) Genesis of the novel epidemic *Vibrio cholerae* O139 strain: evidence for horizontal transfer of genes involved in polysaccharide synthesis. EMBO J:209–216. https://doi.org/10.1002/j.1460-2075.1995.tb06993.x

Bina XR et al (2008) *Vibrio cholerae* RND family efflux systems are required for antimicrobial resistance, optimal virulence factor production, and colonization of the infant mouse small intestine. Infect Immun 76(8):3595–3605

Box AM et al (2016) Functional analysis of bacteriophage immunity through a type I-E CRISPR-Cas system in *Vibrio cholerae* and its application in bacteriophage genome engineering. J Bacteriol:578–590. https://doi.org/10.1128/jb.00747-15

Bridges AA, Bassler BL (2021) Author response: inverse regulation of *Vibrio cholerae* biofilm dispersal by polyamine signals. https://doi.org/10.7554/elife.65487.sa2

Campos J, Martínez E, Suzarte E, Rodríguez BL, Marrero K, Silva Y, Ledón T, del Sol R, Fando R (2003) VGJ phi, a novel filamentous phage of Vibrio cholerae, integrates into the same chromosomal site as CTX phi. J Bacteriol 185(19):5685–5696. https://doi.org/10.1128/JB.185.19.5685-5696.2003

Carpenter MR et al (2017) CRISPR-Cas and contact-dependent secretion systems present on excisable pathogenicity islands with conserved recombination modules. J Bacteriol 199:10. https://doi.org/10.1128/JB.00842-16

Carraro N et al (2016) IncA/C conjugative plasmids mobilize a new family of multidrug resistance islands in clinical *Vibrio cholerae* non-O1/non-O139 isolates from Haiti. MBio. https://doi.org/10.1128/mbio.00509-16

Cash RA et al (1974) Response of man to infection with *Vibrio cholerae*. I. Clinical, serologic, and bacteriologic responses to a known inoculum. J Infect Dis:45–52. https://doi.org/10.1093/infdis/129.1.45

Castillo D et al (2018) Exploring the genomic traits of non-toxigenic strains isolated in southern Chile. Front Microbiol 9:161

Ceccarelli D et al (2006) New cluster of plasmid-located class 1 Integrons in *Vibrio cholerae* O1 and a dfrA15 cassette-containing Integron in Vibrio parahaemolyticus isolated in Angola. Antimicrob Agents Chemother:2493–2499. https://doi.org/10.1128/aac.01310-05

Ceccarelli D et al (2017) Characterization of two cryptic plasmids isolated in Haiti from clinical *Vibrio cholerae* non-O1/non-O139. Front Microbiol. https://doi.org/10.3389/fmicb.2017.02283

Chaand M, Dziejman M (2013) *Vibrio cholerae* VttR a and VttR B regulatory influences extend beyond the type 3 secretion system Genomic Island. J Bacteriol:2424–2436. https://doi.org/10.1128/jb.02151-12

Chakraborty S et al (2009) Assessment of the evolutionary origin and possibility of CRISPR-Cas (CASS) mediated RNA interference pathway in *Vibrio cholerae* O395. In Silico Biol 9(4):245–254

Chatterjee SN, Chaudhuri K (2006) Lipopolysaccharides of *Vibrio cholerae*: III. Biological functions. Biochim Biophys Acta (BBA) - Mol Basis Dis:1–16. https://doi.org/10.1016/j.bbadis.2005.08.005

Che D, Hasan M, Chen B (2014) Identifying pathogenicity islands in bacterial pathogenomics using computational approaches. Pathogens:36–56. https://doi.org/10.3390/pathogens3010036

Cho JY, Liu R, Hsiao A (2022) Microbiota-associated biofilm regulation leads to resistance against intestinal environmental stress. Front Cell Infect Microbiol 12:861677

Cho JY et al (2021) The interface of *Vibrio cholerae* and the gut microbiome. Gut Microbes. https://doi.org/10.1080/19490976.2021.1937015

Chowdhury F et al (2021) A phase I/II study to evaluate safety, tolerability and immunogenicity of Hillchol®, an inactivated single Hikojima strain based oral cholera vaccine, in a sequentially age descending population in Bangladesh. Vaccine:4450–4457. https://doi.org/10.1016/j.vaccine.2021.06.069

Chowdhury G et al (2016) Extraintestinal infections caused by non-toxigenic *vibrio cholerae* non-O1/non-O139. Front Microbiol 7:144

Chun J et al (2009) Comparative genomics reveals mechanism for short-term and long-term clonal transitions in pandemic *Vibrio cholerae*. Proc Natl Acad Sci:15442–15447. https://doi.org/10.1073/pnas.0907787106

Ciglenecki I et al (2013) Cholera in pregnancy: outcomes from a specialized cholera treatment unit for pregnant women in Léogâne, Haiti. PLoS Negl Trop Dis 7(8):e2368

Clemens JD et al (2017) Cholera. Lancet 390(10101):1539–1549

Coetzee JN, Datta N, Hedges RW (1972) R factors from Proteus rettgeri. J Gen Microbiol 72(3):543–552

Colwell RR (1996) Global climate and infectious disease: the cholera paradigm. Science:2025–2031. https://doi.org/10.1126/science.274.5295.2025

Couchoud C et al (2020) Deciphering the role of insertion sequences in the evolution of bacterial epidemic pathogens with software. Microbial Genomics 6:6. https://doi.org/10.1099/mgen.0.000356

D'Costa VM et al (2006) Sampling the antibiotic resistome. Science:374–377. https://doi.org/10.1126/science.1120800

D'Costa VM et al (2011) Antibiotic resistance is ancient. Nature:457–461. https://doi.org/10.1038/nature10388

D'Mello-Guyett L et al (2022) Identifying transferable lessons from cholera epidemic responses by Médecins Sans Frontières in Mozambique, Malawi and the Democratic Republic of Congo, 2015–2018: a scoping review. Confl Heal. https://doi.org/10.1186/s13031-022-00445-1

Dalsgaard A et al (2000) Class 1 integron-borne, multiple-antibiotic resistance encoded by a 150-kilobase conjugative plasmid in epidemic *Vibrio cholerae* O1 strains isolated in Guinea-Bissau. J Clin Microbiol 38(10):3774–3779

Das B (2014) Mechanistic insights into filamentous phage integration in *Vibrio cholerae*. Front Microbiol. https://doi.org/10.3389/fmicb.2014.00650

Das B et al (2010) Molecular keys of the tropism of integration of the cholera toxin phage. Proc Natl Acad Sci:4377–4382. https://doi.org/10.1073/pnas.0910212107

Das B et al (2014) A novel, broad-range, CTXΦ-derived stable integrative expression vector for functional studies. J Bacteriol 196(23):4071–4080

Das B et al (2016) Molecular evolution and functional divergence of *Vibrio cholerae*. Curr Opin Infect Dis:520–527. https://doi.org/10.1097/qco.0000000000000306

Das B et al (2020) Antibiotic resistance in *Vibrio cholerae*: understanding the ecology of resistance genes and mechanisms. Vaccine 38(Suppl 1):A83–A92

Davies BW et al (2012) Coordinated regulation of accessory genetic elements produces cyclic di-nucleotides for V. cholerae virulence. Cell 149(2):358–370

De R (2021) Mobile genetic elements of *Vibrio cholerae* and the evolution of its antimicrobial resistance. Front Trop Dis. https://doi.org/10.3389/fitd.2021.691604

Deen J, Mengel MA, Clemens JD (2020) Epidemiology of cholera. Vaccine:A31–A40. https://doi.org/10.1016/j.vaccine.2019.07.078

Didelot X et al (2015) The role of China in the global spread of the current cholera pandemic. PLoS Genet 11(3):e1005072

Dutta D et al (2013) *Vibrio cholerae*Non-O1, non-O139 serogroups and cholera-like diarrhea, Kolkata, India. Emerg Infect Dis:464–467. https://doi.org/10.3201/eid1903.121156

Dziejman M et al (2002) Comparative genomic analysis of *Vibrio cholerae*: genes that correlate with cholera endemic and pandemic disease. Proc Natl Acad Sci:1556–1561. https://doi.org/10.1073/pnas.042667999

Dziejman M et al (2005) Genomic characterization of non-O1, non-O139 *Vibrio cholerae* reveals genes for a type III secretion system. Proc Natl Acad Sci U S A 102(9):3465–3470

Faruque SM, Mekalanos JJ (2012) Phage-bacterial interactions in the evolution of toxigenic *Vibrio cholerae*. Virulence 3(7):556–565

Faruque SM et al (1998) Induction of the lysogenic phage encoding cholera toxin in naturally occurring strains of toxigenic *Vibrio cholerae* O1 and O139. Infect Immun:3752–3757. https://doi.org/10.1128/iai.66.8.3752-3757.1998

Faruque SM et al (2003) Examination of diverse toxin-coregulated pilus-positive *Vibrio cholerae* strains fails to demonstrate evidence for vibrio pathogenicity Island phage. Infect Immun:2993–2999. https://doi.org/10.1128/iai.71.6.2993-2999.2003

Faruque SM, Mekalanos JJ (2003) Pathogenicity islands and phages in Vibrio cholerae evolution. Trends Microbiol 11(11):505–510. https://doi.org/10.1016/j.tim.2003.09.003

Faruque SM, Bin Naser I, Fujihara K, Diraphat P, Chowdhury N, Kamruzzaman M, Qadri F, Yamasaki S, Ghosh AN, Mekalanos JJ (2005) Genomic sequence and receptor for the Vibrio cholerae phage KSF-1phi: evolutionary divergence among filamentous vibriophages mediating lateral gene transfer. J Bacteriol. 187(12):4095–4103. https://doi.org/10.1128/JB.187.12.4095-4103.2005

Faruque SM et al (2006) Transmissibility of cholera: in vivo-formed biofilms and their relationship to infectivity and persistence in the environment. Proc Natl Acad Sci U S A 103(16):6350–6355

Finkelstein RA (2011) Cholera, O1 and O139, and other pathogenic Vibrios. In: Baron S (ed) Medical Microbiology. University of Texas Medical Branch at Galveston, Galveston (TX)

Folster JP et al (2014) Multidrug-resistant IncA/C plasmid in *Vibrio cholerae* from Haiti. Emerg Infect Dis 20(11):1951–1953

Galen JE et al (1992) Role of *Vibrio cholerae* neuraminidase in the function of cholera toxin. Infect Immun:406–415. https://doi.org/10.1128/iai.60.2.406-415.1992

Gallego-Hernandez AL et al (2020) Upregulation of virulence genes promotes biofilm hyperinfectivity. Proc Natl Acad Sci U S A 117(20):11010–11017

Gao H et al (2020) Direct binding and regulation by Fur and HapR of the intermediate regulator and virulence factor genes within the ToxR virulence regulon in *Vibrio cholerae*. Front Microbiol. https://doi.org/10.3389/fmicb.2020.00709

Garg P et al (2000) Expanding multiple antibiotic resistance among clinical strains of *Vibrio cholerae* isolated from 1992-7 in Calcutta, India. Epidemiol Infect 124(3):393–399

Glass RI et al (1980) Emergence of multiply antibiotic-resistant *Vibrio cholerae* in Bangladesh. J Infect Dis 142(6):939–942

Gupta PK et al (2016) Cholera outbreak caused by drug resistant *Vibrio cholerae* serogroup O1 biotype ElTor serotype Ogawa in Nepal; a cross-sectional study. Antimicrob Resist Infect Control. https://doi.org/10.1186/s13756-016-0122-7

Hacker J, Carniel E (2001) Ecological fitness, genomic islands and bacterial pathogenicity. A Darwinian view of the evolution of microbes. EMBO Rep 2(5):376–381

Hall RM, Collis CM (2006) Mobile gene cassettes and integrons: capture and spread of genes by site-specific recombination. Mol Microbiol:593–600. https://doi.org/10.1111/j.1365-2958.1995.tb02368.x

Hall RM et al (2007) What are superintegrons? Nat Rev Microbiol:1–2. https://doi.org/10.1038/nrmicro1462-c1

Hao Y et al (2015) A case of non-O1/non-O139 *Vibrio cholerae* septicemia and meningitis in a neonate. Int J Infect Dis:117–119. https://doi.org/10.1016/j.ijid.2015.05.004

Harris JB, LaRocque RC, Qadri F, Ryan ET, Calderwood SB (2012) Cholera Lancet 379(9835):2466–2476. https://doi.org/10.1016/S0140-6736(12)60436-X

Hassan F, Kamruzzaman M, Mekalanos JJ, Faruque SM (2010) Satellite phage TLCφ enables toxigenic conversion by CTX phage through dif site alteration. Nature 467(7318):982–985. https://doi.org/10.1038/nature09469

Haycocks JRJ et al (2019) The quorum sensing transcription factor AphA directly regulates natural competence in *Vibrio cholerae*. PLoS Genet 15(10):e1008362

Hazen TH et al (2010) The contribution of mobile genetic elements to the evolution and ecology of Vibrios. FEMS Microbiol Ecol:485–499. https://doi.org/10.1111/j.1574-6941.2010.00937.x

Henderson JC et al (2014) Antimicrobial peptide resistance of *Vibrio cholerae* results from an LPS modification pathway related to nonribosomal peptide synthetases. ACS Chem Biol 9(10):2382–2392

Herrera A et al (2022) A Genomic Island of Vibrio cholerae encodes a three-component cytotoxin with monomer and protomer forms structurally similar to alpha-pore-forming toxins. J Bacteriol 204(5):e0055521. https://doi.org/10.1128/jb.00555-21

Herrington DA et al (1988) Toxin, toxin-coregulated pili, and the toxR regulon are essential for *Vibrio cholerae* pathogenesis in humans. J Exp Med:1487–1492. https://doi.org/10.1084/jem.168.4.1487

Hille F, Charpentier E (2016) CRISPR-Cas: biology, mechanisms and relevance. Philos Trans R Soc Lond Ser B Biol Sci 371:1707. https://doi.org/10.1098/rstb.2015.0496

Hochhut B, Waldor MK (1999) Site-specific integration of the conjugal *Vibrio cholerae* SXT element into prfC. Mol Microbiol:99–110. https://doi.org/10.1046/j.1365-2958.1999.01330.x

Hochhut B et al (2001) Molecular analysis of antibiotic resistance gene clusters in *Vibrio cholerae* O139 and O1 SXT constins. Antimicrob Agents Chemother 45(11):2991–3000

Hsiao A, Zhu J (2020) Pathogenicity and virulence regulation of *Vibrio cholerae* at the interface of host-gut microbiome interactions. Virulence:1582–1599. https://doi.org/10.1080/21505594.2020.1845039

Hu D et al (2016) Origins of the current seventh cholera pandemic. Proc Natl Acad Sci. https://doi.org/10.1073/pnas.1608732113

Huda N et al (2003) Molecular cloning and characterization of an ABC multidrug efflux pump, VcaM, in non-O1 *Vibrio cholerae*. Antimicrob Agents Chemother 47(8):2413–2417

Igere BE, Okoh AI, Nwodo UU (2022) Non-serogroup O1/O139 agglutinable *Vibrio cholerae*: a phylogenetically and genealogically neglected yet emerging potential pathogen of clinical relevance. Arch Microbiol. https://doi.org/10.1007/s00203-022-02866-1

Ikema M, Honma Y (1998) A novel filamentous phage, fs-2, of Vibrio cholerae O139. Microbiology (Reading) 144(Pt 7):1901–1906. https://doi.org/10.1099/00221287-144-7-1901

Imamura D et al (2017) Comparative genome analysis of VSP-II and SNPs reveals heterogenic variation in contemporary strains of *Vibrio cholerae* O1 isolated from cholera patients in Kolkata, India. PLoS Negl Trop Dis 11(2):e0005386

Ismail H et al (2011) Genetic characterization of multidrug-resistant, extended-spectrum-β-lactamase-producing *Vibrio cholerae* O1 outbreak strains, Mpumalanga, South Africa, 2008. J Clin Microbiol:2976–2979. https://doi.org/10.1128/jcm.00293-11

Jain M, Kumar P, Goel AK (2016) Emergence of tetracycline resistant *vibrio cholerae* O1 biotype El Tor serotype Ogawa with classical ctxB gene from a cholera outbreak in Odisha, Eastern India. Journal of pathogens 2016:1695410

Jain M et al (2013) Molecular characterization of *Vibrio cholerae* O1 reveals continuous evolution of its new variants in India. Indian J Microbiol:137–141. https://doi.org/10.1007/s12088-013-0372-5

Jaskólska M, Adams DW, Blokesch M (2022) Two defence systems eliminate plasmids from seventh pandemic *Vibrio cholerae*. Nature 604(7905):323–329

Jensen PKM et al (2021) Historical and contemporary views on cholera transmission: are we repeating past discussions? Can lessons learned from cholera be applied to COVID-19? APMIS:421–430. https://doi.org/10.1111/apm.13102

Jermyn WS, Fidelma Boyd E (2002) Characterization of a novel vibrio pathogenicity island (VPI-2) encoding neuraminidase (nanH) among toxigenic *Vibrio cholerae* isolates. Microbiology:3681–3693. https://doi.org/10.1099/00221287-148-11-3681

Jermyn WS, Fidelma Boyd E (2005) Molecular evolution of vibrio pathogenicity island-2 (VPI-2): mosaic structure among *Vibrio cholerae* and Vibrio mimicus natural isolates. Microbiology:311–322. https://doi.org/10.1099/mic.0.27621-0

Juhas M et al (2009) Genomic islands: tools of bacterial horizontal gene transfer and evolution. FEMS Microbiol Rev 33(2):376–393

Jutla A et al (2013) Environmental factors influencing epidemic cholera. Am J Trop Med Hygiene:597–607. https://doi.org/10.4269/ajtmh.12-0721

Kanungo S et al (2010) Cholera in India: an analysis of reports, 1997–2006. Bull World Health Organ:185–191. https://doi.org/10.2471/blt.09.073460

Kanungo, S. et al. (2014) 'Safety and immunogenicity of a live Oral recombinant cholera vaccine VA1.4: a randomized, placebo controlled trial in healthy adults in a cholera endemic area in Kolkata, India', PLoS ONE, p. e99381. doi:https://doi.org/10.1371/journal.pone.0099381

Kanungo S et al (2022) Cholera. Lancet:1429–1440. https://doi.org/10.1016/s0140-6736(22)00330-0

Kaper JB, Morris JG, Levine MM (1995) Cholera. Clin Microbiol Rev:48–86. https://doi.org/10.1128/cmr.8.1.48

Karaolis DK et al (2001) Comparison of *Vibrio cholerae* pathogenicity islands in sixth and seventh pandemic strains. Infect Immun 69(3):1947–1952

Karaolis DKR et al (1998) A *Vibrio cholerae* pathogenicity island associated with epidemic and pandemic strains. Proc Natl Acad Sci:3134–3139. https://doi.org/10.1073/pnas.95.6.3134

Karaolis DKR et al (1999) A bacteriophage encoding a pathogenicity island, a type-IV pilus and a phage receptor in cholera bacteria. Nature:375–379. https://doi.org/10.1038/20715

Karlsson SL et al (2014) Development of stable *Vibrio cholerae* O1 Hikojima type vaccine strains co–expressing the Inaba and Ogawa lipopolysaccharide antigens. PLoS One:e108521. https://doi.org/10.1371/journal.pone.0108521

Kaysner CA and Hill WE (2014) 'Toxigenic*Vibrio cholerae*O1 in Food and Water', *Vibrio cholerae* and Cholera, pp. 27–39. doi:https://doi.org/10.1128/9781555818364.ch2

Kostiuk B et al (2022) Author correction: type VI secretion system mutations reduced competitive fitness of classical *Vibrio cholerae* biotype. Nat Commun 13(1):916

Kovach ME, Shaffer MD, Peterson KM (1996) A putative integrase gene defines the distal end of a large cluster of ToxR-regulated colonization genes in *vibrio cholerae*. Microbiology:2165–2174. https://doi.org/10.1099/13500872-142-8-2165

Kumar A, Bag S, Das B (2018) 'Novel genetic tool to study the stability of Genomic Islands', recent patents on. Biotechnology:200–207. https://doi.org/10.2174/1872208312666180223113618

Kumar A, Das B, Kumar N (2020) Vibrio pathogenicity Island-1: the master determinant of cholera pathogenesis. Front Cell Infect Microbiol 10:561296

Labbate M et al (2016) A genomic island in *Vibrio cholerae* with VPI-1 site-specific recombination characteristics contains CRISPR-Cas and type VI secretion modules. Sci Rep 6:36891

Laj N et al (2022) Quorum-sensing molecules: sampling, identification and characterization of N-acyl-homoserine lactone in vibrio sp. Saudi J Biol Sci:2733–2737. https://doi.org/10.1016/j.sjbs.2021.12.062

Lantagne D et al (2014) The cholera outbreak in Haiti: where and how did it begin? In: Cholera Outbreaks, pp 145–164. https://doi.org/10.1007/82_2013_331

Letchumanan V et al (2016) Insights into bacteriophage application in controlling vibrio species. Front Microbiol. https://doi.org/10.3389/fmicb.2016.01114

Levade I et al (2021) Predicting *Vibrio cholerae* infection and disease severity using metagenomics in a prospective cohort study. J Infect Dis:342–351. https://doi.org/10.1093/infdis/jiaa358

Li J (2015) 'Cholera', radiology of infectious diseases. 2:75–82. https://doi.org/10.1007/978-94-017-9876-1_6

Lippi D, Gotuzzo E, Caini S (2016) 'Cholera', microbiology spectrum. https://doi.org/10.1128/microbiolspec.poh-0012-2015

Lopez AL et al (2014) Killed oral cholera vaccines: history, development and implementation challenges. Ther Adv Vaccines 2(5):123–136

Lu B, Leong HW (2016) Computational methods for predicting genomic islands in microbial genomes. Comput Struct Biotechnol J:200–206. https://doi.org/10.1016/j.csbj.2016.05.001

Lu W-J et al (2018) ATP-binding cassette transporter VcaM from *Vibrio cholerae* is dependent on the outer membrane factor family for its function. Int J Mol Sci 19:4. https://doi.org/10.3390/ijms19041000

de Magny GC et al (2011) Role of zooplankton diversity in *Vibrio cholerae* population dynamics and in the incidence of cholera in the Bangladesh Sundarbans. Appl Environ Microbiol 77(17):6125–6132

Mahillon J, Chandler M (1998) Insertion sequences. Microbiol Mol Biol Rev:725–774. https://doi.org/10.1128/mmbr.62.3.725-774.1998

Mandal J et al (2012) Third-generation cephalosporin-resistant *Vibrio cholerae*, India. Emerg Infect Dis 18(8):1326–1328

Mandal S, Mandal MD, Pal NK (2011) Cholera: a great global concern. Asian Pac J Trop Med 4(7):573–580

Mashruwala AA, Bassler BL (2020) The *Vibrio cholerae* quorum-sensing protein VqmA integrates cell density, environmental, and host-derived cues into the control of virulence. MBio. https://doi.org/10.1128/mbio.01572-20

McDonald ND et al (2019) CRISPR-Cas systems are present predominantly on mobile genetic elements in Vibrio species. BMC Genomics 20(1):105

Mengel MA et al (2014) Cholera outbreaks in Africa. Cholera Outbreaks:117–144. https://doi.org/10.1007/82_2014_369

Mohanty P, Patel A, Bhardwaj AK (2012) Role of H- and D- MATE-type transporters from multidrug resistant clinical isolates of Vibrio fluvialis in conferring fluoroquinolone resistance. PLoS One:e35752. https://doi.org/10.1371/journal.pone.0035752

Mojica FJM et al (2005) Intervening sequences of regularly spaced prokaryotic repeats derive from foreign genetic elements. J Mol Evol 60(2):174–182

Morita D et al (2020) Genomic characterization of antibiotic resistance-encoding genes in clinical isolates of *Vibrio cholerae* non-O1/non-O139 strains from Kolkata, India: generation of novel types of genomic islands containing plural antibiotic resistance genes. Microbiol Immunol 64(6):435–444

Morita M et al (2013) Horizontal gene transfer of a genetic island encoding a type III secretion system distributed in *Vibrio cholerae*. Microbiol Immunol 57(5):334–339

Morris JG Jr (2014) Non-O group 1 *Vibrio cholerae* strains not associated with epidemic disease. In: *Vibrio cholerae and Cholera*, pp 103–115. https://doi.org/10.1128/9781555818364.ch8

Morris JG Jr et al (1995) Clinical and immunologic characteristics of *Vibrio cholerae* O139 Bengal infection in North American volunteers. J Infect Dis 171(4):903–908
Mosley JF 2nd et al (2017) Vaxchora: the first FDA-approved cholera vaccination in the United States. P T 42(10):638–640
Murphy SG et al (2021) *Vibrio cholerae*'s mysterious Seventh Pandemic island (VSP-II) encodes novel Zur-regulated zinc starvation genes involved in chemotaxis and cell congregation. PLoS Genet:e1009624. https://doi.org/10.1371/journal.pgen.1009624
Nadeem A et al (2021) A tripartite cytolytic toxin formed by proteins with flagellum-facilitated secretion. Proc Natl Acad Sci U S A 118:47. https://doi.org/10.1073/pnas.2111418118
Nair GB et al (2002) New variants of *Vibrio cholerae* O1 biotype El Tor with attributes of the classical biotype from hospitalized patients with acute diarrhea in Bangladesh. J Clin Microbiol:3296–3299. https://doi.org/10.1128/jcm.40.9.3296-3299.2002
Nair GB et al (2006) Cholera due to altered El Tor strains of *Vibrio cholerae* O1 in Bangladesh. J Clin Microbiol:4211–4213. https://doi.org/10.1128/jcm.01304-06
Naser IB et al (2017) Analysis of the CRISPR-Cas system in bacteriophages active on epidemic strains of *Vibrio cholerae* in Bangladesh. Sci Rep 7(1):14880
Nelson EJ et al (2009) Cholera transmission: the host, pathogen and bacteriophage dynamic. Nat Rev Microbiol:693–702. https://doi.org/10.1038/nrmicro2204
Nelson EJ et al (2011) Antibiotics for both moderate and severe cholera. N Engl J Med 364(1):5–7
Ng QX et al (2020) Yemen's cholera epidemic is a one health issue. J Prev Med Public Health:289–292. https://doi.org/10.3961/jpmph.20.154
Nuñez JK et al (2014) Cas1-Cas2 complex formation mediates spacer acquisition during CRISPR-Cas adaptive immunity. Nat Struct Mol Biol 21(6):528–534
O'Hara BJ et al (2017) A highly specific phage defense system is a conserved feature of the *Vibrio cholerae* mobilome. PLoS Genet 13(6):e1006838
Ochi K et al (2021) Recent *Vibrio cholerae* O1 epidemic strains are unable to replicate CTXΦ prophage genome. mSphere 6(3):e0033721
Odevall L et al (2018) The Euvichol story—development and licensure of a safe, effective and affordable oral cholera vaccine through global public private partnerships. Vaccine:6606–6614. https://doi.org/10.1016/j.vaccine.2018.09.026
Olejnik E, Davidovitch S (1951) Action of Terramycin and chloromycetin on cholera vibrio in mice. Nature:654–654. https://doi.org/10.1038/168654a0
Vakrani GP, Nambakam T (2021) Retrospective study on acute kidney injury among cholera patients in an outbreak in Whitefield, Bengaluru. Int J Nephrol 6682838. https://doi.org/10.1155/2021/6682838
van der Oost J et al (2014) Unravelling the structural and mechanistic basis of CRISPR–Cas systems. Nat Rev Microbiol:479–492. https://doi.org/10.1038/nrmicro3279
Pant A, Das B, Bhadra RK (2020b) CTX phage of *Vibrio cholerae*: genomics and applications. Vaccine:A7–A12. https://doi.org/10.1016/j.vaccine.2019.06.034
Pant A et al (2020a) Molecular insights into the genome dynamics and interactions between core and acquired genomes of. Proc Natl Acad Sci U S A 117(38):23762–23773
Partridge SR et al (2001) Family of class 1 Integrons related to In4 from Tn 1696. Antimicrob Agents Chemother:3014–3020. https://doi.org/10.1128/aac.45.11.3014-3020.2001
Pérez-Reytor D et al (2018) Accessory toxins of vibrio pathogens and their role in epithelial disruption during infection. Front Microbiol. https://doi.org/10.3389/fmicb.2018.02248
Pérez-Reytor D et al (2020) Analysis of the Zonula occludens toxin found in the genome of the Chilean non-toxigenic Vibrio parahaemolyticus Strain PMC53.7. Front Cell Infect Microbiol. https://doi.org/10.3389/fcimb.2020.00482
Piret J, Boivin G (2021) Pandemics throughout history. Front Microbiol 11:631736
Pollitzer R, Burrows W (1955) Cholera studies. IV. Problems in immunology. Bull World Health Organ 12(6):945–1107
Prithvisagar KS et al (2021) Whole genome analysis unveils genetic diversity and potential virulence determinants in Vibrio parahaemolyticus associated with disease outbreak among

cultured Litopenaeus vannamei (Pacific white shrimp) in India. Virulence:1936–1949. https://doi.org/10.1080/21505594.2021.1947448

Pundhir S, Vijayvargiya H, Kumar A (2008) PredictBias: a server for the identification of genomic and pathogenicity islands in prokaryotes. In Silico Biol 8:223–234

Qin Z et al (2020) Crosstalks between gut microbiota and *Vibrio cholerae*. Front Cell Infect Microbiol. https://doi.org/10.3389/fcimb.2020.582554

Quilici ML et al (2010) *Vibrio cholerae* O1 variant with reduced susceptibility to ciprofloxacin, Western Africa. Emerg Infect Dis 16(11):1804–1805

Rahaman MH et al (2015) Molecular tools in understanding the evolution of *Vibrio cholerae*. Front Microbiol 6:1040

Rajanna C et al (2003) The Vibrio Pathogenicity Island of epidemic *Vibrio cholerae* forms precise extrachromosomal circular excision products. J Bacteriol:6893–6901. https://doi.org/10.1128/jb.185.23.6893-6901.2003

Ramamurthy T et al (2019) Revisiting the global epidemiology of cholera in conjuction with the genomics of. Front Public Health 7:203

Rapa RA et al (2015) A genomic island integrated into recA of *Vibrio cholerae* contains a divergent recA and provides multi-pathway protection from DNA damage. Environ Microbiol 17(4):1090–1102

Reen FJ et al (2006) The genomic code: inferring Vibrionaceae niche specialization. Nat Rev Microbiol:697–704. https://doi.org/10.1038/nrmicro1476

Reid AJ, Amyes SG (1986) Plasmid penicillin resistance in *Vibrio cholerae*: identification of new beta-lactamase SAR-1. Antimicrob Agents Chemother 30(2):245–247

Rivard N, Colwell RR, Burrus V (2020) Antibiotic resistance in *vibrio cholerae*: mechanistic insights from IncC plasmid-mediated dissemination of a novel family of Genomic Islands inserted at. mSphere 5:4. https://doi.org/10.1128/mSphere.00748-20

Ruimy R et al (1994) Phylogenetic analysis and assessment of the genera vibrio, photobacterium, aeromonas, and plesiomonas deduced from small-subunit rRNA sequences. Int J Syst Bacteriol:416–426. https://doi.org/10.1099/00207713-44-3-416

Sakib SN et al (2018) Environmental role of pathogenic traits in *Vibrio cholerae*. J Bacteriol. https://doi.org/10.1128/jb.00795-17

Samai P et al (2015) Co-transcriptional DNA and RNA cleavage during type III CRISPR-Cas immunity. Cell 161(5):1164–1174

Seed KD, Bodi KL, Kropinski AM, Ackermann HW, Calderwood SB, Qadri F, Camilli A (2011) Evidence of a dominant lineage of Vibrio cholerae-specific lytic bacteriophages shed by cholera patients over a 10-year period in Dhaka, Bangladesh. mBio 2(1):e00334–e00310. https://doi.org/10.1128/mBio.00334-10

Seed KD et al (2012) Phase variable O antigen biosynthetic genes control expression of the major protective antigen and bacteriophage receptor in *Vibrio cholerae* O1. PLoS Pathog:e1002917. https://doi.org/10.1371/journal.ppat.1002917

Seed KD et al (2013) A bacteriophage encodes its own CRISPR/Cas adaptive response to evade host innate immunity. Nature 494(7438):489–491

Shaikh H et al (2020) Current and future cholera vaccines. Vaccine 38(Suppl 1):A118–A126

Shi M et al (2022) ToxT Regulon is nonessential for *Vibrio cholerae* colonization in adult mice. Appl Environ Microbiol. https://doi.org/10.1128/aem.00072-22

Siddique AK, Cash R (2014) Cholera outbreaks in the classical biotype era. Cholera Outbreaks:1–16. https://doi.org/10.1007/82_2013_361

Siguier P et al (2015) Everyman's guide to bacterial insertion sequences. Microbiol Spectrum. https://doi.org/10.1128/microbiolspec.mdna3-0030-2014

Silva AJ, Benitez JA (2016) *Vibrio cholerae* biofilms and cholera pathogenesis. PLoS Negl Trop Dis:e0004330. https://doi.org/10.1371/journal.pntd.0004330

Slipski CJ et al (2021) Characterization of proteobacterial plasmid Integron-encoded efflux pump sequence diversity and quaternary ammonium compound antiseptic selection in Escherichia coli grown planktonically and as biofilms. Antimicrob Agents Chemother 65(10):e0106921

Smillie C et al (2010) Mobility of plasmids. Microbiol Mol Biol Rev:434–452. https://doi.org/10.1128/mmbr.00020-10

Smith KP, Kumar S, Varela MF (2009) Identification, cloning, and functional characterization of EmrD-3, a putative multidrug efflux pump of the major facilitator superfamily from *Vibrio cholerae* O395. Arch Microbiol 191(12):903–911

Soares SC et al (2012) PIPS: pathogenicity Island prediction software. PLoS One:e30848. https://doi.org/10.1371/journal.pone.0030848

Soares SC et al (2016) GIPSy: genomic island prediction software. J Biotechnol:2–11. https://doi.org/10.1016/j.jbiotec.2015.09.008

Stroeher UH et al (1992) Serotype conversion in *Vibrio cholerae* O1. Proc Natl Acad Sci:2566–2570. https://doi.org/10.1073/pnas.89.7.2566

Sun D, Jeannot K, Xiao Y, Knapp CW (2019) Editorial: horizontal gene transfer mediated bacterial antibiotic resistance. Front Microbiol 27(10):1933. https://doi.org/10.3389/fmicb.2019.01933

Svennerholm AM, Holmgren J (1976) Synergistic protective effect in rabbits of immunization with *Vibrio cholerae* lipopolysaccharide and toxin/toxoid. Infect Immun 13(3):735–740

Svennerholm AM et al (1984) Local and systemic antibody responses and immunological memory in humans after immunization with cholera B subunit by different routes. Bull World Health Organ 62(6):909–918

Taviani E et al (2010) Discovery of novel *Vibrio cholerae* VSP-II genomic islands using comparative genomic analysis. FEMS Microbiol Lett 308(2):130–137

Teschler JK et al (2015) Living in the matrix: assembly and control of *Vibrio cholerae* biofilms. Nat Rev Microbiol 13(5):255–268

Thapa Shrestha U et al (2015) Multidrug resistant *Vibrio cholerae* O1 from clinical and environmental samples in Kathmandu city. BMC Infect Dis 15:104

Towner KJ et al (1980) Resistance to antimicrobial agents of *Vibrio cholerae* E1 Tor strains isolated during the fourth cholera epidemic in the United Republic of Tanzania. Bull World Health Organ 58(5):747–751

Varela MF, Kumar S, He G (2013) Potential for inhibition of bacterial efflux pumps in multidrug-resistant Vibrio cholera. Indian J Med Res 138(3):285–287

Verma J et al (2019) Genomic plasticity associated with antimicrobial resistance in. Proc Natl Acad Sci U S A 116(13):6226–6231

Vernikos GS, Parkhill J (2006) Interpolated variable order motifs for identification of horizontally acquired DNA: revisiting the Salmonella pathogenicity islands. Bioinformatics:2196–2203. https://doi.org/10.1093/bioinformatics/btl369

Waldor MK, Mekalanos JJ (1996) Lysogenic conversion by a filamentous phage encoding cholera toxin. Science:1910–1914. https://doi.org/10.1126/science.272.5270.1910

Waldor MK, Tschäpe H, Mekalanos JJ (1996) A new type of conjugative transposon encodes resistance to sulfamethoxazole, trimethoprim, and streptomycin in *Vibrio cholerae* O139. J Bacteriol:4157–4165. https://doi.org/10.1128/jb.178.14.4157-4165.1996

Walsh TR et al (2011) Dissemination of NDM-1 positive bacteria in the New Delhi environment and its implications for human health: an environmental point prevalence study. Lancet Infect Dis:355–362. https://doi.org/10.1016/s1473-3099(11)70059-7

Wang R, Li J, Kan B (2016) Sequences of a co-existing SXT element, a chromosomal integron (CI) and an IncA/C plasmid and their roles in multidrug resistance in a *vibrio cholerae* O1 El Tor strain. Int J Antimicrob Agents:305–309. https://doi.org/10.1016/j.ijantimicag.2016.05.020

Wang R et al (2015) IncA/C plasmids harboured in serious multidrug-resistant *Vibrio cholerae* serogroup O139 strains in China. Int J Antimicrob Agents:249–254. https://doi.org/10.1016/j.ijantimicag.2014.10.021

Wei W et al (2016) Zisland explorer: detect genomic islands by combining homogeneity and heterogeneity properties. Brief Bioinform:bbw019. https://doi.org/10.1093/bib/bbw019

Weil AA, Becker RL, Harris JB (2019) Vibrio cholerae at the Intersection of immunity and the microbiome. mSphere 4(6):e00597–e00519. https://doi.org/10.1128/mSphere.00597-19

WHO (2017) 'Cholera annual report 2017. Wkly Epidemiological Rec 2018′, 93, pp. 489–500. doi:

WHO (2020) 'Cholera Annual Report 2019. Wkly Epidemiological Rec 2020′, 95, pp. 441–448 doi:
WHO Publication (2010) Cholera vaccines: WHO position paper-recommendations. Vaccine 28(30):4687–4688
Williams PCM, Berkley JA (2018) Guidelines for the management of paediatric cholera infection: a systematic review of the evidence. Paediatr Int Child health 38(sup1):S16–S31
Woolley RC et al (2005) Characterization of the *Vibrio cholerae* vceCAB multiple-drug resistance efflux operon in Escherichia coli. J Bacteriol 187(15):5500–5503
Wong E, Vaaje-Kolstad G, Ghosh A, Hurtado-Guerrero R, Konarev PV, Ibrahim AFM et al (2012) The Vibrio cholerae colonization factor GbpA possesses a modular structure that governs binding to different host surfaces. PLoS Pathog 8(1):e1002373. https://doi.org/10.1371/journal.ppat.1002373
Wozniak RAF et al (2009) Comparative ICE genomics: insights into the evolution of the SXT/R391 family of ICEs. PLoS Genet 5(12):e1000786
Xu M, Cao C, Wang D, Kan B (2014) Identifying environmental risk factors of cholera in a coastal area with geospatial technologies. Int J Environ Res Public Health. 12(1):354–370. https://doi.org/10.3390/ijerph120100354
Zhang R et al (2014) Identification of horizontally-transferred Genomic Islands and genome segmentation points by using the GC profile method. Curr Genomics:113–121. https://doi.org/10.2174/1389202915999140328163125
Zhu J, Mekalanos JJ (2003) Quorum sensing-dependent biofilms enhance colonization in *Vibrio cholerae*. Dev Cell 5(4):647–656

Genomic Islands in Marine Bacteria 15

Mohit Kumar, Ashutosh Khaswal, Santosh Kumar Mishra, Abha Vashistha, Indra Mani, and Vijai Singh

Abstract

Bacterial genomes are dynamic in nature having conserved pool genes across taxonomic boundaries. Deep-sea presently emerging as a new source for isolation and screening of new and novel compounds produced by microbes. There is wide importance of marine bacteria is already proven but bacterial genome of these microorganisms is not well explored and studied systemically. Recent findings showed that Horizontal Gene Transfer (HGT) mechanism played most important and significant role in the evolution results in a diversified role in the adaptation. The source of genetic variability remains an important aspect of HGT. In this chapter, different aspects of genomic island bacteria and recent findings have been summarized.

Keywords

Genomic · Actinomyces · Transposons · Island

M. Kumar · A. Khaswal · S. K. Mishra (✉)
Department of Biotechnology, IMS Engineering College, Ghaziabad, U.P, India

A. Vashistha
Department of Biosciences, University institute of Biotechnology, Chandigarh University, Mohali, Punjab, India

I. Mani
Department of Microbiology, Gargi College, University of Delhi, New Delhi, India

V. Singh
Department of Biosciences, School of Science, Indrashil University, Rajpur, Mehsana, Gujarat, India

I. Mani et al. (eds.), *Microbial Genomic Islands in Adaptation and Pathogenicity*,
https://doi.org/10.1007/978-981-19-9342-8_15

15.1 Introduction

Comparative genomics of bacterial systems provides us with a once-in-a-lifetime chance for retrieval of valuable information in the genome structure, evolution and diversity in functional aspects in the microorganisms present in marine ecosystem. This has been observed that the genome of these bacterial communities is dynamic in nature, with a conserved pool of genes shared across taxonomic boundaries. Horizontal Gene Transfer (HGT), these are large genomic areas that cause considerable genetic changes across closely related species, and they have ability to disclose ecologically important characteristics among the genomes. Bacterial comparative genomics offers a rare opportunity to learn about the genome structure, function to show the evolution in the marine microbial community as well as functional diversity. Since genomes of bacteria are dynamic in nature, having conserved groups of genes shared at multiple taxonomic levels at the core genome and a flexible (or adaptable) genome. These microbial entities showed number of taxa specific genes which is not related across characteristically similar (Hacker and Carniel 2001; Ochman et al. 2005). HGT is considered one of the significantly important evolutionary processes that increase the adaptability of bacterial populations' genome pools, allowing them to adjust to new ecological conditions (Boucher et al. 2003; Doolittle 1999). Genomic islands (GIs) are group of genes that is laterally transported and are related to the prokaryotic genome pool's flexible genome pool. Comparative genome analysis was used to examine these highly variable genomic regions in a few bacterial species (Hacker and Carniel 2001). Genes with species-specific origins are generally present in GIs (Hacker and Kaper 2000). These are substantial genomic areas that cause considerable genetic changes across closely related species, and they may disclose specific ecologically relevant genome traits (Cuadros-Orellana et al. 2007; Hacker and Kaper 2000). GIs may include a diverse assortment of genes from various origins. However, by employing a hypothesis-free method in GI identification, certain similar traits may be identified. This suggests that GIs might be considered as superfamily having important mobile components (Vernikos and Parkhill 2008). Genes identified in GIs show variable function and nature. These genes range from critical survival genes to virulence genes as well as antibiotic resistance genes. It has been observed that GIs which have rich cluster virulence genes, as well as DNA sequences made up of succession repeats of 23–47 base pair long and having role in developing resistance to genetic elements, especially in the plasmids as well as phages (Ho Sui et al. 2009). GI content reveals information related to habits of bacteria toward survival methods (Read and Ussery 2006).

15.2 Mining of Genomes in Actinomycetes from Marine Environments

Presently deep sea, provided new and novel compounds from marine environment and emerged as a key source for the identification of novel bacterium having the potential to produce new compounds (Yang et al. 2020; Jagannathan et al. 2021). Since there is an extreme and adverse environmental actinomycetes of deep-sea origin provide some important *Streptomyces* which display some unique metabolic characteristics leading these marine microorganisms in the production of some important biomolecules (Kamjam et al. 2017; Wang et al. 2020). Genome-level studies showed that in *Streptomyces* genomes there are more than 20 biosynthetic gene clusters (BGCs) responsible for the production of natural products of economic importance. However, in laboratory conditions, they showed a small amount of natural products is produced (Baltz 2017). It may be due to hindrances in the activation of BGCs that have reduced expression of desired or sometimes due to no expression of gene. All these strategies emerged as a major activity in the identification as well as production of bioactive compounds with novel properties. In the past few decades due to emergence of bioinformatics tools genome mining activities is now becoming an effective method in genomic study. Genome mining and efficient in silico method for identification of BGCs needed precise and genomes sequences with high quality. Short-read sequencing methods such as Illumina are now considered in genome sequencing and widespread due to low-cost, high coverage, precise as well as delivery of accurate reads (Miller et al. 2010). Recently developed long-reads technologies, i.e., PacBio, have improved the correctness of de novo units in the providing genomic structure (Jayakumar and Sakakibara 2019). However, the use of hybrid strategies for study of complex bacterial genomes when combines with the correctness of short reads provide significantly relevant information on the genomic structure (De Maio et al. 2019). These hybrid strategies have proven to be particularly efficient in the study of repetitive sequences of *Streptomyces*.

15.3 Relative Significance of Genomic Island in Marine Bacteria

IslandViewer has emerged as itself one of the most effective tools and was used in in silico study of 70 manually curated genomes. It is a web-based user interface that combines a number of techniques used for GI recognition as well as visualization. IslandPick (Waack et al. 2006), SIGI-HMM, and IslandPath-DIMOB (Hsiao et al. 2005) verified the GIs prediction methods incorporated in IslandViewer. However, there is a very less number of genomes that have been successfully sequenced and thoroughly edited for marine bacteria. GIs found in prior research were frequently based on a single approach either a comparison between two genomes or variations in tetra nucleotide frequency, or the existence of mobility genes. Because this web-based tool look for numerous features, we may presume a further robust and conservative exposure of genomic islands. Table 15.1 compares the genomic islands

Table 15.1 Comparison of earlier reported GIs with IslandViewer GIs

Bacterial strain	Earlier studies genomic islands: Number of GIs	Earlier studies genomic islands: Genomic islands length identified in control (kb)	IslandViewer tool-based GIs: Genomic islands present (+)	IslandViewer tool-based GIs: Extra genomic islands (+)	IslandViewer tool-based GIs: Genomic islands absent (−)	IslandViewer tool-based GIs: Genomic islands length with IslandViewer estimation (kb)	Comparative analysis of previous versus IslandViewer genomic islands: Overlap percentage (kb)	Comparative analysis of previous versus IslandViewer genomic islands: Sensitivity percentage (%)	Comparative analysis of previous versus IslandViewer genomic islands: Precision (%)
P. marinus str. MIT9312	5	233	3	0	3	44.9	19.3	50	100
S. sp. RCC307	15	271	9	1	6	56.9	18.7	60	90
S. sp. WH7803	11	344	3	2	8	73.7	20.3	27	60
S. sp. CC9605	20	505.3	18	4	2	300.4	54.3	90	82
S. sp. CC9311	24	578.6	9	2	15	125.9	16.9	38	82
A. macleodii “deep ecotype”	13	480	6	5	7	272.7	69.6	46	55
S. ruber DSM13855	3	221.8	3	1	0	168.2	41.1	100	75
Salinibacter ruber M8	2	265.9	2	3	0	144.9	44.5	100	40
Average							**35.58**	**64**	**73**

of eight marine bacteria known as Control Genomes for whose GIs were previously provided with the GIs projected by IslandViewer for the same genomes in this work.

15.4 GIs Quantitative Significance in Marine Bacteriological Genomes

The 70 chosen marine bacterial genomes include the following four main oceanic prokaryotic taxa:Alphaproteobacteria, Bacteroidetes, Gammaproteobacteria, as well as Cyanobacteria. These four bacteria account for nearly 80% of all marine bacterial species (Barberán and Casamayor 2010). On the other hand, out-group Bacteroidetes genomes, comprised 14 Flavobacteria and seven nonmarine Bacteroidetes. Numerous genomes from diligently connected bacterial strains from each taxonomic group were used to investigate the rate of intra-specific variability of GIs and their value as primary contributors to strain-specific genes.

The marine *Bacteroidetes flavobacteria* BBFL7, *Pelagibacter ubique* HTCC1062, *Flavobacterium psychrophilum* JIP02/86 and *Flavobacteria* ALC-, have no GIs found in their genomes. Given that several of these genomes had extremely small genome sizes and that relevant genomes were not available for comparison, the lack of GIs in these genomes may be explained by the GI predictor's poor sensitivity in certain situations. When the data for each of the four groups was evaluated independently, Bacteroidetes (p 0.001, $R = 0.78$) and Cyanobacteria (p 0.05, $R = 0.60$) and revealed noteworthy relationships whereas proteobacteria did not. Greater relationships ($R^2 = 0.9$) between genomic island size and size of the genome for *picocyanobacteria* were seen out of 14 genomes from the relevant genera *Prochlorococcus* and *Synechococcus* (Dufresne et al. 2008). Cyanobacteria had a substantially poorer average ratio than the other three groups. This is most likely owing to IslandViewer's weak GI detection rate for several Cyanobacteria (Table 15.1).

The genomes of *Synechococcus* sp., *Rhodobacter sphaeroides* ATCC1705 *Psychrobacter* sp. *gammaproteobacterium*, *flavobacterium Robiginitalea biformata* HTCC250, and PRwf-1 had the highest ratios for each major bacterial class. Earlier research found that between 10 and 31% of the genomes of *Synechococcus* were made up of GIs, and similar fractions nearly 17%, have also been observed for pathogenic islands in the *E. coli strain* (Ochman et al. 2000). Study of marine bacteria showed that a smaller proportion of their genome in the genomic island.

15.5 Architecture of Marine Bacterial GIs

In line with prior GI investigations, we discovered that 70% of the identified GIs included phage integrase-related genes, MGE, conjugative transposons, predominantly transposases, or integrons (Dobrindt et al. 2004; Juhas et al. 2009). Furthermore, we discovered that at least 27% of the genomic island were bordered by or included tRNAs, suggesting that they acted as GI integration sites (Reiter et al. 1989; Williams 2002).

The *Anabaena variabilis*, a cyanobacterium possesses a gene connected to psaC of photosystem 1 subunit 7, four cas genes involved with the CRISPR system, and three Tn7-like transposition genes in a single 13-kb GI. The *Pseudoalteromonas* atlantica T6 has a genomic island of 62.7 kb, which was largely made up of a prophage with numerous phage-linked genes. In addition, we discovered a 16-kb GI containing several MGE elements and flagellar protein genes in the *Roseobacter denitrificans* OCh 114. Finally, *Leeuwenhoekiella blandensis*, a marine Bacteroidetes has a genomic island of 26.8 kb and many MGE, nitrite reductase genes and cas. This kind of gene has only been discovered in the deep-water flavobacterium which possesses the ability to hydrolyze organic nitrogen (Qin et al. 2010). Second, we discovered several GIs that included tRNAs and site-specific recombinases. These HR-GIs contain genes that frequently have a distinct sequence arrangement from the rest of the genome and some of them are known to encrypt ribosomal proteins, which are acknowledged to be highly expressed (Karlin 2001). As a result, these genome segments could turn out to be false-positive predictions if sequence configuration bias (% GC content) was the only measure used to detect GIs. The three most precise genomic island prediction methods (Hsiao et al. 2005; Langille et al. 2010; Waack et al. 2006) were added to IslandViewer, each utilizing a different methodology to predict GIs, hence, more than one technique detected both HR-GIs. It provides supplementary evidence that these gene cartridges were found in a true GI and were not false positives.

EF-Tu and RpoB gene sequences found in HR1 genomic island, as well as the 16S rRNA, and sequences from 20 Bacteroidetes genomes were used to create phylogenetic trees. If these GIs were false positives, then we would assume the phylogeny of the EF-Tu and RpoB genes to coincide with those of 16S rRNA. We would forecast slightly different phylogeny for the 16S rRNA, EF-Tu, and RpoB, genes if these were real GIs responsive to HR1. There were notable variances between the 16S rRNA phylogenies and the other two genes, despite the fact that the basic topology of the Sphingobacteria, Flavobacteria, and Bacteroides branches were preserved throughout the three genes. This is consistent with the two functional genes sharing a family and having similar evolutionary trajectories (Fig. 15.1).

15.6 Differences in GI Gene Content Among Marine Bacteriological Classes

Each bacterial class possessed a distinct set of functions that were preferentially represented in its GIs, as seen by the significant disparities that were present in all of the instances. Cyanobacteria has by far the biggest and greatest diversified number of augmented GO keywords (18). A vast of these were connected to photosynthesis, especially the electron transport system and antenna or photosystem proteins. These photosynthesis-related genes (30 genes) were discovered within genomic island of 50% of Cyanobacteria, indicating that they are a very prevalent trait across

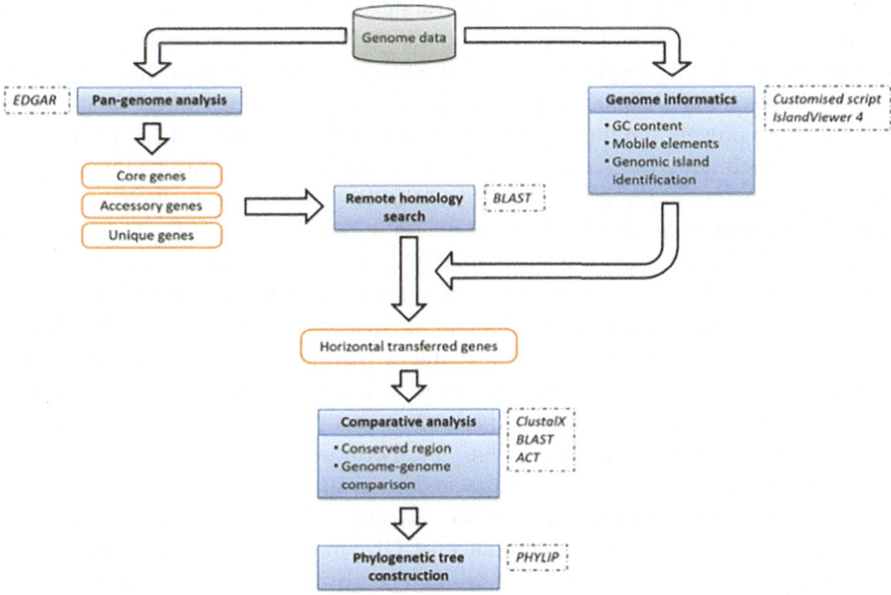

Fig. 15.1 Approaches in the analysis of HGT using pan-genome data. The availability of genome data allows for a series of investigations via pan-genome analysis that can lead to the identification of horizontally transferred genes within the organism

Cyanobacteria GIs. Other GO keywords associated with Cyanobacteria included histidine/cobalamin biosynthesis, hydrolysis/proteolysis activity, and glucose metabolism.

15.7 Biologically Important Genes Found in Marine Bacterial GIs

We divided the genes found in GIs into different biological groups that may enhance bacterial fitness in order to evaluate the ecological importance of each gene. DNA restriction-modification systems, energy metabolism, hydrolysis activity, ribosomal proteins, stress response proteins, transporters, DNA-directed RNA polymerase, two-component systems, and MGE are a few of the biological categories that were present in significant numbers among the bacterial taxa. Although their abundance varied within the GIs, these categories were uniformly distributed across the genomes. By moving DNA sequences to different locations inside or across genomes, transposases, and integrases can change the structure of the genome (Rice and Baker 2001). Recently, transposases have been found to be among the most common and abundant genes (Aziz et al. 2010) and found that 675 genes linked to transposases, such as IS elements or transposons, which make up about 8.2% of the entire database. These genes, which make up around 30% of all transposases, were overrepresented in Alphaproteobacteria in GIs (194 genes). Genes involved in bacteriophage defense, including DNA modification restriction mechanisms, were also present in the GI sample. These defense mechanisms, which include restriction endonucleases and sequence-specific restriction enzymes, have long been recognized to prevent the introduction of external DNA (Arber and Linn 1969).

Another technique to give resistance from phages and this may be due to other ambulant elements is the availability of CRISPR systems due to their linked cas genes (Barrangou et al. 2007; Mojica et al. 2000; Sorek et al. 2008). These genetic features have been discovered in around 90% and 40% of archaea and bacteria genomes, respectively. According to a recent study, CRISPR systems are mostly transmitted by horizontal gene transfer and are overrepresented among GIs (Ho Sui et al. 2009). Due to their potential advantages for the host's lifestyle, several GI genes are quite likely to be selected favorably. It has been shown that various *Prochlorococcus* strains' GIs are expressed in distinct ways in response to light and dietary stresses (Coleman et al. 2006). Furthermore, *Prochlorococcus* GIs play a crucial role in the cohabitation of viruses and hosts since they carry genes important in viral attachment to the host cell surface (Avrani et al. 2011). Since there is some endorsement for the adaptive importance of GI genes amid environmental bacteria, however, it is uncertain how much influence of these genes on the routes for diversification in marine bacterial species.

15.8 Conclusions

The majority of the marine bacterial genomes examined had genomic island. Research data demonstrates that horizontal gene transfer through plasmids; phages, etc. play a vital role in the maneuverability of gene clusters among taxa and within closely related genomes, changing the flexible cluster of the genome. By introducing unique foreign genes as well as altering their regulation, transcription, and/or transduction, GIs have potential to increase bacterial fitness in response to changing environmental circumstances. The results demonstrate the potential role of GIs in altering the composition and accumulative diversity of marine bacteriological genomes. The physiology and ecology of microorganisms were shown to be closely related to a number of GIs, but we also found several significant conserved genes that were theoretically connected to the core genome. Given the importance of GIs in fully comprehending the evolution and ecology of marine bacteria in the ocean, our findings underscore the necessity of establishing a pangenome system for marine bacterial species. Learning about the evolutionary processes that shape the marine bacterial genomes are helpful in investigating the mechanisms that maintain and choose GIs.

References

Arber W, Linn S (1969) DNA modification and restriction. Annu Rev Biochem 38:467–500
Avrani S, Wurtzel O, Sharon I, Sorek R, Lindell D (2011) Genomic island variability facilitates Prochlorococcus–virus coexistence. Nature 474:604–608
Aziz RK, Breitbart M, Edwards RA (2010) Transposases are the most abundant, most ubiquitous genes in nature. Nucleic Acids Res 38:4207–4217
Baltz RH (2017) Gifted microbes for genome mining and natural product discovery. J Ind Microbiol Biotechnol 44(4–5):573–588
Barberán A, Casamayor E (2010) Global phylogenetic community structure and β-diversity patterns in surface bacterioplankton metacommunities. Aquat Microb Ecol 59:1–10
Barrangou R, Fremaux C, Deveau H, Richards M, Boyaval P, Moineau S, Romero DA, Horvath P (2007) CRISPR provides acquired resistance against viruses in prokaryotes. Science 315:1709–1712
Boucher Y, Douady CJ, Papke RT, Walsh DA, Boudreau MER, Nesbø CL, Case RJ, Doolittle WF (2003) Lateral gene transfer and the origins of prokaryotic groups. Annu Rev Genet
Coleman ML, Sullivan MB, Martiny AC, Steglich C, Barry K, DeLong EF, Chisholm SW (2006) Genomic Islands and the ecology and evolution of prochlorococcus. Science 311:1768–1770
Cuadros-Orellana S, Martin-Cuadrado A-B, Legault B, D'Auria G, Zhaxybayeva O, Papke RT, Rodriguez-Valera F (2007) Genomic plasticity in prokaryotes: the case of the square haloarchaeon. ISME J 1:235–245
De Maio N, Shaw LP, Hubbard A, George S, Sanderson ND, Swann J, Wick R, AbuOun M, Stubberfield E, Hoosdally SJ, Crook DW (2019) Comparison of long-read sequencing technologies in the hybrid assembly of complex bacterial genomes. Microbial Genomics 5:9
Dobrindt U, Hochhut B, Hentschel U, Hacker J (2004) Genomic islands in pathogenic and environmental microorganisms. Nat Rev Microbiol 2:414–424
Doolittle WF (1999) Lateral genomics. Trends Biochem Sci 24:M5–M8
Dufresne A, Ostrowski M, Scanlan DJ, Garczarek L, Mazard S, Palenik BP, Paulsen IT, de Marsac NT, Wincker P, Dossat C, Ferriera S, Johnson J, Post AF, Hess WR, Partensky F (2008) Unraveling the genomic mosaic of a ubiquitous genus of marine cyanobacteria. Genome Biol 9: R90
Hacker J, Carniel E (2001) Ecological fitness, genomic islands and bacterial pathogenicity. EMBO Rep 2:376–381
Hacker J, Kaper JB (2000) Pathogenicity islands and the evolution of microbes. Annu Rev Microbiol
Ho Sui SJ, Fedynak A, Hsiao WWL, Langille MGI, Brinkman FSL (2009) The Association of Virulence Factors with Genomic Islands. PLoS One 4:e8094
Hsiao WWL, Ung K, Aeschliman D, Bryan J, Finlay BB, Brinkman FSL (2005) Evidence of a large novel gene pool associated with prokaryotic Genomic Islands. PLoS Genet 1:e62
Jagannathan SV, Manemann EM, Rowe SE, Callender MC, Soto W (2021) Marine actinomycetes, new sources of biotechnological products. Mar Drugs 19(7):365
Jayakumar V, Sakakibara Y (2019) Comprehensive evaluation of non-hybrid genome assembly tools for third-generation PacBio long-read sequence data. Brief Bioinform 20(3):866–876
Juhas M, van der Meer JR, Gaillard M, Harding RM, Hood DW, Crook DW (2009) Genomic islands: tools of bacterial horizontal gene transfer and evolution. FEMS Microbiol Rev 33:376–393
Kamjam M, Sivalingam P, Deng Z, Hong K (2017) Deep sea actinomycetes and their secondary metabolites. Front Microbiol 8:760
Karlin S (2001) Detecting anomalous gene clusters and pathogenicity islands in diverse bacterial genomes. Trends Microbiol 9:335–343
Langille MGI, Hsiao WWL, Brinkman FSL (2010) Detecting genomic islands using bioinformatics approaches. Nat Rev Microbiol 8:373–382

Miller JR, Koren S, Sutton G (2010) Assembly algorithms for next-generation sequencing data. Genomics 95(6):315–327

Mojica FJM, Díez-Villaseñor C, Soria E, Juez G (2000) Biological significance of a family of regularly spaced repeats in the genomes of archaea, bacteria and mitochondria. Mol Microbiol

Ochman H, Lawrence JG, Groisman EA (2000) Lateral gene transfer and the nature of bacterial innovation. Nature 405:299–304

Ochman H, Lerat E, Daubin V (2005) Examining bacterial species under the specter of gene transfer and exchange. Proc Natl Acad Sci 102:6595–6599

Qin Q-L, Zhang X-Y, Wang X-M, Liu G-M, Chen X-L, Xie B-B, Dang H-Y, Zhou B-C, Yu J, Zhang Y-Z (2010) The complete genome of Zunongwangia profunda SM-A87 reveals its adaptation to the deep-sea environment and ecological role in sedimentary organic nitrogen degradation. BMC Genomics 11:247

Read TD, Ussery DW (2006) Opening the pan-genomics box. Curr Opin Microbiol

Reiter W-D, Palm P, Yeats S (1989) Transfer RNA genes frequently serve as integration sites for prokaryotic genetic elements. Nucleic Acids Res 17:1907–1914

Rice PA, Baker TA (2001) Comparative architecture of transposase and integrase complexes. Nat Struct Biol

Sorek R, Kunin V, Hugenholtz P (2008) CRISPR—a widespread system that provides acquired resistance against phages in bacteria and archaea. Nat Rev Microbiol 6:181–186

Vernikos GS, Parkhill J (2008) Resolving the structural features of genomic islands: a machine learning approach. Genome Res 18:331–342

Waack S, Keller O, Asper R, Brodag T, Damm C, Fricke WF, Surovcik K, Meinicke P, Merkl R (2006) Score-based prediction of genomic islands in prokaryotic genomes using hidden Markov models. BMC Bioinformatics 7:142

Wang YN, Meng LH, Wang BG (2020) Progress in research on bioactive secondary metabolites from deep-sea derived microorganisms. Mar Drugs 18(12):614

Williams KP (2002) Integration sites for genetic elements in prokaryotic tRNA and tmRNA genes: sublocation preference of integrase subfamilies. Nucleic Acids Res 30:866–875

Yang Z, He J, Wei X, Ju J, Ma J (2020) Exploration and genome mining of natural products from marine streptomyces. Appl Microbiol Biotechnol 104(1):67–76

Challenges in Eventing Horizontal Gene Transfer

16

Mahak Bhushan, Goutam Kumar Dhandh, Vijayaraghava S. Sundararajan, Amita Sharma, Harshita Bhargava, Meera Indracanti, Sankalp Sharma, Ashkan Dashtban, M. Nidheesh, Jayaraman Valadi, and Prashanth Suravajhala

M. Bhushan
Department of Biological Sciences ISER Kolkata, Kolkata, West Bengal, India

G. K. Dhandh (✉) · S. Sharma
Bioclues.org, Hyderabad, India

V. S. Sundararajan
School of Computing and Information Systems, Singapore Management University, Bras Basah, Singapore

A. Sharma · H. Bhargava
Department of Computer Science, IIS University, Jaipur, Rajasthan, India
e-mail: amita.1983@iisuniv.ac.in; harshita.bhargava@iisuniv.ac.in

M. Indracanti
Department of Life Sciences, School of Allied Health Sciences, Malla Reddy University, Hyderabad, Telangana, India
e-mail: drmeera@mallareddyuniversity.ac.in

A. Dashtban
University College, London, UK

M. Nidheesh
Amrita School of Biotechnology, Amrita Vishwavidyapeetham, Clappana, Kerala, India
e-mail: Nidheesh@am.amrita.edu

J. Valadi
Department of Computer science, FLAME University, Pune, MH, India

P. Suravajhala
Bioclues.org, Hyderabad, India

Amrita School of Biotechnology, Amrita Vishwavidyapeetham, Clappana, Kerala, India
e-mail: prash@Bioclues.org; prash@am.amrita.edu

I. Mani et al. (eds.), *Microbial Genomic Islands in Adaptation and Pathogenicity*,
https://doi.org/10.1007/978-981-19-9342-8_16

Abstract

Horizontal gene transfer (HGT) takes place when two organisms share their genetic material. Contrary to the ancestral transfer of genes, viz. Vertical gene transfer, the HGT, is very commonly found in many species of bacteria even if there is distantly related evidence which is phylogenies, sequence comparison, and genome characteristics. In this work, we use an Artificial Intelligence/ Machine Learning (AI/ML) model which is based on unsupervised learning on the human fungal pathogenesis dataset. The AI/ML model discovers distinctive classification/annotation features which are able to predict a correlation between pathogenesis and horizontal transfer events with unknown queries. A brief perspective of this is enlisted in our chapter.

Keywords

Evolution · Horizontal gene transfer · Pathogenesis · Machine learning · Bioinformatics

16.1 Introduction

The movement of genetic material between species is referred to as horizontal gene transfer (HGT) or lateral gene transfer (LGT). In contrast to the ancestral transfer of genes, viz. Vertical Gene Transfer, HGT event is non-hereditary (Soucy et al. 2015). The HGT is very commonly found in many species of bacteria even if they are distantly related by evidence of sequence and phylogenetic comparison using genome characteristics (Husnik and McCutcheon 2018). The HGT is beneficial to bacteria if the non-inheritable protein or gene of interest, and If the gene has no function, it confers a good function but is harmful. Bacteria being simple, are known to even HGT complexity and they do this by causing numerous diseases. Numerous harmful bacteria have been shown to be crucial not just in relation to illness, but also with regard to any genes that might have transferred from the host to the bacteria (Olszak et al. 2017). The rate of HGT was previously thought to be pretty high in the beginning of bacterial evolution, with HGT events leading to the development and rapid spread of novel genes (Ram and Hadany 2019). This further allows the build-up of large genomes; However, even if the rate of gene loss is lower, the situation is still unfavorable. Monaco has earlier reviewed and proposed a model to check this, in Bioessays, "The selfish environment meets the selfish gene: Coevolution and inheritance of RNA and DNA pools: A model for organismal life incorporating coevolution, horizontal transfer, and inheritance of internal and external RNA and DNA pools.: A model for organismal life incorporating coevolution, horizontal transfer, and inheritance of internal and external RNA and DNA pools" (Monaco 2022).

There are over 130,00 vivid bacteria that have been sequenced, and It is possible that a lot of bacterial proteins implicated in host disease moved to primate hosts like humans, according to microorganisms that have been sequenced (Loman and Pallen 2015; Reddy et al. 2014). Pathogens represent an extremely small fraction of all microbial species, according to estimations of bacterial diversity from multiple resources, as was previously stated (DeLong 1997). While the majority of them do not spread diseases. However, the above argument in lieu of gene transfer from bacteria to humans needs to be thoroughly reasoned keeping in view the host's many physical, cellular, and molecular barriers. With the rise in systems, genomic approaches attempts are made to decipher protein–protein interaction (PPI) networks and determine whether they have been passed down between organisms. Interologs are hypothetical model organisms such as bacteria that researchers use to supplement the interactomes of higher eukaryotes. Various network components have been conserved in this manner (Brown and Jurisica 2007) keeping in view the gene transfer that lays an enormous interest in considering the magnitude of infection of certain organisms like *Streptococcus* spp. (Enright et al. 2001). During evolution, pathogenic bacteria are thought to have lost some virulent genes while gaining virulence factors. In addition, there are two types of experiments: the first tests pathogen genes for their being able to confer avirulence phenotype of such a typically non-virulent strain, and the second tests pathogen genes for their role in virulence via mutational analysis. Many virulence-related genes are specific to pathogenic organisms and have been laterally transferred into other genomes, according to the analysis of sequences recovered using these methods. Efforts in this direction have paved the way, for example, random walking approaches have been used for "synteny index" (SI) with statistical models (Sevillya et al. 2020).

With sequences of entire genomes offering new ways to assess gene content, and much more importantly showing the effect of HGT on pathogenesis and its evolution, there remains a challenge on how organisms are studied experimentally. Therefore, it is suggested to proceed with comparative genomics involving sequences that, for the first time, have homologs shared by phylogenetically distinct, completely sequenced species (Cywes Bentley et al. 2005). It was observed that genes assumed to be necessary for host interactions were exchanged between *Chlamydia trachomatis* and *Rickettsia prowazekii* (Aravind et al. 1998) implying that a shared niche makes it easier to transfer and maintain genes needed for adaptation to a specific host or lifestyle. Keeping in view of the aforementioned views, we foresee the following bioinformatics challenge to identify proteins or genes involved in pathogenesis in the host.

Identifying substitution rates between two protein homologues in different strains. For example, MutS, a DNA repair protein known to be conserved in all bacteria, has several isoforms. In identifying several housekeeping loci of such proteins, we can determine the population genetic structure must be determined first, followed by the nucleotide sequence to check for substitution changes. Such selection in close relatives is always purifying (*see* Fig. 16.1). Given the evidence for recombination's significant impact, it is surprising that HGT in several bacteria, including the evolution of group A Streptococcus (GAS) populations, it appears that

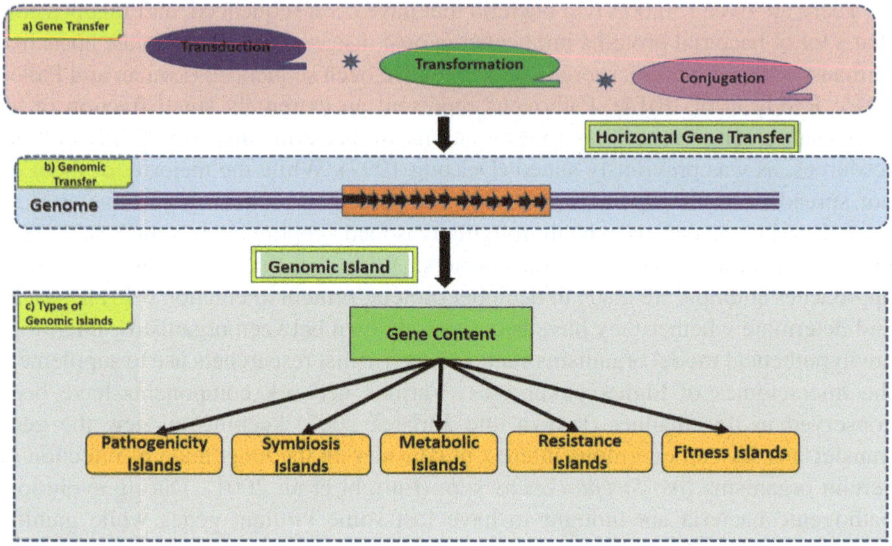

Fig. 16.1 (**a**): A simplest overview of how HGT events take place. (**b**) During the event of gene duplication, if w = dA/dS <1, then it would lead to a purifying selection, while w = 1 also leads to purifying selection but the nucleotide sequences fit in selection and so there is no plausibility of the proteins with allied function. But when w > 1 and > > 1, there is a strong reason to show that there is a change in amino acids that leads to a greater increase in non-synonymous substitutions through which novel functions arise

the presence of some lineages has been rare, which has been proved earlier (Nelson et al. 1999). We, therefore, conceptualize a web server (*See Supplementary Flowchart*) to find if the genes event HGT, localized to the subcellular location of interest, any synonymous or nonsynonymous substitutions in the process. As a first step, only *Streptococcus pyogenes* were considered for the study. In order to find out the genes transferred from *Streptococcus pyogenes* to human mitochondrial proteins BLAST analysis was done yielding many significant hits (data not shown) which underlines the horizontal transfer of genes since it is believed that human mitochondria evolved from bacterial cells (Fig. 16.2).

```
STEP#1: Query
STEP#2: Here, X and X' are orthologs even as Y and Y'.
STEP#3: Use X' and Y' as query and crosscheck if X and Y, respec-
tively, are observed in Blast, then we could say
Y and Y' are genuine interactors. If genuine interactors, then a
similar function between two pairs (X and Y and X' and Y'
STEP#4: CONCLUDE
```

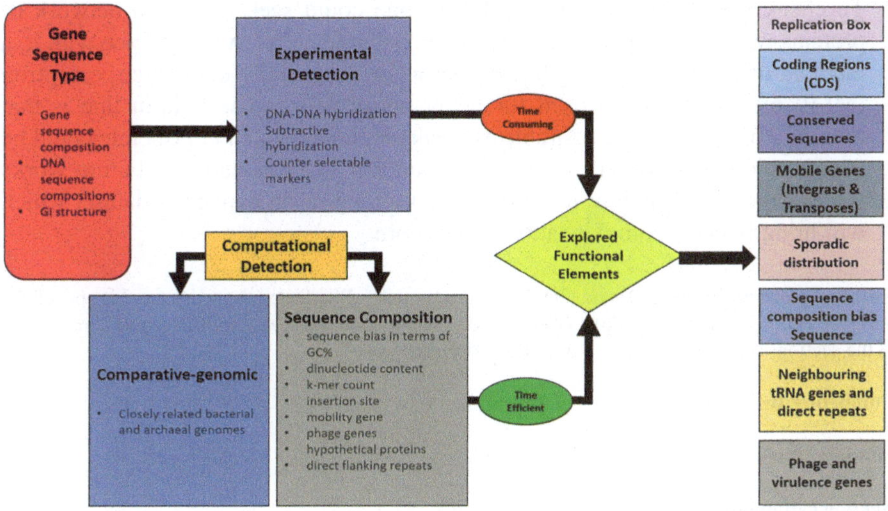

Fig. 16.2 A precise methodology employed to evaluate the machine learning algorithms

16.2 Conclusions and Future Outlook

Horizontal Gene Transfer (HGT) events are common across various domains of life and have come to be known as major driving forces of evolution in both geological timescales and short timescales as an example, consider antibiotic resistance. Current approaches to event HGT employ either phylogeny-based methods or composition-based methods, both of which demand extensive computational labor. We provide a few perspectives and challenges in these directions for better dissemination of HGT data:

1. Establishing a correlation between pathogenesis and horizontal transfer events could predict these events with unknown queries. An ultimate goal carved from this is a web server to infer HGTed events. To check this, we have made some efforts in making a dataset with a total of about 400 annotated proteins (Unpublished). This dataset was split into 2; training and test, which were checked using the WEKA Software and several in-house programs. The datasets were then split into various attributes, viz. HGT likeliness from BLAST results, pathogenicity prediction, number of domains, and sequence associated with any noncoding nucleotides. Cross-validation followed by precision/recall was associated with all machine learning heuristics. What we could observe was that a large number of proteins evented HGT in plants when compared to others while the pathogenic nature of bacteria eventing HGT is compromised. The training efficiency measured a class of various subtypes of bacteria keeping all attributes intact.

2. To check the efficacy of these datasets, one could seek to train a network of proteins with various parameters based on pathogenicity. In accordance with it, a plethora of machine learning based approaches, especially multilayer perceptron (MLP) models could be used for better performance. The bottom line is that training efficiency is not desirable as weak classifiers which in principle could be used to model the system with a small number of proteins used. Future works such as these will require more rigorous training of an expanded dataset which will bring better recommendations to the fore.

Authors' Contributions MB and GD contributed equally. PS ideated the project and proposed the horigene concept. All other authors contributed to the general framework of the chapter and agreed to the sections of the manuscript before PS proofreading it.

Competing Interests None.

Funding None.

Acknowledgments None.

Data Availability Github link: goutamdh/HGT_Website_Server (github.com).

References

Aravind L, Tatusov RL, Wolf YI, Walker DR, Koonin EV (1998) Evidence for massive gene exchange between archaeal and bacterial hyperthermophiles. Trends Genet 14(11):442–444. https://doi.org/10.1016/s0168-9525(98)01553-4. Erratum in: Trends Genet 1999 Jan;15(1):41. PMID: 9825671

Brown KR, Jurisica I (2007) Unequal evolutionary conservation of human protein interactions in interologous networks. Genome Biol 8(5):R95

Cywes Bentley C, Hakansson A, Christianson J, Wessels MR (2005) Extracellular group a streptococcus induces keratinocyte apoptosis by dysregulating calcium signalling. Cell Microbiol 7(7):945–955

DeLong EF (1997) Marine microbial diversity: the tip of the iceberg. Trends Biotechnol 15:203–220

Enright MC, Spratt BG, Kalia A, Cross JH, Bessen DE (2001) Multilocus sequence typing of streptococcus pyogenes and the relationships between emm type and clone. Infect Immun 69(4): 2416–2427

Husnik F, McCutcheon JP (2018) Functional horizontal gene transfer from bacteria to eukaryotes. Nat Rev Microbiol 16(2):67–79. https://doi.org/10.1038/nrmicro.2017.137. Epub 2017 Nov 27

Loman N, Pallen M (2015) Twenty years of bacterial genome sequencing. Nat Rev Microbiol 13: 787–794. https://doi.org/10.1038/nrmicro3565

Monaco AP (2022) The selfish environment meets the selfish gene: coevolution and inheritance of RNA and DNA pools: a model for organismal life incorporating coevolution, horizontal transfer, and inheritance of internal and external RNA and DNA pools.: a model for organismal life incorporating coevolution, horizontal transfer, and inheritance of internal and external RNA and DNA pools. BioEssays 44(2):e2100239. https://doi.org/10.1002/bies.202100239. Epub 2022 Jan 5

Nelson KE, Clayton RA, Gill SR, Gwinn ML, Dodson RJ, Haft DH, Hickey EK, Peterson JD, Nelson WC, Ketchum KA, McDonald L, Utterback TR, Malek JA, Linher KD, Garrett MM, Stewart AM, Cotton MD, Pratt MS, Phillips CA, Richardson D, Heidelberg J, Sutton GG,

Fleischmann RD, Eisen JA, White O, Salzberg SL, Smith HO, Venter JC, Fraser CM (1999) Evidence for lateral gene transfer between Archaea and bacteria from genome sequence of Thermotoga maritima. Nature 399(6734):323–329. https://doi.org/10.1038/20601. PMID: 10360571

Olszak T, Latka A, Roszniowski B, Valvano MA, Drulis-Kawa Z (2017) Phage life cycles behind bacterial biodiversity. Curr Med Chem 24(36):3987–4001. https://doi.org/10.2174/0929867324666170413100136

Ram Y, Hadany L (2019) Evolution of stress-induced mutagenesis in the presence of horizontal gene transfer. Am Nat 194(1):73–89. https://doi.org/10.1086/703457. Epub 2019 May 20

Reddy TBK, Thomas AD, Stamatis D, Bertsch J, Isbandi M, Jansson J (2014) The genomes online database (GOLD) v.5: a metadata management system based on a four level (meta)genome project classification. Nucl Acids Res 43:D1099–D1106

Sevillya G, Doerr D, Lerner Y, Stoye J, Steel M, Snir S (2020) Horizontal gene transfer phylogenetics: a random walk approach. Mol Biol Evol 37(5):1470–1479. https://doi.org/10.1093/molbev/msz302

Soucy SM, Huang J, Gogarten JP (2015) Horizontal gene transfer: building the web of life. Nat Rev Genet 16(8):472–482. https://doi.org/10.1038/nrg3962

Artificial Intelligence and Machine Learning for Prediction and Analysis of Genomic Islands

17

Ankita Shukla, Navnit Kumar Mishra, and Ankur Choudhary

Abstract

The Genomic Islands (GIs) are the groups of gene clusters in bacterial genomes acquired by lateral gene rransfer (LGT). These Genomic Islands are significantly linked with bacterial pathogenicity, adaptations, and evolution. The ancestries of GIs and their association with virulence or pathogenicity factors in bacteriophages and bacteria could provide detailed genetic diversity to identify nucleotide sequences. Increasing evidence suggests that LGT is the prime cause of the transfer of virulence genes through transduction, transformation, and conjugation. However, most of the GIs are the main origin of novel genes for some bacteria. Therefore, the prediction of Genomic Islands and their analysis have gained attention in bacterial genomic sequence research. Although, recently, several bioinformatics tools have been developed for detecting these GIs. Providing researchers with diverse options for effectively identifying these GIs in a bacterial genome. However, rarely any one of them is effectively identifying precisely the complete function of GIs in the bacterial genome. Therefore, advanced algorithms of Artificial Intelligence and Machine Learning approaches are used to process large GIs.

Keywords

Artificial Intelligence · Machine Learning · Genomic Islands · Comparative genomics · Horizontal gene transfer

A. Shukla (✉) · N. K. Mishra
University Institute of Biotechnology (UIBT), Chandigarh University, Mohali, Punjab, India

A. Choudhary
Department of Civil Engineering, Chitkara School of Engineering & Technology, Chitkara University, Himachal Pradesh, Baddi, India

I. Mani et al. (eds.), *Microbial Genomic Islands in Adaptation and Pathogenicity*,
https://doi.org/10.1007/978-981-19-9342-8_17

17.1 Introduction

Genomic Islands (GIs) can be identified through microarray technology (DNA–DNA hybridization experiment), using counter selectable markers, or subtractive hybridization (Dobrindt et al. 2004; Juhas et al. 2009). These processes for the detection of strain-specific GIs would be time consuming and expensive. Therefore, there is a need for Machine Learning and Artificial Intelligence techniques arises for accurately predicting the GIs (Lu and Leong 2016a; De Brito et al. 2016). The prior research on computationally predicting GIs are mainly categorized into two classes: sequence composition and comparative genomics. The comparative genomics-based methods encompass the use of two closely related archaeal and bacterial genomes (Binnewies et al. 2006). A GI from this approach is identified when a cluster of genes present in an organism is not present in any related genomes. Recently, Bertelli and colleagues conducted research and explained that comparative genomic-based methods can identify Genomic Islands regions precisely (Bertelli et al. 2017; Lindblad-Toh 2020; Alföldi and Lindblad-Toh 2013; Oyedara et al. 2018). The most common disadvantage of this approach is its dependence on closely related genomes, along with the variance in the result based on the selection of these genomes. The nucleotide sequence arrangement methods are based on recognizing atypical sequences in the core genome. To achieve this, these methods make use of various structural features previously studied, such as a nucleotide sequence bias in terms of GC%, dinucleotide content, codon usage or k-mer count, presence of an insertion site, mobility gene, phage genes, hypothetical proteins, and direct flanking repeats (Bertelli et al. 2019). To apprehend these features which are associated with gene, sequence annotation is needed. Therefore, these features leave the prediction of unannotated sequences exclusively based on identifying biases in nucleotides, which have a very few features sets, making the prediction models with low precision. On the other hand, this enables the findings on gene-level sequence composition to predict GIs successfully with higher precision and accuracy. The downside of the gene-level research is the dependency on annotated sequences, which may not be available for the naive sequenced genomes or there may be annotation errors (Jungid et al. 2020). In summary, there are certain limitations to the GI prediction techniques: (i) Requirement of closely related genomes in case of comparative genomics-based approach; (ii) Dependency on annotated genomes in case of gene-level sequence-based approach; (iii) Lack of good feature set in nucleotide level sequence-based approach.

Genomic Islands are often present in most of the microbes such as pathogenic and non-pathogenic microbes. Specifically, a Genomic Islands is a large sequence of continuous nucleotides sequence transferred by the process of Horizontal Gene Transfer (HGT) (Lu and Leong 2016a). This Genomic Island consists of a cluster of genes. The length of Genomic Islands fluctuates from 4.5 to 600 kb of nucleotide bases. In other words, horizontally transferred genomic strings briefer than a specific threshold are known as genomic islets (Juhas et al. 2009; Bellanger et al. 2014). Genomic Islands often have aperiodic distribution in their evolutionary relationship. The phylogenetic analysis suggests that they are present in microorganisms, however, lineages linked to Genomic Islands are devoid of phylogenetically related organisms. GIs have many unique features which distinguish them from rest of the genomic regions (Juhas et al. 2009; Hacker et al. 1997; Schmidt and Hensel 2004). These unique features are varying order of sequence composition relative to the principal genome, the occurrences of transposons genes, specific integration sites, and flanking direct repeats (DRs). To illustrate, transfer DNA (T-DNA) is well-known hotspot for the Genomic Islands recombination (Bellanger et al. 2014; Williams 2002). Moreover, all these properties linked to Genomic Islands are seldom present in single Genomic Islands, however, many of the GIs lack these characteristics. As a consequence, Genomic Islands are also considered as a super-family of transposons with essential and fluxional genomic sequence features (Vernikos and Parkhill 2008). GIs are broadly categorized as mobile genetic elements (MGEs) besides the slandered GI definition (Boyd et al. 2009; Langille et al. 2010). They can be further subclassified on their mobility basis. Some genomic sequences of GIs are transposable because they can have the potential to transfer to the new host, in the form of Integrative and Conjugative Elements (ICEs). Other transposable elements in bacterial genome are prophages, and conjugative transposons which persist in their mobility (Juhas et al. 2009). Genomic Islands can also be classified based on their functional gene strings. They are PAIs (Pathogenicity Islands) with DNA sequence encoding for virulence factors. Resistance islands (REIs) with the genomic sequence being transcribed and translated to metabolically active proteins pertain to antibiotic resistance (Fig. 17.1) (Dobrindt et al. 2004). However, the classification based on function may not be certain since the special sequence features related to genes within GIs not being in exercise.

Genomic Islands play significant roles in phylogeny of microbial genomes and their evolution and consequently microbial adaptation to the ecosystem (Hacker et al. 2001). Being involved in the dynamic gene pool, the insertions of GIs induce substantial evolution, permitting microbes to gain large genomic sequences associated to much more evolved adaptive biological functions and thereby persuading evolutionary fitness (Dobrindt et al. 2004; Juhas et al. 2009). Significantly, the recursively genomic sequence (gene) in the Genomic Islands guides various important metabolic traits, such as pathogenicity, symbiotic relation of bacteria to the host, antibiotic resistance, fitness, and metabolism. Particularly, PAIs have many genes causative to pathogen virulence (Hacker et al. 1997; Schmidt and Hensel 2004; Ho Sui et al. 2009), therefore, the potential vaccine Oligos could be located within PAIs (Moriel et al. 2010). Thus, the precisely locating and predicting the GIs is imperative for both evolutionary lineages and their therapeutic potential. The overall flowchart for the GIs prediction has been shown in Fig. 17.2.

17.2 Deficiency in Benchmark for GIs Datasets

Insufficient consistent benchmarks of Genomic Islands datasets are available to validating to supervised prediction methods. There are several available GI databases like Islander (Mantri and Williams 2004), and ICEberg (Bi et al. 2012),

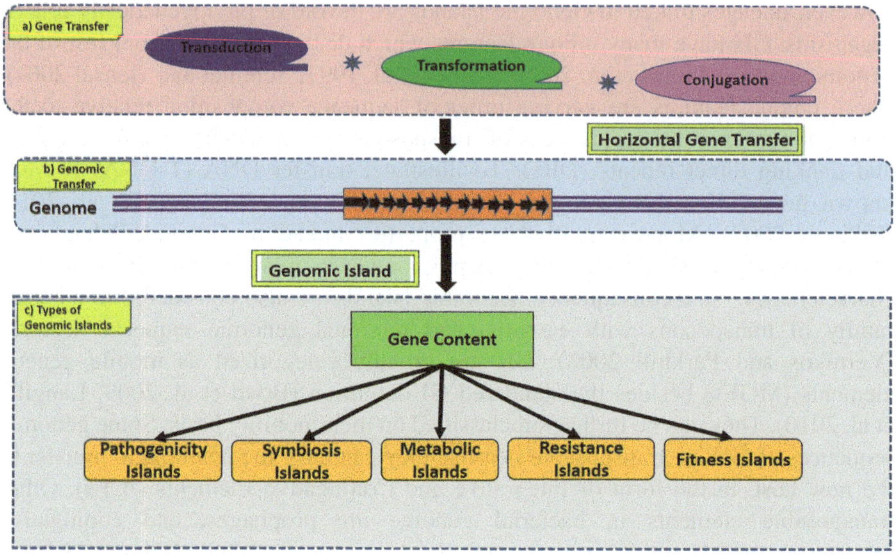

Fig. 17.1 Schematic illustration of Genomic Island insertions through horizontal gene transfer and their metabolic consequences in microbes. (**a**) Transduction, Transformation, and Conjugation are the most common methods of DNA transfer in Horizontal Gene Transfer (HGT) from one bacterium to another. Transduction is the method of DNA sequence transmission from one bacterium to another mediated via bacteriophage. Transformation is a genetic process of Lateral Gene Transfer by which exogenous DNA sequence is transferred to the host bacterium. Conjugation is a genetic process by which one bacterium transfers genetic material to another through pili formation. (**b**) Hybridization of the host genomic material or extrachromosomal material through the aforementioned Lateral Gene Transfer (red color) methods. (**c**) Functional features of Genomic Islands are acquired through the process of HGT. They are Pathogenicity Islands responsible for virulence; Symbiosis Islands consists of conserved region linked to the symbiotic gene; metabolic gene Islands are fortified in genomic sequence related to secondary metabolite biochemical pathways responsible for the functional adaptation; Resistance Islands clustered within the same genetic locus through horizontal accusation, mostly they are antibiotic resistance genes; Fitness Islands increase the bacterial fitness either directly or indirectly

PAIDB (Yoon et al. 2007) which are accustomed to predicting and verifying more GIs. But these GIs databases are limited to a particular group of GIs, like transfer-DNA borne GIs (GIs inserted at ribonucleic acid or transfer ribonucleic acid DNA sequence sites), Pathogenicity Islands, and ICEs. Recently, a machine learning approach has been applied on two constructed GI datasets using whole genome comparison as training datasets (Vernikos and Parkhill 2008; Langille et al. 2008a). However, the dimensions of those datasets are not large enough, and this whole genome dataset's consistency has yet to be confirmed by considerable biological evidence.

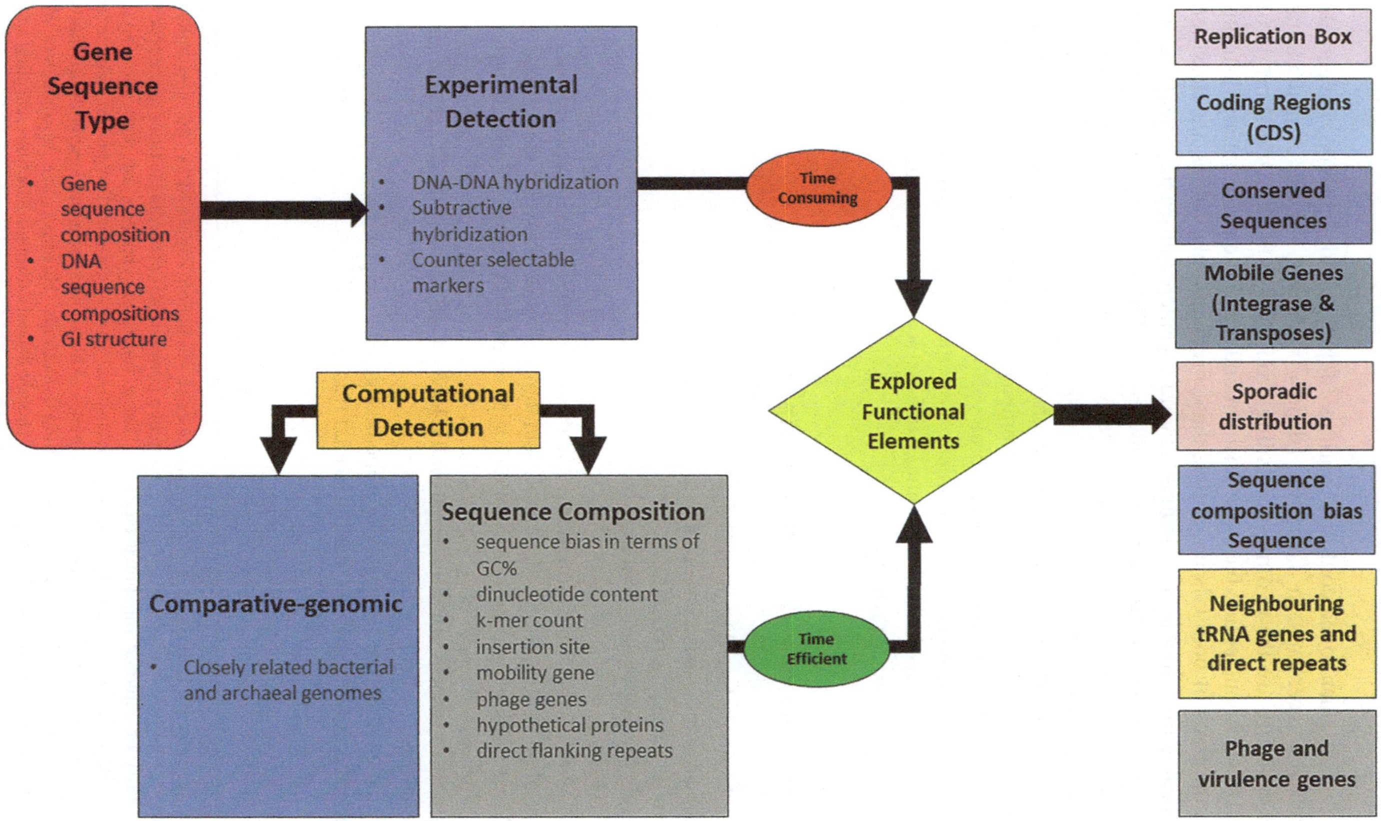

Fig. 17.2 The flowchart illustrates the comparative prediction of Genomic Islands through experimental and computational approaches. The input parameter for GI detection is gene sequence composition, DNA sequence composition, and the GI structure. Experimental Detection: Experimental techniques for the detection of GIs based on gene expressions includes DNA-DNA hybridization, substrate hybridization, and counter selectable markers. Computational

17.2.1 GI Prediction Methods

Despite the aforementioned confronts, various computational algorithms have made significant advancements in Genomic Islands prediction. These computational methods typically use the two most revealing properties of the horizontal origin of the GIs: Biases in nucleotide base composition and aperiodic evolutionary distribution. On the basis of both GIs properties, these prediction algorithms are classified into dual categories: first is comparative genomics-centered procedures, and second is composition center approaches (Langille et al. 2008a). On the basis of input genome numbers, the Genomic Islands prediction algorithms can be categorized into two large groups. They are grouped into single genome approach and multiple genomes datasets approach. The first approaches are often composition based, while the second group of methods are based on comparative genomics. Ensemble learning methods shall be also included that combine various approaches and approaches to predict Genomic Islands in incomplete genomes. These ensemble learning algorithms predict GIs in draft genomes such as scaffolds or contigs rather than WGS (whole genome sequence).

17.2.2 Genomic Islands Prediction Approaches

1. Single Genome-based approaches
2. Multiple Genome-based approaches
3. Ensemble Learning procedures
4. Methods for incomplete genome

GIs have various metabolic functions, such as pathogenicity, antibiotic resistance, heavy metal and nanoparticle resistance, and xenobiotic degradation (Langille et al. 2010). Most fundamental biochemical characteristics of Genomic Islands include:

1. Integration hotspots for GIs are usually located adjacent to RNA genes on a chromosome.
2. Genomic Islands are bordered by repeated genetic sequences (Direct Repeats) that are the consequence of the horizontal transfer of GIs.

Fig. 17.2 (continued) Detection: While computational prediction can be done through either comparative genomics methods or, genomic sequence composition procedures. Both approaches help to identify or explored functional elements. These functional elements are replication box, coding sequence (CDS), conserved sequences, Transpose genes, sporadic distribution, sequence composition bias, flanking T-DNA, and virulence genes. Although experimental methods are more time consuming in comparison to the computational methods. However, the advancement of computational algorithms for the prediction of Genomic Islands through Machine learning and Artificial Intelligence based on these functional elements ensures high precision

3. Lateral transfer-associated genes, such as recombinases, transposases, and integrases, are mostly present at the intersection of Genomic Islands and fundamental (core) genome. The principal genome and GIs (Genomic Islands) are significantly identified on the basis of GC content.

Horizontal gene transfer in a population refers to discrete Genomic segments that can be identified as GIs. This segment of genomic region may integrate into the host chromosome and relocate to new microbes via the method of genetic transformation, transduction, or conjugation. This genomic segment is being transcribed and translated into proteins and represents a novel biological function in the recipient host. In prokaryotes, the Genomic Islands can be easily identified or predicted than in the eukaryotes because of genomes diversity in eukaryotes. The occurrence of GIs in bacterial genomic contents are predicted by GIPSy, IslandViewer 4, and Zisland Explorer computational programmes. The best approach to identifying the Genomic Islands is by combining the predicted results from various programmes considering the various algorithms that every programme implements to detect the GIs. IslandViewer 4 predicts the GIs in bacterial and archaeal Genomes from their EMBL or GenBank sequence format already submitted in the NCBI database. This computer programme consists of four high-precision prediction approaches: Islander, IslandPath-DIMOB, IslandPick, and SIGI-HMM. Combining the various prediction programme reduces the false-positive outcomes and also ensures GIs identification precisely.

Predicting accurate GI segment borders is important to locate the Genomic Islands strings in microbial genomes. Based on reference genome IslandPick can identify the GIs through a comparative genomics approach using Genomic Islands segment boundaries. Otherwise, Islander is recommended hence, Islander is able to precisely identify a repeated genetic sequence that flank the Genomic Island segment region. On the other hand, SIGI-HMM and IslandPath-DIMOB revealed their lower precision while predicting Genomic Islands segment borders and tend to identify fragmented genomic regions of horizontally transferred genes. However, predicting Genomic Islands through bias approaches in genomic regions, both SIGI-HMM and IslandPath-DIMOB tools are favored more discrete genomic segments gained by HGT.

17.3 Genomic Islands Prediction

To predict the Genomic Islands precisely the quality of Genome sequences should be good. Second factor is the sensitivity of the algorithm implemented in GIs predicting computational tools.

17.3.1 Data Quality

Out of the several factors, one of the crucial factors for predicting the GIs is the base quality (the correct bases have been identified in nucleotide sequence reads) of the genome sequence. In comparison to classical sequencing method, the advent of NGS techniques has led to an enormous amount of whole Genome Sequences (small sequencing reads) along with draft genomes (Niedringhaus et al. 2011). The draft genomes obtained through the NGS methods can be used for the prediction of GIs, however, one of the problems related to such predictions is that frequently they are trapped to false-negative or false-positive prediction (de Castro Soares et al. 2016). These statistically significant false results occur as a consequence of the unavailability of DNA strings either in the query or reference genome due to unsolved gaps (de Villemereuil et al. 2014). Hence, this proved that for precisely predicting GIs, whole genome sequence is mandatory as an input (Bertelli et al. 2019). To get the complete genome sequence, combined sequencing approaches have been developed by researchers. These technologies are transformed into PacBio and MinIon for long-read sequencing and Illumina and Ion Torrent improves the base quality required to attain a good excellence of whole genome sequence (Neubert et al. 2021).

It has been noticed that the reads generated by Illumina and Ion Torrent platforms may have a high frequency of substitutions as well as insertion/deletions events, respectively (Frio 2015). Such events lead to substitutions (non-synonymous) and pseudogene formation which in turn will impact the composition of the gene, GC content, and codon usage (Zhang et al. 2002; Radványi and Kun 2021). Therefore, manual curation of genome sequences using genome mapping software programme along with a high-quality genome is also necessary. Gene arrangement also plays a significant role in predicting GIs as well as in determining its biological consequences. Therefore, manual curation is the best practice for whole genome annotation to improve the quality of annotation.

17.3.2 Computational Programme for GIs Identification

The pathogenicity islands (a subset of Genomic Islands) were first identified via molecular biology methods; though, this approach was very time consuming and costly (Gal-Mor and Finlay 2006). The advent of sequencing technologies has led to the programming of some computational tools for GI identification. It has been found through different studies that GIs shares some specific properties such as the presence of an integrase (that provides the ability to recombine), their GC content, connotation with tRNA genes, and occurrence of flanking DNA strings on dual sides (Vernikos and Parkhill 2008; Langille et al. 2010). GIs are also classified depending on their functions such as DNA strings in bacteria contributing to antibiotic resistance are categorized as "resistance islands (RI)"; while islands with their capability to metabolize new organic sources are categorized as "metabolic islands (MI)" and the GIs that enable bacteria to cause infection is categorized as "virulence islands

(PAI)" (da Silva Filho et al. 2018; Hentschel and Hacker 2001). Most of the software tools developed for GIs prediction primarily utilize the common genetic properties such as high GC content genomic strings or triplet codon biases rather than considering the entire genome sequencing (Langille et al. 2008b) (Table 17.1). Therefore, some software tools predict GIs based on its unique genomic signature such computational tools are SIGI-HMM, IGIPT, GI-SVM, and PAI-IDA (Langille et al. 2010; Lu and Leong 2016b; Jain et al. 2011; Waack et al. 2006). Likewise, there are tools such as EGID, Islander, and Islandpath for predicting GI features (homogeneous genomic signature) like flanking regions (i.e., tRNAs), mobile elements, and insertion sites (Che et al. 2011; Hudson et al. 2015; Hsiao et al. 2003). It has been found through various studies that genomic comparison is one of the significant features to find the absence of the region in closely related organisms (de Castro Soares et al. 2016; Rajashekara et al. 2004). Tools such as GIHunter, GIST, GIPSy, GI-POP, INDeGenIUS, PIPS, IslandViewer, RPGFinder, and PAIDB make use of the comparative genomics and mobile genetic contigs to make the GIs predictions more efficient (Che and Wang 2013; Hasan et al. 2012; Soares et al. 2016; Lee et al. 2013; Shrivastava et al. 2010; Soares et al. 2012; Langille and Brinkman 2009; Ogier et al. 2010; Yoon et al. 2015). Studies have also shown the importance of genomic signatures in terms of correctly predicting the GIs as an increasing number of genomic variables used by prediction tools precisely and efficiently identified the GIs. Therefore, ensemble methods were designed that combine different software tools to achieve a comprehensive correlation of all genomic characteristics.; it includes GIPSy, EGID, IslandViewer, GIST, and PIPS. As discussed, there are other subclasses of GIs (such as PAI, RI, MI, and SI) also but very little is known about others than PAIs (da Silva Filho et al. 2018). Although some of the GIs prediction software such as IslandViewer, PAIDB, and InDeGenIUS partially address predictions for these subclasses of GIs (da Silva Filho et al. 2018). This shows that though lot of algorithms have been designed so far for predicting GIs. However, still there is a need to develop a unique and precise genetic algorithm to specifically recognize all four classes of GIs (Table 17.1).

17.4 Future Improvements

There is a need for approaches that may involve the effective use of overall features that are present in the genomic region and thus can be predicted via machine learning-based classification methods. The features include the distribution of genes with variance in GC percentage, codon usage, dinucleotide bias, k-mer frequency, flanking tRNA genes, transposase genes, and so on (Hsiao et al. 2005). It has been seen through various studies that GIs shows a mosaic arrangement in in vivo and individual GIs might show a diversity in genomic features (Jani et al. 2016). An example, some GIs show features such as variation in GC percentage, flanking strings of tRNA contigs, and transposable genes, while others may only have information related to deviation in triplet codon biases, and virulence factors. The mosaic composition of the GIs structure predicts false-negative even though

Table 17.1 List of the Genomic Island Prediction tools

Tool	Strong feature in GI	Significance	Prediction
DeepHGT	GC contento, ATGC ratio, K-mer frequency	Bacterial genome	HGT insertion sites
GIPSy	Triplet codon usage and GC content, flanking strings of tRNA contigs, transposable genomic region, antibiotic resistance/ symbiosis, metabolism, virulence factors, absence in closely related species or other organisms in the same genus	*Escherichia coli* CFT073, *A. baumannii* strain AYE, *Burkholderia pseudomallei* K96243,, *Mesorhizobium loti* MAFF303099	Pathogenicity Islands, Symbiotic Islands, Resistance Islands, and Metabolic Islands
IslandViewer4	Codon usage, dinucleotides	Bacterial and Archaeal genomes	Pathogen-associated genes
Zisland Explorer	G + C%, codon usage	Bacterial and Archaeal genomes	Similarity within an island
Predict Bias	G + C%, codon usage, dinucleotides	Prokaryotes	Pathogenicity Islands
GI Hunter	IVOM, k-mers	Prokaryotic genomes	GIs in whole genome
Alien Hunter	IVOM, k-mers	*E. coli*, *Salmonella* lineages	Higher order motifs
COLOMBO SIGI-HMM	G + C%, codon usage	Microbial genomes	Putative donor gene
INDeGenIUS	k-mers	Proteobacteria	Mobile islands, Metabolic islands, virulent islands, Symbiotic islands, Secretion islands, and Resistance islands,
IslandPath	G + C%, dinucleotides	Prokaryotes	Multiple DNA signal analyses and annotation features
PAI-IDA	G + C%, codon usage, dinucleotides	Bacterial genomes	Pathogenicity islands, Anomalous gene clusters
(IGIPT (Integrated Genomic Island Prediction Tool)	G + C%, unique Genomic signature, Triplet codon biases, k-mer Distribution,	Bacterial genomes	Anomalous nucleotide composition
Wn-SVM	Single genome, gene annotation	Viruses, Archaeal and Bacterial genomes	Conserved Regions
Centroid	Partitioning entire genome into non-overlapping	Bacterial genomes	Compositionally distinct regions

(continued)

Table 17.1 (continued)

Tool	Strong feature in GI	Significance	Prediction
	histograms of equal window		
IslandPick	Comparative genomics based	Bacterial genomes	Multiple sequence composition
MobilomeFInDER	tRNA, whole genome alignments	Bacterial genomes	Similarity w.r. t closest sequence

using the combinational approaches that are ensemble methods (Waack et al. 2006). That is why the computational tools with machine learning algorithms proved to be best approaches in identifying all probable genomic states using different features during the classification of GIs. One more thing that needs to be addressed in predicting the GI's origin this is because GIs blend with core genome and acclimatize their unique genomic characteristics with time. Therefore, this is impractical to predict the evolutionary time of the GIs integration by comparing it with the unique genomic characteristics of other organisms (Juhas et al. 2009). Various studies have shown that two distantly converging organisms could have similar triplet codon biases, as a result of tRNA bioavailability (Koski et al. 2001). The alternative method to identify the GI origin is the comparison through phylogenetic methods wherein syntenic genes inside one organism can be compared to their diverged orthologous genes in other species. GIs prediction through advanced pan-genomics research effectively predicts the GIs in microbes considering non-redundant clusters of genes. The algorithm uses the divergence prediction in orthologous genomic contigs, and identifies commonly shared genes among all strains; those shared between two or more strains, and, unique ones that are present in a single strain only. It has been found from various studies that movable accessory genes shared in the conserved core genome are found to be significant in drug and vaccine designing. While the accessory genes in the common genomes and the singletons (that are present in a single strain only) are found to be significant in case of acclimatization to new loci and hence account for the GIs (Bazin et al. 2020; Trost et al. 2012). Impending approaches in pan-genomics might aim at detecting GIs in all possible strains and thereafter comparing them to find the degree of mosaicism. Thereafter, GIs identified for all sets of strains follow epidemiological analyses that can be computed through phylogeny of genomes (phylogenomics based). The final step involves finding the GI's origin between distantly related species via gene synteny conservation methods. Finding the evolutionary time of the GIs integration with pan-genomics studies aids in determining the acquired set of genes that influences the host adoptive environment of bacteria to new traits. The method proved to be efficient in determining the basis of these new clonal complexes and creating novel diagnostic approaches for identifying evolving pathogenic strains and also knowing their role in epidemiological analyses (Bertelli et al. 2019; Escher et al. 2016). GIs account for inducing considerable variance in genomic strings specifically if we talk about bacterial strain, so there is the possibility of creating new GIs comparative methods for each and every field of comparative genomics.

References

Alföldi J, Lindblad-Toh K (2013) Comparative genomics as a tool to understand evolution and disease. Genome Res. https://doi.org/10.1101/gr.157503.113

Bazin A, Gautreau G, Médigue C, Vallenet D, Calteau A (2020) panRGP: a pangenome-based method to predict genomic islands and explore their diversity. Bioinformatics 36(Suppl_2): i651–i658. https://doi.org/10.1093/bioinformatics/btaa792

Bellanger X, Payot S, Leblond-Bourget N, Guédon G (2014) Conjugative and mobilizable genomic islands in bacteria: evolution and diversity. FEMS Microbiol Rev. https://doi.org/10.1111/1574-6976.12058

Bertelli C, Laird MR, Williams KP, Lau BY, Hoad G, Winsor GL, Brinkman FSL (2017) IslandViewer 4: expanded prediction of genomic islands for larger-scale datasets. Nucleic Acids Res. https://doi.org/10.1093/nar/gkx343

Bertelli C, Tilley KE, Brinkman FSL (2019) Microbial genomic island discovery, visualization and analysis. Brief Bioinform. https://doi.org/10.1093/bib/bby042

Bi D, Xu Z, Harrison EM, Tai C, Wei Y, He X, Jia S, Deng Z, Rajakumar K, Ou HY (2012) ICEberg: a web-based resource for integrative and conjugative elements found in bacteria. Nucleic Acids Res. https://doi.org/10.1093/nar/gkr846

Binnewies, Motro Y, Hallin PF, Lund O, Dunn D, La T, Hampson DJ, Bellgard M, Wassenaar TM, TT Ussery DW (2006) Ten years of bacterial genome sequencing: comparative-genomics-based discoveries. Funct Integr Genomics 6:165–185

Boyd EF, Almagro-Moreno S, Parent MA (2009) Genomic islands are dynamic, ancient integrative elements in bacterial evolution. Trends Microbiol. https://doi.org/10.1016/j.tim.2008.11.003

Che D, Hasan MS, Wang H, Fazekas J, Huang J, Liu Q (2011) EGID: an ensemble algorithm for improved genomic island detection in genomic sequences. Bioinformation 7:311

Che D, Wang H (2013) GIV: a tool for Genomic Islands visualization. Bioinformation 9:879–882

da Silva Filho AC, Raittz RT, Guizelini D, De Pierri CR, Augusto DW, dos Santos-Weiss ICR, Marchaukoski JN (2018) Comparative analysis of genomic island prediction tools. Front Genet 9:619

De Brito DM, Maracaja-Coutinho V, De Farias ST, Batista L V, Do Rêgo TG (2016) A Novel Method to Predict Genomic Islands Based on Mean Shift Clustering Algorithm. https://doi.org/10.1371/journal.pone.0146352

de Castro Soares S, de Castro OL, Jaiswal AK, Azevedo V (2016) Genomic Islands: an overview of current software tools and future improvements. J Integr Bioinform 13:82–89

Dobrindt U, Hochhut B, Hentschel U, Hacker J (2004) Genomic islands in pathogenic and environmental microorganisms. Nat Rev Microbiol. https://doi.org/10.1038/nrmicro884

de Villemereuil P, Frichot É, Bazin É, François O, Gaggiotti OE (2014) Genome scan methods against more complex models: when and how much should we trust them? Mol Ecol 23:2006–2019

Escher R, Brunner C, von Steiger N, Brodard I, Droz S, Abril C, Kuhnert P (2016) Clinical and epidemiological analysis of Campylobacter fetus subsp. fetus infections in humans and comparative genetic analysis with strains isolated from cattle. BMC Infect Dis 16:198. https://doi.org/10.1186/s12879-016-1538-7

Frio TR (2015) High-Throughput Technologies: DNA and RNA sequencing strategies and potential. In: Pan-Cancer Integr. Mol. Portrait New Paradigm Precis. Med. Springer, pp 47–68

Gal-Mor O, Finlay BB (2006) Pathogenicity islands: a molecular toolbox for bacterial virulence. Cell Microbiol 8:1707–1719

Hacker J, Blum-Oehler G, Mühldorfer I, Tschäpe H (1997) Pathogenicity islands of virulent bacteria: structure, function and impact on microbial evolution. Mol Microbiol. https://doi.org/10.1046/j.1365-2958.1997.3101672.x

Hacker J, Carniel E, Achtmann M et al (2001) Ecological fitness, genomic islands and bacterial pathogenicity a Darwinian view of the evolution of microbes. EMBO Rep 2

Hasan MS, Liu Q, Wang H, Fazekas J, Chen B, Che D (2012) GIST: genomic island suite of tools for predicting genomic islands in genomic sequences. Bioinformation 8:203

Hentschel U, Hacker J (2001) Pathogenicity islands: the tip of the iceberg. Microbes Infect 3:545–548

Ho Sui SJ, Fedynak A, Hsiao WWL, Langille MGI, Brinkman FSL (2009) The association of virulence factors with genomic islands. PLoS One. https://doi.org/10.1371/journal.pone.0008094

Hsiao W, Wan I, Jones SJ, Brinkman FS (2003) IslandPath: aiding detection of genomic islands in prokaryotes. Bioinformatics 19:418–420

Hsiao WWL, Ung K, Aeschliman D, Bryan J, Finlay BB, Brinkman FSL (2005) Evidence of a large novel gene pool associated with prokaryotic genomic islands. PLoS Genet 1:e62

Hudson CM, Lau BY, Williams KP (2015) Islander: a database of precisely mapped genomic islands in tRNA and tmRNA genes. Nucleic Acids Res 43:D48–D53

Jain R, Ramineni S, Parekh N (2011) IGIPT-integrated genomic island prediction tool. Bioinformation 7:307

Jani M, Mathee K, Azad RK (2016) Identification of novel genomic islands in Liverpool epidemic strain of Pseudomonas aeruginosa using segmentation and clustering. Front Microbiol 7:1210

Juhas M, Van Der Meer JR, Gaillard M, Harding RM, Hood DW, Crook DW (2009) Genomic islands: tools of bacterial horizontal gene transfer and evolution. FEMS Microbiol Rev. https://doi.org/10.1111/j.1574-6976.2008.00136.x

Jungid H, Venturaid T, Chungid JS, Kimid W-J, Nam B-H, Kongid HJ, Kimid Y-O, Jeon M-S, Eyunid S-I (2020) Twelve quick steps for genome assembly and annotation in the classroom. https://doi.org/10.1371/journal.pcbi.1008325

Koski LB, Morton RA, Golding GB (2001) Codon bias and base composition are poor indicators of horizontally transferred genes. Mol Biol Evol 18(3):404–412. https://doi.org/10.1093/oxfordjournals.molbev.a003816

Langille MG, Brinkman FS (2009) IslandViewer: an integrated interface for computational identification and visualization of genomic islands. Bioinformatics 25:664–665

Langille MG, Hsiao WW, Brinkman FS (2008b) Evaluation of genomic island predictors using a comparative genomics approach. BMC Bioinformatics 9:1–10

Langille MGI, Hsiao WWL, Brinkman FSL (2008a) Evaluation of genomic island predictors using a comparative genomics approach. BMC Bioinformatics. https://doi.org/10.1186/1471-2105-9-329

Langille MGI, Hsiao WWL, Brinkman FSL (2010) Detecting genomic islands using bioinformatics approaches. Nat Rev Microbiol. https://doi.org/10.1038/nrmicro2350

Lee C-C, Chen Y-PP, Yao T-J, Ma C-Y, Lo W-C, Lyu P-C, Tang CY (2013) GI-POP: a combinational annotation and genomic island prediction pipeline for ongoing microbial genome projects. Gene 518:114–123

Lindblad-Toh K (2020) What animals can teach us about evolution, the human genome, and human disease. Ups J Med Sci. https://doi.org/10.1080/03009734.2020.1722298

Lu B, Leong HW (2016a) Computational methods for predicting genomic islands in microbial genomes. Comput Struct Biotechnol J 14:200–206

Lu B, Leong HW (2016b) GI-SVM: a sensitive method for predicting genomic islands based on unannotated sequence of a single genome. J Bioinforma Comput Biol 14:1640003

Mantri Y, Williams KP (2004) Islander: a database of integrative islands in prokaryotic genomes, the associated integrases and their DNA site specificities. Nucleic Acids Res. https://doi.org/10.1093/nar/gkh059

Moriel DG, Bertoldi I, Spagnuolo A et al (2010) Identification of protective and broadly conserved vaccine antigens from the genome of extraintestinal pathogenic Escherichia coli. Proc Natl Acad Sci U S A. https://doi.org/10.1073/pnas.0915077107

Neubert K, Zuchantke E, Leidenfrost RM, Wuenschiers R, Grützke J, Malorny B, Brendebach H, Al Dahouk S, Homeier T, Hotzel H (2021) Testing assembly strategies of Francisella tularensis genomes to infer an evolutionary conservation analysis of genomic structures. BMC Genomics 22:1–21

Niedringhaus TP, Milanova D, Kerby MB, Snyder MP, Barron AE (2011) Landscape of next-generation sequencing technologies. Anal Chem 83:4327–4341

Ogier J-C, Calteau A, Forst S, Goodrich-Blair H, Roche D, Rouy Z, Suen G, Zumbihl R, Givaudan A, Tailliez P (2010) Units of plasticity in bacterial genomes: new insight from the comparative genomics of two bacteria interacting with invertebrates, Photorhabdus and Xenorhabdus. BMC Genomics 11:1–21

Oyedara OO, Segura-Cabrera A, Guo X, Elufisan TO, Cantú González RA, Rodríguez Pérez MA (2018) Whole-genome sequencing and comparative genome analysis provided insight into the predatory features and genetic diversity of two Bdellovibrio species isolated from soil. Hindawi Int J Genomics. https://doi.org/10.1155/2018/9402073

Radványi Á, Kun Á (2021) Phylogenetic analysis of mutational robustness based on codon usage supports that the standard genetic code does not prefer extreme environments. Sci Rep 11:1–9

Rajashekara G, Glasner JD, Glover DA, Splitter GA (2004) Comparative whole-genome hybridization reveals genomic islands in Brucella species. J Bacteriol 186:5040–5051

Schmidt H, Hensel M (2004) Pathogenicity Islands in bacterial pathogenesis. Clin Microbiol Rev. https://doi.org/10.1128/CMR.17.1.14-56.2004

Shrivastava S, Reddy SK, Mande SS (2010) INDeGenIUS, a new method for high-throughput identification of specialized functional islands in completely sequenced organisms. J Biosci 35: 351–364

Soares SC, Abreu VA, Ramos RT, Cerdeira L, Silva A, Baumbach J, Trost E, Tauch A, Hirata R Jr, Mattos-Guaraldi AL (2012) PIPS: pathogenicity island prediction software. PLoS One 7:e30848

Soares SC, Geyik H, Ramos RT, de Sá PH, Barbosa EG, Baumbach J, Figueiredo HC, Miyoshi A, Tauch A, Silva A (2016) GIPSy: genomic island prediction software. J Biotechnol 232:2–11

Trost E, Blom J, Soares SC, Huang I-H, Al-Dilaimi A, Schröder J, Jaenicke S, Dorella FA, Rocha FS, Miyoshi A, Azevedo V, Schneider MP, Silva A, Camello TC, Sabbadini PS, Santos CS, Santos LS, Hirata R Jr, Mattos-Guaraldi AL, Efstratiou A, Schmitt MP, Ton-That H, Tauch A (2012) Pangenomic study of Corynebacterium diphtheriae that provides insights into the genomic diversity of pathogenic isolates from cases of classical diphtheria, endocarditis, and pneumonia. J Bacteriol 194(12):3199–3215. https://doi.org/10.1128/JB.00183-12

Vernikos GS, Parkhill J (2008) Resolving the structural features of genomic islands: a machine learning approach. Genome Res. https://doi.org/10.1101/gr.7004508

Waack S, Keller O, Asper R, Brodag T, Damm C, Fricke WF, Surovcik K, Meinicke P, Merkl R (2006) Score-based prediction of genomic islands in prokaryotic genomes using hidden Markov models. BMC Bioinformatics 7:1–12

Williams KP (2002) Integration sites for genetic elements in prokaryotic tRNA and tmRNA genes: sublocation preference of integrase subfamilies. Nucleic Acids Res. https://doi.org/10.1093/nar/30.4.866

Yoon SH, Park Y-K, Kim JF (2015) PAIDB v2. 0: exploration and analysis of pathogenicity and resistance islands. Nucleic Acids Res 43:D624–D630

Yoon SH, Park YK, Lee S, Choi D, Oh TK, Hur CG, Kim JF (2007) Towards pathogenomics: a web-based resource for pathogenicity islands. Nucleic Acids Res. https://doi.org/10.1093/nar/gkl790

Zhang L, Vision TJ, Gaut BS (2002) Patterns of nucleotide substitution among simultaneously duplicated gene pairs in Arabidopsis thaliana. Mol Biol Evol 19:1464–1473

The manufacturer's authorised representative in the EU is Springer Nature Customer Service Centre GmbH, Europaplatz 3, 69115 Heidelberg, Germany. If you have any concerns regarding our products, please contact ProductSafety@springernature.com

Printed and bound by CPI Group (UK) Ltd, Croydon, CR0 4YY
30/09/2026
02225242-0016